Ecological and Environmental Physiology of **Amphibians**

Ecological and Environmental Physiology of **Amphibians**

Stanley S. Hillman
Biology, Portland State University
Philip C. Withers
Animal Biology, University of Western Australia
Robert C. Drewes
Herpetology, California Academy of Sciences
Stanley D. Hillyard
School of Dental Medicine, University of Nevada, Las Vegas

OXFORD
UNIVERSITY PRESS

OXFORD
UNIVERSITY PRESS

Great Clarendon Street, Oxford OX2 6DP

Oxford University Press is a department of the University of Oxford.
It furthers the University's objective of excellence in research, scholarship,
and education by publishing worldwide in

Oxford New York

Auckland Cape Town Dar es Salaam Hong Kong Karachi
Kuala Lumpur Madrid Melbourne Mexico City Nairobi
New Delhi Shanghai Taipei Toronto

With offices in

Argentina Austria Brazil Chile Czech Republic France Greece
Guatemala Hungary Italy Japan Poland Portugal Singapore
South Korea Switzerland Thailand Turkey Ukraine Vietnam

Oxford is a registered trade mark of Oxford University Press
in the UK and in certain other countries

Published in the United States
by Oxford University Press Inc., New York

First published 2009

British Library Cataloguing in Publication Data

Data available

Library of Congress Cataloging in Publication Data

Data available

Typeset by Newgen Imaging Systems (P) Ltd., Chennai, India
Printed in Great Britain
on acid-free paper by
CPI Antony Rowe, Chippenham, Wiltshire

ISBN 978–0–19–857031–8 (Hbk.)
ISBN 978–0–19–857032–5 (Pbk.)

10 9 8 7 6 5 4 3 2 1

We dedicate this volume to two colleagues and friends who have had strong and lasting influence on our careers: Professor Vaughan H. Shoemaker and Professor Rodolfo Ruibal.

Ecological and Environmental Physiology Series (EEPS)

Series Editor: Warren Burggren, University of North Texas

This authoritative series of concise, affordable volumes provides an integrated overview of the ecological and environmental physiology of key taxa including birds, mammals, reptiles, amphibians, insects, crustaceans, mollusks and fish. Each volume provides a state-of-the-art review and synthesis of topics that are relevant to how that specific group of organisms have evolved and coped with the environmental characteristics of their habitats. The series is intended for students, researchers, consultants and other professionals in the fields of physiology, physiological ecology, ecology, and evolutionary biology.

A Series Advisory Board assists in the commissioning of titles and authors, development of volumes, and promotion of the published works. This Board comprises more than 50 internationally recognized experts in ecological and environmental physiology, providing a combination of both depth and breadth to proposal evaluation and series oversight.

The reader is encouraged to visit the EEPS website for additional information and the latest volumes (http://www.eeps-oxford.com/index.htm). If you have ideas for new titles in this series or just wish to comment on EEPS, please do not hesitate to contact the Series Editor, Warren Burggren (University of North Texas; burggren@unt.edu).

Volume 1: Ecological and Environmental Physiology of Amphibians
Stanley S. Hillmen, Philip C. Withers, Robert C. Drewes, Stanley D. Hillyard

Contents

Acknowledgements

We would first like to thank our mentors Rodolfo Ruibal, Vaughan Shoemaker, Lon McClanahan, Malcolm Gordon, Park Nobel, George Bartholomew, Tony Lee, Otto Sokol, and George Gorman who provided the enthusiasm, opportunity and tolerance for our development as frog doctors. We feel privileged to follow in the footsteps of a host of colleagues in both a real and proverbial sense, who have contributed immeasurably to our understanding of how amphibians work in particular: D. Jones, M. Hedrick, L. Crawshaw, A. Smits, W. Burggren, C.B. Jorgensen, B. Boutilier, W. Weathers, G. Snyder, C. Carey, B. Brattstrom, P.J. Bentley, V. Hutchison, S. Yokota, A. Romspert, H. Harlow, R. Putnam, H. Lillywhite, C. Talbot, D. Toews, K. Olson, D. Jackson, G. Malvin, G. Ultsch, J. Stinner, and G. Thompson. There are so many who have passed before, whom we never got a chance to know except through their work. We hope we have been fair in acknowledging contributions and accept all blame for any mistakes of omission. We owe our own students a deep debt of gratitude for bringing their energy to our lives and contributing to our understanding of how amphibians work in particular: M. Hedrick, S. Landrey, P. Kimmel, A. Zygmunt, M. Baustian, W. Palioca, T. Solberg, D. Crossley, M. Bolke, E. DeGrauw, T. Hancock, T. Hoagland, J. Word, J. Martin, Z. Harlow, A. Truitt, V. Cartledge, L. Wilson, R. Lucas, R. Stoelting, H. Constable and K. McMaster. Other colleagues have shored up weak spots of our knowledge bases: Ronn Altig, Mark Wilkinson, Simon Loader, and John Measey. L. McClanahan, M. Hedrick and C. Cooper provided invaluable assistance and insight through reading drafts of chapters. There is so much more to be learned and we hope this effort will help in passing the baton to the next generation of amphibian physiologists. We also wish to thank our families for their understanding and tolerance of our lifelong pursuit of understanding how amphibians work.

Foreword

The amphibia represent one of the most diverse vertebrate groups. They have successfully adapted to almost every habitat available on planet earth. Given their physiological vulnerabilities, this would seem to be a conundrum. As a group they generally have a limited tolerance to temperature variation. They have very high cutaneous evaporative water loss compared to other vertebrates. Their kidneys are not capable of retrieving water that is lost in urine production, as are many other vertebrates. Hence they produce an extremely dilute urine. With few exceptions, as a group they are tied to water for reproduction. Eggs are deposited in water and tadpoles develop and metamorphose in an aquatic medium. Knowing only these facts, a sagacious physiological ecologist would likely predict that all amphibians would be limited to aquatic or tropical environments.

The authors of this text have combined their talents to amass a large body of literature that amply demonstrates why the sagacious physiological ecologist would be wrong. The text is well written and the chapters are organized in logical succession. For the graduate student who is interested in amphibian physiology, this text should serve as a springboard to a world of potential problems to solve. As everyone who has done research in this area knows, if one door is opened, then many more doors will become apparent and each one hides important information that fills in the missing gaps.

The authors of this text are esteemed scientists who have made seminal contributions to this field as a result of their research. The connecting thread that runs throughout this book is that there is a continuum of information in amphibian physiological ecology from the molecular level to the whole animal level, and no one piece of information or particular area of research is more important than another. The key is understanding how it all fits together to give a whole picture.

Another important point that can be made for this text is that it reveals there is no room for dogma in this field of endeavor, or any other scientific field for that matter. Unfortunately there are times when a researcher becomes overzealous and makes broad predictions based on findings for one or more species. A good example would be the research that demonstrated that bullfrog tadpoles lacked key ornithine cycle enzymes to produce urea as a nitrogenous end product and therefore were obligated to ammonotelism. Unfortunately, that research finding was generalized to all amphibian tadpoles. Now it is known that many species of amphibians have tadpole stages that are fully capable of producing urea throughout their development.

This is an important text because not only does it summarize existing literature in the field of amphibian physiological ecology, it also gives future directions and potential research projects for those who are entering the field. Given the diversity, complexity, and beauty of amphibians it is easy to understand why they have been idolized and revered historically by so many different cultures.

Lon McClanahan
Emeritus Professor of Biology
California State University, Fullerton

1 Introduction

Amphibians are the only descendents of the first terrestrial vertebrates to invade the land; they are morphologically and physiologically radically different from other terrestrial vertebrates. Lacking cleidoic eggs and impermeable skins, they have followed a separate evolutionary trajectory from those of the Amniota, the reptiles, birds and mammals. Amphibians present unique evolutionary models for our understanding of the rigors and solutions to the aquatic–terrestrial transition for all terrestrial vertebrates and insights into how the Amniota have been more successful in exploiting the land.

There are nearly 6000 currently recognized, extant species of amphibians contained within three orders: Anura (frogs), Caudata (salamanders) and Gymnophiona (caecilians) included within the group Lissamphibia ("smooth amphibians"). The larger class Amphibia also incorporates the extinct tetrapods that transitioned from fish to all terrestrial vertebrates. In the following, the term "amphibian" is meant to refer to the extant forms exclusively.

1.1 Amphibian Characteristics and Monophyly

There is little doubt that the orders of lissamphibians, Anura, Caudata and Gymnophiona, together represent a monophyletic group (clade) with respect to other tetrapods, the Amniota; this concept is well supported by morphological, molecular, and studies using combined datasets. However, the relationship between modern amphibians and extinct forms (notably the temnospondyls; Bolt 1977) is less clear. While it is believed that all three lissamphibian orders had a unique common ancestry distinct from the amniotes, no known fossil can be recognized as either an immediate common ancestor, or a plausible sister-taxon of such a group (Carroll 2001). Interordinal relationships of the three major groups of amphibians are discussed below.

The skin of adult amphibians lacks an external layer of epidermal scales or similar protective layer, although dermal osteoderms may be present (Ruibal and Shoemaker 1984); the skin is smooth (except in some anurans and salamanders), highly glandular and with a reduced stratum corneum; most importantly, amphibian skin is permeable to water and respiratory gasses.

With some fascinating exceptions, fertilization is external in most anurans, internal in most salamanders, and internal in all caecilians; the males of all caecilians have a protrusible intromittent organ known as the phallodeum. All lissamphibian eggs are unshelled and lack the specialized membranes (amnion, chorion and allantois) present in other tetrapod eggs; amphibian ova are surrounded by various layers of jelly and are highly susceptible to desiccation. Eggs are almost always deposited either in moist terrestrial environments, adjacent to or directly in fresh water in a number of ovipositional modes (Altig and McDiarmid 2007), singly, in clutches, or masses. Except for the few species in which true viviparity has evolved (a few frog groups, some salamanders and caecilians) or those that exhibit direct development, the majority of modern amphibians have a posthatching, gilled, aquatic larval stage, except caecilians (external gills are usually resorbed immediately after hatching in species with free-swimming aquatic larvae). The tadpoles of frogs are largely herbivorous, while the aquatic larvae of salamanders and caecilians are carnivorous. In all three groups, the larval period usually involves various degrees of developmental change (metamorphosis) to the air-breathing, fully terrestrial, reproductive adult stage; however, there are many exceptions. Frogs and their relatives undergo the most radical metamorphic changes by gradually acquiring limbs, adult jaw structures and the capacity to breathe air. Our current knowledge of anuran larvae has been compiled by McDiarmid and Altig (1999). Larval salamanders and caecilians more closely resemble the adult forms and in most groups metamorphosis centers around loss of gills and assumption of terrestrial life style. Some salamanders such as proteids and sirenids are gilled as adults, and individuals of some other groups such as the ambystomatid salamanders may reach sexual adulthood while retaining the larval gills through neoteny; in these groups, neoteny is probably triggered by environmental factors and is thus facultative. Additional, unique characters point to the monophyly of modern amphibians. Skin permeability, lack of shelled eggs, larval stage and metamorphosis from aquatic to air-breathing terrestrial only partly define the Lissamphibia from other tetrapods, but susceptibility to evaporative water loss and, conversely, water uptake has fundamentally driven the evolution and diversity of the entire group.

There are additional unique characters (synapomorphies) that point to the monophyletic nature of modern amphibians. All adult amphibians possess pedicellate teeth, which include a bicuspid crown (lateral and lingual faces) associated with an underlying pedicel; there is a fibrous zone of weakness between the two (Parsons and Williams 1962). All amphibians possess a thin sheet of muscle on the roof of the mouth beneath the eye, the musculus levator bulbi, that allows the eye to be elevated, a function also unique to the amphibians (Walls 1942; Pough *et al.* 1998). Walls (1942) discussed a unique retinal structure present in all salamanders and anurans, known as a "green rod," perhaps intermediate between cones and the normal "red" rod. Green rods are absent in caecilians but within this group, eyes and associated structures are greatly reduced.

The inner ear of all extant amphibians contains two sensory patches: the papilla amphibiorum, a unique area of epithelium sensitive to sound vibrations of less that 1000 Hz, and the papilla basilaris which is found in other tetrapods. The sound-conducting elements of the middle ear of all amphibians are the columella (or stapes, common to most vertebrates) and the operculum, a structure unique to extant amphibians; it is connected by muscles to the suprascapula (except in caecilians which lack a pectoral girdle); low-frequency sound is conducted to the papilla amphibiorum of the inner ear by the operculum, and higher frequencies are conducted to the basilar papilla by the columella (stapes). In amphibians, the columella is directed dorsolaterally from the fenestra ovalis; in all other tetrapods, the columella (stapes) is directed ventrolaterally (Parsons and Williams 1962; Duellman and Trueb 1986; Pough *et al.* 1998).

Fat bodies in modern amphibians are associated with the gonads and develop from the germinal ridge in all three groups, an origin unique to the Lissamphia. Cutaneous and subcutaneous fat deposits have been identified in some anuran species (Wygoda *et al.* 1987). As mentioned above, all extant amphibians have semipermeable skin and can exhibit various degrees of cutaneous gas exchange with the environment; otherwise, air intake is by means of a unique buccal force pump that involves raising and lowering of the floor of the mouth and the action of valvular nostrils. The skin in amphibians contains both mucous and serous (granular) glands that are considered homologous in the three orders.

Additional synapomorphies linking the modern amphibians include ribs (if present) not encircling the body cavity, the presence of two occipital condyles for articulation of the skull to the vertebral column (craniovertebral joint; most other tetrapods have a single condyle), and similar patterns of reduction of skull elements.

Recent molecular and combined studies have supported the monophyly of modern amphibians with respect to the Amniotes (Hillis 1991; Hedges *et al.* 1993; Frost *et al.* 2006).

1.2 Amphibian Orders

1.2.1 Caecilians (Gymnophiona)

Caecilians are the least understood of the major amphibian groups; they are restricted to the tropical regions of both hemispheres excluding Madagascar and the Australo-Papuan region east of Wallace's Line. Isolated taxa occur on the Seychelles Islands in the Indian Ocean and São Tomé in the Gulf of Guinea. All species are fossorial, semifossorial or aquatic. Caecilians are vermiform in appearance and lack limbs and limb girdles; the tail is reduced or absent, the eyes are reduced and located beneath the skin or the bones of the skull, and the body has distinct annular grooves similar in external appearance to the costal grooves of salamanders. Each primary groove is associated with a single trunk vertebra. Many species have secondary and tertiary grooves, and in many species dermal "scale-like" structures within the grooves are unique to caecilians among the Lissamphibia. Other gymnophionan synapomorphies include the presence of the male phallodeum (an intromittent organ, all fertilization in the caecilians is internal), a specialized, protrusible sensory tentacle located between the mouth and the eye, a partially roofed, somewhat kinetic (zygokrotaphic) or completely roofed (stegokrotaphic) skull, and a unique dual jaw-closing mechanism involving the usual mandibular adductor muscles but also specialized elements of the musculus interhyoideus. The left lung is reduced or absent in some species, and one unique, enormous (700+ mm) aquatic species, *Atretochoana* is lungless. Aquatic eggs and larvae (external gills lacking after hatching but gill slits present) characterize a number of species, others exhibit direct development and about 75% of species are viviparous. The female provides direct sustenance to the developing young. Until recently, about 100 species of caecilians were recognized in five (Hedges *et al.* 1993) or six (Nussbaum and Wilkinson 1989) families; Frost *et al.* (2006) recognized three families based on a combination of morphological and molecular data, and the following is based upon their analysis.

The family Rhinatrematidae (two genera/nine species) is basal to the remaining groups of caecilians; these caecilians retain a number of characters considered primitive among the Gymnophiona, such as a true tail, mouth at the tip of the snout, and positioning of the tentacular opening. Skulls are kinetic (zygokrotaphic). This group of small Neotropical caecilians is poorly known but suspected to be semifossorial or perhaps even a surface dweller. All rhinatrematids are presumed to be oviparous; larvae are known for some species.

The Ichthyophiidae (three genera/44 species), including the formerly recognized Uraeotyphlidae, exhibit a number of characters that appear intermediate between the primitive rhinatrematids and more derived caecilians, including retention of a true tail but more solidly roofed skulls (stegokrotaphy), although the southern Indian *Uraeotyphlus* have somewhat kinetic skulls. Ichthyophiids are oviparous with aquatic larvae, although *Uraeotyphlus* may have direct development (larvae have never been observed). Ichthyophiid species are widely distributed from India and Sri Lanka, through Southeast Asia and Indonesia, west of Wallace's Line. The Caeciliidae (23 genera/120 species), including the formerly recognized Scolecomorphidae and Typhlonectidae (Frost *et al.*, 2006), is the largest and most diverse of caecilian families. They are widely distributed in the Neotropics, in the western and eastern tropics of Africa, and the west coast of India and Bangladesh. Three genera are endemic to the Seychelles Islands of the Indian Ocean, and a single endemic species (*Schistometopum thomense*) is known only from the island of São Tomé, off the West African coast. All caeciliids lack a true tail, most are burrowers with fully roofed, stegokrotaphic skulls; exceptions are the South American typhlonectines, most of which are aquatic, and some of the African species. Aquatic species are laterally compressed in the posterior parts of the body with tail fins and fins on the posterior part of the dorsum. The reproductive diversity of caeciliids ranges from oviparous with aquatic larvae, oviparous with direct development, to a number of species with viviparity. Parental attendance of egg clutches has been observed in some species.

1.2.2 Salamanders (Caudata)

The majority of salamanders are rather generalized vertebrates possessing four limbs (except sirenids) that are roughly equal in size and a long tail, in both adult and aquatic larval forms; only the Nearctic Sirenidae lack a pelvic girdle and hind limbs; and limbs are greatly reduced in all three species of the Amphiumidae (*Amphiuma*) and some species of plethodontid salamanders. All salamanders lack the middle ear cavity, and external auditory apparati are absent. The majority of salamander species are terrestrial but many return to the aquatic environment for reproduction; some New World salamandrids such as *Taricha* may travel up to 1000 m to breed (Twitty 1966), and adults of this genus undergo epidermal changes during aquatic breeding periods. Fertilization is internal in all except cryptobranchids, hynobiids and sirenids; in the remainder, males deposit spermatophores that females pick up with their cloacal lips and store in the spermatheca. Larvae have true teeth, gill slits and external gills but lack eyelids and mineralized maxillae; four limbs and a finned tail are present at hatching and elements of the palate are wholly reorganized during metamorphosis. Incomplete metamorphosis (obligate neoteny) is a characteristic of all cryptobranchids, sirenids, amphiumids and proteids; facultative metamorphosis is present in some groups in the Ambystomatidae (absent in rest of the Lissamphibia). Many species of plethodontids lay eggs within which the larvae develop directly to hatchlings, without a gilled larval stage, and females of some members of the salamandrid genera *Salamandra* and *Mertensiella* retain larvae within the body to advanced stages of development (ovoviviparity) or give birth to fully metamorphosed young (viviparity). In the salamandrid genus, *Notophthalmus,* a terrestrial, immature (eft) stage may occur for some years during which most development occurs. Once efts undergo a second "metamorphosis" to the adult stage they remain permanently aquatic; in other populations, development may be through the more standard aquatic larva mode.

There are over 550 known species of salamanders, in at least 63 genera in nine families. The majority of lineages are Holarctic in distribution, ranging as far north as the Arctic Circle in Siberia, but the most speciose family, the lungless salamanders (Plethodontidae), exhibit a significant radiation in the Neotropics, south to the Tropic of Capricorn.

The families Cryptobranchidae (two genera/three species) plus Hynobiidae (19 genera/50 species) form a clade that is basal to the rest of the extant families of the Caudata (Frost *et al.* 2006). Most members of both families have external fertilization; the cryptobranchids are wholly aquatic, exhibit incomplete metamorphosis and have a disjunct Palearctic distribution; the Asian hynobiids are largely terrestrial but the family includes one genus (*Onychodactylus*) that is aquatic. The Proteidae (two genera/six species) and Sirenidae (two genera/four species) are highly paedomorphic and retain many juvenile characteristics as adults. Members of both families are fully aquatic and retain external gills and well developed caudal fins as adults; eggs are aquatic and fertilization is presumed to be external. The Proteidae are unique among the Caudata in having a diploid chromosome number of 38.

The Nearctic Ambystomatidae (two genera/35 species) have internal fertilization, as do members of their sister taxon, the family Salamandridae (20 genera/74 species), and all remaining families in the Caudata. Most ambystomatids are primarily terrestrial, returning to water only to breed, although some species deposit eggs on land; some species in this group exhibit facultative metamorphosis and retain larval characteristics such as gills into adulthood. Ambystomatids are usually robust and are characterized by costal grooves on the trunk. The salamandrids of Europe, eastern Asia and North America are predominantly aquatic as adults, many species having fins on the tail and mid-dorsum. While most groups have aquatic larvae, some also have a prolonged terrestrial juvenile phase (*Notophthalmus*), and a few others exhibit ovoviviparity (*Mertensiella, Salamandra*). The Salamandridae include a number of species with dermal toxins and bright coloration, presumed to be aposematic. These toxins are very similar to tetrodotoxin, originally described from pufferfish; it is the most poisonous non-protein substance known (Stebbins and Cohen 1995).

The Rhyacotritonidae (one genus/four species) of the northwestern United States occur in small cold-water streams. They have greatly reduced lungs that lack alveoli and a narrow tolerance for warm temperatures (CTM = 28°C).

The Amphiumidae (one genus/three species), the smallest family within the Caudata, is the sister-group to the largest, the lungless salamanders of the Family Plethodontidae (24 genera/377 species). Amphiumids, restricted to the southeastern United States, are aquatic and somewhat paedomorphic in such features as absence of eyelids and retention of gill slits; both girdles and fore- and hind limbs are present although vestigial. Eggs are laid in burrows near water and attended by the parents. The Plethodontidae of the Palearctic, Nearctic and Neotropical regions is the largest and most diverse of the salamander families. All plethodontid species are lungless (a condition that seems to have evolved independently in some stream-dwelling hynobiids and salamandrids) and have nasolabial grooves in metamorphosed adults, which may function in chemoreception. The lungless salamanders show a general trend toward terrestriality; the diversity of reproductive modes ranges from aquatic eggs and larvae to direct development, although several species are permanently aquatic and several subterranean taxa are obligate neotenes. Many Neotropical taxa are arboreal with prehensile tails and webbed feet. The most advanced feeding mechanism among the Caudata, which involves projection of the tongue, is found among the bolitoglossine plethodontids, although apparently similar mechanisms have evolved independently in two lungless groups of the Salamandridae.

1.2.3 Frogs (Anura)

Anurans are certainly the most successful and diverse of the lissamphibian orders; there are over 5300 species in 42 families (Frost *et al.* 2006) found in diverse habitats throughout the world, except above very high latitudes in both hemispheres, most oceanic islands (but see Loument 1992; Vences *et al.* 2003; Measey *et al.* 2007) and the most low-lying, arid regions.

A large suite of synapomorphies supports the monophyly of the Anura within the Lissamphibia; some include reduction and fusion of bones of the skull, for example the frontal and parietal bones are fused, dentition is absent from the lower jaw except for one species, reduction in the number of presacral vertebrae from nine in basal groups to as few as five in some highly derived groups (although the most common number of presacral vertebrae is eight); the caudal vertebrae in all extant species of frogs are fused into a single element known as the urostyle (or coccyx, in older literature), and the ilia of the hind limbs are elongate and directed anteriorly. Hind limbs and feet are greatly elongated in most species except for some fossorial and semifossorial species, and a true tail is absent. The typical anuran tongue is unique among vertebrates in being attached to the anterior part of the floor of the jaw (except in rhinophrynids) and free posteriorly (the tongue is absent in the basal Aglossa).

Fertilization is most commonly external, and eggs are usually deposited in or near water; internal fertilization is known in some species such as *Ascaphus*, two species of *Eleutherodactylus* and in small African bufonids of the genera *Mertensophryne* and *Nectophrynoides.* Anuran larvae are unique within the Lissamphibia; they are limbless, lack true teeth, have a large gill basket and internal gills, and, unlike caecilians and salamanders, undergo a radical reorganization of the *Bauplan* during metamorphosis to terrestriality: fore- and hind limbs complete development, and the tail and gills are lost. Aquatic anuran larvae (tadpoles) are usually filter-feeders or herbivorous. Tadpoles have been grouped into a number of basic morphotypes (Orton 1953, 1957; Starrett 1973), and our accumulated knowledge of all aspects of tadpole biology has recently been reviewed by McDiarmid and Altig (1999). A number of frogs exhibit direct development, such as species of the African arthroleptid *Arthroleptis* and the Asian rhacophorid, *Philautus*; ovoviviparity, here defined as eggs remaining associated with the body of the adult during development, has evolved independently in a number of lineages including Neotropical Pipidae (*Pipa)*, Cyclorhamphidae (*Rhinoderma*) and

Amphignathodontidae (*Gastrotheca*) and the Australian myobatrachid, *Rheobatrachus*. True viviparity (maternal nourishment of developing embryos) appears in the Africa bufonid genus, *Nectophrynoides.*

The large and varied clade Anura has been the subject of numerous and contested phylogenetic analyses; the advent of molecular sequence data and analysis has added significant breadth and depth to modern phylogenetic constructs but the enormous data set is not nearly comprehensive. Studies presented so far remain contentious, especially with regard to methodology. Some of the more family-inclusive studies of the order Anura in recent years include Ford and Cannatella (1993), Haas (2003), San Mauro *et al.* (2005) and Frost *et al.* (2006). Some major non-molecular characters that have traditionally been used to define the subordinal groups of frogs include a variety of features of the skull, number of vertebrae and shape of vertebral centra, the configuration of the pectoral girdle, musculature, tadpole morphology and certain aspects of behavior (Pough *et al.* 1998). Extant frogs have been divided into two, sometimes three, suborders by various authors; most authors accept that these groups may not be phylogenetically valid (paraphyletic). Those basal lineages which evidently diverged earliest in geologic time (San Mauro *et al.* 2005) and exhibit many plesiomorphic (primitive) character states have been referred to as the Archaeobatrachia (Reig 1959) and usually include the tailed frog (*Ascaphus*), the unique species of the genus *Leiopelma* of New Zealand, the fire-bellied toad group Bombinatoridae and the Discoglossidae of Europe and Asia. An intermediate group, suborder Mesobatrachia, was defined by Laurent (1979) to include a number of families that exhibit intermediate or "transitional" characters and character states. This group usually includes the disjunct, aquatic Pipidae of the Afrotropics and Neotropics, the monotypic, burrowing family Rhinophrynidae, the largely fossorial pelobatids and scaphiopodids of arid Palearctic regions, the Pelodytidae of Europe and the Near East, and the Megophryidae of the Oriental Region (Ford and Cannatella 1993). The remaining group, suborder Neobatrachia (Reig 1958), contains the majority of families, the preponderance of species and the greatest diversity among the extant anurans, and a number of authors have suggested that there is support for the monophyly of this group.

The assemblage of derived groups (Neobatrachia), comprised of some 33 families (Frost *et al.* 2006), was divided into two major groups based on E. D. Cope's (1865) recognition of two basic types of pectoral girdle configuration: the arciferal condition exhibited by the great majority of species in which epicoracoidal cartilages overlap ventromedially, and the firmisternal condition, in which the epicoracoidal cartilages are fused with the coracoids at the midline and the sternum is fused directly to the pectoral arch (Pough *et al.* 1998). Cope (1865) termed the former group the Arcifera, later referred to as the Bufonidea (Duellman 1979; Laurent 1979), Hyloidea (Dubois 1983) and Hyloides (Frost *et al.* 2006). Frog groups with the firmisternal girdle condition were initially grouped as the Raniformia by

Cope (1865) but subsequent authors have suggested two lineages of firmisternal frogs usually referred to as the Ranoidea and the Microhyloidea (Duellman and Trueb 1986). Although a number of intermediate conditions between the two major types of pectoral suspensoria have been described, it is widely accepted that there are probably three broad lineages among the "higher frogs," with the arciferal girdle arrangement representing the plesiomorphic condition of the structure. Among the many neobatrachian frog families with the arciferal pectoral girdle condition are toads and their relatives (Bufonidae, *sensu lato*), all of the Neotropical, Holarctic and Australian treefrog groups and the majority of remaining neobatrachian Neotropical families, including the most speciose vertebrate genus, *Craugaster* (formerly *Eleutherodactylus* [in part, see Frost *et al.* 2006). Firmisternal families traditionally grouped under the subfamily Ranoidea include the "typical" frogs of the genus *Rana*, widely studied in Europe and North America, Asian treefrogs (Rhacophoridae, *sensu lato*), and all of the neobatrachian anuran families of the Afrotropical Region. Members of families traditionally placed in the Microhyloidea, which are fermisternal but retain a cartilaginous sternum, include several families and over 60 genera of mostly terrestrial species with distinctive aquatic larvae (see McDiarmid and Altig 1999). Species of this group are broadly distributed in southeastern United States south to below the Tropic of Capricorn in the Neotropical Region, sub-Saharan Africa and Madagascar, southern Asia, and Indo-Australia.

After a century and a half, the template for our knowledge of basic anuran anatomy is still *Anatomie des Frosches*, an exhaustive illustrated investigation of the European frog, *Rana esculenta*, written, revised and expanded over a 40-year period (1864 to 1904) by three distinguished German anatomists, Alexander Ecker, Robert Wiedersheim and Ernst Gaupp of the University of Freiburg (Ecker 1864; Gaupp 1904). Similarly, much of our knowledge of anuran physiology has come from observation of and experimentation with a relatively few species, and these are from distantly related groups: the Bufonidae (many species but most frequently *Chaunus* (*Bufo*) *marinus*, the Ranidae (in Europe, *Rana esculenta*; in North America, *Lithobates* (*Rana*) *catesbeiana* or *L. pipiens*), the Pipidae (*Xenopus* spp.) and the arid-adapted Scaphiopodidae (*Scaphiopus* spp. and *Spea* spp.) (nomenclature of Frost *et al.* 2006). A major systematic revision of the amphibians by Frost *et al.* (2006) has resulted in a number of proposed nomenclatural changes; among these are the scientific names of the American bullfrog, the leopard frog and the marine toad, species that have been widely used for experimentation in laboratories. In order to avoid confusion in the following chapters we will retain the scientific names as used in the original publications cited; thus although they are correctly *Lithobates catesbeiana* and *Chaunus marinus*, the American bullfrog and marine toad will be usually referred to by the traditional names, *Rana catesbeiana* and *Bufo marinus*, respectively.

1.3 Environmental Biophysics of Gas, Water and Ion Exchange: Aquatic–Terrestrial Transition

Amphibians present an interesting vertebrate model system to study the transition from an aquatic to a terrestrial environment. This transition was a significant event from a vertebrate evolutionary perspective. Modern day amphibians are the closest tetrapod descendants of the fish that crawled out on land and most are still tied to water for reproduction. A physical understanding of the characteristics of each environment will provide unique insights into the stresses amphibians faced in this transition as well as providing an understanding of how the environment influences their physiology. As physiology is the study of the processes of life and life is dependent upon exchanges of energy, water and gasses with the environment it is essential to understand how these fluxes are quantitatively described.

The flux of any material can be described in the general format:

Flux/area = driving force/resistance to flux

or

Flux/area = conductance to flux × driving force

Obviously, resistance = 1/conductance. We will use both conductance and resistance in the following chapters to understand fluxes. There is no correct answer as to which is more appropriate. Resistance is a simpler concept in a description of a flux that entails series resistance elements, as the total resistance is simply the addition of individual resistances ($R_{total} = R_1 + R_2 + R_3$ etc.) while a series conductance solution would be $1/C_{total} = 1/C_1 + 1/C_2 + 1/C_3$ etc. Conductance is a simpler solution to fluxes in parallel as total conductance is again the addition of individual conductances, $C_{total} = C_1 + C_2 + C_3$ etc. where parallel resistances would sum as $1/R_{total} = 1/R_1 + 1/R_2 + 1/R_3$ etc.

Consequently to describe any flux we need to understand both the driving force and what physical, morphological and physiological variables are involved in determining the resistance (or conductance) to that flux. Let us start with the driving forces available to move gasses, water and/or solutes. We will use an irreversible thermodynamic approach, non-equilibrium, where the driving forces will be assumed to be linear and the forces can interact. An excellent synthesis of these processes can be found in Nobel (1999). The energy available to move gasses, water or solutes between the organism and environment can entail some combination of the following forces: concentration, pressure, and voltage.

1.3.1 Gas Flux

The basic gas law is:

$$PV = nRT$$

This can be rearranged to

$$1/RT \times P = n/V$$

Since n/V is synonymous with concentration (mol/V), it is clear that concentration is dependent upon the product of pressure and a capacitance coefficient, 1/RT. The capacitance coefficient is mol/l P, that is the concentration of gas per unit of partial pressure. In aerial situations the capacitance coefficient is essentially the same for all gasses only varying inversely with absolute temperature (Fig. 1.1). This is not true for aquatic situations as R is intertwined with the physical solubility and chemical binding of the particular gas with the water and its chemical constituents. Consequently the capacitance coefficient in aquatic situations is particular for both a specific gas species and a specific temperature and solute concentration (see Fig. 1.1). The product of the diffusion coefficient for any molecule and its capacitance coefficient is known by physiologists as Krogh's constant of diffusion. Diffusion in tissues and water is described by the product of Krogh's constant of diffusion (K) and the driving force for the molecule.

Increasing temperature and solute concentration decrease the capacitance coefficient for any gas species. In the two gas species CO_2 and O_2, the solubility and thus capacitance for CO_2 is about 25–30 times greater than O_2 at the same temperature and solute concentration in water (see Dejours 1975). Therefore at equivalent pressures the concentration of CO_2 will be 25 times the concentration of O_2 in water. The capacitance coefficient for CO_2 in water is greater than in air when water is less than 20°C, for O_2 it is always greater in air. This means that metabolic O_2 consumption and CO_2 production by an animal will more markedly alter the environmental gas concentration in water, and this will be most dramatic for O_2 in water (25–30 times). This can be substantial in waters rich with plant, animal and microbial respiration characteristically driving down O_2 concentrations at night and concentrations rising during the day with photosynthetic

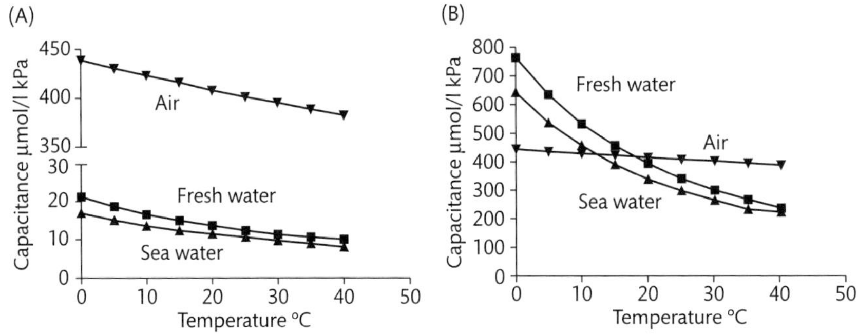

Fig. 1.1 Capacitance coefficients for (A) O_2 and (B) CO_2 in air and water of different salinities at different ambient temperatures.

activity. An aerial environment from the perspective of O_2 provides an environment where an animal's metabolism will have less effect on environmental O_2 concentrations, while an aquatic environment can serve as a CO_2 sink for the CO_2 produced during metabolism because of the greater CO_2 capacitance. One advantage of an aquatic to terrestrial transition is entering an environment where the concentration of O_2 is greater and less fluctuating, hence there should be less interference with aerobic metabolism on a daily basis. Even at the top of Mt Everest (8850 m) there is a 24 times higher concentration of O_2 than in fully saturated water at 10°C (see Fig. 1.2).

Since the highest known distributional limit for amphibians is 4–5000 m, the concentration of O_2 is still 14 times the saturated concentration of O_2 in water at sea level. Therefore altitude should be less of an environmental stress on aerobic metabolism in terrestrial amphibians but could present major issues for aquatic amphibians.

Another consequence of the difference in capacitance coefficients leading to differences in air and water concentrations of O_2 and CO_2 is that concentration differences will not describe the driving force for gas movement. The driving force for gas movement is based on a difference in pressure. This means that P = concentration × RT. From Dalton's law of partial pressures, the pressure of any gas species is equal to its fractional percentage of the gas mixture. Since O_2 makes up 20.95% of the total gasses in the atmosphere, its partial pressure at sea level would equal 0.2095 × 103.4 kPa for dry air or 21.7 kPa. The SI unit for pressure is the pascal (Pa). The literature is rich with other terms for pressure. The medical and physiological literature commonly uses units of height for either mercury (Hg) or water. An mmHg = 1 torr = 133 Pa, while a cmH_2O = 99 Pa. The plant and soil literature frequently uses the terms bars or atmospheres. An atmosphere = 101.3 kPa, while a bar = 100 kPa. We will attempt to use SI units throughout the

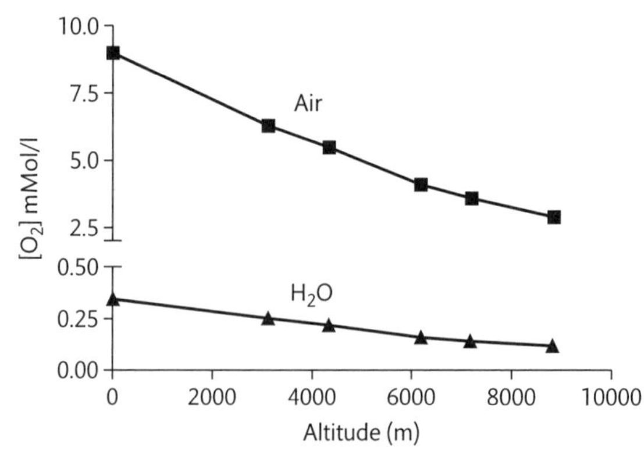

Fig. 1.2 Concentrations of O_2 in air and water at different elevations at 10°C.

book, even though much of the literature has been tardy in adopting this international convention.

The flux of any gas in air (as opposed to water or tissue) from point 1 to point 2 can be described from Fick's Law of planar diffusion as:

Number of Moles/time = D A ($P_1 - P_{2)}$/distance

where: D = Diffusion Coefficient for the gas at a specified temperature, A = surface area being monitored, $P_1 - P_2$ = the difference in pressure from point 1 to point 2, and distance = the linear distance from point 1 to point 2.

The media air and water (or tissue) are functionally different in that water contains both the diffusion coefficient and capacitance coefficient, while air has only the diffusion coefficient for O_2 and CO_2. The diffusion coefficient for air compared to Krogh's diffusion coefficient for water for both O_2 and CO_2 are about 7000–8000 times greater with $DO_2 > DCO_2$; as CO_2 is a larger molecule it diffuses slower. A good rule of thumb for estimating the effect of molecular weight on diffusion coefficients for small molecules is that $D \propto 1/\sqrt{\text{mol weight}}$. Diffusional gas fluxes will be about 7500 times faster in an aerial environment for the same pressure difference, another potential selective advantage for the evolutionary and developmental transitions from aquatic to terrestrial environs by amphibians.

1.3.2 Water Flux

The driving force for water flux into or out of an organism is a function of the water potential gradient where water potential refers to the free energy of water. The water potential (ΔG_{H_2O}) of pure water or water in a solution is a function of the mole-fraction (X_A) which is the ratio of the number of water (nW) molecules to the total number of solute and water molecules (nS + nW) in a given volume of solution (Marshall 1978):

$$\Delta G_{H_2O} = RT \ln X_A = RT \ln nW/(nW - nS)$$

For a molal solution a given amount of solute is added to 1 kg of water and osmotic concentration is expressed as mOsmoles/kg while for a molar solution less water is needed to make more concentrated solutions up to a 1 liter volume and the osmotic concentration is expressed as mOsmoles/l. The mole fraction for pure water is 1 and the water potential (ln 1) is zero. As more solute is added the mole faction becomes less than 1 and the water potential (ln <1) is negative so that water moves passively down a free energy gradient. Osmolality is theoretically a preferable term since the free energy of water varies only as a function of the added solute. However, for dilute solutions, in the order of amphibian body fluids, the difference between osmolar and osmolal solutions is negligible.

Historically, the force generated by a difference in solute concentration across a membrane was termed osmotic pressure, a concept that Pfeffer (1877) developed using artificial membranes. Basically, osmotic water movement from pure water to concentrated solutions generates a hydrostatic pressure that can be measured with a manometer. When the hydrostatic pressure balances the osmotic pressure there is no net water flux, so osmotic pressure for a given solution is measured as a function of the equivalent hydrostatic pressure. Note in this convention that a greater solute concentration will produce a greater osmotic pressure but at the same time the free energy of water will become more negative. Van't Hoff (1887) showed that the ideal gas law could be applied to dilute solutions and related the osmotic pressure (π) exerted by a solution relative to pure water in terms of the ideal gas law:

$$\Pi = CRT \quad \text{or} \quad C = \Pi/RT$$

where C is the concentration, R is the gas constant and T the absolute temperature. Note that for pure water C and thus $\Pi = 0$. Historically, values for π have been measured in units of atmospheres or bars but more recently the SI units of kPa is used (1 bar = 1.01 atm = 100 kPa). Concentrations determined from osmotic pressure are expressed as (milli)osmolar or (milli)osmolal and refer to the ideal behavior of the solutes. When estimating (milli)osmolar concentrations from molar concentrations it is important to correct for non-ideal behavior. For example, the osmotic coefficient of NaCl is 0.93 so a 100 mM solution will have an osmotic concentration of 186 mOsm. Conversely, to make a 200 mOsm solution requires 107.5 mM NaCl per liter of solution. Instruments currently used to calculate osmotic concentration evaluate the free energy of water based on the colligative properties of freezing point depression and reduced vapor pressure. However, the same considerations regarding osmolarity, osmolality and ideality apply to these techniques and may become important for studying osmotic balance of animals in brackish water habitats.

1.3.3 Soil Water

The relationship becomes a little more complicated when it involves water fluxes between an organism and water in the soil. Because soil has both air and water interfaces and because of the surface tension between water and air, the surface area of the soil particles will interact with the water molecules and effectively hold onto them, preventing their interaction with solutes. This binding of water is dependent upon the surface area of the soil particles. Small particles will have greater surface area/volume hence the matric potential (τ) will tend to be greater in these soils and water will be less likely to move from these soil surfaces, and there will be less water available for solutes to dissolve in this interstitial water. Less water raises the osmotic concentration of the water between the soil particles. Clay soils are described as having particle sizes $<2\,\mu m$ across, silt with particles 2–50 μm across and sand having particles 50–2000 μm across. This difference in particle size can lead to a two to three orders of magnitude greater surface area/mass in clay compared to sandy soils. Hence the ability to hold water is greater in clay versus sandy soil. From an amphibian perspective it will be harder to extract water from clay versus sandy soil. The overall water potential of soil then becomes:

$$\Psi_{soil} = -(\Pi_s + \tau)$$

where τ = matric potential.

Actual soil is obviously a mixture of various particle sizes, so practically soil water potentials are determined by exerting external air pressures on the soil until water is released. This is characteristically determined over a wide range of soil water contents, so by simply knowing soil water content one can extrapolate what the equivalent water potential would be if one has the water potential–soil water curve (Fig. 3.5). The water contents for soils at a $\Psi = -1.5\,\text{MPa} = 615\,\text{mosm/l}$ would be 15% for clay soils and 3% for sandy soil. Consequently, water potential is frequently expressed in units of pressure. This means one needs to be able to relate osmotic concentration to pressure. This relationship is:

$$\Psi = -2.44\ \text{MPa}\ C_{osm}$$

Fresh water has an osmotic concentration of about 0.02 while sea water is 1 osm/l, and these represent the extreme environmental aquatic environment they face. Amphibian osmotic concentrations run from about 0.2 to almost 0.9 osm/l. This would correspond to water potentials of −0.49 to −2.2 MPa.

1.3.4 Air

The water content of air is usually expressed as % relative humidity = $100\ C_{wv}/C_{wv\ saturated}$. The $C_{wv\ sat}$ to air temperature is exponential (Figs. 1.3 and 3.7).

The water potential of air is described by:

$$\Psi_{air} = 135\ \text{MPa}\ \ln(\text{RH}/100)$$

This gives us the ability to compare both aquatic and aerial environments from a water potential perspective. Normally, aerial environments can run from 5% RH (−404 MPa) in a desert to almost 99% RH (−1.36 MPa) when it rains. When compared to sea water (−2.44 MPa) and fresh water (−0.05 MPa) it is clear the driving force for water loss ($\Psi_{animal} - \Psi_{environment}$) is extreme in a terrestrial environment in comparison to the most hostile aquatic situation, as sea water is only the terrestrial equivalent of 98.2% RH. It also points out that increasing the osmotic concentration of body fluids will not substantially alter the driving force for water loss in terrestrial situations even though this is an important adaptational strategy in dealing with acclimation either to increasing salinities for aquatic species or extracting water from soil for terrestrial species (Chapter 4). The transition from

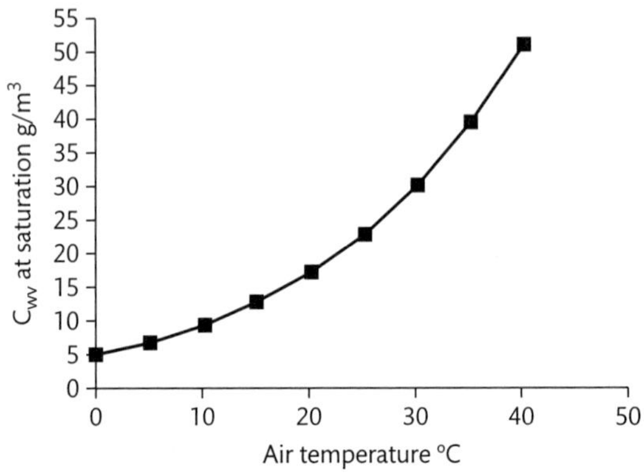

Fig. 1.3 The concentration of water vapor in air when fully saturated (100% relative humidity) at different air temperatures.

an aquatic to a terrestrial environment developmentally, seasonally or evolutionarily is certainly an osmotically hostile one.

Normally, from an organismal perspective, the diffusional flux is not quite this simple as the medium and diffusion coefficients change from cell to environment and we have the potential of convective processes (wind, ventilation, perfusion) in both the skin and immediate environment. This then expands the number of variables and brings us to the more general equation of:

$$\mathbf{Flux} = \Delta \mathbf{P}/\mathbf{R}_{\text{total}} \quad \textbf{or} \quad \mathbf{C}_{\text{total}} \times \Delta \mathbf{P}$$

If we consider the diffusional movement of a gas across a respiratory membrane it entails a diffusional movement from the environment to the surface of the membrane (boundary layer) and a diffusional movement across the cells of the membrane to the capillaries that will either absorb the O_2 or release the CO_2. Hence rather than a single D and distance, we have two separate "media" hence two separate conductances or resistances in series for the diffusional process. This leads to a consideration of the resistance associated with the environment or boundary and the resistance inherent to the physiology and morphology of the organism.

1.3.5 Boundary Layers

Between any organism and the medium in which it resides, air or water, is an unstirred boundary layer. Diffusional movements by molecular movement of gasses, solute or solvent flow across this boundary layer in accordance with Fick's law of planar diffusion described above. The resistance to diffusion can be described as:

$$\mathbf{R} = \mathbf{distance/DA}$$

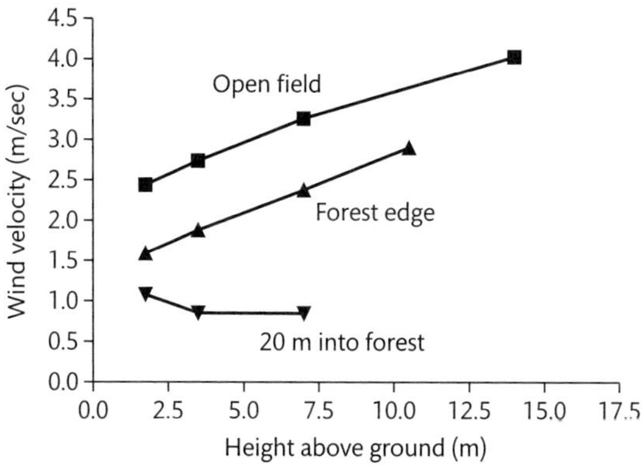

Fig. 1.4 Mean wind velocities over 5 years at various heights, in different habitats, at the interface of an open field and pine forest; data from Raynor (1971).

From an environmental perspective, boundary layers of equivalent thickness in air versus water will be only 1/7500 as resistive as the diffusion coefficients are 7500 times greater in air. Boundary layers are established as a consequence of the frictional resistance between the organism and the fluid, essentially stopping the moving fluid in contact with their surface. Boundary layers are not uniform across the surface of an organism, and are generally thinner at the leading and trailing edges of an object. What are the environmental variables that determine the thickness of a boundary layer? The principal variables are air or water velocity, a characteristic dimension of the organism in the direction of the moving fluid and the shape of the organism. The general equation for the thickness is:

Boundary Layer Thickness = "Factor" X length$^{0.3-0.5}$/wind velocity$^{0.5-0.7}$

The actual values of the factor and exponents are normally empirically determined and are influenced principally by shape (see Chap. 3.2). As most amphibians are small, their actual boundary layer may be more determined by the object they are perched upon than their own boundary layer: that is they sit within the boundary layer of the object. Wind velocity in the environment is characteristically variable and influenced by the objects in the environment. Objects tend to increase the frictional form drag of moving fluids, hence slow down wind velocity. Wind velocity tends to increase with distance from the substrate. A representative example of wind measurements for a New York pine forest from Raynor (1971) has been replotted from 5 years of data in Fig. 1.4.

Consequently, boundary layers will tend to be largest for larger animals on soil in a forest and least for small species perched on small objects above the ground. Boundary layers will tend to be larger for equivalent objects in air compared to water, as air has a kinematic viscosity (ordinary viscosity divided by density) about 15 times greater than water at 20°C (1.5×10^{-5} versus $1 \times 10^{-6}\,m^2/s$).

Let us take a practical example of the diffusion of gas and water across the boundary layer to the surface of the epidermis in the amphibian integument. First let us consider what shape we wish to investigate. The two extremes might be a cylindrical shape approximating many urodeles and caecilians versus a planar state like many arboreal tree frogs. The equations for calculating the mean boundary layer thickness (see Nobel 1999) are:

planar

$$\textbf{Boundary Layer Thickness (mm)} = \textbf{4 length (m)}^{0.5}\textbf{/wind velocity (m/s)}^{0.5}$$

cylinder

$$\textbf{Thickness} = \textbf{5.8 diameter}^{0.5}\textbf{/wind velocity}^{0.5}$$

If we assume the length and diameter are 1 cm, the effective boundary layer thicknesses are 0.82 mm and 0.57 mm for the cylinder and planar surfaces respectively at wind velocity of 0.5 m/s, and 0.47 and 0.33 mm at a wind velocity of 1.5 m/s. The next distance that the O_2 must traverse is the epidermis of the skin, then it must traverse the non-vascular dermis until it reaches the capillaries. We then have the $R_{total} = R_{bl} + R_{epi} = R_{der}$. The amphibian epidermis is two to six layers thick and can vary developmentally, interspecifically and with body region. Each of these epidermal cells is about 5 µm tall so the total epidermal thickness, diffusional distance, is 10–30 µm. If we assume 20 µm as an average, then

$$\mathbf{R_{epi} = 2 \times 10^{-5}\,m/Diffusion\ Coefficient\ O_2 \times capacitance\ coefficient\ O_2}$$

See Chap 2.4.3 for empirical solutions.

1.3.6 Evaporative Water Loss

Evaporative water loss (EWL) is measured by flowing dry air over an animal in a confined chamber and measuring the water content of the incurrent and excurrent air. EWL can be calculated from the equation:

$$\mathbf{EWL = [(C_{wv} \times RH\ of\ the\ incurrent\ air) - (C_{wv} \times RH\ of\ the\ excurrent\ air)] \times F}$$

In this equation, C_{wv} is the water vapor density of water-saturated air (mg/cm^3) and relative humidity is the fractional saturation (0–1) at the experimental temperature. C_{wv} values are generally obtained from reference sources, many of which are available from internet sites. F is the flow rate over the animal. When F is expressed as cm^3/h, EWL is in units of mg/h. In some cases the area-specific rates of water loss are obtained to better characterize the properties of the skin. This can be accomplished by calculating the surface area from the body mass

(Hutchison *et al.* 1968) or applying a small chamber of known area to the surface of the skin (Gomez *et al.* 2006). In this study a total resistance value, R_t, was calculated from the driving force for evaporation and the rate of water loss:

$$R_t = (C_{wv} \text{ of the skin surface} - C_{wv} \times \text{RH of the excurrent air}) \times \text{EWL}^{-1}$$

When EWL is expressed in area-specific units (mg water/$cm^2 \times s$), R_t values are in units of s/cm. In addition, boundary layers can affect evaporation so flow rates are critical and contribute to the resistance to water loss. R_t is then the sum of the skin resistance (R_s) and boundary layer resistance (R_b) and $R_s = R_t - R_{bl}$. R_{bl} can be estimated from the EWL of a saturated tissue of the same surface area as the animal under the same conditions of flow and temperature (Gomez *et al.* 2006). Lillywhite (2006) estimated R_s values from a large selection of literature values for EWL (which they termed transepithelial water loss, TEWL). They ranged from 0–5 s/cm for semiaquatic ranids and bufonids to over 200 s/cm for cocoon-forming anurans such as *Cyclorana austrailis* with a formed cocoon in place and as high as 400–900 s/cm for *Chiromantis* spp.

1.3.7 Ion Exchange

The aquatic to terrestrial transition is dramatically different from the perspective of ion movement. The aquatic world has ions in solution. In fresh water ions are in lower concentration in the environment than in the amphibian. In saline environments the ions are in greater concentration than in the amphibian's body fluids. Hence ions will tend to be passively lost in fresh water and gained in saline environments. We shall see that amphibians generally have active transport systems in their skin to move both sodium and chloride against their concentration difference in fresh water environments, counteracting the dilution effect of fresh water. Since ions do not exist in air, ion movement from the environment to the animal or out across the skin is impossible. Consequently, ion exchange with the environment across the skin is only possible in aquatic situations.

One of the confusing issues in the literature on cutaneous ion exchange is that most of the biophysical equations deal with delineating standard "Ussing cell" conditions, which entail equal concentrations on both sides of the skin or bladder to study ion transport. This has proved an exciting model for the study of active transport mechanisms (see Chapter 3) but has very limited applicability to actual physiological conditions in which environmental concentrations for amphibians are normally very much lower than body fluid concentrations. Hence the normal textbook delineation is presented in terms of equilibrium thermodynamics rather than an irreversible thermodynamic approach where actual fluxes are involved. This standard equilibrium approach (Ussing and Zerahn 1951) is delineated as:

$$J_{out}/J_{in} = C_{in}/C_{out}\, e^{-zFE/RT}$$

where J = flux, C = concentration, Z = charge of the ion involved, F = Faraday's constant, E = voltage, R = gas constant, and T = temperature (°K).

This equation is useful in both equilibrium analyses and has been more recently modified to incorporate a work term to account for active ion transport (Ussing 1978; Sten-Knudsen and Ussing 1981). This equilibrium approach does not incorporate any relationship between the coupling of ion and water flux via mechanisms like solvent drag or solute drag.

1.4 Biophysics of Thermal Balance

Temperature is a function of the mean kinetic energy of the molecules of a material. It determines the direction of heat movement as heat moves from high to low kinetic energy. Heat is the form of energy that moves between materials of different temperature and is a measure of the total kinetic energy. The actual quantity of heat exchanged is not only dependent upon the temperature difference, but also the heat capacities of the materials in question. Heat capacity of a material is equal to the Δ heat/Δ temperature × mass.

1.4.1 Conduction/Convection

This is the transfer of energy from high to low temperature between molecules in the environment and molecules of the organism. Convection is the transfer between the skin surface and a fluid, either air or water, while conduction is the transfer between the skin surface and a solid. Obviously the proportion of skin in contact with the substrate will determine the proportion that can be ascribed to conduction. The general form of the equation describing conductive/convective processes is:

$$J_{heat} = A \times K \times \Delta T/d_{bl}$$

where A = surface area, K = thermal conductivity, d_{bl} = boundary layer thickness, ΔT = temperature difference between the organism and environment. Obviously heat is gained from the environment if $T_{environment} > T_{skin}$, and lost from the animal if $T_{skin} > T_{environment}$. The other variable to appreciate is thermal conductivity. Thermal conductivity of any liquid or substrate increases with temperature and varies with material. For instance the values of K are 0.0257 and 0.599 W/m °C for air and water at 20°C. Soil K varies directly with water content and has values between 0.2 and 2 W/m °C. Hence the transition from water to air would retard conductive/convective heat exchange as K is 23 times lower and boundary layers are thinner in water.

1.4.2 Solar Radiation

This is the transfer of energy by photons in the visible spectrum. Visible radiant energy can impinge upon the animal directly, or be scattered by molecules in the atmosphere (clouds and sky). Hence some fraction of incident solar radiation is direct, some is cloudlight, and some is skylight. Obviously the wavelengths that make up each pathway are different based on the colors represented by each (sky = blue, clouds = white black). Solar radiation also can impinge upon and be reflected onto the surface of the animal by the surroundings. The overall amount of solar radiation absorbed is described by:

$$J_{solar} = a_d A_d S_d + a_c A_c S_c + a_s A_s S_s + a_r A_r S_r$$

where a = absorptivity for solar radiation (direct, cloudlight, skylight, reflected light), A = surface area impinged by direct, cloudlight, skylight and or reflected, and S = amount of solar radiation impinging of each type. The maximal S impinging on the earth's atmosphere is about 1360 W/m^2. On a clear, cloudless day S is only about 1000 W/m^2 at sea level based on absorptive and reflective processes in the atmosphere. As water will absorb and reflect incident solar radiation the importance of solar radiation in an overall energy balance equation is less important in aquatic versus terrestrial environments for amphibians.

1.4.3 Infrared Radiation

This is the transfer of energy by photons in the long wavelength but non-visible portion of the electromagnetic spectrum. All objects above 0°K (−273°C) emit infrared radiation. Consequently energy is both lost and gained via this avenue. The overall amount of infrared radiation absorbed from the environment is described by:

$$J_{infrared} = a_{ir}\, \sigma\, [(A_{sur}\, T_{surr})^4 + (A_{sky}\, T_{sky})^4]$$

where a_{ir} = absorptivity for infrared radiation, σ = Stefan–Boltzmann constant = 5.67×10^{-8} W/m^2°K^{-4}, A = area facing surroundings or sky and T = temperature (°K) of either the sky or surroundings. Since amphibians are 78% water the a_{ir} is about 0.97. In an aquatic situation the animal is totally surrounded by water hence input would be dependent upon water temperature. In an aerial environment, the effective temperature of the sky will depend upon the amount of water vapor present as clouds or fog. Cloudy environments can have a T_{sky} of around 270°K (292 W/m^2), while on a clear night it might be only 220°K (129 W/m^2). This is why you feel colder on clear versus cloudy nights at the same ambient temperature, there is less than half the infrared radiation energy to be absorbed from the sky. Activity by aquatic amphibians is dependent upon the water being liquid rather than solid (273°K).

The overall amount of infrared radiation lost by the animal is described by:

$$J_{infrared} = e_{ir}\,A\,\sigma\,T^4_{skin}$$

where e_{ir} = emissivity for infrared radiation, A = surface area of the animal, σ = Stefan–Boltzman constant = $5.67 \times 10^{-8}\,W/m^2$ and T_{skin} = temperature (° K) of the skin. The emissivity for infrared energy is about 0.97.

The net infrared energy balance can be positive or negative, dependent upon an amphibians body temperature and its environmental conditions. As aquatic animals are characteristically close the temperature of the water, the loss and gain by infrared radiation tend to be equal.

1.4.4 Latent Heat

Both evaporation (loss) and condensation (gain) are potential avenues of heat exchange. They are considered together as they represent a change of state of the water molecules, that is water goes from a liquid to a gas requiring heat input (evaporation) or it goes from a gas to a liquid releasing heat (condensation). The heat flux associated with evaporation is described by:

$$J_{ewl} = J_{wv}\,H_{vap} = H_{vap}\,D_{wv}\,\Delta\,c_{wv}/d_{bl} = H_{vap}\,\Delta\,c_{wv}/R_{bl}$$

where H_{vap} = 44.2 kJ/mol at 20°C, D_{wv} = diffusion coefficient of water vapor in air = $1.95 \times 10^{-5}\,m^2/s$ at 20°C, $\Delta\,c_{wv}$ = water vapor concentration difference between skin and environment, d_{bl} = boundary layer thickness and $R_{bl} = d_{bl}/D_{wv}$. An important thing to note is that we are describing in this formulation a situation where the cutaneous resistance to evaporation is like a free surface of water at an [osmotic] = 250 mOsm. This is normative for the vast majority of amphibians studied to date. There are exceptions (see Lillywhite 2006 for review) and actual cutaneous resistance to evaporation will be dealt with in later chapters. To correct for cutaneous resistance, the cutaneous resistance would be added to the boundary layer resistance in a new formulation.

Condensation is a special situation where air temperature is greater than the skin temperature, and the humidity is such that when the air is cooled by the skin, the humidity reaches 100% RH and water condenses. This heat transfer process is described by:

$$J_{condensation} = H_{sub}\,A\,D_{wv}\,\Delta\,c_{wv}/d_{bl} = H_{sub}\,A\,\Delta\,c_{wv}/R_{bl}$$

where the terms are the same as in evaporation, A = surface area and H_{sub} = 51 kJ/mol at 0°C. It is not clear whether this naturally occurs in amphibians, but behavior might allow condensation to contribute to wetting the skin and simultaneously adding heat to the organism.

1.4.5 Metabolism

As animals breakdown energy and since they do no net work on the environment, all this metabolism will come off as heat energy. This flux is formulated as:

$$J_{met} = VO_2 = 21.4\,kJ/l\,O_2$$

If VO_2 is expressed as l/s, and a W = 1 J/s then rates of O_2 consumption can be used to calculate the overall heat contributed by metabolism. We ignore anaerobic metabolism in these calculations since vertebrates generate only 2 ATPs/glucose anaerobically versus 36 aerobically. Therefore a glucose molecule metabolized anaerobically will generate only 1/18 of the heat energy as the same molecule oxidized aerobically. Given the extremes in metabolism for a 100 g amphibian at 20°C at rest $1.4 \times 10^{-6}\,l\,O_2/s$ and $14 \times 10^{-6}\,l\,O_2/s$ during activity, metabolism generates 0.03 and 0.3 W during rest and activity, respectively. If we assume a 100 g amphibians has a surface area of $0.157\,m^2$ (SA (cm^2) = 7.2 Body $mass^{0.67}$; Jorgensen 1997), then this translates to 0.2 and 2 W/m^2. These values are very low in comparison to the other avenues of heat exchange.

1.4.6 Heat Storage

When energy input does not equal energy output there will be a change in the temperature of the animal or environmental situation. This rate of change of temperature can be described by the following relationship:

$$J_H/\Delta t = m\,C_m\,\Delta T/\Delta t$$

or

$$J_H/\Delta t = v\,C_v\,\Delta T/\Delta t$$

where m = mass, C_m = mass specific heat capacity, v = volume, C_v = volumetric heat capacity, ΔT = temperature change and Δt = change in time. The difference in the two equations is whether you are interested in the rate of temperature change per volume or mass. Air has a C_m of 1 J/g°C or 1.2 J/g°C, while water has a C_m of 4.19 kJ/g°C or 4190 J/g°C. From an environmental perspective, aerial environments will register temperature changes 4200 times faster that aquatic environments for the same net energy input or loss on a volumetric basis. Consequently the transition from aquatic to terrestrial environments would characteristically expose an organism to a more variable temperature regime.

1.5 Habitats and Morphotypes

As we have indicated, the presence of amphibian species in their varied habitats is wholly dependant upon the availability of water and the organism's physical and behavioral ability to deal with the universally present risk of evaporative water loss at all life stages. In this section, we attempt to identify external morphological characteristics that are indicative of habitat use. Although the array of such morphological adaptations is large and the character states are often variable, there are combinations of physical features that may allow us to predict the nature of habitat to which an amphibian is adapted with somewhat more accuracy than is reflected by the traditionally used broad terms "aquatic," "fossorial", "terrestrial" and "arboreal."

1.5.1 Gymnophiona

Morphologically, the caecilians are the most conservative among the Lissamphibia, although when compared with frogs, salamanders are rather conservative as well. Caecilians evidently diverged early from other Lissamphibian orders (Estes and Wake 1972; San Mauro *et al.* 2005); early loss of fore- and hind limbs seems to have restricted the variety of habitats that can be exploited by caecilians to aquatic, fossorial and semifossorial life modes—there are no arboreal caecilians nor adults that retain gills. With the exception of one group among the aquatic lineage (South American species of the caeciliid genus, *Chthonerpeton*), caecilians are restricted to the world's tropical regions and seemingly have been unable to expand into the temperate, more arid areas of higher latitudes. Although anatomically conservative, caecilians have some morphological characteristics that may allow characterization of some of the groups into basic morphotypes; these characters include condition of the skull, presence or absence of fins and compression of the posterior part of the body, number of annular grooves, condition of the eyes and perhaps tentacle position relative to the snout.

Aquatic

The posterior regions of the body of many aquatic species (Typhlonectidae and Ichthyophiidae) are compressed to some degree, have well-developed tail fins (if the tail is present) and fins on the posterior portion of the dorsum (Fig. 1.5). Many aquatic forms have only partially roofed, semikinetic skulls (zygokrotaphic), while burrowing caecilians largely exhibit the more compact stegokrotaphic skull condition. Toledo and Jared (1993) suggested that the costal grooves of salamanders may be an adaptation for the capillary transport of water to parts of the body not in contact with moist substrate. If true, aquatic species of the caecilians might be expected to have reduced numbers of annuli when compared with fossorial and semifossorial species. Such seems to be the case as in many aquatic species, secondary and tertiary annular grooves are absent in many aquatic forms, and primary annuli may be indistinct, while the converse seems to be true among non-aquatic caecilians (Taylor 1968; Wilkinson and Nussbaum 1997).

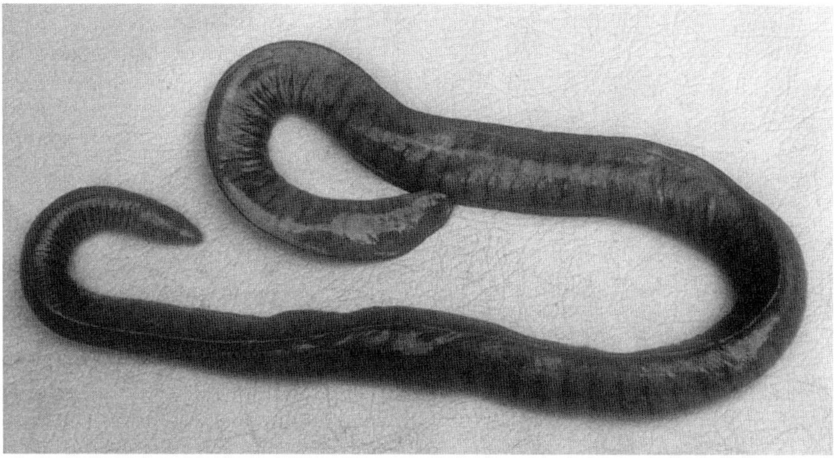

Fig. 1.5 *Potomotyphlus kaupii*, an aquatic typhlonectine (Caeciliidae) from the Amazon and Orinoco drainage systems of the Neotropical Region; note posterior lateral body compression, dorsal fin and presence of primary annulae only. Photo: J. E. Cadle.

Fossorial and Semifossorial

Gower *et al.* (2004) reported an excellent field study of two sympatric, non-aquatic species of caecilians in a forest reserve in Tanzania, the endemics *Boulengerula boulengeri* and *Scolecomorpus vittatus* (Caeciliidae, *sensu* Frost *et al.* 2006). Their results indicate that *B. boulengeri* is primarily fossorial, whereas (based in part on the frequency of capture in buried can traps), *S. vittatus* is more active in leaf litter on the surface and thus semifossorial. While both species have distinct primary annuli (Taylor 1968), Gower *et al.* (2004) noted a number morphological differences between the two species that may be correlated with habitat type, which may prove useful for inferring morphotypes in other caecilian species.

The more fossorial *B. boulengeri* has a blunt, subconical head, a slender and subcylindrical body, and a strongly stegokrotaphic skull, all of which may be associated with efficient burrowing (Fig. 1.6). Eyes are strongly reduced and overlain by bone, and the tentacles are short, globular and cannot be strongly protruded; the condition of the eyes and tentacle suggest that these sensory systems are better adapted for a subterranean existence.

On the other hand, the head of the semifossorial, *S. vittatus*, is slightly depressed and the body is rather slender and tapered anteriorly. The temporal region of the skull is not roofed (zygokrotaphic) and less well-suited for burrowing. The eye of *S. vittatus* is better developed than in *B. boulengeri* and lies within a tentacular groove covered by translucent, pigment-free skin, indicating that the visual system is more important in *S. vittatus*. The tentacle is closer to the snout and more

Fig. 1.6 *Dermophis mexicanus*, a fossorial caeciliid from Mesoamerica; note lack of lateral compression of posterior portion of the body, and presence of secondary and tertiary annulae. Photo: J. H. Tashjian.

ventrally oriented when protruded, suggesting a greater utility above ground than beneath it.

1.5.2 Caudata

Like caecilians, salamanders do not have as broad a range of habitat adaptations as do anurans (see below). For instance, there are no specializations for digging, and thus no truly fossorial species; we exclude members of the aquatic Sirenidae that burrow into mud and estivate for months within cocoons formed from multiple layers of shed skin (Stebbins and Cohen 1995). There appear to be no obvious dermal modifications for rapid uptake of water such as those found in many frog species; however, modifications of the skin do occur that may serve different functions (see Cryptobranchidae, below) such as the costal grooves in most terrestrial forms. Habitat can be inferred from other morphological characteristics such as presence or absence of gills or lungs, limb loss and modifications of hands, feet and tail, presence or absence of costal grooves on the trunk, tail shape and prehensility, and specializations of the tongue. Many semiterrestrial and terrestrial species have toxins in the skin and exhibit bright, aposematic coloration. Egg deposition and larval development is also closely linked to habitat type.

Aquatic

We define truly aquatic salamanders as those species that retain gills as adults; reproduction, feeding and all other life processes occur in fresh water environments (Fig. 1.7 A, B). Eggs are deposited in the aquatic environment; a larval stage is present. There are no specialized feeding mechanisms involving the tongue. We exclude the species of some salamanders that are facultative neotenes (i.e. individuals that attain sexual maturity while retaining gills in response to environmental factors). True aquatic salamanders include all species of the Nearctic Sirenidae and the Proteidae of North America and Europe. Largely inhabitants of still waters, Sirenids lack the hind limbs and the forelimbs are greatly reduced, while proteids show a reduction in the of numbers of digits of the fore- and hind limbs; tails are compressed, and caudal fins are sometimes present as are dorsal fins in some species; costal grooves are absent or indistinct. There are a number of species of eastern Nearctic lungless salamanders (Plethodontidae) that inhabit aquatic cave environments and retain gills as adults: a number of species of *Eurycea*, at least three of the genus *Gyrinophilus* and members of *Typhlomolge* (Duellman and Trueb 1986; Larson 1996e; Larson *et al.* 2006b).

Semiaquatic

Semiaquatic salamanders live almost exclusively in fresh water habitats but are air breathers that lack gills. All biological activity, including reproduction, occurs in the aquatic milieu. Costal grooves are usually absent or not distinct. In most semiaquatic species, the tail is compressed, and a caudal fin is present (Fig. 1.8A, B). The three eel-like species of the Amphiumidae of the southeastern United States are inhabitants of still or slow-moving waters and are the most specialized of the semiaquatic Caudata; fore- and hind limbs are greatly reduced, although they are moveable.

The massive Asian giant salamanders (*Andrias*) and *Cryptobranchus* of eastern North America, Family Cryptobranchidae, are inhabitants of higher elevations in well-oyxgenated rivers and streams. The head and body are depressed, but their tails are compressed with strong caudal fins. In addition, cryptobranchids have large folds of highly vascularized skin along the lateral surfaces of the body (Fig. 1.8B) which may enhance subaerial cutaneous respiration (Harlan and Wilkinson 1981; Stebbins and Cohen 1995). We include the four species of the torrent salamanders, Family Rhyacotritonidae, endemic to the northwestern United States in this category. While sometimes found beneath stones in the spray zone, these small salamanders usually occur under rocks within fast moving, extremely cold streams (6 to 10°C); they are nearly lungless, the tails are compressed but lack well-developed caudal fins. A number of Eurasian species of the Salamandridae (*Euproctus, Pleurodeles* and *Triturus*) are semiaquatic, as are members of the genus *Pachytriton* of China; most dwell in brooks and streams at higher elevations; tails are compressed and the fins, if present, are not strongly developed. Several genera of the Old World

Fig. 1.7 Aquatic salamanders. (A) *Typhlomolge rathbuni*, the blind cave salamander of central Texas (family Plethodontidae). Photo: J. H. Tashjian. (B) *Siren intermedia*, a sirenid of the Southeastern United States and Mexico; note gills and caudal fins. Photo: J. H. Tashjian.

Fig. 1.8 Semiaquatic salamanders. (A) *Pleurodeles waltl*, a salamandrid of the Iberian Peninsula and Morocco; note caudal fin and absence of costal grooving. Photo: J. H. Tashjian. (B) The Japanese giant salamander, *Andrias japonicus*, family Cryptobranchidae, an inhabitant of cold water streams; note caudal fin and loose lateral folds of highly vascularized skin. Photo: J. H. Tashjian.

Hynobiidae, *Onychodactylus* and *Batrachuperus,* have species that inhabit environments very similar to *Rhyacotriton* and are semiaquatic as adults (Duellman and Trueb 1986; Larson 1996b, 1996c, 1996d; Larson *et al.* 2006a, 2007).

Semiterrestrial

We define the semiterrestrial category as including those salamanders which may spend considerable amounts of time in the terrestrial environment, up to several years in some cases, but must return to water in order to breed; semiterrestrial species deposit eggs in water which hatch into gilled larvae, then metamorphose into air-breathing adults. Many species have prolonged terrestrial periods in their development and have a number of characteristics associated with terrestrial life, especially with respect to skin texture. Morphological changes may occur upon return to the aquatic breeding environment. Many semiterrestrial salamanders are brightly colored as adults (usually interpreted as aposematic or defensive), such as the European fire salamander—in fact all members of the Salamandridae have toxic skin secretions. Still others, such as the Eastern newt, *Notophthalmus viridescens* (Nearctic Salamandridae) have a brightly colored terrestrial phase commonly known as the "eft" but assume more cryptic coloration as aquatic, breeding adults. Western American newts of the genus *Taricha* are brightly colored beneath (Fig. 1.9A) and, when threatened, expose the ventral surface by arching the back in an "Unken reflex". A number of species are capable of facultative gill retention at sexual maturity as in the New World mole salamanders, family Ambystomatidae. Mole salamanders, widely distributed in the Nearctic region, are rather robust as adults with short blunt heads; most of the year is spent in rodent burrows or under detritus on the ground, but adults return to ponds and slow-moving streams to breed. Ambystomatids have prominent costal grooves on the body (Fig. 1.9B) which has been suggested as a mechanism for the capillary transport of water to parts of the body not in contact with the substrate (Toledo and Jared 1993); in most species, the tail is slightly compressed and never fin-like after metamorphosis. In semiterrestrial members of the Salamandridae, costal grooves are usually not distinct and, like the mole salamanders, the tails are not markedly compressed or fin-like. The lungless salamanders, Plethodontidae, are broadly distributed in the Holarctic and Neotropical Regions; although mostly true terrestrial organisms (see below), a number of species have the biphasic reproductive pattern (with aquatic larvae) typical of this category; examples include most members of the large genus *Desmognathus.* Costal grooves occur in adult plethodontids, and in semiterrestrial species the tail is not usually modified and fin-like. (Larson 1996a; Larson *et al.* 2006b, 2007)

Terrestrial

Terrestrial salamanders are those species that do not deposit eggs in water and lack a gilled, aquatic larval stage; reproduction is independent of free water. Most

Fig. 1.9 Semiterrestrial salamanders. (A) A Nearctic salamandrid, *Taricha torosa*, with bright ventral coloration and dry skin. Photo: J. Vindum. (B) A mole salamander of the family Ambystomatidae, *Ambystoma macrodactylum* of the western United States, with a blunt head and strong costal grooving. Photo: R. C Drewes.

terrestrial salamanders have direct developing eggs or give birth to young that resemble miniatures of the adults. Members of this group have no obvious adaptations for climbing (see below) and lack compressed tails and caudal fins. Costal grooves are distinct in most species (Fig. 1.10A, B). The great majority of terrestrial species belong to the largest salamander family, the lungless salamanders (Plethodontidae), of the Palearctic and Nearctic regions. With a number of exceptions, the majority of species in this huge family deposit direct-developing eggs on land- the aquatic gilled larval stage is absent. The family Salamandridae also contains three groups that belong in the terrestrial category; the genus *Lyciasalamandra*, the species *Salamandra atra* and *S. lanzai,* and some populations of *S. salamandra* and *S. algira* are viviparous (Larson *et al.* 2007). The physics of feeding in the semiterrestrial and terrestrial environments is very different from that in the aquatic milieu, and terrestrial plethodontids species have a range of adaptations for tongue projection: simple protrusion in *Aneides* and *Plethodon* where the tongue is still attached to the anterior part of the lower jaw to tongue pads that are completely free of the floor of the mouth and can be rapidly projected a distance of up to half the snout-vent length of the animal. The free projectile tongue occurs in at least 11 genera of bolitoglossine salamanders and is shared by two salamandrid genera, *Chioglossa* and *Salamandrina*. (Larson *et al.* 2006b, 2007). The evolution of projectile tongue mechanisms appears to be linked to the loss of lungs in these groups (Lombard and Wake 1977, 1986).

Scansorial

We have substituted the term "scansorial" (from the Latin "to climb") for the widely used term, "arboreal"; while the latter term, implying tree-dwelling, is appropriate for an important group of salamanders, it is misleading when applied to amphibians in general. For instance, many frogs with striking adaptations for climbing (greatly elongate hind limbs, strongly modified finger and toe-tips, etc.) are not inhabitants of trees or shrubs. All of the scansorial salamanders are members of the largest salamander family, the lungless Plethodontidae, and as such they share many characteristics of the terrestrial group defined above, including the absence of aquatic larvae, non-aquatic egg deposition sites and direct-developing ova, costal and nasolabial grooves, and the same specializations for tongue projection. In addition, members of the scansorial group may also exhibit extremely slender, elongate bodies, tails of great length that may be fully prehensile and hands and feet that are modified for climbing (i.e. widely-spaced digits and complete webbing of the feet group; Fig. 11A, B). The scansorial salamanders are most numerous in Central America; there are at least 11 genera and 140 species described so far from the New World tropics, all related to the supergenus *Bolitoglossa* (Larson *et al.* 2006b; Wake 1987). This group has been a major focus of study by David B. Wake and his students at the University of California, Berkeley, for several decades.

Fig. 1.10 Terrestrial salamanders. (A) A Nearctic lungless salamander of the family Plethodontidae, *Aneides lugubris*; note costal grooving and lack of lateral compression of the tail. Photo: R. C. Drewes. (B) A Neartic plethodontid, *Ensatina eschscholtzi*. Photo: J. H. Tashjian.

Fig. 1.11 Scansorial salamanders. (A) *Chiropterotriton magnipes*, a scansorial plethodontid from high elevations in central Mexico; note prehensile tail, broad, palmate, webbed hands and feet and costal grooving. Photo: G. Velo-Antón. (B) A bromeliad dweller from Guatemala, *Dendrotriton bromeliacius* (Plethodontidae). Photo: S. Rovita.

1.5.3 Anura

The anurans are by far the most successful of the amphibian orders in terms of geographic distribution, numbers of species, and habitats and ecological niches exploited. Although most numerous in the moist tropical regions, frog species are also quite successful in mesic midlatitude regions of all continents, and they have also successfully exploited xeric environments such as arid semideserts and high elevation montane areas.

Below, we delineate combinations of physical features that reflect the nature of the habitat to which a frog species is adapted with somewhat more accuracy than the traditionally used terms "aquatic," "fossorial", "terrestrial" and "arboreal."

For instance, the extent of webbing between the toes (Character A, Table 1.1) often indicates degree of aquatic behavior, a species with full, non-emarginate webbing to the toe tips can be presumed to be an efficient swimmer such as the wholly aquatic African clawed frogs (*Xenopus*, *Silurana*, *Hymenochirus*); these species also retain a functioning lateral line system (Character state B1, Table 1.1) and lack tongues (aglossan), feeding strictly underwater. In the same way, finger and toe tips (Character E, Table 1.1) that are obviously modified and enlarged into flattened pads usually indicate species that are efficient climbers, such as the scansorial hylid treefrogs, and numerous Old World species of the families Rhacophoridae and Hyperoliidae. Some arboreal and semiarboreal species do have extensive webbing, even between the fingers; in some species of the treefrog genera *Rhacophorus*, *Agalychnis* and some *Hyla,* extensive webbing provides sufficient surface area to create an air foil, allowing the animals to glide or parachute (Davis 1965). However, while both aquatic and many arboreal frogs have extensive webbing of the feet, no aquatic, fossorial or semifossorial species possess finger and toe tips that are expanded into obvious climbing digital pads.

Most fossorial and semi fossorial species spend much of their lives beneath the ground below the boundary layer (Chapter 1.2) and are efficient diggers. With few exceptions, frogs burrow by utilizing the hind legs in a ventrolateral shuffling motion, which moves the animal backwards and downwards into the soil, an activity unique to the Anura among terrestrial vertebrates (Emerson 1976). Substrate displacement is accomplished by a modified metatarsal tubercle (Character D, Table 1.1) on the plantar surface of the foot; in the most efficient burrowers the metatarsal tubercle is a greatly enlarged, edged, spade-like structure that may be cornified (Character D3, Table 1.1; Fig. 1.12). The only frogs known to burrow head-first are the South American microhylids *Myersiella subnigra* and *Dasypops schichi*, the Southeast Asian microhylid, *Glyphoglossus molossus*, the African shovel-snouted frogs *Hemisus marmoratus* and *H. guineensis,* and the fossorial Australian sand-dune frog *Arenophryne rotunda* and turtle frog *Myobatrachus gouldi.* Emerson (1976) studied anuran burrowing and indicated that a shortened tibia (see Section Semiterrestrial, below) functions as a powerful out lever in *Glyphoglossus* but that a greatly shortened tibia may also be an adaptation for walking.

Table 1.1 Anuran morphotype indicators

Family (Frost *et al.* 2006) (# species)	Taxa	Character							
		A	B	C	D	E	F	G	H
Leiopelmatidae (2)	*Ascaphus, Leiopelma*	2	2	1	1	2	1	38–55	2
Pipidae (4)	*Hymenochirus, Pipa, Xenopus*	1	1	1	1	2	1	38–44	2
Rhinophrynidae	*Rhinophrynus dorsalis*	2	2	1	3	2	2	26	2
Alytidae (2)	*Alytes, Discoglossus*	2	2, 4	1, 2	1	2	1	38–51	2
Bombinatoridae (2)	*Barbourula, Bombina*	1, 2	2	1	1	2	1	40–42	2
Pelodytidae (1)	*Pelodytes caucasicus*	2	4	2	1	2	1	52	2
Scaphiopodidae (2)	*Scaphiopus, Spea*	1, 2	2	1, 2	3	2	2	29–36	2
Pelobatidae (3)	*Pelobates*	1, 2	2	1	3	1, 2	2	33–40	2
Megophryidae (4)	*Leptobrachium, Megophrys, Scutiger*	2	2, 4	1	1	1, 2	1, 2	34–59	2
Heleophrynidae (1)	*Heleophryne natalensis*	2	4	1	1	1	1	52	2
Sooglossidae (1)	*Nesomantis thomasetti*	2	4	2	1	2	1	47	1
Batrachophrynidae (1)	*Caudiverbera caudiverbera*	2	4	2	3	2	2	40	2
Limnodynastidae (4)	*Helioporus. Neobatrachus, Notaden, Limnodynastes*	2	2, 3	1, 2	1, 3	2	1, 2	29–48	2
Myobatrachidae (4)	*Rheobatrachus, Crinia, Mixophyes, Uperoleia*	1, 2	2, 4	1	1, 2	2	1, 2	38–64	X, 2
Hemiphractidae (1)	*Hemiphractus proboscidea*	2	4	1	1	1	2	57	2
Brachycephalidae (2)	*Brachycephalus, Craugaster*	2	3, 4	1, 2	1	1, 2	1, 2	41–53	1
Amphignathodontidae (1)	*Gastrotheca riobambae*	2	4	1	1	1	1	45	X
Hylidae (14)	*Cyclorana, Lysapsus, Acris, Hyla, Litoria, Osteopilus, Phyllomedusa, Pternohyla, Smilisca, Triprion*	1, 2	2, 4	1, 2	1, 3	1, 2	1, 2	38–75	2
Centrolenidae (1)	*Centrolenella fleischmanni*	2	4	1	1	1	1	50	2
Leptodactylidae (1)	*Leptodactylus pentadactylus*	2	3	1	1	2	1	52	2
Ceratophryidae (3)	*Ceratophrys, Lepidobatrachus, Telmatobius*	2	3	1, 2	1, 3	2	1, 2	28	2

Table 1.1 (*cont.*)

Family (Frost *et al.* 2006) (# species)	Taxa	Character							
		A	B	C	D	E	F	G	H
Dendrobatidae (3)	*Colostethus, Dendrobates*	2	2, 3	1	1	1, 2	1, 2	41–48	2
Cycloramphidae (2)	*Eupsophus, Rhinoderma*	2	2, 4	1	1	2	1	48	2
Bufonidae (12)	*Ansonia, Atelopus, Anaxyurus, "Bufo", Chaunus, Mertensophryne, Nectophrynoides, Phrynoidis, Poyntonophrynus, Schismaderma*	1, 2	4	1, 2	1, 2	1, 2	1, 2	31–47	2
Microhylidae (8)	*Callulops, Cophixalus, Gastrophryne, Glyphoglossus, Kaloula, Microhyla, Phrynomantis, Sphenophryne*	1, 2	2, 4	1, 2	1, 3	1, 2	1, 2	35–69	1, 2
Brevicipididae (4)	*Breviceps, Probreviceps*	2	2, 4	1, 2	1, 2, 3	2	1, 2	18–25	3
Hemisotidae (1)	*Hemisus marmoratum*	2	2	1	3	2	1	44	2
Hyperoliidae (16)	*Afrixalus, Cryptothylax, Hyperolius, Kassina, Kassinula, Phlyctimantis, Tachycnemis*	1, 2	2, 4	1, 2	1	1, 2	1	31–54	2
Arthroleptidae (4)	*Arthroleptis, Cardioglossa, Leptopelis, Schoutedenella*	2	2, 3, 4	1, 2	1, 3	1, 2	1, 2	33–56	1
Ptychadenidae (2)	*Hildebrandtia, Ptychadena*	1, 2	3, 4	1	1, 3	2	1, 2	43–63	2
Ceratobatrachidae (2)	*Ceratobatrachus, Palmatorappia guentheri*	2	2, 4	1	1	1, 2	1	45–51	1
Micrixalidae	*Micrixalus borealis*	2	2	1	1	2	1	46	2
Phrynobatrachidae (2)	*Phrynobatrachus*	2	2, 3	1	1	2	1	51–56	2
Petropedetidae (4)	*Arthroleptides, Conraua, Petropedetes*	1, 2	2, 4	1, 2	1	1, 2	1, 2	48–56	2
Pyxicephalidae (7)	*Aubria, Cacosternum, Pyxicephalus, Strongylopus, Tomopterna*	2	2, 3, 4	1, 2	1, 2, 3	2	1, 2	31–52	2
Dicroglossidae (3)	*Altirana, Fejervarya, Occdiozyga*	1, 2	2	1	1	2	1, 2	40–46	2

Table 1.1 (*cont.*)

Family (Frost *et al.* 2006) (# species)	Taxa	Character							
		A	B	C	D	E	F	G	H
Mantellidae (1)	*Mantidactylus ulcerosus*	2	3	1	1	2	1	44	2
Rhacophoridae (11)	*Buergeria, Chirixalus, Chiromantis, Nyctixalus, Philautus, Polypedetes*	1, 2	2, 4	1, 2	1	1	1, 2	34–60	2
Ranidae (6)	*Amnirana, Lithobates (Rana), Platymantis, Staurois*	1, 2	2, 3, 4	1, 2	1	1, 2	1, 2	48–70	1, 2

A. Toe webbing
   1. toes fully webbed to tips, webbing not emarginated
   2. webbing absent , or if present, it is emarginated (deeply incised between the toes)
B. Tongue, lateral line system, ventral/ lateral skin modification
   1. tongue absent; functional lateral line system present; ventral and lateral skin not modified for water uptake
   2. tongue present; lateral line system absent; skin unmodified
   3. tongue present; lateral line system absent; ventral skin "sculptured" with verrucae hydrophilicae in pelvic patch only
   4. tongue present; lateral line system absent; ventral dermal sculpturing extending beyond pelvic region onto ventral and lateral surfaces
C. Dorsal skin modification
   1. dorsal skin not modified to enhance capillarity (rugose, ridged or glandular but not pebbled)
   2. dorsal skin sculptured (pebbled) or otherwise modified to enhance capillarity
D. Metatarsal tubercle
   1. tubercle unmodified, not obviously enlarged
   2. tubercle enlarged but not spade-like
   3. metatarsal tubercle edged, spade-shaped, may be cornified
E. Digital tips
   1. tips of fingers and toes clearly expanded, relative to body size—modified for adhesion (climbing)
   2. digital finger and toetips unmodified to slightly enlarged
F. Dorsal lymph sac
   1. dorsal lymph sac extending onto dorsal surface of the head to at least mid-point of eyes
   2. dorsal lymph sac not extending to head; skin overlying skull is non-compliant
G. Hind limb length
   % tibia length of snout-vent length
H. Egg deposition/ larval development
   1. eggs deposited in terrestrial environment (but not underground); direct development or endotrophic, non-aquatic larvae
   2. eggs deposited in, near, above water; aquatic exotrophic tadpole phase
   3. eggs deposited underground, endotrophic larvae or direct development

X = parental involvement with eggs or larva

As stated previously, there are osmotic challenges inherent in all of the habitats occupied by frogs with respect to evaporative water loss (EWL) (with the exception of the freshwater aquatic environment). Given that the integument is the interface between the environment and the lymphatic sac system of the living frog, it is not surprising that most anurans that utilize osmotically hostile environments at least

Fig. 1.12 Posterior aspect of cornified, spade-like metatarsal tubercles (Character D3, Table 1.1) in the Southeast Asian microhylid, *Glyphoglossus molossus*. Photo: R. C. Drewes.

part of the time exhibit various modifications of the skin (Characters B and D, Table 1.1). The anuran "pelvic patch" has been demonstrated to be a primary site of rapid water uptake for terrestrially adapted anurans (see Chapter 3). Elias and Shapiro (1957) described and categorized the various elevations of the skin (verrucae), which have been and often still are traditionally called "warts." Drewes *et al.* (1977) in a detailed study of the skin of the African northern foam-nest treefrog (*Chiromantis petersii*) provided the description of verrucae hydrophilicae, pelvic and ventral dermal sculpturing that is clearly specialized for rapid water uptake (Figs 1.13 and 1.14). It may be noted that this species is capable of a maximum rate of water uptake averaging 75% of body weight per hour in the first 10 minutes of rehydration. Continuous ventral and lateral areas of closely-spaced, polygonal verrucae hydrophilicae are common in frogs that spend substantial amounts of time in osmotically hostile environments; the ability to rehydrate rapidly with temporarily available free water (in the form of dew, rainfall, etc.) would be of obvious advantage to these species (Toledo and Jared 1993). Such dermal sculpturing may be absent in fully aquatic and fully fossorial species (Fig. 1.15); it may be restricted to the pelvic region only (Fig. 1.16), or it may include much of the entire ventral and lateral surfaces (Fig. 1.17).

The skin of the dorsal surfaces of many non-aquatic frogs is rough, rugose or tuberculate; in some species, the subequal size and close proximity of dorsal

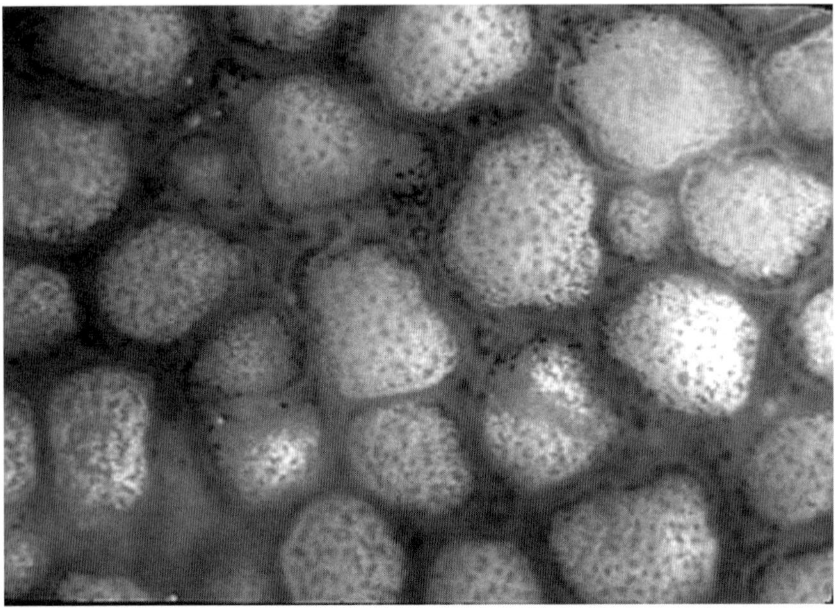

Fig. 1.13 Ventral close-up of verrucae hydrophilicae on the ventral surface of *Chiromantis petersii*, an African scansorial rhacophorid treefrog. Photo: R. C. Drewes.

ornamentation (pebbling) may enhance the rapid mobilization of free water by capillary action (Character state C2, Table 1.1; Fig. 1.18).

The anuran lymphatic system is unique among terrestrial vertebrates, consisting of a series of interconnected lymph sacs beneath the skin rather than a discrete lymphatic circulatory system. The means by which lymph is returned from these sacs to the blood circulatory system via the dorsally-located lymph hearts has been the subject of recent investigation (see Chapter 2; Hillman *et al.* 2004, 2005; Drewes *et al.* 2007; Hedrick *et al.* 2007). Carter (1979) figured the position and dimensions of the anuran lymph sacs in an aquatic (*Xenopus*), semiaquatic (*Rana*), scansorial (*Litoria*), and semifossorial (*Neobatrachus*) species and found that the dorsal lymph sac of *Neobatrachus* does not extend anteriorly beneath the skin on to the top of the head, whereas in the first three species, the dorsal lymph sac is continuous to the snout. The absence of the cranial portion of the dorsal lymph sac (Character state F2, Table 1.1) is fairly widespread among anuran species that are frequently exposed to hostile osmotic conditions in xeric to mesic environments.

Overall body shape is often used as a Gestalt indicator of anuran lifestyle, but body shape has only partial utility, as there are a number of different morphological characters that are seemingly adapted to the same limiting environmental factors. Most fossorial and some semifossorial species are usually stout, squat or

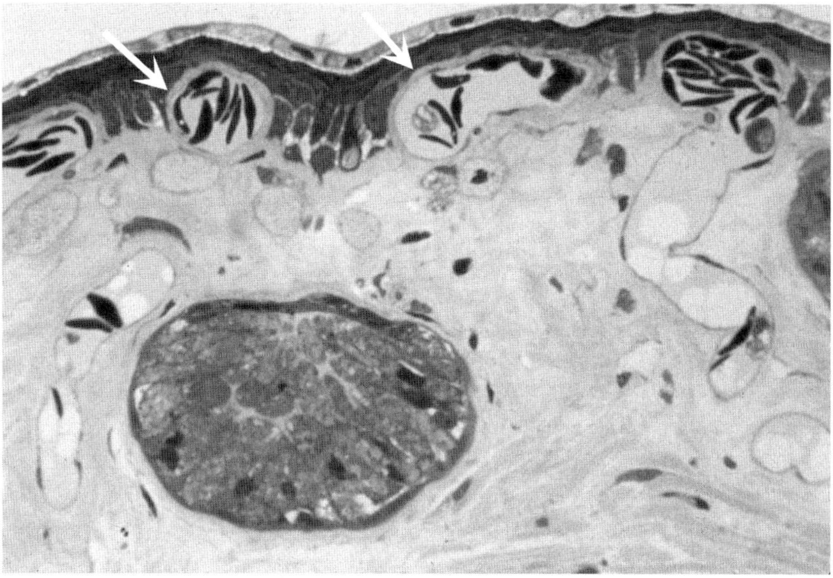

Fig. 1.14 Epoxy cross-section through a verruca hydrophilica—ventral surface of *Chiromantis petersi*; note capillaries (arrows) embedded in basement membrane of the epidermis, close to the ventral surface. Photo: R. C. Drewes.

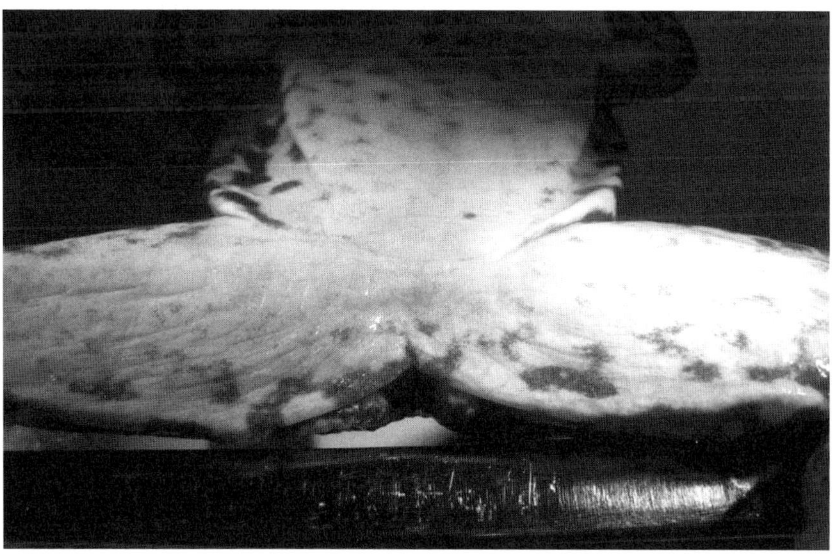

Fig. 1.15 Ventral aspect of *Lithobates* (*Rana*) *ridibunda*, a semiaquatic ranid; note absence of verrucae hydrophicae (Character state B2, Table 1.1). Photo: R. C. Drewes.

Fig. 1.16 Ventral aspect of *Rana muscosa*, a semiterrestrial (Mesic) ranid; verrucae hydrophilicae are restricted to the pelvic region (Character state B3, Table 1.1). Photo: R. C. Drewes.

even globose (e.g. South African rain frogs of the genus *Breviceps*) with short forelimbs and short, powerful hind legs. Emerson (1976) studied anuran burrowing and indicated that a shortened tibia functions as a powerful out lever in burrowing forms such as *Glyphoglossus* but that a greatly shortened tibia may also be an adaptation for walking.

In aquatic and semiaquatic frogs, some species may have smooth, ovoid, hydrodynamic body shapes (*Xenopus, Barbourula*), some may be conspicuously flattened (*Pipa, Hymenochirus*), and still others are rather robust and may have striking ancillary dermal structures such as the extensive loose skin folds of the giant Lake Titicaca frog, *Telmatobius culeus*. However, the aquatic morphotype almost always includes moderately long, but powerful hindlimbs and extensive webbing (Character A, Table 1.1) . Most scansorial species have elongate, gracile bodies with long hind limbs; however a number of climbing frogs have hind limbs of only moderate length relative to SVL (snout to vent length; Character G, Table 1.1) with rather globular bodies, such as the warty frogs of the microhylid genus, *Callulina*; the Malayan climbing toad, *Phrynoidis* (formerly *Pedostibes*) *hosii* has the normal stocky build associated with the familiar, largely terrestrial toads (formerly of the genus *Bufo*). In such cases, these species almost always possess modifications of the finger and toe tips that aid in climbing and/or grasping. The terrestrial environment has produced the greatest number of species with

Fig. 1.17 Ventral aspect of *Pternohyla dentata*, a semifossorial hylid; verrucae extend to all ventral and lateral surfaces (Character state B4, Table 1.1). Photo: R. C. Drewes.

the fewest obvious adaptations (excepting dermal ornamentation [Characters B and C, Table 1.1]) and is inhabited by most of the species usually considered "typical" frogs and toads in the temperate latitudes. Body shape is extremely difficult to quantify; lacking ribs, frog body shapes are highly compliant, and body width can vary with hydration status, lymph sac content, lung expansion, etc. Moreover, body shape is not necessarily an accurate predictor of lifestyle, as noted briefly above. In Table 1.1 below, we attempt to correlate habitat utilization with hind limb length relative to body (SVL) length (Character G). Hind limb length

Fig. 1.18 *Hyla gratiosa*, a Nearctic scansorial treefrog; note sculptured (pebbled) dorsal skin and enlarged digital tips (Character states C2 and E1, respectively, of Table 1.1). Photo: R. C. Drewes.

is estimated by expressing the length of the tibia (technically the tibiofibula) as a percentage of SVL (G).

While most anuran species deposit eggs in free water and have aquatic larvae, there are a number of exceptions to this rule that are related to the nature of the habitat utilized (Character H, Table 1.1). Wholly fossorial species (see definition below) deposit eggs underground (Character state H3, Table 1.1), and these eggs bypass the larval stage (direct development); a number of other species utilize terrestrial deposition sites in the absence of free water. In these species, an aquatic larval stage is absent, the eggs either undergo direct development as above or hatch into endotrophic larvae (Character state H1, Table 1.1).

We have plotted the various states of these characters in Table 1.1 which is based on the phylogeny proposed by Frost *et al.* (2006). These same external characters have been combined with a number of myological characters that are associated with water uptake and lymphatic flow (Drewes *et al.* 2007). The species chosen for Table 1.1 represent all but three of the anuran families recognized in Frost *et al.* (2006), the Cryptobatrachidae, Thoropidae and Nyctibatrachidae, and include both typical and atypical members where possible. Below, we attempt to categorize habitat utilization using combinations of these external morphological features.

Fossorial (Fig. 1.19)

Toes not fully webbed or if present emarginate (A2); functional lateral line system absent; ventral and lateral skin unsculptured; tongue present (B2); dorsal skin not modified to enhance capillarity or not (C1, 2); metatarsal tubercle modified for digging—enlarged, spade-like, may be cornified; (D3); digital finger and toe tips not obviously modified/expanded to enable climbing or grasping (E2); dorsal lymph sac may or may not extend anteriorly on to the head (F1, 2); hind limbs short—tibia length less than 30% SVL (G); eggs deposited underground, and undergo direct development—no aquatic larval stage (H3).

We define truly fossorial frogs as including only those groups that actually breed underground; as such we recognize as truly fossorial only the Australian turtle frog (*Myobatrachus*) and the brevicipitid genera *Breviceps* and *Probreviceps* of East and South Africa. Members of all three genera may appear above ground especially during and after heavy rainfall, and species of *Breviceps* and *Probreviceps* feed above ground, but the majority of the life of the frog (including reproductive processes) occurs in a subterranean setting below the boundary layer.

Semifossorial (Fig. 1.20)

Toes not fully webbed, or if webbing present, it is emarginate (A2); functional lateral line system absent; ventral and lateral skin unsculptured; tongue present (B2); dorsal skin not modified to enhance capillarity (C1); metatarsal tubercle enlarged, spade-like, may be cornified; metatarsal tubercle spade-like, modified for digging (D3); digital finger and toe tips not obviously modified/expanded to enable climbing or grasping (E2); dorsal lymph sac not extending anteriorly on to the head (F2); hind limbs short—tibia length less than 40 SVL (G); eggs deposited in or near water, aquatic, exotrophic larval stage present (H2).

A large number of frog species survive in xeric environments by estivating underground during extended dry periods; they emerge briefly during or after significant rainfall in order to breed in free water, burrowing shortly thereafter to underground retreats. This reproductive mode is typified by the North American spadefoots *Scaphiopus* and *Spea*, the monotypic *Rhinophrynus dorsalis*, European pelobatids, the African bullfrog *Pyxicephalus*, species of the genus *Tomopterna*, and several limnodynastine species in Australia. A number of such species form cocoons as a means of reducing the flux of water between the animal and the surrounding soil during aestivation. For others, behaviorally pursuing appropriate soil conditions allowing rehydration also reduces evaporative water loss during estivation. Since these species are active only in the presence of ephemeral water sources they are only exposed to higher rates of EWL for short periods in spite of the overall aridity of the habitat; thus specialized dermal sculpturing to effect rapid water uptake is not needed. Among those studied here, there are two species that do not wholly fit the parameters above: in *Pyxicephalus*, the dorsal lymph sac extends anteriorly to include the top of the head and in the Australian *Heleioporus*,

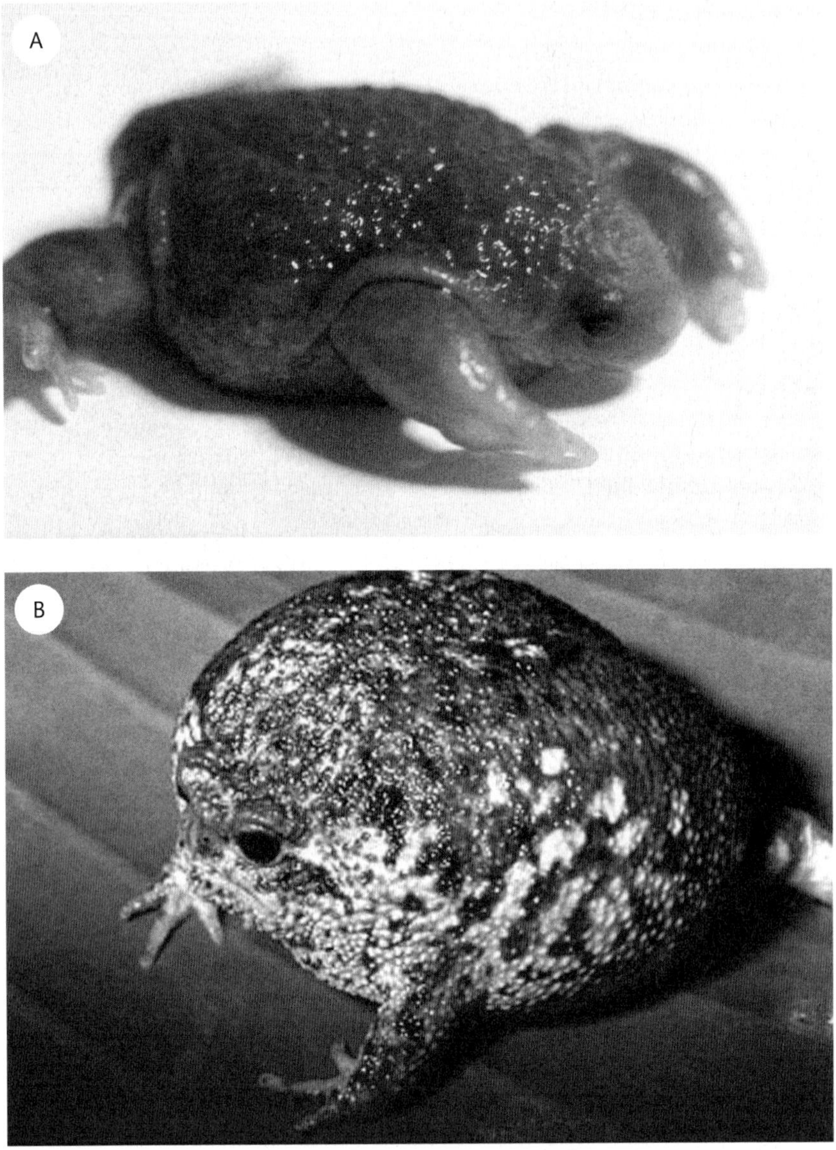

Fig. 1.19 Fossorial frogs. (A) *Myobatrachus gouldii*, the wholly fossorial Australian turtle frog (Myobatrachidae); note extremely short forelimbs. Photo: M. J. Tyler. (B) The South African Cape rain frog, *Breviceps gibbosus* (Brevicipitidae). Photo: R. C. Drewes.

Fig. 1.20 Semifossorial frogs. (A) *Rhinophrynus dorsalis*, a termitophilous endemic to Central America (Rhinophrynidae); note extremely short limbs and visible spade-like metatarsal tubercle (Character state D3, Table 1.1). Photo: J. H. Tashjian. (B) *Spea hammondi*, a spadefoot "toad" from xeric areas of western North America (Scaphiopodidae). Photo: J. H. Tashjian.

the dorsum appears to be modified for capillarity. It may be inferred that these species spend more time in the terrestrial environment than other frogs in this category before retreating underground.

Fossorial/Terrestrial (Fig. 1.21)

Toes not fully webbed, webbing if present emarginate (A2); functional lateral line system absent; pelvic, ventral and lateral skin sculptured for water uptake; tongue present (B3, 4); dorsal skin may or may not be modified to enhance capillarity (C1, 2); metatarsal tubercle, spade-like (D3); digital finger and toe tips not obviously modified/expanded to enable climbing or grasping (E2); dorsal lymph sac not extending anteriorly on to the head (F2); hind limbs short—tibia length moderate, 35–55 SVL (G); eggs deposited in or near water, aquatic, exotrophic larval stage present (H2).

Fossorial/terrestrial frog species are all capable of digging but spend more time exposed to environmental dehydration gradients than do fossorial and semi-fossorial species. Such species all exhibit some dermal modifications (verrucae hydrophilicae) for rapid uptake of free water. These modifications may be in the pelvic area only (B3) indicating either a more mesic environment or less time exposed to EWL; more arid-adapted species exhibit dermal sculpturing that extends well onto the belly and ventrolateral surfaces of the frog (B4). Some species have dorsal skin that is apparently modified for enhancement of capillary action (C2). This group is diverse, ranging from the rather gracile *Pternohyla fodiens*, a cocoon-forming hylid treefrog, and *Leptopelis bocagei,* one of only two members of this African arthroleptid genus to be non-arboreal, to larger, more robust fossorial/terrestrial frogs such as the large-headed South American horned frogs, *Ceratophrys*, the cocoon-forming helmeted water toads, *Caudiverbera*, African ornate frogs of the genus *Hildebrandtia,* various Australian lymnodynastid (*Notaden*) and hylid (*Cyclorana*) species, as well as a number of species of the genus formerly known as *Bufo* (Frost *et al.* 2006).

Aquatic (Figure 1.22)

Toes fully webbed (A1); functional lateral line system present; ventral and lateral skin unsculptured; tongue absent (B1); dorsal skin not modified to enhance capillarity (C1); metatarsal tubercle if present not obviously modified (D1); digital finger and toe tips not obviously modified/expanded to enable climbing or grasping (E2); dorsal lymph sac extends anteriorly on to the head (F1); hind limbs moderate—tibia length usually around 40 to 60% SVL, with some exceptions (G); eggs deposited in or near water, aquatic, exotrophic larval stage present (H2).

So far as is known, the lateral line system, characteristic of fish, is only retained in functional form in members of the Pipidae: the frogs of the genus *Pipa* of South America, and the African clawed frog group, *Xenopus, Hymenochirus* and *Silurana* (Fritsch *et al.* 1987), and this family also includes the only adult anurans known to lack tongues and to feed underwater.

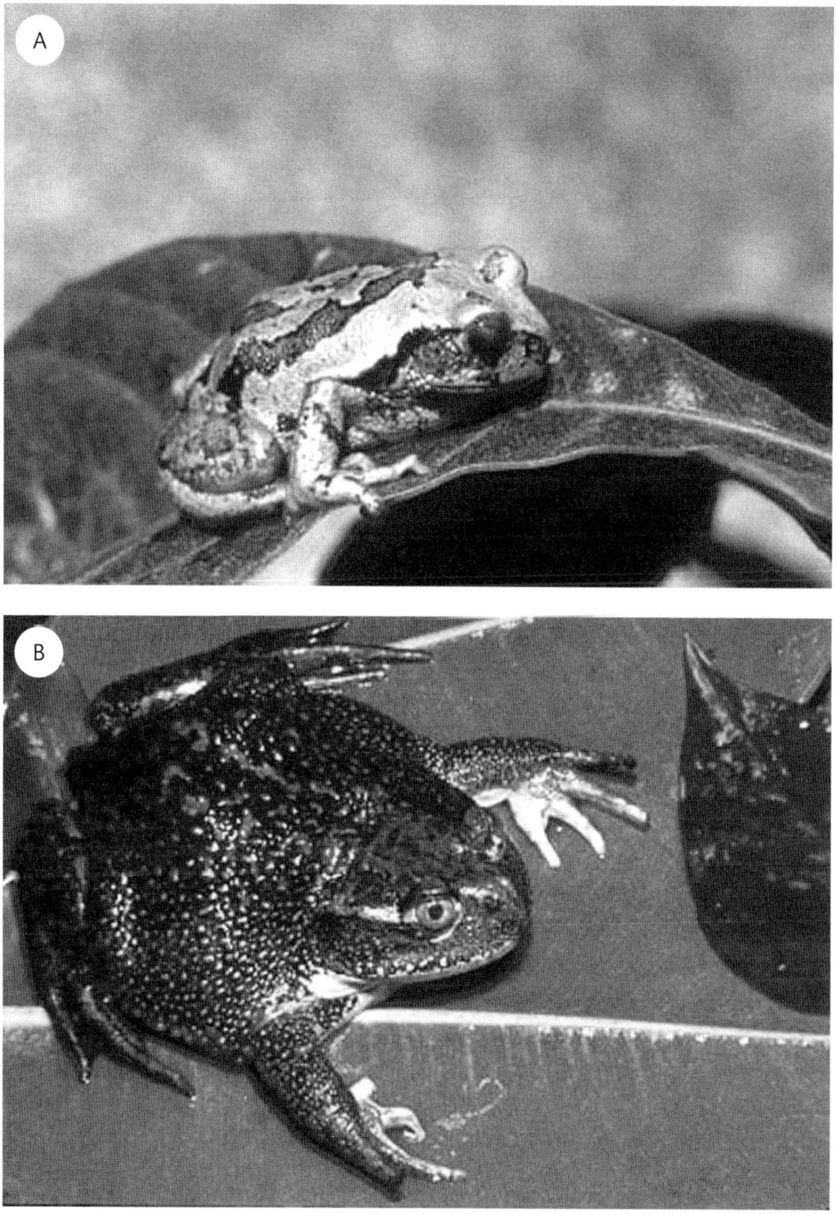

Fig. 1.21 Fossorial/ terrestrial frogs. (A) An African treefrog, *Leptopelis bocagei* (Arthroleptidae), one of only two members of the genus in this category. Photo: R. C. Drewes. (B) *Caudiverbera caudiverbera*, a fossorial/ terrestrial frog from Peru (Batrachophrynidae). Photo: J. H. Tashjian.

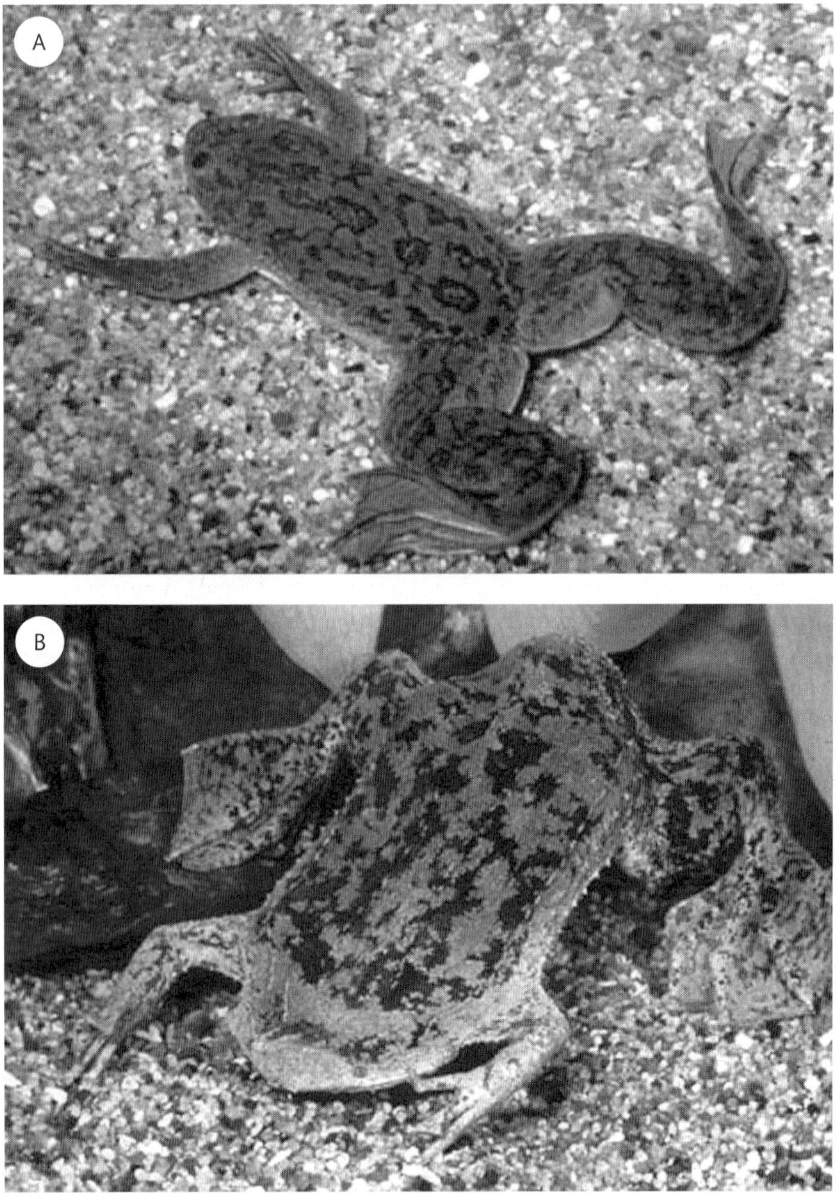

Fig. 1.22 Aquatic frogs. (A) The African clawed frog, *Xenopus sp.* (Pipidae); note full, non-emarginate toe webbing; Photo: J. H. Tashjian. (B) *Pipa pipa*, the Surinam toad of northern South America; note full, non-emarginate toe webbing; tongue is absent, functional lateral line system is present. Photo: J. H. Tashjian.

Semiaquatic (Figure 1.23)

Toes fully webbed (A1); functional lateral line system absent; ventral and lateral skin unsculptured; tongue present (B1); dorsal skin usually not modified to enhance capillarity (C1); metatarsal tubercle not obviously modified (D1); digital finger and toe tips not obviously modified or expanded to enable climbing or grasping (E2); dorsal lymph sac extends anteriorly on to the head (F1); hind limbs moderate—tibia length usually around 40 to 60% SVL, with some exceptions (G); eggs deposited in or near water, aquatic, exotrophic larval stage present (H2).

There is a large group of anurans with fully webbed toes that live almost exclusively within fresh water but lack the lateral line system; these species all possess a tongue and feed in the phaneroglossan manner typical of the majority of the Anura; this group includes frogs such as the Philippine species *Barbourula busuangensis,* the famous Lake Titicaca frog, *Telomatobius culeus* and some of its relatives, the Asian rice paddy frogs, *Occidozyga,* the Neotropical hylids *Pseudis* and *Lysapsus*, and the Australian myobatrachids, *Mixophyes* and the probably extinct mouth-brooding frog, *Rheobatrachus silus* (Tyler 1982; Cogger 1992). An even larger group of aquatic anurans lives primarily in water or in very close proximity to it such as on stream and river banks or on the shores of ponds. A number of species formerly included in the genus *Rana*, such as the North and South American bullfrogs, *Lithobates catesbeiana and L. palmipes* respectively, fit into this category, as do the African groove-crowned frog, *Dicroglossus occipitalis*, and the African giant frogs (*Conraua,* although members of this genus are atypical in lacking the cephalic extension of the dorsal lymph sac, a character we associate with a more terrestrial lifestyle). It is important to note that although tongued aquatic frogs may spend significant amounts of time on land, the immediate proximity of free water has obviated the necessity for the evolution of any dermal adaptations for rapid water uptake.

Semiterrestrial (Mesic) (Figure 1.24)

Toes not fully webbed (A2); lateral line system absent; ventral and lateral skin unsculptured; tongue present (B2); dorsal skin not modified to enhance capillarity (C1); metatarsal tubercle not obviously modified (D1); digital finger and toe tips not obviously modified or expanded to enable climbing or grasping (E2); dorsal lymph sac extends anteriorly on to the head (F1); hind limbs moderate—tibia length usually around 35 to 55% SVL, with some exceptions (G); eggs deposited in or near water, aquatic, exotrophic larval stage present (H2).

A number of species may spend significant time in the terrestrial habitat removed from free water but live in rather mesic environments such as moist leaf litter in forested areas or near water courses. Such species do not exhibit the extensive webbing associated with efficient swimming, neither are there any apparent dermal modifications for rapid water uptake; the dorsal lymph sac usually

Fig. 1.23 Semiaquatic frogs. (A) *Pseudis paradoxa*, a Neotropical hylid frog of northern South America. Photo: J. H. Tashjian. (B) *Barbourula busuangensis* (Bombinatoridae), a semiaquatic species from the Philippine Islands. Photo: Steinhart Aquarium.

Fig. 1.24 Semiterrestrial (Mesic) frogs. (A) *Gastrophryne carolinensis*, a Nearctic microhylid from the southeastern United States. Photo: R. C. Drewes. (B) *Oophaga* (*Dendrobates*) *lehmanni*, a dart poison frog from Colombia; note lack of toe webbing and aposematic coloration. Photo: J. H. Tashjian.

extends on to the top of the head. We consider such disparate anurans as the tailed frog *Ascaphus* of Northwestern US, the Eurasian fire bellied toad, *Bombina,* the European painted frog, *Discogolossus pictus* and the crab-eating frog, *Fejervarya* (formerly *Rana*) *cancrivora* to be members of the semiterrestrial (mesic) group (see Table 1.1). Many species of forest leaf-litter frogs of the African genus *Phrynobatrachus* probably also belong in this category, as does the African rubber frog (*Phrynomantis*) and species of the microhylid genus, *Gastrophryne.* Some species of the Neotropical dart poison frog family Dendrobatidae fall into this category (although some do have moderately expanded digital tips) as do a number of the anurans of the former genus *Rana.*

Semiterrestrial (Xeric) (Figure 1.25)

Toes not fully webbed (A2); lateral line system absent; ventral and lateral skin sculptured to aid capillary action; tongue present (B3, 4); dorsal skin may be modified to enhance capillarity (C1, 2); metatarsal tubercle not modified (D1); digital finger and toe tips not obviously modified/expanded to enable climbing or grasping (E2); dorsal lymph sac not extending anteriorly on to the head (F2); hind limbs moderate—tibia length usually around 40 to 60% SVL, with some exceptions (G); eggs deposited in or near water, aquatic, exotrophic larval stage present (H2).

A large diverse group of frogs that lack burrowing mechanisms to escape EWL are adapted to life on the ground through the ability to absorb free moisture up through the skin rapidly. In some species specialized dermal verrucae hydrophilicae are restricted to the pelvic region only (B3). Such species are assumed to live in less hostile environments and include many of the former genus *Rana* such as *Lithobates pipiens,* African rocket frogs of the genus *Ptychadena,* many dendrobatids, the Neotropical chicken frog, *Leptodactylus pentadactylus,* many ground-dwelling African species of the genera *Phrynobatrachus, Cacosternum, Aubria,* the arthroleptid *Cardioglossa,* and Malagasy mantellids. The African ranid, *Amnirana albolabris* has enlarged digital tips (B1) but these are small, relative to the mass of the frog, and are not deemed to be climbing adaptations *per se.* Sinsch (1991) was able to correlate the rates of water reabsorption with degree of terrestrialism in three European species of *Rana;* we assume a similar correlation exists between the rate of water uptake and the extent of specialized areas of verrucae hydrophilicae.

In other species of this group, dermal sculpturing extends well onto the ventrolateral surfaces of the body (B4), and in a few, the dorsal epidermis appears modified for capillary spreading of water as well (C2). This more arid-adapted group includes many toads (formerly *Bufo*), many terrestrial species of the African genus *Kassina;* Darwin's frog, *Rhinoderma darwini,* the Neotropical *Atelopus,* the famous obstetrical toad of Europe, *Alytes,* Asian and Australian ground-dwellers, such as *Leptobrachium* and *Uperoleia* (See Table 1.1).

Fig. 1.25 Semiterrestrial (Xeric) frogs. (A) *Amietophrynus* (formerly *Bufo*) *pardalis* (Bufonidae) a toad of coastal Cape Province, South Africa. Photo: R. C. Drewes. (B) A widespread species in sub-Saharan Africa, *Kassina senegalensis* (Hyperoliidae). Photo: R. C. Drewes.

Terrestrial (Figure 1.26)

Toes not fully webbed (A2); lateral line system absent; ventral and lateral skin either sculptured to aid capillary action, or skin modifications absent; tongue present (B2, 4); dorsal skin usually not modified to enhance capillarity (C1); metatarsal tubercle not modified (D1); digital finger and toe tips usually not obviously modified/expanded to enable climbing or grasping (E2); dorsal lymph sac usually extending anteriorly on to the head (F1); hind limbs moderate—tibia length variable from 35 to 65% SVL (G); eggs deposited in terrestrial environment, not underground; direct development or endotrophic larval stage present (H1).

We define true terrestriality as the lifestyle of only those taxa that deposit eggs in the terrestrial environment (on land, rock crevices, tree roots, leaf litter, etc.), not in underground burrows and not in free water. Eggs may either hatch into endotrophic larvae which continue development nourished by yolk or undergo direct development into miniature adult morphology (Modes 15 and 16 respectively of Duellman and Trueb 1986). Development in either mode does not include parental involvement, such as evidenced by some marsupial frogs, *Gastrotheca,* which produce endotrophic larvae but carry them in specialized dorsal pouches. Mode 15 is known for but a few species of the families Bufonidae, Cyclorhamphidae, Microhylidae, and African Phrynobatrachidae and Pyxicephalidae. Direct development (larval stages absent) has evolved many times in at least nine anuran families. It is the dominant reproductive mode among members of the Neotropical brachycephalid genus *Craugaster* (formerly *Eleutherodactylus* in part—one of the most speciose of all vertebrate genera [Frost *et al.* 2006]); it is also the dominant mode of the Solomon Islands Ceratobatrachidae, the Seychelles Islands Sooglossidae, and two major groups of Australo-Papuan microhylids (Duellman and Trueb 1986). Full terrestriality is difficult to define morphologically. Most terrestrial species with direct development inhabit areas that are perpetually humid; in Table 1.1 there are a number of species that reflect this, lacking dermal modifications for water uptake (B2), such as the African *Arthroleptis*, the endemic sooglossid *Nesomantis* of the Seychelles Islands and the New Zealand endemics of the genus *Leiopelma*. Other frogs in this category exhibit modified digital tips (E1), longer hind limbs (G) or dermal adaptations (B4) that suggest they are scansorial in nature in spite of reproductive mode (*Ceratobatrachus, Palmatorappia, Platymantis, Philautus schmackeri).*

Scansorial (Figure 1.27)

Toes usually not fully webbed (A2); lateral line system absent; ventral and lateral skin may or may not be sculptured to aid capillary action; tongue present (B2, 4); dorsal skin usually not modified to enhance capillarity (C1); metatarsal tubercle not modified (D1); digital finger and toe tips obviously modified to enable climbing and/or grasping (E2); dorsal lymph sac usually extending anteriorly on to the head (F1); hind

Fig. 1.26 Terrestrial frogs. (A) *Leiopelma hochstetteri* (Leiopelmatidae), endemic to New Zealand; Photo: D. M. Green. (B) *Ceratobatrachus guentheri* endemic to the Solomon Islands. All members of the family Ceratobatrachidae deposit terrestrial, direct-developing eggs. Photo: J. H. Tashjian.

Fig. 1.27 Scansorial frogs. (A) *Hyperolius thomensis* (Hyperoliidae) is endemic to São Tomé Island, Gulf of Guinea, West Africa; note greatly expanded digital tips, and B4 condition of ventral skin. Photo: D. Lin. (B) *Triprion spatulatus*, a scansorial hylid from coastal Mexico; note digital expansions; skin of head is co-ossified with skull (dorsal lymph sac condition F2). Photo: J. H. Tashjian.

limbs moderate to long—35 to 70% SVL (G); eggs deposited in, near or above water; aquatic exotrophic larva present (H2).

We define climbing frogs as those possessing obvious modifications of the digital tips to enable climbing and grasping, meaning expansions of the digits that appear large enough to support the body mass of the frog in near-vertical situations—many species with noticeable dilations of the digital tips do not climb; in many of such cases the digital expansions are rather small relative to the mass of the frog. While many of these species do dwell in trees and bushes, we avoid the use of the familiar term "arboreal" because there are a great many anuran species that are adapted for climbing but do not inhabit trees and bushes. Examples of such species in Table 1.1 include torrent-dwellers such as *Arthroleptides martiensseni* and *Heleophryne natalensis*, the presumed rock-dweller of the arid Somali Horn, *Kassina maculifer, Cryptothylax greshoffi,* an inhabitant of swamps, lakes and streams in the Congo Basin and the stream toad, *Ansonia muelleri* of the Philippines. *Staurois natator* of Southeast Asia is another. A number of climbing frogs inhabit very mesic conditions; *Centrolenella fleishmanni* has no dermal modifications for water uptake, and the dorsal lymph sac extends on to the surface of the head to the snout. All of the remaining climbing species studied have sculptured skin in the pelvic, ventral and, lateral regions for rapid water uptake (B4, Table 1.1); some species have, in addition, dorsal skin modified to enhance capillary transport of water (C2). The latter include the Nearctic *Hyla versicolor*, the African frogs, *Leptopelis flavomaculatus* (Arthroleptidae) and *Petropedetes newtoni* (Petropedetidae), African foam-nest frogs, *Chiromantis xerampelina,* and the Asian treefrog, *Polypedetes leucomystax*. In still other species the compliance of the skin of the head is reduced by the absence of the underlying dorsal lymph sac (D2): the arboreal toad, *Phrynoidis* (*Pedostibes*) *hosei*, the New World treefrog, *Osteopilus septentrionalis* and the aforementioned *Polypedates leucomystax*. The dorsal lymph sac is also absent in the casque-headed *Hemiphractus proboscidea* and the hylid, *Triprion spatulatus*; in both species the skin is co-ossified with the skull.

1.6 Phylogeny of the Amphibia

The diversity of the nearly 6000 species of modern amphibians (Lissamphibia) of three major lineages (caecilians, salamanders and anurans) reflects a long phylogenetic history, dating to at least the early Triassic Period. The earliest lissamphibian fossil records are from the early Triassic (*Triadobatrachus*, Rage and Rocek 1989; *Czatobatrachius*, Evans and Borsuk-Bialynicka 1998) but in general the lissamphibian fossil record is poor. Consequently the origin and interrelationships of the three monophyletic orders of lissamphibians is unclear (Meyer and Zardoya 2003; Zardoya and Meyer 2001; San Mauro *et al.* 2005). Recent molecular phylogenetic analyses of San Mauro *et al.* (2005) suggest that the caecilians diverged from the common batrachian ancestor about 367 MYBP, and the divergence of salamanders and anurans occurred soon after at about 357 MYBP. These old divergence times suggest that the three lissamphibian lineages originated in the Palaeozoic soon after the transition to land by sarcopterygian fishes and in parallel with the radiation of other non-lissamphibian amphibians such as *Acathostega* and *Ichthyostega* (Benton 1990; Milner 1993; Carroll *et al.* 2004; San Mauro *et al.* 2005).

It is likely that the three lisamphibian orders diverged rapidly in their evolutionary histories (San Mauro *et al.* 2005; Frost *et al.* 2006; see Figure 1.28), and this ancient radiation makes the early evolutionary nodes short, so reconstruction of phylogenetic histories is difficult (especially because the fossil record is poor) and application of "phylogenetic methods" to physiological data analysis (see below) can be problematic.

The divergence of the major caecilian lineages probably occurred before the breakup of Pangaea, since living caecilians are found in South America, Africa, the Seychelles and India. A subsequent dispersal of caecilians from the Indian tectonic plate after its collision with Asia explains the southeast Asian distribution of icthyophiid caecilians (Duellman and Trueb 1994; Wilkinson *et al.* 2002; San Mauro *et al.* 2005). Most ancient caecilian clades (rhinotrematids, ichthyophiids/uraeotyphlids) may have been widespread in Gondwana before its breakup and now persist as relict groups, and the more recently derived scolecomorphid and typhlonectid caecilians have evolved *in situ* (Duellman and Trueb 1994; San Mauro *et al.* 2005).

Salamanders presumably arose in the Laurasian part of Pangaea since they now have a mostly Laurasian distribution, although sirenid fossils are known from both South America and Africa (Duellan and Trueb 1994; Evans *et al.* 1996). San Mauro *et al.* (2005) suggest that modern salamanders diverged during the late Palaeozoic (273 MYBP) when sirenids and the hynobiid/cryptobranchid clade separated from other salamanders (Fig 1.28). Plethodontid salamanders separated from ambystomatids at about 253 MYBP, and ambystomatids from salamandrids at about 230 MYBP. Thus, the main salamander divergences all occurred before the breakup of Pangaea and the fragmentation of Laurasia.

The anurans probably occurred throughout Pangaea, for example *Triadobatrachus* in Madagascar (Rage and Rocek 1989) and *Czatobatrachius* in Poland (Evans and Borsuk-Bialynicka 1998) in the early Triassic. The splitting of the leiopelmatids occurred at about 263 MYBP, then pipids at 245, discoglossids at 234 and pelobatoideans at 216 MYBP. Thus, all of the major archaeobatrachian (Ascaphidae, Leiopelmatidea, Discoglossidae, Bombinatoridae) and mesobatrachian (Pipidae, Rhinophrynidae, Pelobatidae, Megophyridae, Pelodytidae) lineages existed before the breakup of Pangaea (San Mauro *et al.* 2005). The Archaeobatrachia and Mesobatrachia do not appear to be monophyletic (San Mauro *et al.* 2005; Frost *et al.* 2006). Most of the remaining families (Neobatrachia), which constitute about 96% of the modern species, are in two well-defined clades,

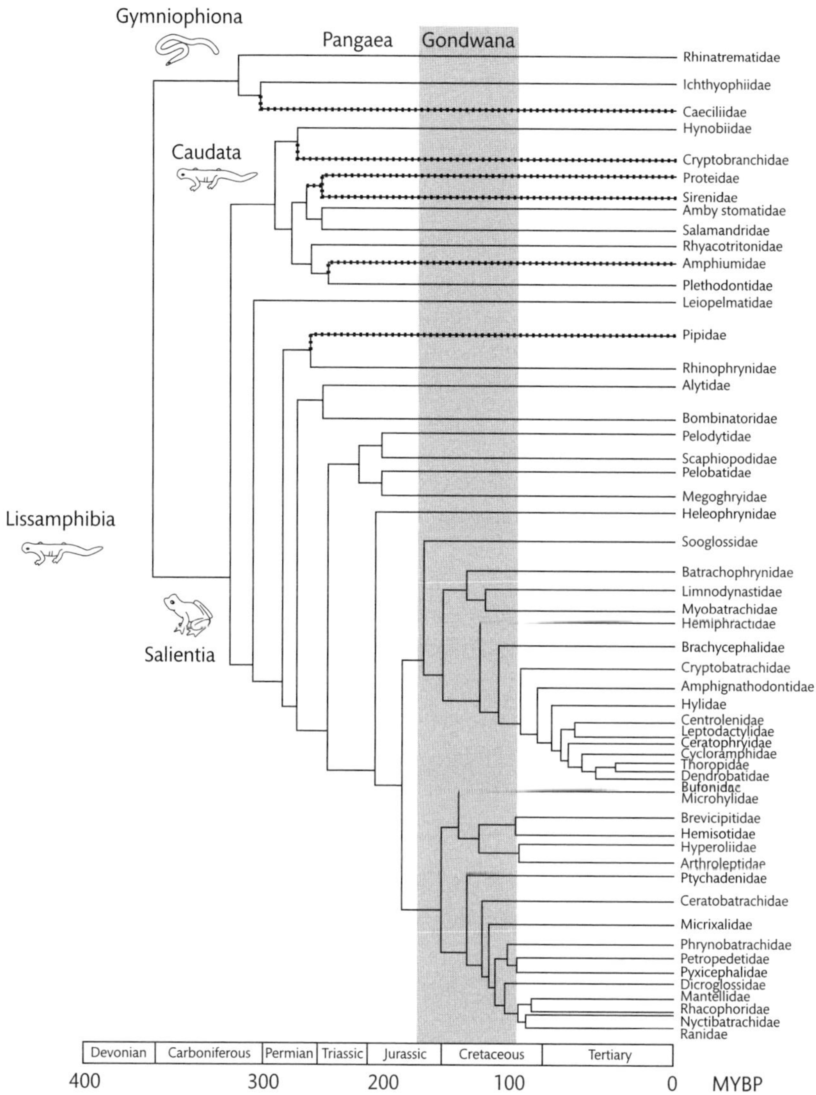

Fig. 1.28 Family tree of the Amphibia with approximate geological time scale, and indication of presence of Pangaea and Gondwana. Based on the phylogenetic analysis of Frost *et al.* (2006; their Fig. 50 and Appendix 4) and San Mauro *et al.* (2005). Ancestral states were reconstructed using MacClade (version 4). Lineages that are partially or completely aquatic are indicated by dotted lines and partially or completely terrestrial lineages by continuous lines.

Hyloidea and Ranoidea (San Mauro *et al.* 2005). These clades radiated extensively around 65 and 95 MYBP respectively, after the splitting of Gondwana (Pitman *et al.* 1993). Some species-poor neobatrachian groups, such as the South African heleiophrynids, Australian myobatrachids, Seychelles sooglossids and Neotropical *Caudiverbera,* were present before the breakup of Gondwana, and may have once been more diverse and widespread but have been reduced by competition with modern hyloids/ranoids.

There are many more species of living anurans (about 5400) than salamanders (about 560), and more salamander species than caecilians (about 170). This may reflect the geographic distribution of amphibians. Most anurans are tropical, and there appear to have been multiple invasions of the tropics by temperate zone anurans, whereas there was a single and more recent invasion of the tropics by salamanders (Wiens 2007). The species-poor caecilians are tropical but are also specialized as limbless fossorial or aquatic species. Tropical invasions are associated with increased rates of species diversification, and so this explains at least in part the species diversity of anurans compared to salamanders and caecilians (Wiens 2007).

1.7 Phylogeny and Environmental Physiology

One role of "modern" phylogenetic methods is to quantitatively analyze biological traits in a phylogenetic context and infer adaptive significance. Such analyses can indicate whether there is a strong phylogenetic pattern for the evolution of traits (is there a phylogenetic signal?), can indicate the ancestral state for traits and estimate trait values for ancestors, and can "remove" phylogenetic effects to allow comparison of different groups of taxa based on other biotic influences such as diet or life history parameters, or abiotic influences such as environment or climate. Accounting for phylogeny when analyzing biological data is not a particularly new concept, but the availability of phylogenies and complex computer software for analysis are relatively new. Physiologists have often accounted for phylogeny when analyzing allometric data using analysis of covariance (ANCOVA). A variety of other taxonomic (rather than phylogenetic) statistical approaches have also been used, for example nested ANOVA to partition variability at various taxonomic levels (see Harvey and Pagel 1991). A variety of more recent phylogenetic methods quantitatively account for the phylogeny of a group of species. Phylogenetic methodologies are essential tools for comparative biologists because they allow us to interpret patterns in biological traits in a phylogenetic context. However, the nature of the questions that we can ask after phylogenetic correction is different from the questions we can ask for data before phylogenetic correction. After phylogenetic correction, we can only ask questions pertaining to the phylogenetically corrected data; we have lost the direct connection to the real data.

Many (or most) biological traits will have some phylogenetic signal based on evolutionary inertia, but traits do not necessarily have a phylogenetic signal. A phylogenetic signal can be detected for numerical traits (k/k* test, Blomberg *et al.* 2003; C test, Abouheif 1999) and categorical traits (Vanhooydonck and Van Damme 1999). Some evolutionarily plastic traits (e.g. related to behavior or physiology) will not have a strong phylogenetic signal. For example, the evolution of cocoon formation during aestivation (see Chapters 3 and 4) appears to have independently evolved a number of times, at least once in salamanders (*Cryptobranchus/Siren*) and more than once in anurans (*Neobatrachus*, *Cyclorana*, *Pternohyla/Smilisca*, *Pyxicephalus*, *Leptopelis*, *Lepidobatrachus*). At the family level, there is no significant phylogenetic clustering of the non-cocooning/cocooning trait (categorical clustering test; Vanhooydonk and Van Damme, 1999; P = 0.866). Trait values can be extrapolated for ancestral forms from values for living species, based on phylogeny. Non-cocooning would be predicted to be the ancestral state from the phylogenetic pattern of non-cocooning/cocooning. It is also possible to estimate ancestral values for numerical traits (e.g. body mass, metabolic rate, urine flow rate, etc.) from phylogeny.

It is inevitable that the most closely related species are likely to be the most similar, in all aspects of their biology (see Chapter 5 Phylogenetic Methods). This can easily lead to an inflated estimation of the significance of a phylogenetic pattern for the trait. Including all of the species in a genus in a phylogenetic analysis of a trait would over-estimate the occurrence of that trait if the trait had evolved only once in the common ancestor of the genus rather than separately in each species. For example, all species of *Neobatrachus* form a cocoon and presumably inherited the trait from a common cocooning ancestor. Here, it would be more appropriate to consider *Neobatrachus* as a single datum in phylogentic-based analysis of the trait, so that this genus contributes n = 1 to the analysis, not n = 8 species that inherited the trait. Distantly-related species can, of course, converge in aspects of their biology in response to adaptation to particular aspects of their environment. It is these examples of independent evolution of traits that provide strong evidence for the functional significance of traits.

Recent molecular phylogenies of the Lissamphibia (e.g. San Mauro *et al.* 2005; Frost *et al.* 2006; Wiens 2007) provide a framework for phylogenetic analysis at the family to species level. There are now many phylogenies for various groups of amphibians in the ever-expanding literature. Branch lengths connecting the three lissamphibian orders are short, indicating an ancient and rapid divergence of caecilians, salamanders and anurans (e.g. Figure 1.28); this makes it difficult to infer relationships between the major clades. However, many genera have diverged relatively recently so it is easier to infer relationships at this level.

Amphibians are a general transitional group from completely aquatic, to semi-aquatic, to completely terrestrial (Feder 1992). Marine/estuarine osteolepiform fishes were the immediate ancestors of earliest amphibians (late Devonian), which radiated into freshwater and terrestrial environments in the next 20–30 MY, but the origin of the modern lissamphibians is uncertain because of the poor fossil record for freshwater deposits in this period (Carroll 2001). The general consensus is that the ancestral lissamphibian had a salamander-like form and a biphasic lifestyle, with an aquatic larval stage that metamorphosed into a more terrestrial adult form (Reilly *et al.* 2000; Reiss 2002; Bruce 2007). Each of the three lissamphibian lineages includes partially or completely aquatic families. Some caecilians are completely aquatic (e.g. *Typhlonectes*) whereas some terrestrial species are viviparous and do not need to enter water even to breed (e.g. *Schistometopum*). The terrestrial species are generally fossorial or under leaf litter in humid tropical environments. Some salamanders are completely aquatic to the point of retaining larval characteristics (neoteny; e.g. the axolotl) whereas others are completely terrestrial or scansorial (e.g. some *Plethodon*). Some anurans are so aquatic that they can hardly move on land (e.g. *Xenopus*) whereas others never enter water (e.g. *Eleutherodactylus*) or even have direct development of eggs in moist sand (e.g. *Myobatrachus*). Terrestrial anurans have radiated into various environments ranging from mountains to deserts, ocean beaches, forests to tundra, and niches from fossorial to terrestrial and scansorial. There is a phylogenetic pattern amongst amphibians for aquatic/terrestrial habits (see Fig 1.28) where aquatic lineages are more basal. However, the ancestral adult lissamphibian was terrestrial and so the reversion to an aquatic adult habit has independently occurred in caecilians (Caeciliidae), salamanders (Cryptobranchidae, Proteidae/Sirenidae, Amphiumidae and some Plethodontidae and Hynobiidae) and anurans (Pipidae).

We might expect many basic biological traits of amphibians to show strong phylogenetic relationships. Some lissamphibian morphological traits have a phylogenetic pattern (categorical clustering test; Vanhooydonk and Van Damme 1999). Absence of a tongue and presence of a lateral line system is related to an aquatic habitat. A suctorial feeding strategy is similarly related to an aquatic habitat. There is no phylogenetic pattern for some important morphometric characteristics for anuran amphibians, including dorsal skin sculpturing, expansion of digit tips, extent of the dorsal lymph sac, relative length of the hind limbs or pattern of egg deposition, at least at the family level. There is no phylogenetic pattern for cocoon formation amongst amphibians; each family that includes cocoon-forming species has apparently independently derived the capacity. As we shall see throughout this book, the physiological data that are available for amphibians are often sparse, and concentrated on particular taxa. For example, there are very few physiological data available for caecilians (Table 1.2). More data are available for salamanders, and even more for anurans. Interestingly, the distribution of literature citations for the three orders of amphibians parallels their species numbers. However, the citations (data) are from a narrow selection of families. For instance, for anurans the data are concentrated in relatively few families, particularly the Ranidae, Bufonidae and Hylidae. This lack of

Table 1.2 Summary of total citations for the three major lissamphibian lineages in Biological Abstracts (1965–2007), and citations for physiology and major physiological topics (the search for anuran included frog and toad)

	Caecilian	Salamander	Anuran
Physiology	21	499	11,994
metabolism	12	497	2,671
respiration	11	103	795
cardiovascular	1	2	447
solutes	0	3	86
ions	0	36	1,070
water balance	0	5	63
thermoregulation	0	9	52
Total	217	3,785	36,380

physiological data for many of the diverse taxa of amphibians makes comprehensive physiological comparisons difficult, and often precludes phylogentic analysis and interpretation.

The nature of the gas exchange organ is strongly related to habitat; gills occur in aquatic species and lungs in terrestrial species, although the skin can be an important avenue for gas exchange in both aquatic and terrestrial species (Chapter 2). Significant cutaneous exchange occurs in eggs, larvae and most adult amphibians ("waterproof" and cocooned frogs being possible exceptions). Branchial (gill) gas exchange only occurs in aquatic amphibians—gills are essentially non-functional in air because the branchial surfaces must be kept moist, and water tension adheres adjacent gill filaments in air. Lungs occur in each of the three lissamphibian lineages (presence of paired lungs was presumably an ancestral trait), but are reduced, non-functional or even absent in some caecilians (e.g. *Atretochoana*), many salamanders (e.g. plethodontids), and highly reduced in aquatic frogs from fast-flowing streams (e.g. *Ascaphus*) or lakes (e.g. *Telmatobius*).

Skin permeability might be expected to be related to terrestriality. Thick, armored, and likely impermeable (to gases, water and solutes) skin was presumably the ancestral state for amphibians (Lillywhite 2006) so the thin, vascular skin that is characteristic of most extant amphibians is presumably a derived trait, but the ancestral state for lissamphibians. Thin, vascular skin promotes cutaneous O_2 and CO_2 exchange, as well as water and ion exchange, and would be expected to be favored in aquatic environments (but at the expense of physical protection against skin damage). For aquatic amphibians there is no disadvantage of a high evaporative water loss (EWL) since their skin is not usually exposed to dry air. Terrestrial amphibians, apart from a few scansorial species, have retained a thin, vascular skin despite the disadvantage of a high EWL, presumably to retain a significant cutaneous component of O_2 and CO_2 exchange, as well as water and solute exchange, especially through the ventral "pelvic patch" region when exposed to water. The disadvantage of a high cutaneous EWL is typically avoided by behavioral strategies (nocturnality, burrowing, use of the water conserving posture, etc.). A few scansorial anurans have skin modified to reduce EWL (and as a consequence gas exchange as well). *Chiromantis xerampelina* (an African rhacophorid frog) was the first frog described to be "waterproof" (and to excrete uric acid as its primary nitrogenous waste product; Loveridge 1970); its congener *C. petersii* also has a very low cutaneous EWL (Drewes *et al.* 1977). *Phyllomedusa sauvagei* (a southern American hylid frog) was similarly found to be waterproof (and uricotelic; Shoemaker *et al.* 1982). Subsequently, reduced EWL has been described for some African hyperoliid frogs (Withers *et al.* 1982a, b) and Australian hylid treefrogs (Withers *et al.* 1984; Buttemer 1990). "Waterproof" skin has clearly evolved infrequently and independently in a diverse selection of frogs, and is not a "mainstream" evolutionary trend amongst terrestrial amphibians. That so many amphibians nevertheless occur in arid environments is testament to the efficacy of their behavioral, physiological and reproductive strategies.

Thermal biology also might be expected to change with the transition between aquatic and terrestrial habitats. The high thermal conductivity of water and animal tissues (see above) effectively precludes physiological thermoregulation in an aquatic environment (at least for ectotherms such as amphibians) so it isn't surprising that aquatic amphibians generally thermoconform to water temperature. Some nevertheless are able to behaviorally thermoregulate by selecting from the available range of water temperatures (e.g. some tadpoles) or by basking in sunlight (e.g. some dense aggregations of melanistic tadpoles). Terrestrial amphibians are potentially more able to behaviorally thermoregulate by heliothermy (sun basking) or thigmothermy (using warm substrates), but relatively few species do because of the potential for desiccation. Scansorial frogs, particularly the waterproof species, do sun bask and elevate their body temperature to 35°C or even more, and are adept thermoregulators at higher temperatures by using evaporative cooling (see Chapter 3). It is apparent then that low cutaneous EWL and tolerance or active selection of high body temperature are not "mainstream" evolutionary trends for terrestrial amphibians.

Similarly, aspects of water and solute balance tend not to show any particular evolutionary patterns. There appears to be considerable ionic and osmotic flexibility in even aquatic amphibians. For example, *Xenopus laevis* is one of the most highly aquatic anuran amphibians but has considerable hypersaline tolerance, involving urea accumulation (Romspert 1976). Dehydration tolerance, hence tolerance to accumulated solutes, tends to be higher for more terrestrial species (see Chapter 3), as is their bladder volume (Chapter 2) which ameliorates the ionic/osmotic effects of dehydration.

Finally, resting and active metabolic rate might be expected to differ for caecilians, salamanders and anurans (Gatten *et al.* 1992) and be higher for terrestrial species (since O_2 is more readily available in air) than for aquatic species (for which the low O_2 solubility in water restricts O_2 availability for metabolism). Each of the three major lissamphibian clades includes aquatic and terrestrial species. Anurans and salamanders differ in standard metabolic rate (SMR is 2.5 × higher for anurans at 5°C and 1.5 × higher at 25°C; Gatten *et al.* 1992) and the allometric relationship for anurans is parallel to, but higher than, that for salamanders (Gatten *et al.* 1992). The allometric relationships for SMR of salamanders and anurans, based on $\log_{10}$ mass (M) and $\log_{10}$ SMR of genus averages, are:

salamanders: $\log_{10} \text{SMR} = 0.861 \log_{10} \text{M} - 1.197$

anurans: $\log_{10} \text{SMR} = 0.834 \log_{10} \text{M} - 1.005$

(see Fig. 1.29). The slopes of these lines are not significantly different, but the elevations of the lines are; anurans have a higher SMR than salamanders (cf Gatten *et al.* 1992). It is not reasonable to simply pool the data for salamanders and anurans because their elevations differ; the pooled relationship would be weighted by the numbers and masses of salamanders relative to anurans in the dataset. Phylogenetic analysis of these data "re-scale" their SMR according to phylogeny, and salamanders and anurans are then found to fit the same allometric relationship with

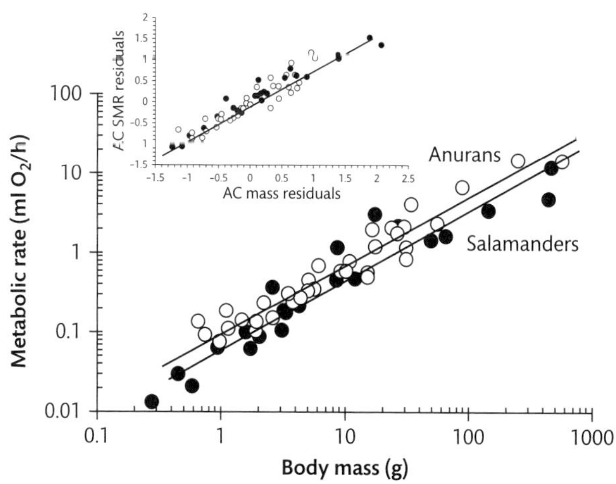

Fig. 1.29 Allometric relationship for standard metabolic rate of salamanders and anurans, based on genus averages calculated from data summary of Gatten *et al.* (1992) using the phylogenetic relationships from Wiens (2007) and Frost *et al.* (2006); the regression lines for anurans and salamanders do not differ in slope but differ significantly in elevation (see text). Inset shows phylogenetically-corrected mass and SMR residuals (using autocorrelation); there is no significant difference in slope or elevation after phylogenetic correction.

respect to slope and elevation (Fig. 1.29 inset). We don't conclude that anurans and salamanders do not differ in standard metabolic rate (SMR)—in fact they do (by conventional ANCOVA analysis). Rather, we conclude that the difference in SMR between anurans and salamanders can be accounted for by their phylogenetic history. There is no correlation of the residuals from these phylogentically-corrected data with aquatic/terrestrial habits, or reliance on gills, lungs, skin (lungless terrestrial species). Body mass and SMR of the ancestral salamander is calculated to be 9.4 g and 0.70 ml O_2/h (8.7 nmol/s), and for the ancestral anuran is 34.3 g and 1.31 ml O_2/h (16.2 nmol/s); for the ancestral amphibian, body mass and SMR are 18.5 g and 0.97 ml O_2/h (12 nmol/s).

2
General Physiological Principles

This chapter will focus on the physical principles and anatomic variables involved in heat, gas, water and solute fluxes in amphibians. The purpose of this chapter is to quantitatively delineate what is known about water, ion, chemical energy, heat and gas fluxes in amphibians. The data will be organized from a budgetary outlook when sufficient data are available. Physiological control systems (neural, endocrine and physical) will be introduced and delineated, especially as they relate to environmental adaptations.

2.1 Water and Ion Budgets

Budgets are a simple yet rigorous methodology for delineating both the inward and outward fluxes of molecules and energy. Budgets frame the variables involved in a flux and apportion importance based on the relative contribution to the total flux and, if influx and efflux do not balance, it indicates storage or loss. In this way, they help direct investigators to the most important variables to be investigated. Budgets also allow comparisons between species and taxonomic groups based on the total flux that might relate to environmental variation.

2.1.1 Water in the Environment

Water has a very significant role in the distribution and ecology of amphibians for a variety of reasons. First, water availability in the environment has a major role in the reproductive ecology of most amphibian species since most have an aquatic larval stage. Adult amphibian densities and the likelihood of successful metamorphosis are correlated with rainfall and hydroperiod (Inger 1980; Toft 1980; Pechmann *et al.* 1989). Water availability influences the behavior, distribution and activity of many amphibian species (Schwarzkopf and Alford 1996; Grover 2000; Seebacher and Alford 2002; Rohr and Madison 2003). Finally, water seems to be a primary determinant of amphibian species diversity (Pianka and Schall 1981; Scott 1976; Woinarski *et al.* 1999; Zhao *et al.* 2006; Buckley and Jetz 2007). For example there is a strong positive relationship of rainfall on amphibian species diversity both globally (Buckley and Jetz 2007) and continentally (Fig 2.1, Scott 1976; Pianka

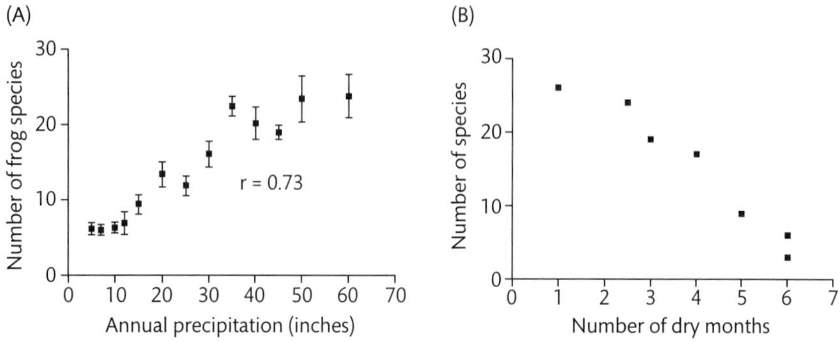

Fig. 2.1 (A) The effect of rainfall on the number of amphibian species in various quadrates of Australia (Pianka and Schall 1981) and (B) the number of amphibian species in Central American lowland forests as a consequence of the number of rainfall-free months (Scott 1976).

and Schall 1981; Zhao *et al.* 2006). Global climate change has been implicated in creating drier conditions in many environments and these drier conditions have been implicated as the cause of many of the declines and extinctions of amphibians (Pounds *et al.* 1999; Kiesecker *et al.* 2001; Daszak *et al.* 2005).

Free water can occur as surface water (drips, puddles, ponds, streams, lakes) or as moisture associated with soil. Amphibians have access to both types of free water. For exchange with surface water, the driving force for water uptake is related to the difference in water potential between the external environment and amphibian body fluids. Soil water has both an osmotic and a matric potential. The matric potential arises from the surface tension of water in the interstitial spaces between the soil particles. This interstitial holding capacity is higher for small soil particles, so the matric potential of clay soils with small particle size make it more difficult to extract water than from sandy soils with large soil particle size (see Chapter 1).

2.1.2 Water Budget and Daily Turnover

The water budget of a generalized amphibian is based on the water balance equation:

$$\textbf{In} \pm \textbf{Storage} - \textbf{Out} = 0$$

The components of the equation are:

- Inputs:
  preformed food
  metabolic

oral drinking
skin uptake

- Storage:
  urinary bladder
  body water
  1. intracellular
  2. extracellular
     (a) plasma
     (b) lymph
     (c) interstitial
- Outputs:
  fecal
  urinary
  evaporative
  1. skin
  2. lungs

The overall water flux is generally described as some fraction of body mass that is either lost and regained (terrestrial) or gained and then excreted (aquatic) on a daily basis to maintain body mass (% body mass/day). Amphibians generally have high water turnovers in comparison to other ectothermic vertebrates (Nagy and Peterson 1988).

Water turnover rates, in most vertebrate groups, are usually assessed by measuring the turnover of tritiated water. The high turnover rates of amphibians make the application of tritiated water turnover techniques very difficult. Where have these estimates for amphibian water turnover come from? One method is to actually measure mass changes on a daily basis, before and after hydration periods. The first such turnovers in a terrestrial environment (pet treefrogs allowed to free roam in an apartment) were measured for *Hyla arborea* at 20–80%/day (Towson 1799, graphically presented by Jorgensen 1994a). Turnovers on the order of 4%/day were measured for *Bufo bufo* in seminatural conditions with access to shelter and allowed to behaviorally aggregate to minimize surface area in a low convective laboratory setting (Jorgensen 1991). Using urine flow rates for the aquatic *X. laevis,* a water turnover of 29%/day was calculated (Henderson *et al.* 1972). Turnovers of about 26%/day have been measured for *Pachymedusa dachnicolor* and *Agalychnis annae* in a laboratory setting with some convection and dry air (Shoemaker and McClanahan 1975; Shoemaker 1987). The question is why are water turnovers in an aerial environments so high? To answer that question we have to look at the individual components of the water budget.

In Fig. 2.2 are presented two water budgets for a typical anuran, *Pachymedusa dachnicolor,* and a specialized waterproof species, *Phyllomedusa sauvagei.* In

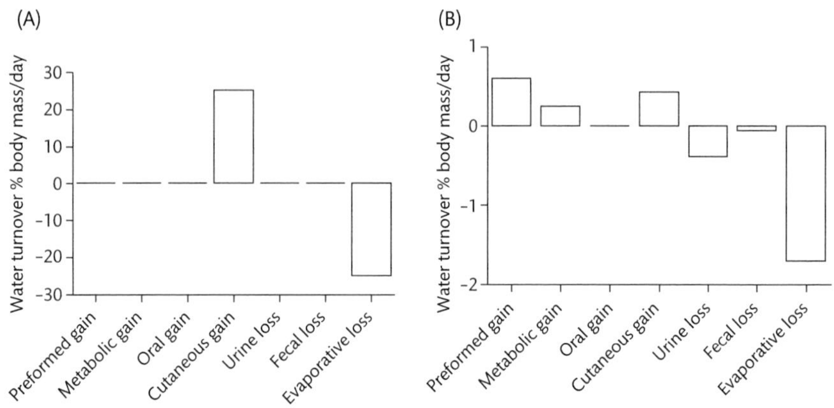

Fig. 2.2 The rates of input and loss expressed as a percent of body mass per day for (A) a typical anuran *Pachymedusa dachnicolor* and (B) a species with low water loss *Phyllomedusa sauvagei.*

the typical anuran the water budget essentially simplifies to a high cutaneous evaporative water loss being balanced by a high cutaneous water uptake. The other avenues of water exchange are so small they are essentially trivial variables. The waterproof species has an order of magnitude lower daily water turnover; the other avenues of water exchange are now more important participants in balancing the water budget.

Water Inputs

Preformed Water in Food

The amount of preformed water assimilated in food is dependent upon two variables; the water content of the food ingested and the amount of food ingested. Amphibians as adults are predators that are generally insectivorous, with a few species being carnivorous. Insects and other vertebrates vary in water content from 49 to 81% of body mass (Smith 1976; Dimmit and Ruibal 1980; Secor and Boehm 2006); this variation is related to insect developmental stage, fat content and amount of chitin. The amount of food consumed also varies from 5 to 29% of body mass per meal (Dimmit and Ruibal 1980) and seems to vary interspecifically. Ingesting food also stimulates the drive to cutaneously absorb water, as about 0.75 ml H_2O is absorbed per gram of food ingested if water is available (Jorgensen 1992). Passage time for food is generally 2–4 days but can take up to a week. Given the seasonal nature of activity patterns for most amphibians, there will be protracted periods where food will be unavailable. A comparison of different species (Table 2.1) indicates that preformed water contributes only a small fraction of the total daily water turnover for amphibians.

Table 2.1 A summary of the preformed water, volume of food eaten, number of feeding bouts and preformed water flux for two species of amphibian

Species	Food	Preformed H_2O % H_2O	Volume eaten % body mass	Number of feeding bouts/year	Preformed H_2O flux % body mass/day
B. cognatus	Ant	65	6	22	0.23
	Cricket	67	6	22	0.24
S. couchii	Mealworm	64	29	2	0.10
	Termite	49	29	2	0.08

Metabolic Water

Oxidation of food to generate ATP via aerobic metabolism leads to the production of water. For the aerobic oxidation of glucose the general formula is:

$$C_6H_{12}O_6 + 6\,O_2 \leftrightarrow 6\,H_2O + 6\,CO_2 + 30\text{–}38\,\text{ATP}$$

The type of substrate being oxidized influences the amount of water that is produced. Metabolic water production is 6 moles of water per mole of glucose or 0.56 ml of H_2O per gram; protein yields 0.4 ml H_2O/g and lipids yield 1.07 ml H_2O/g. This might lead to the simple conclusion that lipid metabolism is the most water productive substrate from a metabolic perspective. However, this ignores the loss side of the water budget, in particular the evaporative water loss associated with O_2 exchange. Aerobic oxidation of 1 g of carbohydrate requires 0.83 l of O_2. For lipids 2.0 l of O_2 is required per gram and 0.97 l of O_2 is required per gram of protein oxidation. Hence the ml H_2O produced/ml O_2 consumed is high for carbohydrates compared to lipids and proteins. Since respiratory evaporative water loss is proportional to respiratory ventilation, which is proportional to V_{O_2}, then carbohydrate metabolism is more water productive (0.67 ml H_2O/l O_2) than lipid (0.54) and protein (0.41). Carbohydrate yields more stored water than lipid and protein metabolism. The water content of stored carbohydrate (e.g. glycogen) of 2.7 ml/g is much higher than for protein of 0.4 ml/g and lipid of 0.09 ml/g (Sherman *et al.* 1982; Digirolamo and Owens 1976; Rupley and Careri 1991).

The amount of metabolic water produced is directly related to metabolic rate (V_{O_2}) of the organism. Amphibians as ectotherms have a V_{O_2} that is dependent upon body mass, body temperature, level of activity, season, and whether they are digesting a meal (see Gatten *et al.* 1992). For a 100 g anuran at 20°C, the metabolic rate is about 100 ml O_2/day (Gatten *et al.* 1992); if the metabolic substrate is lipid (the best case scenario from a water production perspective) then 0.07 ml H_2O is produced by metabolism on a daily basis. This input is less than 0.1% of body mass/day, which is an insignificant fraction of the total water turnover.

Glycolytic anaerobic metabolism also occurs in amphibians and can only use glucose as a substrate. The anaerobic breakdown of glucose follows this general formula:

$$C_6H_{12}O_6 \leftrightarrow 2\,\text{Lactate}- \;+ 2\,H_2O + 2\,H^+ \;+ 2\,\text{ATP}$$

Consequently there is more water produced/ATP breaking glucose down anaerobically (1.0) versus aerobically (0.16), but there is three times as much water produced/glucose aerobically compared to anaerobic metabolism (Hochachka and Somero 1984). Given that anaerobic metabolism is usually only a significant component of intense activity (see Chapter 3), its role in metabolic water production is usually ignored.

Oral Drinking

Drinking and/or osmotic water uptake have to be important sources of water to balance the water budget of amphibians as two of the four avenues for water input (metabolic and preformed water) are small fractions of the known water turnover rates. This makes it particularly unusual that amphibians normally do not orally drink (see Jorgensen 1997 for review). However, drinking in a fresh water environment would exacerbate the dilution caused by the passive water influx across the skin (see below). Very low rates of drinking in fresh water (1–2% body mass/day) have been reported in a few species (Mullen and Alvarado 1976; Shoemaker and Nagy 1977; Bentley and Yorio 1979). Salamanders and anurans have been reported to drink in hypersaline environments. This behavior hastens electrolyte loading and their eventual death in this environment (Krogh 1939; Bentley and Schmidt-Nielsen 1971). The only amphibian where oral drinking represents a significant source of water is the waxy waterproof frog, *Phyllomedusa sauvagei,* with maximal rates of 21% body mass/h (McClanahan and Shoemaker 1987). This arid environment scansorial treefrog is able to orally ingest raindrops that have coalesced and then drip from leaves in its habitat (McClanahan and Shoemaker 1987).

No one has posed the question as to why amphibians are the only vertebrate class that does not orally drink to balance their water budget when water is available. However, it is clear that most amphibian species do not drink when pooled water is available. This phenomenon is quite odd since oral drinking would speed rehydration and decrease the time necessary to fill urinary bladders. There may be selective disadvantages from the perspective of the transmission of water born parasites and infective agents. As *P. sauvagei* is the only species described to drink, and they drink drips of condensed water from leaves, this water source should be relatively free of infective agents.

Cutaneous Uptake

Water can osmotically enter the skin from either pooled water or interstitial soil water. This process is driven by a difference in water potential, which is osmotic in nature for pooled water, and a combination of the osmotic and matric potential components for soil. The rate of flux is proportional to the permeability of the skin in contact with the water. There are interspecific differences in both the normal permeability of amphibian skin to water and the ability to vary permeability. Aquatic species generally have a lower permeability than terrestrial species (Schmid 1965). This is interpreted as an adaptation to lower the rate of passive water uptake in aquatic species and speeding the rate of water uptake and rehydration in terrestrial species. There are also regional differences in the morphology and permeability of amphibian skin to water. The pelvic patch region of anuran skin is the most permeable (McClanahan and Baldwin 1969; Bentley and Main 1972; Marrero and Hillyard 1985). The pelvic and other ventral skin of terrestrial species tends to be granular and contain papillae known as verrucae hydrophilicae (Drewes *et al.* 1977) compared to aquatic species (Toledo and Jared 1993). This granulation or "epidermal sculpturing" would assist water uptake by increasing surface area and making better contact with irregularities in the soil as well as wicking water by capillarity in the interstices between the granulations (Christensen 1974; Lillywhite and Licht 1974; Hillyard 1976). A stereotypic posture is associated with utilizing the pelvic patch for cutaneous drinking, characterized by abducting the hind limbs and adpressing the pelvic patch to the substrate (initially described by Towson (1792); brought into current context by Stille (1958), and recently extensively reviewed by Hillyard *et al.* (1998)). This cutaneous drinking behavior, termed the water absorption response (WR), was induced by angiotensin II, injected either intraperitoneally or intraventricularly. Barometric pressure and the osmotic concentration of the bathing solution seem to modulate the intensity of the behavior (Hoff and Hillyard 1993). The permeability of the pelvic patch skin is influenced by circulation (Christensen 1974; Parsons and Mobin 1991; Parsons and Swartz 1991) and hormones arginine vasotocin (AVT), hydrins, catechols, and atrial natriuretic peptide (ANP) that can alter the distribution of aquaporins (water channels) (McClanahan and Baldwin 1969; Bentley 1971; DeSousa and Grosso 1982; Segura *et al.* 1982; Chauvet *et al.* 1990; Michel *et al.* 1993; Berman *et al.* 1994; Uchiyama *et al.* 1998; Toop and Donald 2004). Once the water enters the skin it appears to enter cutaneous capillaries (Word and Hillman 2005), although there are some that suggest uptake into the subcutaneous lymph sacs (Carter 1979; Parsons and Swartz 1991; Wentzell *et al.* 1993). The literature on aquaporin localization in the capillaries of amphibian skin supports a capillary entry for cutaneously absorbed water (Willumsen *et al.* 2003).

Water Outputs

Fecal Water Loss

The undigested food is eventually eliminated as feces. The amount of water lost via this route is dependent upon the amount of food eaten, the amount that is undigested and the water content of the fecal material. In describing the preformed water input the mass of fecal material produced can be derived from the non-assimilated food and seems to be generally in the 10–15% range (Shoemaker and

McClanahan 1975). Assimilation efficiency, (energy ingested–energy in feces)/ energy ingested, of the chemical potential energy present in these prey items varies with the relative amount of indigestible chitin that is present and can run from 57 to 94%. The eventual water content of the fecal material of amphibians is not documented. In reptiles the lowest water content of feces seems directly dependent upon the plasma colloid osmotic concentration (Murrish and Schmidt-Nielsen 1970). Consequently, 60–70% water content would be about the best expectation for drying feces. Since about 15% of the ingested food is excreted as feces, a reasonable estimate for fecal water loss is about 15% of the preformed water in the ingested food or 0.01–0.06 % body mass/day, a very small volume in comparison to the other avenues for water loss.

Urinary Water Loss

The first step in understanding the potential role of urine to water loss is to understand renal anatomy. The two kidneys produce a urine less concentrated than the body fluids. Urine is moved from the kidneys by peristaltic contractions of the ureters, which open to the dorsal wall of the cloaca. The cloaca has a posterior opening to the exterior, an anterior opening to the large intestine, and a ventral opening to the urinary bladder. The anatomy would argue for the potential of urine either being excreted to the environment or stored in the urinary bladder or intestine. Physiological data suggests that urine is normally moved into the urinary bladder by cloacal peristaltic contractions (Martin and Hillman 2009) rather than to the environment from the cloaca (see Jorgensen 1997). Consequently, as water is lost from the extracellular fluid compartment by filtration in the kidneys, it generally is stored in the urinary bladder rather than being lost to the animal. Hence the prevailing wisdom is that urine represents a shuttle of extracellular water from the plasma to the urinary bladder (Jorgensen 1994, 1997). A potentially different strategy is seen for fully aquatic anurans, like *Xenopus*, that has a urinary bladder only capable of storing a volume equivalent to 1% of its body mass. Given rates of urine formation of 12 ml/kg h (Henderson *et al.* 1972), and a similar continual osmotic uptake of fluid, a 1% urinary bladder capacity would only account for 1 hour's worth of urine production; the rest would have to be excreted to the environment.

Ordinarily, the rate of urine production is generally closely coupled intraspecifically to the rate of cutaneous osmotic uptake (Shpun and Katz 1995; Jorgensen 1991) and maximal urine flow rates tend to be higher for terrestrial species of apodans, urodeles and anurans which have intrinsically higher cutaneous uptake rates (Vondersaar and Stiffler 1989; Stiffler *et al.* 1990; Pruett *et al.* 1991; Hillman and Schimpf 1994). Anurans tend to have greater glomerular filtration rates than urodeles (Pruett *et al.* 1991).

The concentration of the urine is generally hypo-osmotic to the body fluid concentration, and at its most concentrated is iso-osmotic. Consequently urine will always represent a loss of water for an amphibian. Urine in its most concentrated state would be the equivalent of excreting plasma, obviously an untenable state from a homeostatic perspective. Generally, when this state is reached urine formation ceases.

Evaporative Water Loss

Evaporative water loss (EWL) represents the most significant and consistent avenue of water loss for terrestrial amphibians. It is driven by the difference in water potential between the animal and its environment. Normally, evaporative water loss of vertebrates is partitioned into a flux occurring across the respiratory surfaces and a flux occurring across the cutaneous surfaces. Since all amphibians exchange gases to a significant extent across their skin, all evaporative water loss is associated with gas exchange. Aquatic amphibians would only lose a small amount of water by evaporation as inspired air would have to be humidified to 99.6% RH from the ambient relative humidity (see Chapter 1).

Evaporative water loss has not been partitioned into its pulmonary and cutaneous components in amphibians. The closest attempt has been measuring the cutaneous flux of evaporative water loss alone, by separating pulmonary flux via a mask on the nostrils where V_{O_2} was measured in two species with low cutaneous permeabilities, *P. sauvagei* and *C. xerampelina* (Stinner and Shoemaker 1987). Unfortunately they did not simultaneously measure the pulmonary EWL, although as we will see below this can be roughly calculated. This partitioning would represent the most extreme situation in amphibians in terms of maximizing the relative importance of pulmonary loss to the overall rate of evaporative water loss, as these two species represent two of the highest specific resistances to cutaneous evaporative water loss. The pulmonary V_{O_2} was about 0.1 ml O_2/g h at 24°C. Assuming an extraction efficiency of 20% for O_2, and the fact that O_2 represents 21% of the total gas ventilated, then about 2.4 ml air/g h would need to be ventilated to support that metabolic rate. The c_{wv} at 24°C is 21 g H_2O/m^3 air. The maximal pulmonary evaporative loss rate would be $(2.4 \times 10^{-6}\,m^3/g\,h) \times 21\,g/m^3 = 50 \times 10^{-6}\,g\,H_2O/g$ per h assuming breathing dry air and humidifying it by pulmonary evaporation to 99.6% RH. The lowest rates of total EWL reported by Shoemaker and McClanahan (1975) were $0.5 \times 10^{-3}\,g\,H_2O/g$ h, for a variety of phylomedusine frogs and $10 \times 10^{-3}\,g\,H_2O/g$ h for typical hylid frogs, hence the maximal contribution of pulmonary EWL to total EWL, in probably its most significant situation, is about 10%, and pulmonary percentage is normally less than 0.5% in species with a low cutaneous resistance to EWL. Generally then, EWL is assumed to be entirely cutaneous in the calculation of skin resistances to EWL.

Water Storage

Urinary Bladder

One of the unique features of all amphibians is the presence of a urinary bladder, which develops as a ventral outpocketing of the cloaca. The bladder shape varies from cylindrical (salamanders) to bilobate (salamanders, caecilians and anurans). The bladder lies within the pleuroperitoneal cavity, hence it will be subjected to pressure variations associated with ventilatory events and will always be above ambient pressure. Amphibians generally produce a dilute urine during their bouts of cutaneous osmotic uptake, which moves from the kidneys to the cloaca, via the ureters, and then from the cloaca to the urinary bladder where it is stored. Generally this dilute urine is not expelled to the environment from the bladder unless the animal is disturbed. The ionic concentration of the urine in the bladder can be modified by active transport of Na^+ and Cl^- across the bladder wall from mucosal to serosal surfaces (Sinsch 1991). The concentration of urine varies seasonally in both the lab and field (Katz *et al.* 1986). There is generally some volume of urine present in the urinary bladder, which has led to the conclusion that cutaneous uptake of water and its movement to the bladder is anticipatory of deficits and hence represents a true storage compartment of dilute fluid that can be used to replenish body fluids lost via evaporation (Jorgensen 1991, 1994, 1997).

Movement of urine stored in the bladder back into the body fluids has been documented for a variety species (Adolph 1927; Steen 1929; Sawyer and Schisgall 1956; Ruibal 1962; Shoemaker 1964; McClanahan 1967; Shoemaker and Bickler 1979; Pough *et al.* 1983; Sinsch 1991). The presumed mechanism for the movement of water from the bladder to the extracellular space has been the release of AVT (also hydrin, see Acher *et al.* 1997) with dehydration increasing the permeability of the bladder wall and allowing water to osmotically move (Ewer 1952; Sawyer and Schisgall 1956; Bentley and Heller 1965; Heller and Bentley 1965; Alvarado 1972; Dicker and Elliott 1970). This implies that the amphibian must first be dehydrated to stimulate the release of AVT and hydrins, which is inconsistent with the data for freely behaving anurans utilizing bladder reserves and periodically returning to water to add fluid to their bladders before the urine stores are depleted (Jorgensen 1991, 1994). These amphibians are clearly not dehydrated before they add urine to their bladders. This raises some interesting regulatory issues as it relates to controlling both filling and emptying the bladder. Parsons *et al.* (1993) found that pelvic patch permeability was inversely related to bladder volume, implying a mechanism for sensing bladder volume that regulates the rate of water entry across the skin. They present convincing evidence that the compound regulating the influx is angiotensin II (AII); as the bladder empties, AII is released and it increases pelvic patch water flux. How bladder volume is sensed and how it is coupled to AII release is unclear.

The volume of the urine that can be stored in the urinary bladder varies interspecifically (see Jorgensen 1997 for review), and the maximal urine volume has been loosely correlated with habitat, that is more terrestrial species with a greater likelihood of evaporative water loss would tend to store more water (Bentley 1971; Jorgensen 1997). Problems with obtaining a definitive correlation are that a variety of techniques have been used to measure bladder capacity and there is a lack of a phylogenetic analysis. Some studies use repeated measures of individuals over time and report maximal values for freely behaving individuals. Other studies have used *in vitro* filling of isolated bladders. Both techniques have intrinsic difficulties. The *in vivo* (see Martin and Hillman 2009) measurements ignore known effects of the tendency to store more urine in the bladder under arid conditions or repeated draining of the bladder (Schmajuk and Segura 1982; Sinsch 1987; Jorgensen 1991), and these variables are uncontrolled in reporting the bladder volume results. Second, there is no control for secondary variables such as anatomic structures that would take up space in the pleuroperitoneal cavity, like eggs and fat bodies, and limit the space for bladder filling. The *in vitro* measurements ignore behavior and limitations placed on filling by the pleuroperitoneal cavity itself. Clearly an objective comparison of the bladder is still needed. One approach is to measure bladder compliance (Δ volume/Δ pressure) *in vivo*. As compliance is the product of distensibility and original volume it incorporates interspecific differences in bladder distensibility, interspecific differences in volume and inherent differences in the bladder pleuroperitoneal environment. Such measurements would have to be analyzed by phylogenetic methods to conclude that high bladder capacity was a terrestrial adaptation

Body Water

The total body water is composed of the fluid contained within the intra- and extracellular fluid compartments. Amphibians generally have a total body water content of between 77 and 83% of body mass (see reviews of Claussen 1973; Heatwole 1984; Alvarado 1979). This is higher by about 10–12% of body mass than for other vertebrates (Thorson 1961, 1968). Amphibians also have total body fluid concentration between 200 and 250 mosm/l, which is 50–100 mosm/l lower than for bony fishes and other tetrapods.

When bladder water is exhausted, the continued evaporative loss of water will lead to an increase in osmotic concentration of both the intracellular and extracellular water spaces. As the extracellular and intracellular spaces are in osmotic equilibrium, the actual increase in osmotic concentration of both spaces can be predicted from:

$$C_i V_i = C_f V_f$$

where C_i = concentration initial; C_f = concentration final; V_i = % water content initial; V_f = % original water content –% water lost by evaporation (Shoemaker 1964).

What does this greater diluted body water content of amphibians mean from a regulatory perspective? Consider two model organisms, one 80% water and 200 mosm/l ("amphibian like") and the other 70% water and 300 mosm/l ("other tetrapod like") and each loses 20% of its body mass by evaporation. Assuming both species are incapable of producing a hyperosmotic urine and lack salt glands (the amphibian situation), the osmotic concentration during the 20% dehydration, will increase by 66 mosm/l to 266 mosm/l in the "amphibian like" model and 120 mosm/l to 420 mosm/l for the "other tetrapod" model. Obviously, by starting with both a lower initial body fluid osmotic concentration and greater body water content an amphibian ameliorates the absolute change in osmotic concentration associated with dehydration and also has more fluid to redistribute to the circulatory system.

Intracellular Space

The intracellular space is defined as the fluid contained within cell membranes. It is usually measured as the difference between total body water and extracellular fluid volume, and is estimated to be between 40 and 60% of body mass in amphibians.

Cells have two extremes of strategies during dehydration when water loss exceeds gain and bladder water is exhausted. One extreme is to defend cell volume via the accumulation of solutes, raising their internal osmotic concentration to equal that of the extracellular space and preventing a cell volume change via osmotic water loss. The other extreme is the osmometric strategy whereby water is lost to the extracellular fluid space in achieving osmotic equilibrium. Obviously in both extremes the intracellular osmotic concentration increases. In the volume regulatory strategy the cell presumably selects specific solutes to be involved in increasing the osmotic concentration, that will minimize consequences on the intracellular solute environment that the cell proteins face. This strategy also has the added benefit of maintaining the intracellular and intercellular architecture since the cell volume does not change, and hence is presumably more protective of cellular function (see review of Lang *et al.* 1998). The osmometric strategy has the benefit of passive equilibration, hence regulatory mechanisms in handling water imbalances are not necessary and it translocates fluid from the cells to the extracellular compartment, which will help maintain circulatory system function.

How do amphibian cells actually react to increased extracellular osmotic concentration? The answer seems to depend on the specific organs involved. Some cells, like those associated with skeletal muscles and the skin act more as simple osmometers, losing water in proportion to the increase in extracellular osmotic concentration (Smith and Jackson 1931; Alvarado 1972; Hillman 1978, 1982). Since skin and skeletal muscle are two of the largest organ systems, this could free substantial water to maintain circulatory function without losing much in the way of tension generation or integrity. Other cells, like those associated with the nervous system, appear to volume regulate by accumulating both ions and amino acids (Baxter and Ortiz 1966; Tachiki and Baxter 1980; Baxter *et al.* 1993; Hillman 1988). Three dimensional cellular connections are likely to be significant for nervous system function, and nervous system dysfunction by changes in cell volume would ramify throughout all sensory and control systems. It makes intuitive sense that maintenance of function in these cells would be most critical.

Physical forces that modify cell volume can influence the cellular architecture and subsequently gene expression of the cell (Watson 1991).

Extracellular Space

The extracellular space is divided into three intercommunicating compartments; the plasma space is contained within the boundaries of the circulatory system but outside of cells, the interstitial space is the fluid outside the boundaries of both the circulatory system and cells; and the lymphatic space is that volume of the interstitial space that can be mobilized by the lymphatic system back into the plasma space. The actual volumes of plasma and interstitial space can be measured via the use of dyes or radioactive markers and their kinetic exchange. The total extracellular space of amphibians has been estimated by a variety of techniques and falls within 20–30% of body mass (Thorson 1964; Jorgensen *et al.* 1978; DeGrauw 1998). There tends to be quite a bit of variation; however, it is not clear whether this reflects real interindividual/interspecies variation or measurement variation.

Maintenance of the plasma space is critical during periods of water loss. The volume contained within a circulatory system will influence the pressure within that circulatory system, defined by the circulatory compliance (Δ volume/Δ pressure) of that system. Maintenance of plasma volume by the lymphatic system mobilization of interstitial fluid appears to be an important element of enhanced tolerance to dehydration (see Chapter 3).

2.1.3 Physiological Control of Water Exchange

Five main hormones (arginine vasotocin, hydrin, angiotensin II, aldosterone, atrial natriuretic peptide) as well as the sympathetic nervous system are involved in the short-term control of amphibian water balance. The role of mesotocin is still enigmatic. Prolactin seems to have a significant long-term role decreasing cutaneous permeability. Effects of hormones on transport mechanisms in specific tissues is discussed in Chapter 3.

Arginine Vasotocin and Hydrins

The negative feedback control of water balance by arginine vasotocin (AVT) and hydrins is delineated in Fig. 2.3 when water loss exceeds water gain. Changes in cerebrospinal fluid osmolality are sensed in the hypothalamus and stimulate the release of AVT and hydrin from the posterior pituitary. In conjunction, decreases in the volume of blood are sensed by atrial receptors and arterial baroreceptors, stimulating the release of AVT and hydrins. The AVT, and potentially the hydrins, increase skin and water permeability, stop urine formation and stimulate the strength of lymph and cardiac heart contraction, all tending to increase plasma volume and maintenance of cardiac output.

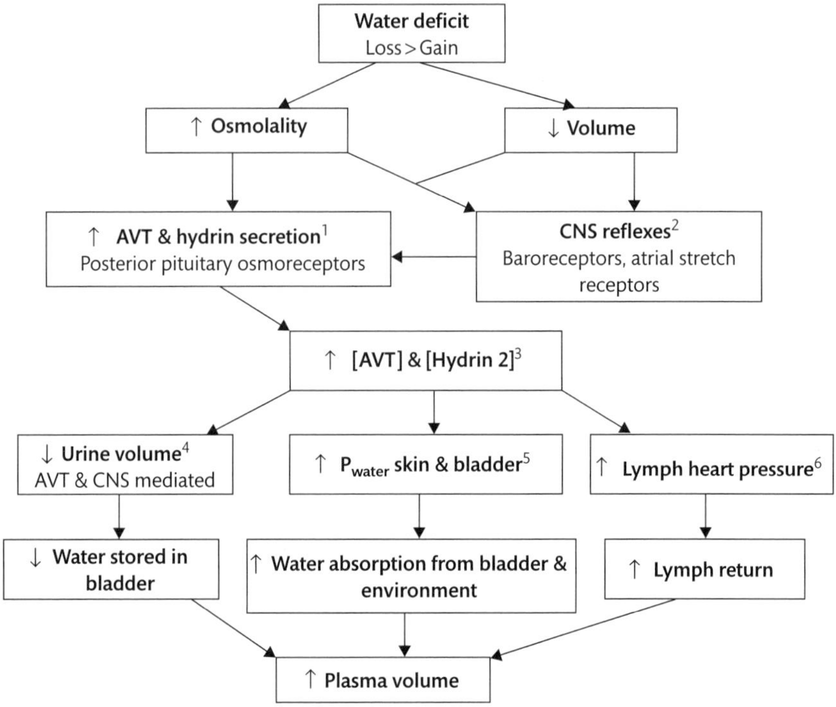

Fig. 2.3 The negative feedback control loops mediated by AVT and hydrins involved in the regulation of water balance when loss exceeds gain: [1]Pang *et al.* 1980, Nouwen and Kuhn 1985, Chauvet *et al.* 1990, Michel *et al.* 1993; [2]Nouwen and Kuhn 1985, Biewald 1980; [3]Nowen and Kuhn 1985, Konno *et al.* 2005; [4]Sawyer 1957, Henderson *et al.* 1972, Gallardo *et al.* 1980, Pang *et al.* 1980; [5]Bentley 1971; [6]DeGrauw and Hillman 2004. AVT, arginine vasotocin.

Angiotensin II and Aldosterone

The negative feedback control elements involved in the release and effects of angiotensin II and aldosterone when water loss exceeds water gain are delineated in Fig. 2.4. A decrease in blood volume decreases renal arterial pressure and blood flow, stimulating the renin–angiotensin–angiotensin II cascade. This stimulates aldosterone release and a series of behavioral and physiological responses to retain fluid in the kidney and mobilize fluid uptake mechanisms.

Atrial Natriuretic Peptide

The apparent negative feedback control of atrial natriuretic peptide (ANP) when water uptake exceeds water loss and its physiological effects are delineated in Fig. 2.5. An increased blood volume stimulates, via mechanical stretch, the release of ANP from principally the atria but also the lungs and kidneys. ANP favors the

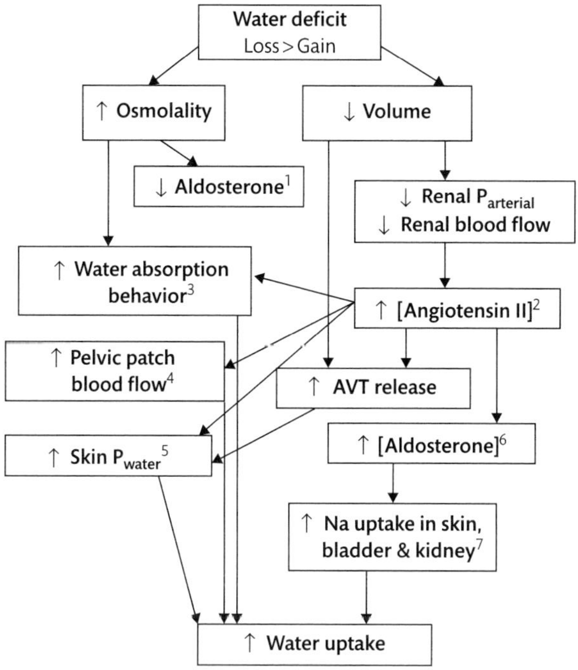

Fig. 2.4 The negative feedback control of water balance mediated by angiotensin II and aldosterone when water loss exceeds gain: [1]Garland and Henderson 1975; [2]West *et al.* 1998; [3]Hoffman and Katz 1999, Hillyard *et al.* 1998; [4]Slivkoff and Warburton 2001; [5]Coviello *et al.* 1974, Bruno and Coviello 1996, Parsons *et al.* 1993; [6]Nolly and Fasciola 1971, Hanke *et al.* 1990; [7]Crabbe 1961, Nagel and Crabbe 1980. AVT, arginine vasotocin.

loss of fluid from the circulatory system via the kidneys and systemic capillaries to the bladder and interstitium, respectively. ANP also antagonizes AVT effects on the skin as well as limiting contraction of the cardiac heart, potentially limiting Frank–Starling-induced work by the heart.

Sympathetic Nervous System Control

The negative feedback control of water and solute metabolism by the sympathetic nervous system when water loss exceeds water gain is delineated in Fig. 2.6. An increase in osmolality and, more importantly, a decrease in blood volume stimulate atrial and arterial baroreceptors to stimulate central nervous system mediated sympathetic nervous system responses that increase cardiac output.

2.1.4 Ions and Ion Budgets

The ion concentration differences between intracellular and extracellular fluids is a source of electrochemical energy that drives all electrical transmission of

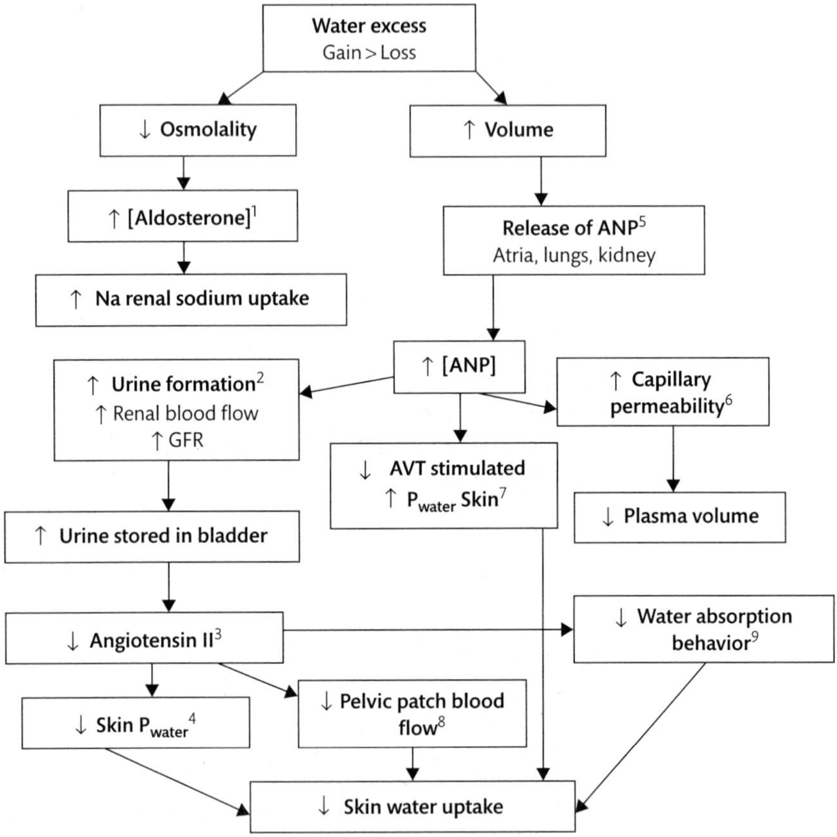

Fig. 2.5 The role of atrial natriuretic peptide in water and solute metabolism: [1]Garland and Henderson 1975; [2]Donald *et al.* 2002; [3]Parsons *et al.* 1993; [4]Coviello *et al.* 1974, Parsons *et al.* 1993; [5]Sakata *et al.* 1988, Lazure *et al.* 1988, Glass *et al.* 1996, Meier *et al.* 1999; [6]Meyer and Huxley 1990; [7]Berman *et al.* 1994, Uchiyama *et al.* 1998, Toop and Donald 2004; [8]Slivkoff and Warburton 2001; [9]Hoff and Hillyard 1991, Propper *et al.* 1995, Hillyard *et al.* 1999. AVT, arginine vasotocin; ANP, atrial natriuretic peptide; GFR, glomerular filtration rate.

information via the nervous system as well as controlling voltage-sensitive ion channels that are important in regulating cell function. The problem that uniquely faces amphibians is that given their high water turnover, the potential for change in these concentrations is arguably faster and greater than for any other vertebrate class. Aquatic amphibians face chronic dilutional stress, while terrestrial species face dilutional stress when cutaneously absorbing water and concentration stress when in aerial situations. Since these concentration differences are established by the active transport of ions, environmental conditions that may compromise

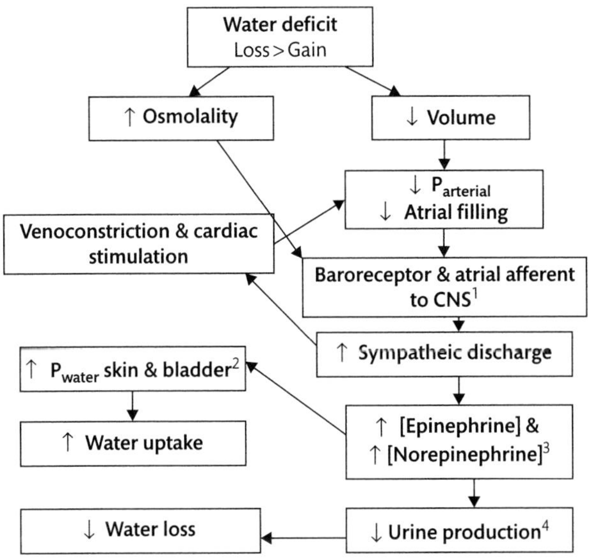

Fig. 2.6 Sympathetic control of water and solute metabolism: [1]West and Van Vliet 1992, 1994; [2]Segura *et al.* 1982; [3]Withers *et al.* 1988, Hillman *et al.* 1998; [4]Gallardo *et al.* 1980.

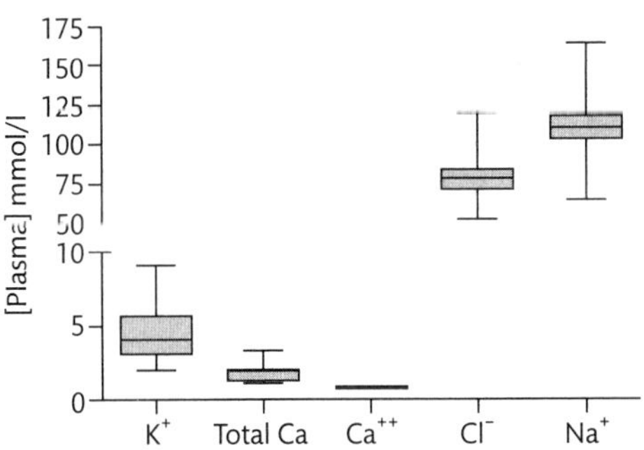

Fig. 2.7 A summary of the concentration of the primary plasma electrolytes present in amphibian plasma. Values are mean, 95% confidence interval of the mean and the range from values in the literature.

metabolism such as hypoxia, low temperature and freezing are of significant interest (see Stinner and Hartzler 2000). Consequently, regulation of ion concentrations is of particular interest for amphibians.

A representative sample of some major plasma ions for a variety of amphibian species is presented in Fig. 2.7 and Table 2.2. Two patterns are obvious, some ions have less variance in their concentrations (K^+, total Ca, and Ca^{++}) hence are regulated more closely. Sodium and chloride are the dominant solutes of the extracellular space.

Table 2.2 A selected summary of plasma ion concentrations in amphibians

Species	$[Na^+]$ mmoll/l	$[K^+]$ mmol/l	$[Cl^-]$ mmol/l	Total [Ca] mmol/l	$[Ca^{++}]$ mmol/l	Reference
Anurans						
A. truei	106	-	81	-	-	Mullen and Alvarado 1976
X. laevis	106	-	-	-	-	Bentley 1973
X. laevis	120	3	80	-	-	Hillman 1978
X. laevis	120	6	120	-	-	Balinsky 1981
X. laevis	101	2	84	-	-	McBean and Goldstein 1970
X. laevis	-	-	-	1.39	-	McWhinnie and Scopelliti 1978
R. catesbeiana	119	-	109	2.1	-	Coppo *et al.* 2005
R. catesbeiana	102	4.0	52	1.08	-	Hillman 1988
R. catesbeiana	-	-	-	2.55	-	Oguro *et al.* 1975
R. catesbeiana	-	-	-	1.39	0.73	Hillman *et al.* 1987
R. muscosa	114	-	-	-	-	Hillman 1980
R. pipiens	116	-	-	-	-	Hillman 1980
R. pipiens	100	-	-	-	-	Bentley 1973
R. pipiens	112	-	68	-	-	Mullen and Alvarado 1976
R. pipiens	104	2.1	74	2	-	Conway 1957
R. pipiens	95	3.8	74	2.1	-	Stiffler *et al.* 1997
R. pipiens	104	5	77	-	-	Campbell *et al.* 1967
R. pipiens	-	-	-	1.47	0.73	Sasayama and Clark 1984
R. pipiens	97	3.5	-	1.2	-	Miller *et al.* 1968
R. pipiens	104	2.5	-	2	-	Conway 1945
R. pipiens	115	7	-	1.7	0.9	Robertson 1985
R. cancrivora	129	9	98	-	-	Shoemaker *et al.* 1992
R. temporaria	-	-	-	2.5	-	Simkiss 1968
R. temporaria	105	2	83	-	-	Balinsky 1981
R. ridibunda	115	6	83	-	-	Balinsky 1981
R. esculenta	104	-	-	-	-	Balinsky 1981
H. regilla	110	-	78	-	-	Mullen and Alvarado 1976
S. couchi	133	-	-	-	-	Hillman 1980

Table 2.2 (*cont.*)

Species	[Na⁺] mmoll/l	[K⁺] mmol/l	[Cl⁻] mmol/l	Total [Ca] mmol/l	[Ca⁺⁺] mmol/l	Reference
B. marinus	107	5.9	-	2.1	-	Cole 1968
B. marinus	118	4	85	-	-	Middler *et al.* 1968
B. marinus	123	4	84	-	-	Hillman 1988
B. marinus	-	-	-	3.3	-	Tufts and Toews 1985
B. marinus	101	-	-	-	-	Bentley 1973
B. marinus	-	-	-	1.26	0.89	Hillman *et al.* 1987
B. cognatus	128	-	-	-	-	Hillman 1980
B. boreas	127	-	-	-	-	Hillman 1980
B. boreas	109	-	77	-	-	Mullen and Alvarado 1976
B. boreas	64	4	61	-	-	Balinsky 1981
B. bufo	107	4	68	-	-	Balinsky 1981
B. bufo	117	2	78	-	-	Ferreira and Jesus 1973
B. viridis	113	5	99	-	-	Balinsky 1981
T. miliaris	165	3.5	75	-	-	Abe and Bicudo 1991
Caecilians	-	-	-	-	-	
Typlonectes	111	5.4	70	-	-	Stiffler *et al.* 1990
Ichthyophis	112	2.7	80	-	-	Stiffler *et al.* 1990
Urodeles						
B. attenuatus	110	-	-	-	-	Shoemaker *et al.* 1992
A. tigrinum	106	-	98	-	-	Shoemaker *et al.* 1992
A. tigrinum	-	-	-	1	-	Stiffler 1991
A. mexicanum	-	-	-	1.62	-	Kingsbury and Fenwick 1989
A. gracile	94	9	75	-	-	Stiffler and Alvarado 1974
N. maculosus	93	2.5	82	-	-	Stiffler *et al.* 1983
D. monticola	-	-	-	1.6	-	Wittle 1983
N. viridescens	-	-	-	2.13	-	Wittle and Dent 1979
C. pyrrogaster	-	-	-	2.45	-	Uchiyama 1980
T. andersoni	-	-	-	1.93	-	Oguro and Sasayama 1978

The balance of ions can be summarized by a budgetary approach, as was done for water. Some of the avenues for input, storage and loss are the same for solutes and water:

- Inputs
  food
  cutaneous
  drinking
  gills
- Storage
  bladder
  intracellular
  extracellular
  anatomic structures
- Outputs
  feces
  urine
  glands
  cutaneous.

Inputs

Food

Adult amphibians are generally insectivorous. There is an extensive literature on the ion contents of arthropods (see Reichle *et al.* 1969). Ion content characteristically varies seasonally and interspecifically. Sodium ion content varies from 2.8 to 9.2 mg Na/g ash free dry mass of arthropods, potassium content from 0.9 to 9.4 and calcium content from 0.3 to 4.7. An average 1 g of insect that is 90% water would tend to raise the extracellular fluid concentration by 29 mmol (4 mmol Ca, 18 mmol Na, 7 mmol K) in a 100-g amphibian assuming all the ions are absorbed and the extracellular fluid compartment represents 30% of the body water. This increase in ion uptake with feeding also characteristically leads to water uptake by the animal in a ratio of 0.75 ml H_2O/g food ingested (Jorgensen 1993).

Cutaneous

There are transport proteins responsible for the active transport of sodium, chloride and calcium in the skin of amphibians. These transport proteins are capable of actively transporting these ions from the environment into the organism (see Chapter 3). Cutaneous ion uptake and ingested food are the two major input sources for ions in adult amphibians.

Outputs

Feces

The feces are associated with some electrolyte loss, both in the remnants of the undigested food as well as the bacterial flora associated with the feces. We are unaware of any systematic, direct measurements of electrolyte loss via this route and have estimated its magnitude as 15% of the ingested electrolyte intake.

Urine

Amphibians normally produce a dilute urine, with low electrolyte concentrations of less than 10 mM (Hillman and Schimpf 1994). These concentrations generally increase if the high urine flow rates are stimulated via hypo-osmotic volume expansion (Hillman and Schimpf 1994). Estimating the actual urinary loss of electrolytes is complicated by the fact the urine is stored in the urinary bladder, which itself is capable of further uptake of sodium, chloride and calcium.

Glands

Amphibians do not possess glands that can produce a fluid more concentrated than their own body fluids, such as salt glands in other vertebrates. They do possess numerous mucous and poison glands that contain electrolytes. We are unaware of any studies quantifying the contribution of glandular secretions to overall electrolyte output in amphibians.

Cutaneous

Amphibians in fresh water aquatic environments will characteristically lose electrolytes to the water by diffusion based on the concentration difference between their body fluids and that of the water. These rates of efflux can be hormonally and physically modulated, as indicated below. Characteristically aquatic species, with lower cutaneous permeabilities will tend to lose fewer electrolytes than terrestrial species in water.

Storage

Urinary Bladder

The urinary bladder has been demonstrated to actively transport sodium, chloride and calcium from its lumen to the body fluids. Consequently, electrolytes excreted by the kidney can be recovered if need be. The size of this reserve is related to the concentration of the urine contained within the urinary bladder and its overall volume. Normally the bladder fluid osmotic concentration is low unless the fluid content is being recovered during periods when water output exceeds water input, where it maximally will be in equilibrium with the body fluid concentration.

Intracellular

The intracellular environment contains mostly potassium as a mobile electrolyte, hence it is a major source of the potassium storage in the body. There is a total body store of potassium of 49 mmol/kg, with 98% of it being intracellular. The intracellular fluid compartment represents about two-thirds of the overall fluid in the body and potassium concentration is approximately 90 mmol/l. The extracellular fluid compartment is a third of the total body fluid content and the potassium concentration is approximately 3 mmol/l (Conway 1957).

Extracellular

The extracellular environment has the highest concentrations of both sodium (110 mmol/l) and chloride (80 mmol/l), with intracellular concentrations of 7 mmol/l for chloride and 4 mmol/l for sodium. The total body stores of chloride would be 24.9 mmol/kg with 88% extracellular and of sodium 31.1 mmol/kg with 93% extracellular.

Anatomic Structures

The principal anatomic reserves exist for calcium and represent the endolymphatic sac system (calcium carbonate) and bones (calcium phosphate).

Sodium Budget

Sodium (Na^+) is the principal cation of the extracellular fluid and characteristically is regulated. Plasma sodium concentration can vary interspecifically and seasonally, but generally falls within the 90–120 mmol/l range. A partial reason for the variation may be activity, as a 15 mmol/l swing in plasma Na^+ was documented 24 h after activity in fasting toads (Andersen and Wang 2003). A Na^+ budget for *Bufo boreas*, a terrestrial toad, and *Rana pipiens*, a semiaquatic frog is presented in Fig. 2.8.

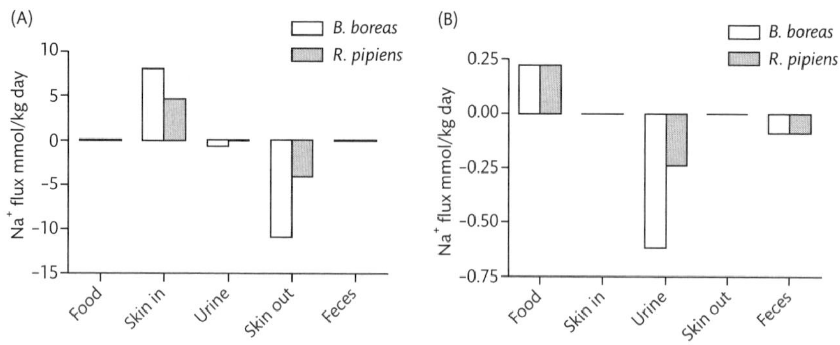

Fig. 2.8 A sodium budget for two species of anurans in both water (A) and air (B). The *Bufo boreas* is normally more terrestrial than the more aquatic *Rana pipiens*.

For adult urodeles and anurans in an aerial environment, the major route of sodium input is dietary and the major route of loss is urinary. In an aquatic situation a much larger flux occurs across the skin, both in and out. This flux can be up to 6% of the total body sodium content on a daily basis (Greenwald 1972). For the two budgets presented, *B. boreas*, (a normally terrestrial species) has a much greater flux of sodium, and efflux exceeds influx. Over time, this would indicate that plasma sodium should decline (hyponatremia) in *B. boreas* compared to *R. pipiens* (a more aquatic species). This is exactly what happens to members of the genus *Bufo* when maintained in fresh water (Bentley 1973; Schmid 1965; Mullen and Alvarado 1976). Plasma sodium concentrations are generally invariant in fasted aquatic species, like the data presented for *R. pipiens* above, based on a decreased rate of renal excretion and a matching active cutaneous uptake (McAfee and Kalckar 1972). Aquatic species, both urodele and anuran, generally have lower rates of Na^+ efflux across the skin, lower urinary efflux, lower transport capacity and a higher affinity transport system for Na^+ than terrestrial species maintained in water (Greenwald 1972; Mullen and Alvarado 1976). This is enigmatic given that the short circuit current across amphibian skin, normally reflective of sodium uptake, does not seem to vary for species from aquatic and terrestrial environments (Schmid and Underhill 1964). The role of the skin and bladder uptake systems (see Chapter 3) is equivocal in the maintenance of long-term sodium balance for aquatic amphibians maintained in water. For fasted amphibians bathed in tap water with cutaneous sodium uptake systems blocked with amiloride (a blocker of sodium uptake), plasma sodium does not change except for terrestrial species (Bentley 1973). If fasted *R. pipiens* are maintained in deionized water, neither plasma sodium concentration nor whole animal sodium content change (McAfee and Kalckar 1972). Blockage of the cutaneous transport system or providing no sodium to transport has no influence on sodium concentrations in aquatic species, indicating that cutaneous sodium transport may play little role in the long-term maintenance of sodium balance in most aquatic species even though cutaneous uptake systems can transport up to 5 mmol/kg day. The long-term acclimation to deionized water also leads to very low efflux of sodium. This is not the case for terrestrial species, when maintained in deionized water, hyponatremia and death may occur within days, mainly as a result of sodium loss from the kidney and across the skin.

Chloride Budget

Chloride (Cl^-) is the main anion of the extracellular fluid. Its concentration tends to be more variable than the other ions, and falls within the 70–100 mmol/l range. Part of the higher variability of chloride concentration is associated with its redistribution to the stomach as HCl secretion during digestion. Immediately post-feeding, acidification of the stomach contents has been calculated to potentially decrease the extracellular [Cl^-] by half (see Jorgensen 1992), although

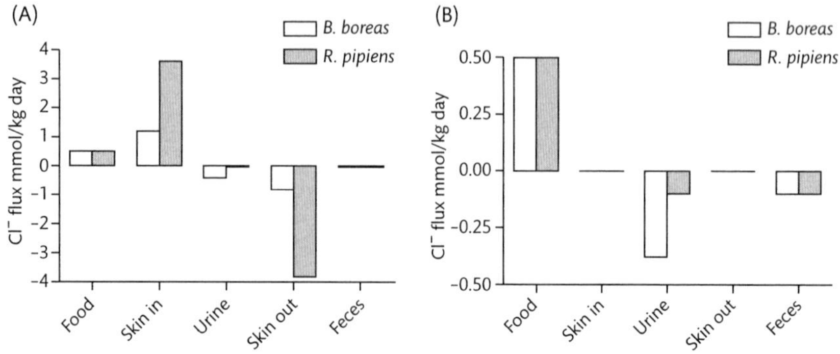

Fig. 2.9 A chloride budget for two species of anurans in both water (A) and air (B). The *Bufo boreas* is normally more terrestrial than the more aquatic *Rana pipiens*.

experimentally it appears closer to a 10 mmol/l decrease (Andersen and Wang 2003). This Cl^- redistribution is accompanied by an accumulation of HCO_3^-, the "alkaline tide" (Jorgensen 1992). Aquatic species, both urodele and anuran, generally have lower rates of Cl^- efflux across the skin, lower urinary efflux and a higher affinity transport system for Cl^- than terrestrial species (Greenwald 1972; Mullen and Alvarado 1976).

A chloride budget for *Bufo boreas* and *Rana pipiens*, presented in Fig. 2.9, illustrates the aquatic–terrestrial species differences. In aquatic situations there is both a large active transport influx and passive cutaneous efflux of chloride, just as we saw for sodium. Consequently, for fasting animals there is the capacity to balance whole body chloride via cutaneous regulation of influx (increase) and cutaneous efflux (decrease). Amphibians provided with food can easily balance their chloride via regulation at a renal level.

Potassium Budget

Potassium (K^+) is principally intracellular in localization. The extracellular concentration runs from about 2–4 mmol/l and is tightly regulated. Exercise has been shown to raise plasma K^+ from 2 to 4.5 mmol/l in *R. pipiens*, hence some of the variation noted between studies may result from handling stress and the loss of K^+ that accompanies skeletal muscle contractions (Warren and Jackson 2005). Concentrations of potassium above 4 mmol/l are generally attributed to extracellular fluid contaminated by intracellular release of K^+ due to cell injury (e.g. red blood cell hemolysis). The cell membrane electrical potential is principally determined by K^+ efflux from the cell, which is governed by the potassium concentration difference across the cell membranes, along with K^+ permeability. Consequently, changes in extracellular K^+ markedly influence both resting membrane potential and action potentials of excitable cells.

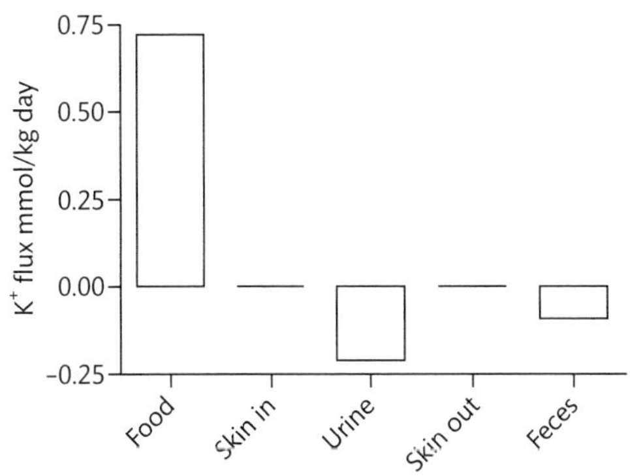

Fig. 2.10 A potassium budget for *Pachymedusa dachnicolor* in air, its characteristic environment.

A potassium budget for a typical scansorial hylid frog (*Pachymedusa dachnicolor*) in air (Fig. 2.10) shows that there is a large dietary input of potassium. This is expected for insectivorous or carnivorous frogs, as two-thirds of the body mass of prey items is intracellular and hence has a high potassium concentration. Urinary losses are low and potentially can increase if intake remains high. We are unaware of a potassium budget for an amphibian in water.

Calcium Budget

Calcium is present in both the extracellular and intracellular compartments, although it is generally at a much higher concentration in the extracellular space, but may be high in the endoplasmic reticulum of muscle cells. Calcium is present in both an ionic form (Ca^{++}) and a bound form associated with anionic proteins. The physiologically active form is Ca^{++}. Measures of total calcium concentration may not accurately reflect the physiologically relevant concentration, although most reported calcium concentrations are for total calcium not the free ionic form. The reported ionic values conclude that 50–70% of the total calcium is in its ionic form (Sasayama and Clark 1984; Robertson 1985; Hillman *et al.* 1987). The most recent review of amphibian calcium metabolism is by Stiffler (1993). Calcium concentrations vary seasonally (Robertson 1977, 1985) and increase with a decrease in pH induced by activity or hypercarbia (Hillman *et al.* 1987). Consequently, there are reservoirs of calcium, both bone and the endolymphatic sac, where Ca can be quickly mobilized by pH shifts. Calcium is actively taken up across the skin of both anurans and urodeles (Stiffler 1995; Guirguis *et al.* 1997; Stiffler *et al.* 1997; Zerella and Stiffler 1999). This cutaneous calcium uptake is enhanced by parathyroid hormone and calcitonin (Stiffler *et al.* 1998). Calcium is also actively secreted into the urinary bladder and can be absorbed from the colon (Rosoff *et al.* 1983). We are unaware of a calcium budget for amphibians in either air or water.

A unique feature of calcium metabolism in amphibians is the presence of a well-developed, highly vascularized endolymphatic sac. It actually surrounds much of the brain in the interdural space, and in anurans extends down the vertebral column with protrusions between the vertebrae known as paravetebral lime sacs, that overlie spinal ganglia. The endolymphatic sac and the paravertebral lime sacs are filled with a form of calcium carbonate known as aragonite (Dempster 1930; Schlumberger and Burk 1953; Carlstrom 1963). This calcium carbonate is readily exchangeable; deposition is enhanced by exposure to elevated calcium in the bathing medium and calcitonin released from the ultimobranchial bodies (embryonic relatives of the parathyroid glands) (Kawamata 1990; Robertson 1969; Oguro *et al.* 1981; Srivastav and Rani 1989). A protein, otoconin-22, seems to be intimately involved in the formation and growth of the calcium carbonate crystals (Pote *et al.* 1993; Yaoi 2003). The need for such an extensive, mobilizable calcium reserve and its intimate anatomic location on the central nervous system raises many interesting questions.

The main physiological function described for this system is to store calcium for bone repair (Sclumberger and Burk 1953) or to support periods of bone growth and mineralization, especially during the non-feeding metamorphic climax where calcium is moved from the endolymphatic sacs to ossify bone (Fig. 2.11, modified from Pilkington and Simkiss 1966). Anurans undergo one of the most dramatic larval to adult vertebrate transformations, which may explain the selective advantage of such an extensive calcium storage system. Another function may be to buffer acidic conditions during respiratory acidosis in the following manner (Simkiss 1968):

$$CaCO_3 + CO_2 + H_2O \leftrightarrow Ca^{++} + 2HCO_3^-$$

The endolymphatic system in conjunction with bone may also partially buffer lactate following exercise (Warren and Jackson 2005).

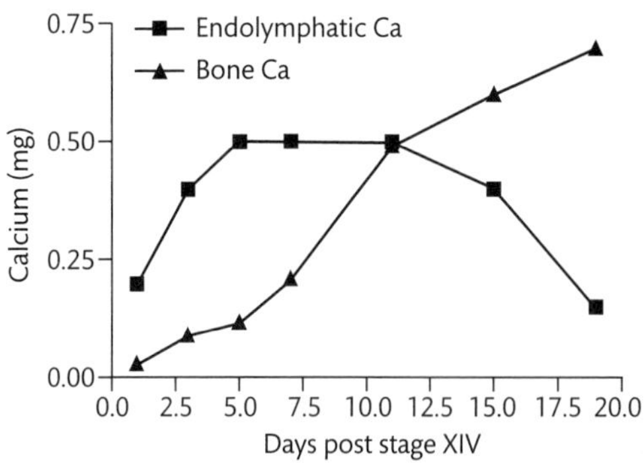

Fig. 2.11 The changes in the quantity of calcium present in the endolymphatic system and in bone during metamorphosis in *Rana temporaria*, after Pilkington and Simkiss 1966.

2.2 Temperature and Thermal Energy Budgets

An organism's body temperature is a physical variable that will impact the rate of essentially every physiological function. Besides its effect on rates of physiological processes, extremes in temperature set both low and high limits to life processes. Thermal tolerance can play a role in the geographic distribution of species. Consequently, an understanding of the processes involved in heat exchange are critical to understanding how and what an amphibian can do to establish a body temperature consistent with normal physiological function. There are excellent reviews available for modeling thermal balances and delineating climate spaces for amphibians (Tracy 1975,1976: Spotilla *et al.* 1992).

Budget

The avenues of heat energy exchange are discussed in Chapter 1. Here we examine the relative magnitude of heat flux through each avenue:

- Inputs
  solar radiation
  infrared (IR) radiation
  convection/conduction
  condensation
  metabolism
- Storage
  Δ body temperature
- Outputs
  evaporative
  IR radiation
  convection/conduction.

Any thermal energy budget is specific for particular environmental conditions of temperature, wind, solar and IR conditions. The budget delineated in Fig. 2.12 is for a model frog where body temperature (20°C) equals air and surface temperature; hence there is no convective/conductive exchange, there is moderate solar input striking the dorsal surface (a_{solar} = 0.85), and the RH = 50% so there is evaporative water loss.

There are some important things to note in this budget. First, the largest amount of energy being exchanged both in and out is via IR radiation. Second, metabolic energy input is essentially zero. Third, solar radiation can be a substantial source of heat gain during the day. Finally, evaporation is a large component of heat loss.

Solar Radiation

Solar radiation can be a significant source of energy input for an amphibian. The magnitude of solar gain depends on the amount of incident solar radiation

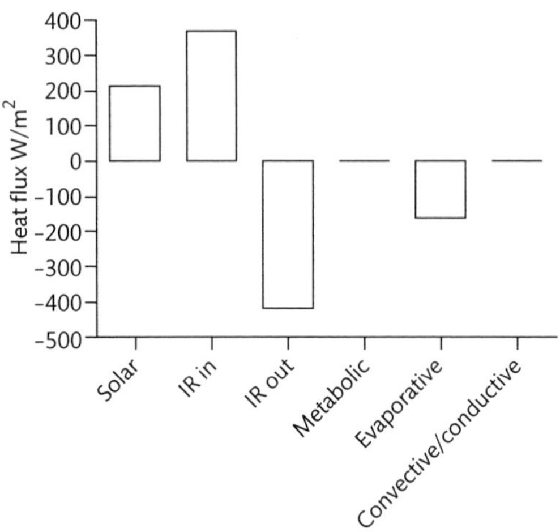

Fig. 2.12 An energy budget for a model frog where body temperature (20°C) equals air and surface temperature, hence there is no convective/conductive exchange, and there is moderate solar input striking the dorsal surface (a_{solar} = 0.85), and the RH = 50%.

available and the absorptivity for solar radiation. The ability to exploit solar radiation has been a selective advantage invoked to explain the evolution of tetrapods from aquatic ancestors (Carroll *et al.* 2005). Figure 2.13 gives the range of solar radiation absorbed over the range of absorptivities reported for amphibians (Spotila *et al.* 1992). On a typical sunny day, radiative heat gain would be about 100–400 W/m^2 dependent upon absorptivity.

Infrared Energy

In a normal environment infrared radiation is one of the predominant forms of energy exchanged, typically in a range of 350–450 W/m^2 (Fig. 2.13). Obviously, behavior can alter the amount of IR energy absorbed by selecting the surrounding thermal conditions and controlling the proportion of surface area facing the sky. Emissivity and absorptivity for infrared radiation (Eir and Air) can be between 0.95 and 1 (Gates 1980). Some tree frogs have special pigments that reflect near IR radiation, known as pterorhodin (Schwalm *et al.* 1977; see Bagnara 2003 for review). The function of pterorhodin is not clear but it has been speculated that the IR reflectance may make it more difficult for predators to detect these species on background vegetation but it may also decrease a_{ir} for the species that possess pterorhodin.

Conduction/Convection

Conduction and convective heat transfer can either be a gain or loss of energy based on the temperature difference between the amphibian and the air or fluid it is in contact with. In our example of a thermal energy input, we purposefully set up an example where body, air and substrate temperatures were equal so there

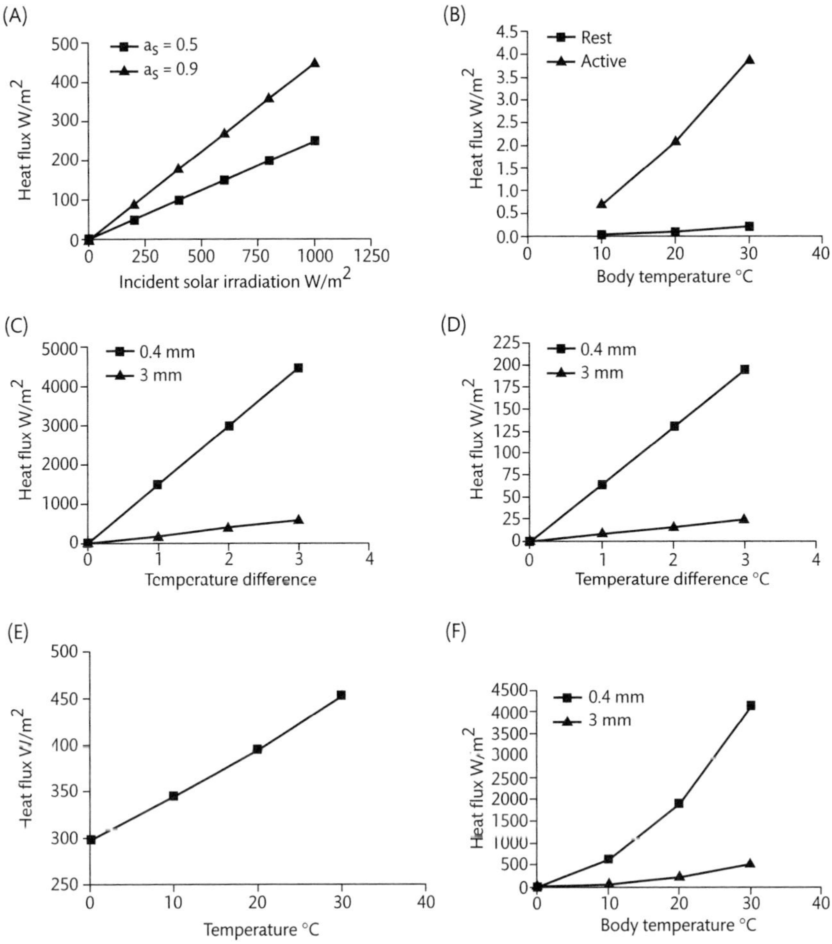

Fig. 2.13 Quantitative estimates of the magnitude of heat fluxes via the various pathways for heat exchange using minimum and maximum conditions described in the text. (A) Solar irradiation gain; (B) metabolism gain; (C) conduction/convection H_2O gain or loss; (D) conduction/convection air gain or loss; (E) infrared radiation gain and loss; (F) evaporation loss.

would be no heat transfer in or out via conduction/convection. In Fig. 2.13 we examine the relative magnitude of heat that would move in response to a difference in temperature in both air and water under what would represent two boundary layer conditions, both a minimum (0.4 mm) and maximum (3 mm) (see Chapter 1.3.5).

Two things are obvious—an amphibian will have very high heat fluxes (200–1500 W/m^2) associated with a 1°C thermal difference in both boundary layer conditions given the high thermal conductivity of water and corresponding large heat flux; and so it will very quickly equilibrate with water temperature. Boundary layers are very important in both air and water, but a large boundary layer in air is actually associated with a very modest net heat flux (<10 W/m^2 at $\Delta T = 1°C$). Consequently it is theoretically possible for an amphibian to maintain a temperature difference between itself and air in low flow (thick boundary layer) conditions. However, a high associated rate of EWL (hence evaporative heat loss) will compromise the potential thermoregulatory capability for normal amphibians. Waterproof species, with low rates of EWL, are able to exploit elevated body temperatures relative to air temperatures.

Evaporation

Evaporative heat exchange is actually relatively unique to amphibians in terms of its magnitude. In the example presented in Fig. 2.13, the condition at 0°C is 100% RH, hence there is no evaporation. As temperature increases in this example for minimum and maximum boundary layer conditions, the ambient concentration of water vapor was held constant (equal to 100% RH at 0°C) and the concentration difference increased in proportion to the difference between the skin surface concentration of water vapor at 99.6% RH (0.3 osm) at body temperature. The diffusion coefficient also increases with temperature and the heat of vaporization decreases in proportion to standard values at these temperatures.

The heat loss due to evaporation is enormous (1000–3000 W/m^2); equivalent to conduction/convection exchange in water. If ambient conditions lead to a net heat gain of the animal, then the increase in body temperature will increase the temperature of the adjacent air boundary layer, which will increase the driving force and diffusivity for water vapor and consequently increase the rate of evaporative heat loss. The high cutaneous permeabilities (low cutaneous resistance) to water movement essentially prevents most amphibians from maintaining a body temperature much above ambient temperatures. This prevents the exploitation of solar heat gain as a significant source of thermal energy that would allow the maintenance of elevated body temperatures over air temperature. Models that allow early tetrapods to thermoregulate, assume low rates of evaporation (Carroll *et al.* 2005), a condition atypical of their extant descendents.

Metabolism

Metabolic heat production is directly proportional to the rate of O_2 consumption; 21.4 J/ml O_2 for carbohydrate, 19.5 J/ml O_2 lipid and 18.8 J/ml O_2 for protein. The heat production of glucose is about 16 kJ/g when burned aerobically, yet it is only 1.1 kJ/g when anaerobically broken down to lactate. Consequently, about 15 times as much heat energy is released with aerobic metabolism of carbohydrates than with anaerobic metabolism. The magnitude of metabolic heat energy input is dependent upon temperature and the metabolic increment associated with activity. Figure 2.13 describes the heat flux associated with resting and maximal activity for *Bufo boreas* (Carey 1979).

Two things are apparent from these data. First, the magnitude of heat flux ($<3\,W/m^2$) is almost at least two orders of magnitude lower than most other avenues of heat exchange. Second, the pattern of metabolism and temperature describes an ectothermic response to temperature, for example metabolism decreases as temperature decreases. Metabolic heat production is therefore not a viable strategy for thermoregulation as it is in endotherms, given the low absolute amount of heat energy associated with even the maximal metabolic rate of amphibians. Even though it is a small amount of energy, it can present a difficulty for amphibians as far as evaporative water loss is concerned. Consider an amphibian in thermal equilibrium bounded by a boundary layer that has the same C_{wv} as the animal. Normally without a driving force for evaporation, one would assume evaporation would then stop and water loss via this avenue would be zero. Metabolic heat production will increase the temperature of an adjacent boundary layer, decreasing the relative humidity and establishing a driving force for evaporation (see Adolph 1932). As long as metabolism continues, there will always be evaporation, and water uptake from near saturated air will be impossible.

2.3 Chemical Energy Budgets

There are many variables involved in understanding food and growth in amphibians. The simplest approach is looking at total mass in (g) and total mass out (g). The mass balance approach has been used by ecologists interested in interspecific feeding capacities and physiologists interested in the effects of variables like temperature (Lillywhite *et al.* 1973; Claussen and Layne 1983). The mass conversion approach assumes energetic equivalency of food, feces and biomass, and ignores specific nutritional requirements required for growth. A slightly more refined approach is energy in versus energy out (the energetic unit in the SI system is the joule (J), but much of the literature is in units of calories, $1\,J = 0.2388\,cal.$) This energy approach accounts for the energetic differences between body mass, food and feces but does not address specific nutritional needs, that is the mixture of fats, carbohydrates and protein that make up the total energy that influence metabolism and growth. Unfortunately the literature on amphibians entails a composite of all three approaches, so caution needs to be exercised about what variables are actually measured in order to interpret them. For instance the term gross conversion efficiency is common to both the mass and energetic approaches but has completely different definitions.

The importance of energy in an evolutionary sense for an amphibian entails both survival and reproduction. The ability to balance an energy budget is crucial to supply the energy required to fuel metabolism supporting life functions.

Extra energy, above the bare necessity required to support life, in part supplies the energy to fuel both the behaviors and gametes involved in reproduction. This combination of factors, survival and reproduction, reinforces the importance of understanding the energy balance of amphibians in an evolutionary sense. On an ecological level, understanding the amphibian energy expenditure has also proved useful. In the three amphibian species in which energy flow has been quantified (*Plethodon cinereus,* Burton and Likens 1975a, b; *Eleutherodactylus coqui,* Stewart and Woolbright 1996 and Beard *et al.* 2002; and *Ambystoma,* Regester *et al.* 2006 and Regester and Whiles 2006) the species are a significant component of both energy flow and nutrient cycling in their particular ecosystem. In the case of *P. cinereus*, they consume only a fifth of the primary production compared to endothermic birds and mammals in their forest ecosystem yet they have a biomass production (kg) as high as birds and small mammals. The terrestrial frog, *E. coqui*, has the greatest density and the greatest biomass of any terrestrial vertebrate in the subtropical wet forests of Puerto Rico. They consequently have a significant impact on insect numbers, reducing rates of herbivory by the insects and increasing primary productivity and availability of nutrients for plants and micro-organisms. The oviposition, larval production and emergence of three species of *Ambystoma,* studied in ponds in southern Illinois, contribute significantly to the carbon, nitrogen and biomass production of these ponds and cycling of energy from the pond to terrestrial ecosystem. Energy is consequently an important element in both the ecology and evolutionary biology of amphibians.

The energy budget of a typical amphibian has an input (food), two material outputs (feces and urine), metabolic energy loss, and energy lost through doing work in its external environment. Storage involves forms of tissue production (such as somatic growth, gamete production for reproduction, and body reserves such as fat bodies), and utilization of body energy stores (e.g. liver glycogen, fat bodies, muscle and liver protein):

- Input
  food
- Storage
  production
  (a) somatic growth
  (b) reproduction
  (c) reserves
- Outputs
  feces
  urine
  metabolism
  work

Input

Food

The amount of energy derived from food is:

$$\textbf{Energy} = \textbf{mass of food ingested} \times \textbf{energy/mass of food} \times \textbf{assimilation efficiency}$$

On the surface this appears to be a fairly straightforward set of variables to measure. The first difficulty is determining the amount of food ingested.

Mass of Food Ingested

Post metamorphic amphibians do not characteristically consume energy at a consistent rate on either a daily, weekly or monthly basis. On a weekly basis in the laboratory there is a pattern characterized by high mass ingestion on the first day followed by much lower rates on the following days (Larsen 1984). The pattern is repeated on 3 to 4-day intervals. Toads will eat about the same number of mealworms/week whether they are fed once a week or three times/week (Jorgensen 1989). After an animal has fed it will cutaneously absorb a volume of water proportional to the mass of food ingested at a ratio of 0.75 ml H_2O/g food ingested (Jorgensen 1992). On an annual basis, long periods with little or no ingestion are followed by intense periods of feeding that last 1–4 months for temperate species. The periods of aphagia correlate to times of the year where above ground activity is constrained due to low water availability or low temperatures. The annual feeding rhythm appears to be endogenous (Jorgensen 1983) and controlled by growth hormone. Growth hormone stimulates feeding (Zipser *et al.* 1969; Wind-Larsen and Jorgensen 1987; Jorgensen and Wind-Larsen 1987) as well as somatic growth (Zipser *et al.* 1969; Farmer *et al.* 1977; Larsen 1984; Nyboe *et al.* 1985). This periodicity in feeding essentially means that chemical energy budgets are most instructive when viewed on an annual cycle or broader time scale, since for temperate species the periods of aphagia are always characterized by outputs exceeding input.

Feeding rates are positively related to temperature (Lillywhite *et al.* 1973). Therefore during the feeding period of one year there are physiological control mechanisms that couple the rate of feeding to metabolism. The cues for feeding and the components of its control are extensively summarized in two excellent reviews by Larsen (1984, 1992).

Anurans also vary interspecifically in the voluntary mass of food ingested given *ad lib* opportunities. An interesting example (Fig. 2.14) of the magnitude of this variation, derived from data of Dimmit and Ruibal (1980), is the two species constrained to the fewest above-ground activity days by living in arid environments and presumably under a greater selective pressure to acquire and process energy quickly. The spadefoot toads *Scaphiopus couchi* and *S. multiplicatus*, on average,

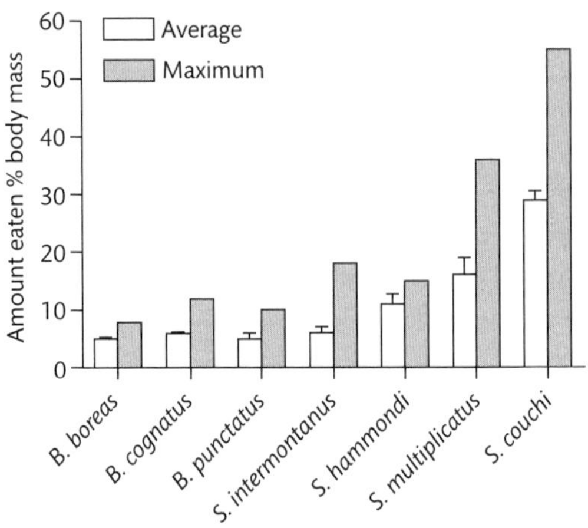

Fig. 2.14 The mean and maximal masses of food voluntarily eaten by seven species of anurans. The data are taken from Dimmit and Ruibal (1980).

eat the greatest amount when given the opportunity to feed. Closely-related congeners, with longer activity cycles (*S. intermontanus* and *S. hammondi*), have mean ingestion rates equivalent to species of *Bufo* that have long activity seasons due to longer availability to free water. This indicates the pattern of ingested energy is not a consequence of phylogenetic inertia but rather due to natural selection.

Energy/Mass Ingested

Adult amphibians, as insectivores and carnivores, generally eat food of higher energy content than herbivorous larval amphibians. The energetic equivalent varies with the type of substrate oxidized; 24 kJ/g for protein, 17.2 kJ/g for glycogen and 37.5 kJ/g for fat (see Jorgensen 1988). Consequently, the energy available in food depends upon its relative proportions of protein, carbohydrate and lipid in that food item. Water content of an insect can vary markedly; this influences the relative proportion of protein, carbohydrate and lipid and hence an insect's energy content/wet mass. Insects range from 5.7 (Scarab beetle) to 26.5 (maggot) kJ/g wet mass, with the variation related to water and lipid content of the prey item (Fitzpatrick 1973; Dimmit and Ruibal 1980).

Assimilation Efficiency

The energy assimilation efficiency of food energy is defined as 100 $(J_{food} - J_{feces})/J_{food}$. Values range from 65 to 95% and average around 90%. For insects, the highest values are for larvae with low amounts of chitin and the lowest values are for adults with a high percentage chitin (Dimmit and Ruibal 1980; Fitzpatrick 1973). There do not appear to be interspecific differences in assimilation efficiency for the same prey item (Dimmit and Ruibal 1980). Adult reed frogs had higher assimilation efficiencies than juveniles (Grafe *et al.* 1992). Another source of variation in

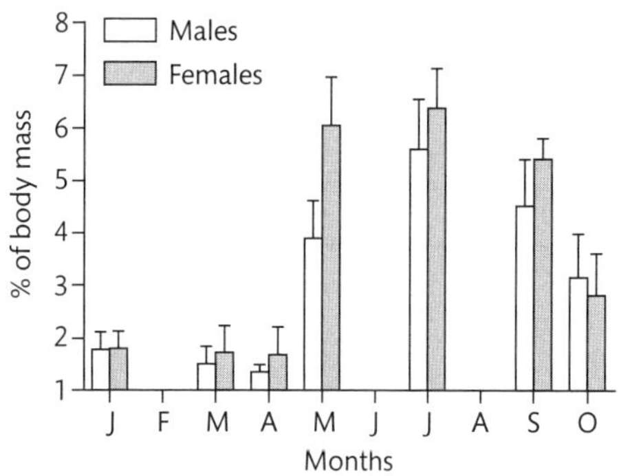

Fig. 2.15 The relative mass of the digestive system of *Rana temporaria* from natural populations in Poland over a 3-year period (data from Juszczyk *et al.* 1966).

assimilation efficiency is in the relative assimilation of the three principle substrates present in insects; fat, protein and carbohydrates. Fat was totally assimilated, whereas only 92% of the protein and 74% of the carbohydrate was assimilated by toads fed mealworms (Jorgensen 1988). As chitin is a complex carbohydrate, it has a low assimilation efficiency by amphibians. This explains why insects with lots of chitin have lower assimilation efficiencies. The digestive system characteristically goes through periods of growth and breakdown, correlating to periods of eating and aphagia respectively (Juszczyk *et al.* 1966; Cramp *et al.* 2005). These morphological fluctuations in intestinal structure do not seem to alter digestive efficiency in amphibians (Cramp and Franklin 2003). An example of the annual periodicity in relative mass of the digestive system is presented in Fig. 2.15 for natural populations from Poland over a 3-year period (data from Juszczyk *et al.* 1966). Note that this species normally breeds upon emergence from hibernation in March and April, and before feeding and intestinal mass begins to increase.

Outputs

Feces

Not all ingested material is assimilated (85–90%), especially the complex carbohydrate known as chitin, which is an important component of the structure of the exoskeleton of insects. As most of the feces are indigestible inorganic materials and chitin, feces will have a lower energy content than the prey item, as carbohydrates have the least energy/mass compared to protein and lipid (Fitzpatrick 1973).

Urine

Amphibians excrete three forms of nitrogenous waste (ammonia, urea and uric acid) in their urine. The predominant nitrogenous waste product varies interspecifically and developmentally (see Chapter 3). Ammonia requires no energy for its biosynthesis, and no energy can be biologically extracted from it. The nitrogenous waste products urea and uric acid require biochemical synthesis, and consequently contain useful chemical energy. Hence there is an energetic cost to their use as nitrogenous waste products. The cost to eliminate the nitrogenous waste of protein/purine metabolism is energy that reduces the chemical energy available to the animal. These costs have been estimated to be 2 ATP/N for urea and 6 ATP/N for uric acid (Atkinson 1977). We are unaware of any study besides Jones (1980) that attempts to incorporate these costs in an amphibian chemical energy budget. Nitrogen excretion rates for amphibians are 16.8 mmol N/g day (Chapter 3). The daily costs would be 1 J/g day for excreting the N totally as urea and 3 J/g day for excreting the N totally as uric acid, assuming 30 kJ/mole ATP.

Work

Mechanical work is defined as the product of force and distance. A one newton force displacing the body one meter is equal to a joule of work (energy). As 1 g = 9.78×10^{-3} N, the distance traveled from natal locations is limited and work can be both positive and negative, net work tends to be low and consequently ignored in most analyses of chemical energy budgets.

Metabolism

Aerobic metabolism is dependent upon temperature, body mass, activities of various kinds, breeding condition and digestive condition to name a few of the numerous variables involved. An integrative measure incorporating all these variables in proportion to the time spent at each temperature in each condition is the goal in estimating field metabolic rate (FMR). One approach to measure FMR is to create a time–energy budget by estimating the time spent in each activity via behavioral observation and then quantify in the laboratory the rate of metabolism associated with each activity. Given the amount of time that amphibians spend in subsurface and nocturnal conditions, there are large gaps in accurate estimates of both the behaviors and time spent in these out of sight behaviors. A technique that circumvents these limitations is the application of the doubly labeled water turnover. Water molecules with O^{18} and tritium (H^3) labels are injected into a captured animal and allowed to equilibrate; a blood sample is taken, and the animal is released into the environment. Periodically the animal is recaptured and a sample of body fluid recovered and analyzed for the concentration of O^{18} and H^3. The O^{18} is lost over time via the carbonic anhydrase catalyzed reaction of $CO_2 + HH^3O^{18} \leftrightarrow HH^3CO^{18}O_2$. Some the O^{18} will rearrange in the reverse reaction as COO^{18}. The COO^{18} is lost in proportion to the CO_2 produced in metabolism and exhaled. Hence one avenue of loss of the O_{18} label is proportional to metabolic production of CO_2. The other avenue for exchange of both the O^{18} and H^3 labels is as water lost then replaced with unlabelled water. Hence the O^{18} is lost in proportion to both metabolism and water turnover, while the H^3 is lost in proportion to only water turnover. The difference in turnover of both labels is due to metabolism. This technique has provided useful estimates of field metabolic rates for a variety of free-ranging species of mammals, birds and reptiles (see Nagy *et al.* 1999). Reptiles generally have field metabolic rates two to three times greater than standard metabolic rates measured at the same temperature and annual metabolic rates that are 1/30 of equivalent-sized endotherms (Nagy *et al.* 1999; Nagy 2005). The doubly labeled water technique has not been applied to amphibians because of their high rate of water turnover. The two labels turnover so quickly they do not provide an integrated estimate over a long enough time period to be useful for free-ranging amphibians. Consequently, the only metabolic estimates available for amphibians are based on time budgets coupled to laboratory estimates of energy expenditure.

Excess Energy

When energy input exceeds output the excess energy can be channeled into somatic growth, growth of reproductive systems, or general storage to support either metabolism or growth when input is less than output. All three of these options can be considered a synthesis of biomass and hence growth. The importance of growth to the overall energy balance of amphibians can be assessed by looking at integrated annual energy budgets. Figure 2.16 summarizes estimated annual energy budgets for three species of temperate zone amphibians (Fitzpatrick 1973; Jorgensen 1988; Dimmit and Ruibal 1980; Jones 1980).

Growth is an important component of the energy budget for two of these three species. In fact, Jorgensen (1988) has elegantly demonstrated that growth was linearly related to ingested energy for juvenile female toads, with 63% of ingested energy being incorporated as biomass energy. The physiological cost of this growth was about a third of the energy deposited. Synthesis of macromolecules (especially

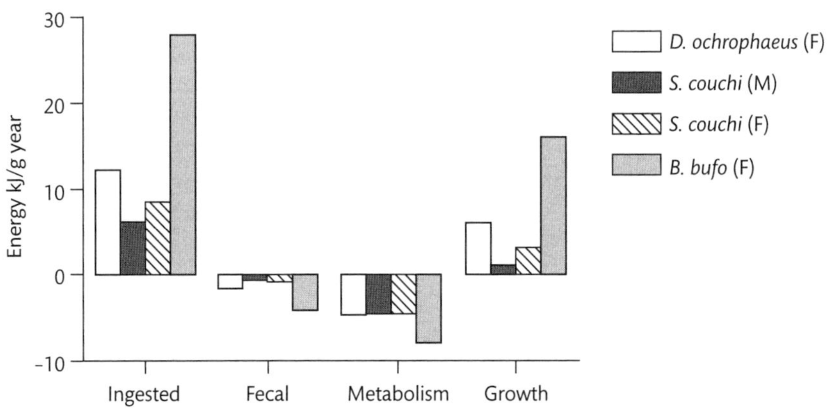

Fig. 2.16 Estimated annual energy budget for three species of temperate zone amphibians (data from Fitzpatrick 1973, Jorgensen 1988, Dimmit and Ruibal 1980 and Jones 1980).

proteins) represented 20–30% of the physiological cost. The remaining 70% of the physiological cost would result from synthesis of protein that is turned over, costs of digestion (specific dynamic effect) and feeding. The data presented above are for an annual budget. How does the budget look on a monthly basis? Monthly energy budgets for two species are presented in Fig. 2.17. The 1-g *Plethodon cinereus* females are from Fitzpatrick 1973 and the 25-g *Scaphiopus couchii* female is a composite of data from Seymour (1973), Dimmit and Ruibal (1980) and Jones (1980), assuming that mass lost is proportional to the metabolic estimates of Seymour (1973).

There are two crucial points concerning growth; for large parts of the year growth is negative and associated with periods of aphagia, and growth represents a significant fraction (about 60%) of assimilated energy. This high production to assimilation (P/E) ratio (0.6) is characteristic of ectotherms. Endotherms have ratios on the order of 0.005–0.03 versus amphibian ratios of 0.5–0.8 (see Pough 1980, 1983). Pough (1983) presents an interesting synthesis of the advantages that ectotherms accrue as a consequence of having low metabolic costs, including release from body shape constraints, low maintenance metabolic costs, increased productivity, and exploitation of low energy flux environments.

If we take the most extreme example of mass loss (negative growth) in *S. couchii,* we see that the energy comes from three sources: fat, protein and carbohydrate. The total energy available in the fall when dormancy begins for *S. couchii* is principally fat reserves (Fig. 2.18; Jones 1980).

Females characteristically enter dormancy with more stored energy than males as a consequence of having eggs/ovaries that can weigh up to 12% of their body mass before they enter dormancy (Seymour 1973); females need these ova to breed immediately upon emergence from dormancy (McClanahan 1967). Most of the energy that fuels metabolism during dormancy in toads comes from fats (Jones 1980; Bush 1963; Chapman and Chapman 1958), although protein metabolism

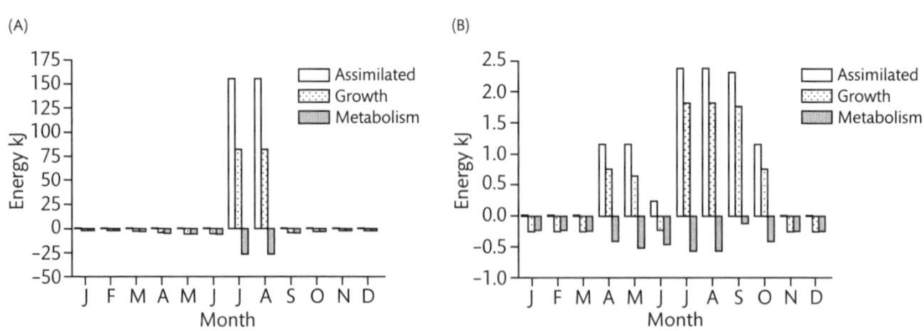

Fig. 2.17 Monthly energy budgets for two amphibians; (A) *Scaphiopus couchii* female (data from Seymour 1973b, Dimmit and Ruibal 1980 and Jones 1980b) and (B) *Plethodon cinereus* females (data from Fitzpatrick 1973).

is also a significant energy source (Jones 1980). The proportion of protein that metabolic energy is derived from can increase if the toads are exposed to dry soil conditions, which necessitates increased urea synthesis to raise their osmotic concentration and absorb water from the drying soil (McClanahan 1972; Jones 1980). The proportion of protein metabolism is more likely to increase later in dormancy when soil is at its driest (Jones 1980). This pattern of substrate utilization appears to vary interspecifically. Ranid frogs may store and utilize more carbohydrate as glycogen than toads (Pasanen and Koskela 1974). This pattern may reflect the predominant type of metabolism utilized during dormant periods. Frogs from cold climates may be more subject to anoxia and freezing during dormancy, hence glycogen would support both anaerobic metabolism and the synthesis of glucose, to assist freeze tolerance (see Chapter 4).

What mechanistically controls the partitioning of growth into reproductive organs and gametes, somatic growth or fat/glycogen storage is a complex question that will vary intersexually. The mechanisms of this partitioning are not totally understood. What does seem clear, at least for seasonal breeders, is that coordination occurs via the hypothalamopituitary axis, with temperature apparently a more significant control than photoperiod (Saidapur and Hoque 1995). Growth hormone seems to play a significant role in somatic growth and gonadotropins in reproduction. Given the great diversity of reproductive modes in amphibians (Duellman and Trueb 1986) and the database for physiological control mechanisms focusing on temperate seasonal breeders, it is clear that we still have a lot to learn about the control of this energy partitioning (see reviews by Wake and Dickie 1998 and Wake 2003), especially from a comparative endocrinology perspective.

Energy Storage

Amphibians characteristically store fat in their gonadal fat bodies (≈ 50% of total fat stores), subcutaneous fat deposits, dermal fat, and in their liver (Wygoda *et al.*

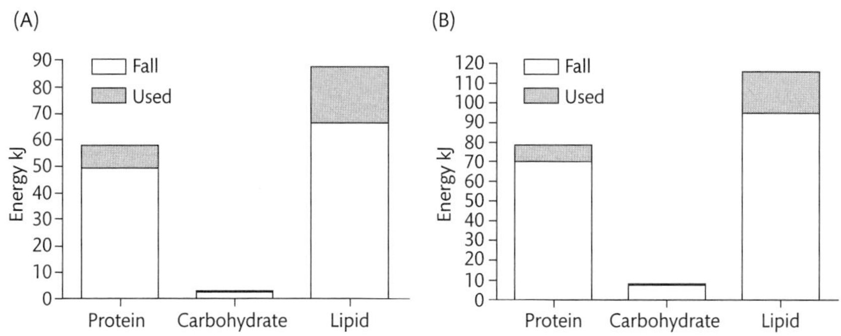

Fig. 2.18 The substrates available at the start of dormancy and their use during dormancy in *Scaphiopus couchii* males (A) and females (B). Data are from Jones (1980).

1987; Zancanaro *et al.* 1996). In the extreme, 24% of body mass is comprised of the gonadal fat bodies for the Australian burrowing frog *Cyclorana platycephala* (van Beurden 1980). Protein is stored in the liver and other cells throughout their body (Jones 1980). Carbohydrate is principally stored in the liver and muscle tissues, but can also be stored in the gonadal fat bodies (Schlaghecke and Blum 1978). The important role of the liver for energy storage is evidenced by its mass being correlated positively with anuran habit (burrowing), body shape (globose) and habitat aridity (Withers and Hillman 2001). Stored energy is mostly used to support metabolism but also fuel the last stages of reproductive maturity in females and frequently reproductive activity in the spring. Temperate species with long periods of aphagia have been the most studied in regards to patterns of energy storage and utilization. Spadefoot toads from arid deserts of southwest North America represent an extreme example of the importance of energy storage in their biology. Spadefoot toad's above ground activity may only encompass one or two months, hence their survival for the other 10–11 months is dependent upon stored energy (McClanahan 1967). What have we learned from the spadefoot toads? First, they demonstrate a metabolic depression during dormancy in comparison to awake toads at the same temperatures, a presumptive adaptation to minimize metabolic costs (Seymour 1973; see Chapter 4). Second, they appear to have the behavioral drive and morphologic capacity to consume and process large amounts of food in a short time period (see above Dimmit and Ruibal 1980). Third, the metabolic substrate of energy during dormancy is coupled to their water budget and ability to generate organic solutes to raise their osmotic concentration (Jones 1980). Many other dormant/aestivating frogs share the trait of metabolic depression (see Chapter 4), though cocoon-forming frogs do not share the production and accumulation of urea to balance their water potential with the soil (see Chapter 3).

Somatic Growth

The growth in snout–vent length in amphibians characteristically occurs when feeding resumes after dormancy in temperate amphibians (Jorgensen 1983; Wind-Larsen and Jorgensen 1987; Jorgensen 1989). The rate of growth increases with increasing temperature and prey ingestion rate (Lillywhite *et al.* 1973). As the feeding season continues, somatic growth slows and fat deposition predominates. Late season feeding experimentally leads to fat deposition and no somatic growth in toads, although if given growth hormone the snout–vent length will increase in the Fall (Wind-Larsen and Jorgensen 1987). Growth hormone supplementation tends to increase the proportion of energy deposited as protein (Jorgensen and Wind-Larsen 1987; Wind-Larsen and Jorgensen 1987). These data imply that growth in length is dependent upon growth hormone, and growth hormone is normally most influential in the early periods after dormancy (Spring) and least influential preceding dormancy (Fall). Jorgensen (1989) argues that growth hormone is required for the feeding, growth and fat deposition but the basic pattern is inherent in the organism. Recent experiments show that transgenic *Xenopus laevis* that overexpress frog growth hormone, grow at accelerated rates and develop skeletal abnormalities akin to acromegaly, the overexpression of growth hormone in humans (Huang and Brown 2000).

Reproduction

The investment in growth of reproductive structure is generally more substantial for females than males, principally because there is so much more mass of eggs than sperm. Reed frog females invest about 30% of net conversion energy into eggs (Grafe *et al.* 1992). Gametic tissue is only one of the energetic costs of reproduction. Males will frequently have higher behavioral metabolic costs than females associated with calling and frequency and time spent at breeding sites (Ryan *et al.* 1983; Woolbright 1985; MacNalley 1981; also see Pough *et al.* 1992). Gametogenesis in both sexes is driven by gonadotropins, but is more dependent upon net energy balance in females than males (Jorgensen *et al.* 1979; Guha *et al.* 1980; Chieffi *et al.* 1975). The effects of low energy stores can result in atresia of vitellogenenetic oocytes (Jorgensen *et al.* 1979; Prasadmurthy and Saidapur 1987) and to reduced egg mass and number for less extreme energy depletion (Girish and Saidapur 2000). There is an indirect, inverse functional relationship between fat bodies and ovarian cycles (Girish and Saidapur 2000). The number of eggs and their size varies interspecifically. The energy content of eggs seems to be a relatively constant 25–26 J/mg dry mass (Crump and Kaplan 1979; Mitchell 2001). Intraspecifically the size of eggs can vary between geographic sites and year-to-year in natural populations (Cummins 1986), consistent with the premise that an individual's energy status is an important component of reproductive energy investment.

2.3.1 Larval Chemical Energy Budgets

There is much interest in larval energetics since variation in larval environment and growth can ramify to the terrestrial phase and influence juvenile growth and survival (see Alvarez and Nicieza 2002a, b) Larval amphibians are characteristically divided into two categories, endotrophic or exotrophic, based on the sources of energy that support their growth. Endotrophic larvae derive their energy exclusively from parental sources while exotrophic larvae derive energy from eating non-parentally supplied food. Endotrophic larvae, with all their energy supplied in the egg metamorphose as feeding toadlets are the easiest to construct an energy budget for, as there is no further energy input during development. The total production efficiencies (gut free body energy of metamorph/(ovum energy—residual yolk energy)) of two endotrophic anurans are between 56 and 61% (Mitchell and

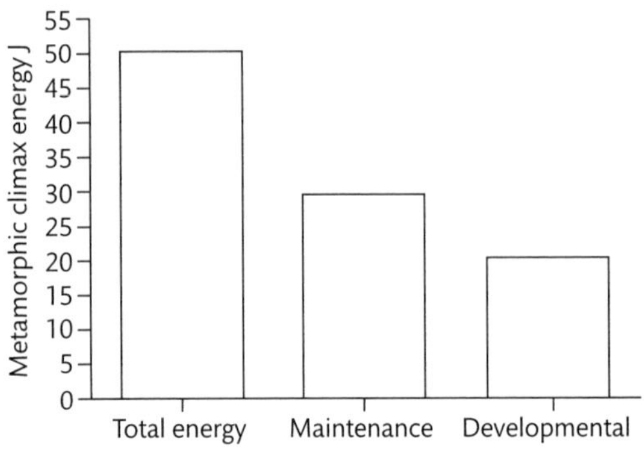

Fig. 2.19 The energetic costs of metamorphic climax for *Bufo terrestris*, data from Beck and Congdon (2003).

Seymour 2000; Mitchell 2001). This is very similar to adult conversion efficiencies (see above).

Exotrophic anuran larvae have two distinct phases of growth. The first is an increase in mass with the feeding larval phase. The second climax phase is a loss of mass with the cessation of feeding and reabsorption of the tail during metamorphosis. The proximate factors influencing the size of larvae at the onset of climax are food quality, density of larvae, water quality and depth, and temperature (Pandian and Marian 1985; Berven *et al.* 1979; Dash and Hota 1980; Kupferberg *et al.* 1994; Skelly and Golon 2003). Assimilation of food varies in part with the type of ingested materials (Altig and McDearman 1975; Waringer-Loschenkohl and Schagerl 2001; Skelly and Golon 2003). Larval feeding rate and energy content are positively correlated to froglet energy content (Pandian and Marian 1985). The cost of metamorphic climax was 2.6 and 6.8 J/g h for *Rana tigrina* and *Bufo terrestris*, respectively (Pandian and Marian 1985; Beck and Congdon 2003). The costs of metamorphic climax in *B. terrestris* can be partitioned as principally maintenance, hence metabolic rate only accounts for 60% of the total energy expended during metamorphosis (Fig. 2.19; Beck and Congdon 2003).

The substrate fueling the metamorphic climax appears to vary interspecifically with little phylogenetic inertia. Some species use lipids (*R. tigrina,* Sawant and Varute 1973; *B. bufo,* Urbani 1962) while others use very little lipid (*X. laevis,* Dodd and Dodd 1976; *B. terrestris,* Beck and Congdon 2003).

2.4 Respiratory Oxygen and Carbon Dioxide Exchange

2.4.1 Metabolism

The aerobic utilization of the chemical potential energy contained in glucose to produce phosphorylated intermediates that drive biological processes follows this

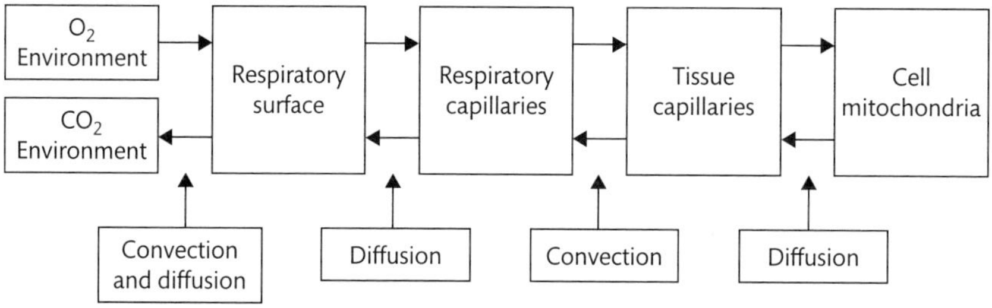

Fig. 2.20 The flux of O_2 from the environment to the mitochondria, and the corresponding efflux of CO_2.

general overall formula:

$$C_6H_{12}O_6 + 6\,O_2 \rightarrow 6\,CO_2 + 6\,H_2O + 30\text{–}38\,ATP$$

The rate of metabolism could be measured by the rate of O_2 consumption (V_{O_2}) or the rate of CO_2 production (V_{CO_2}). In the case of carbohydrates, like glucose, the ratio $V_{CO_2}/V_{O_2} = 1$. This ratio is known as the "Respiratory Quotient" or RQ. The RQ for lipids is about 0.71 and for proteins about 0.83. Therefore by measuring both V_{O_2} and V_{CO_2}, one can determine whether fat or carbohydrate is being used as the sole substrate for metabolism if the RQ was either 0.71 or 1. An RQ between these values would indicate some combination of substrates being oxidized, including protein. Obviously CO_2 efflux and O_2 influx are coupled, dependent upon the substrate being oxidized. The final step of oxidation takes place in the mitochondria contained within cells, along with the synthesis of ATP from $FADH_2$ and $NADH/H^+$.

The flux of O_2 from the environment to the mitochondria, and the corresponding efflux of CO_2 move through the series pathway are shown in Fig. 2.20. The first two steps are considered the function of the respiratory system, the last two steps are considered the function of the cardiovascular system. We will deal with both O_2 and CO_2 fluxes. The efflux of CO_2 has to match O_2 influx, based on the substrate oxidized, or imbalances in pH will occur via the accumulation or loss of H^+ ions via the following pathway catalyzed by the enzyme carbonic anhydrase contained within erythrocytes:

$$CO_2 + H_2O \leftrightarrow H_2CO_3 \leftrightarrow H^+ + HCO_3^-$$

A focus solely on O_2 flux, which is the traditional approach, can lead to some erroneous conclusions about the capacity of various steps. These pitfalls will be dealt with in Chapter 3.

The first step is to move O_2 from the environment and CO_2 to the environment from the respiratory exchange surface. The complication from an amphibian perspective is that there are four respiratory exchange surfaces arranged in parallel: lungs, buccal cavity, skin and gills. The relative contribution of each pathway to the total flux can vary developmentally (larval–adult), can differ on taxonomic bases and can also differ based on environmental conditions. Jorgensen (2000) provides a marvelous review on the history of studies in amphibian respiration. There is extensive information on the partitioning of standard V_{O_2} and V_{CO_2} for amphibians at various temperatures (Whitford and Hutchison 1965; Hutchison *et al.* 1968; Guimond and Hutchison 1972, 1973a,b, 1974, 1976). As can be seen from the summary in Fig. 2.21 of the data for anurans in air normalized to per unit area of skin, O_2 is principally exchanged across the lungs for terrestrial species while CO_2 is principally exchanged across the skin at rest. Aquatic species also exchange the majority of CO_2 across the skin at rest. Increased metabolic fluxes of CO_2 and O_2 associated with elevated temperature or activity are achieved through higher fluxes across the lung surface. The O_2 fluxes display a consistent pattern; lung O_2 flux consistently exceeds skin O_2 flux during rest, with temperature increases and during activity. There is then an enigma associated with CO_2 fluxes at rest. Why

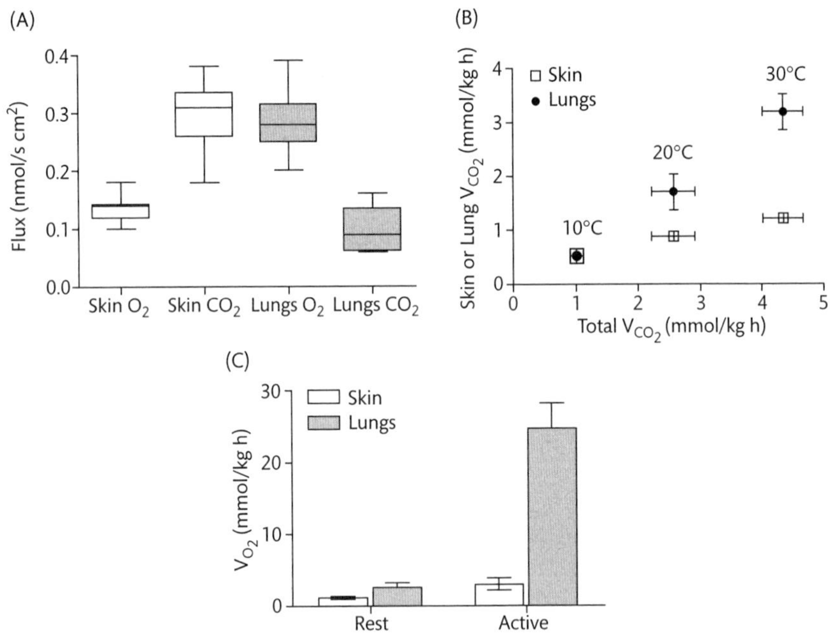

Fig. 2.21 Summary of the partitioning of gas exchange for anurans between the skin (unshaded) and lungs (shaded); (A) in an aquatic environment and air (standard metabolism at 25°C; Whitford and Hutchison 1965; Hutchison *et al.* 1968; Guimond and Hutchison 1972, 1973a, b, 1974), and during increased metabolism associated with (B) elevated temperature (*R. catesbeiana;* MacKenzie and Jackson 1978) or (C) activity (*X. laevis* at 18°C; Hillman and Withers 1981).

should there be greater skin than lung CO_2 fluxes at rest, as increased CO_2 fluxes occur across the lungs? We will answer this when we consider the driving forces for CO_2 diffusion in the lung and skin later in this section.

Fluxes at standard rates of O_2 and CO_2 exchange do not reflect the capacity of each avenue of exchange when these fluxes increase as a consequence of increased temperature or activity as we can see above. The focus simply on standard or "resting" fluxes has led to a variety of literature on the limitations to CO_2 efflux by the skin during the aquatic–terrestrial transition in amphibians (Gans 1970) and even surface area limitations as it relates to body size (Ultsch 1973, 1976). Obviously a more realistic approach to assess the capacity of the pathways for exchange is to look at the exchanges during periods of greatest flux, that is V_{O_2max} and V_{CO_2max} (Chapter 3). However, resting V_{O_2} and V_{CO_2} could be used as a proxy for V_{O_2max} and V_{CO_2max} if the factorial aerobic scope was consistent for species; unfortunately this is not the case.

Whether this first step in gas exchange is described as convection will depend on which respiratory surface we are referring to. Convection is a bulk movement of molecules dependent upon a hydrostatic pressure difference, generally created by muscular work. Diffusion of a gas is a movement of molecules based upon partial pressure differences of that gas (Chapter 1). Movement of the medium (air or water) to the skin would generally be considered diffusion unless the skin is moved by behaviors of the animal. Movement of medium to the lungs, gills or buccal cavity is convection since muscles are responsible for the force necessary to move the medium, or occasionally the gills through the medium.

2.4.2 Lungs

Lungs are present in all anurans, and all but one apodan (Nussbaum and Wilkinson 1995), but varies in urodeles from completely absent in the entire family Plethodontidae (60% of urodele species) and the genus *Onychodactylus* in the family Hynobiidae, to various degrees of reduction in Salamandridae and Ambystomatidae (Duellman and Trueb 1986). The evolutionary loss of lungs in tetrapods is unusual and seems to be restricted to urodele amphibians and the single apodan species; and most species of urodeles are lungless. The selective driving force for lung reduction and loss has historically been attributed to the occupation of cold, fast-flowing streams, where oxygen would be abundant, metabolism would be low, and lungs would increase buoyancy that would increase the likelihood of individuals being swept away in the current (Dunn 1928; Bruce *et al.* 1994). It has been demonstrated that experimentally increased body density decreases stream drift in *Ambystoma* larvae (Bruce *et al.* 1994). Ruben and Boucot (1989) present an alternative explanation for the evolution of lunglessness. The time and region of origin for plethodontid salamanders, the Mesozoic Appalachian region of the US, was not consistent with much vertical topography and was warm and subtropical, not a region of cold mountain streams of swift currents (Ruben *et al.* 1993). Second, cephalic narrowness, a character of many salamanders, presents morphological and physiological constraints on the ability to ventilate lungs. A variety of data suggest that increased reliance on cutaneous respiration is correlated with cephalic narrowness in ambystomatid salamanders (see Ruben and Boucot 1989). A narrow head may open up habitation of narrow spaces, but makes it difficult to effectively ventilate lungs via the buccal pump mechanism (Whitford and Hutchison 1965). The ventilatory dynamics of caecilians offers evidence of the difficulty that narrow heads presents from a ventilatory perspective. Caecilians universally have cephalic narrowness and the respiratory frequency necessary to inflate the lungs of caecilians is four to eight times greater than that of anurans (Carrier and Wake 1995; Gardner *et al.* 2000; KC *et al.* 2000). Hence the selection for lunglessness in urodeles may be a byproduct of selection for cephalic narrowness rather than an ancestral adaptation to prevent drift in aquatic currents. A third hypothesis for the evolution of lunglessness revolving around the use of buccal muscles for feeding in terrestrial plethodontids has also been presented (Reagan and Verrell 1991).

Lungs, besides exchanging gases in amphibians, also play a role in buoyancy (see Jorgensen 2000), dorsal movement of lymph to the lymph hearts (Hedrick *et al.* 2007), creating a hydrostatic turgidity to assist locomotion (Carrier and Wake 1995), and inflating an animal to a large size making it difficult to be swallowed by predators.

The total lung volume of anurans is large in comparison to other tetrapods (Tenney and Tenney 1970). Lungs present a unique problem as a respiratory surface, since it is an interface between air and the watery surface of the lung. Water, because of its polar nature, has a strong tendency to be attracted to other water molecules. Consequently a surface tension force develops at an air–water interface as a result of the attraction between water molecules. The problem for a sac-like amphibian lung is that a force equivalent to surface tension needs to be applied to the lungs' walls or the walls will collapse on themselves based on the surface tension. A collapsed lung will have no aerial surface area for gas exchange. Part of the solution to the problem of pulmonary collapse is the secretion of pulmonary surfactant by pneumocytes of the alveoli lining the lung (see Daniels *et al.* 1995). Surfactants are a complex mixture of phospholipids (mainly phosphotidylcholine in amphibians), neutral lipids and proteins that lie on the inner surface of the lung and serve to decrease the surface tension of the fluid on the lung surface as well as preventing the opposing inner lung surfaces from sticking to one another if they were to contact (Daniels and Orgeig 2003). Amphibians generally have a lipid mixture that is fluid with a low phase transition temperature (Daniels *et al.* 1994), but there is variation in the makeup of surfactant within the amphibia (Daniels *et al.* 1994; Daniels and Orgeig 2003). Aquatic species have lungs more likely to collapse than those of terrestrial species since they normally have higher pressures on their bodies based on both the weight of standing water as well as the atmosphere on top of the water. Aquatic species have high ratios of disaturated phospholipids to phospholipids and cholesterol to phospholipids, while the terrestrial anuran had phosphotidylglycerol and less cholesterol than the aquatic salamanders (Daniels *et al.* 1994). The mixture of surfactants present in urodeles suggests an important antiadhesive function, while the anuran surfactant mixture suggests both an antiadhesive function and fluidity over a broad temperature range. Do we have evidence for lung collapse in amphibians? The answer is yes for species that dive in aquatic environments (Martin and Hutchison 1979; Ultsch *et al.* 2004), hence the antiadhesive function would be important during lung reinflation with aquatic environmental compression forces favoring collapse. It also appears that contraction of smooth muscle in the lung will contribute to lung collapse (Stark-Vancs *et al.* 1984).

As we saw in Chapter 1, terrestrial species are much more likely to see a greater range of temperatures than aquatic species, hence surfactant function over a broad temperature range would present a selective advantage for species likely to be exposed to this broader range of temperatures. The evolutionary origin of the surfactant system in vertebrates seems to be a single event based on similarity of surfactant protein A and its mRNA in all vertebrate groups (Sullivan *et al.* 1998). The available data, though sparse, seem to indicate that selection is operating within amphibians to create a variety of surfactant profiles whose significance remains to be detailed.

The lung can be a simple sac or broken into "sac alveoli" by primary, secondary and/or tertiary septa, which all serve to increase the lung surface area (Goniakowska-Witalinska 1978, 1986; Stark-Vancs *et al.* 1984; Maina 1989; Maina and Maloiy 1988). The degree of septation has been quantified in three ways. The first is to describe and/or document photographically the gross anatomic appearance (see Guimond and Hutchison 1976). The second is to measure the radius of the lung at its midpoint (R_2) and also the central lung radius that has no septa (R_1) (Tenney and Tenney 1970). The ratio of R_1/R_2 will reflect the degree of separation within the lung, a value of 1 indicates no septa, and a value of 0 indicates complete separation. In Fig. 2.22, replotting the data of Tenney and Tenney (1970), gives a rough idea of the variation in lung cross-section occupied by septa present in anurans.

The third, more quantitative approach has been the histological determination of the surface area of the lung walls and the lung septa, then calculating the ratio of septal surface area to non-septal lung wall surface area (Fig. 2.23). Obviously no septa would give a ratio of 0, while the maximal ratio reported for amphibians is about 7 (see Czopek 1965), indicating a large elaboration of pulmonary surface area via increased septation. The degree of alveolization in vertebrates seems to be associated with amplification of parathyroid hormone-related protein (PTHrP) pathway signal (Torday and Rehan 2004).

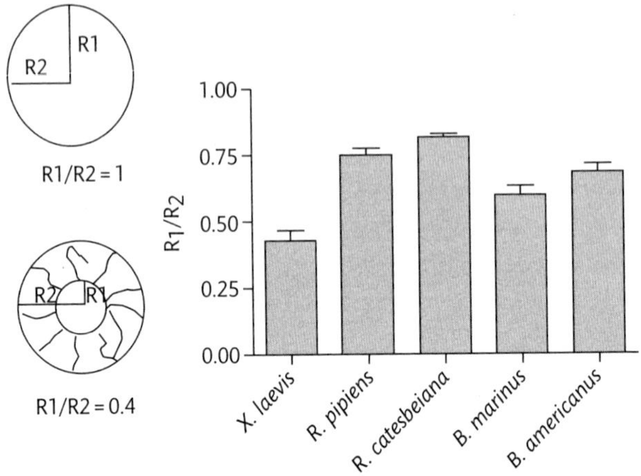

Fig. 2.22 A summary of the degree of lung septation in selected anurans; data are from Tenney and Tenney (1970).

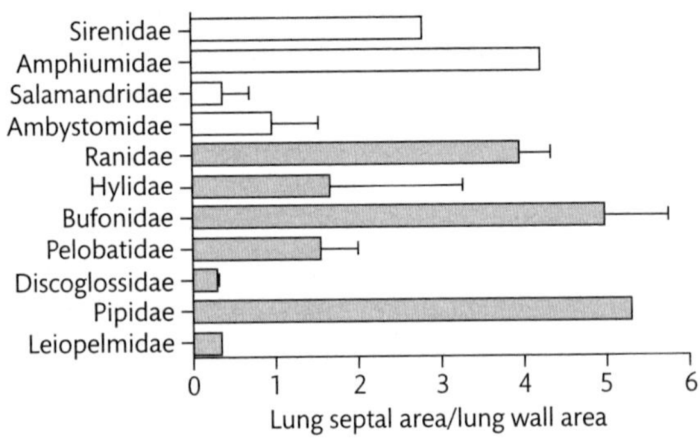

Fig. 2.23 A summary of the total lung septal surface area/total lung surface area in various amphibians; data are from Czopek (1965).

Diffusional Exchange

The diffusive process from the lung interior to the hemoglobin contained within the erythrocyte in the pulmonary capillary can be quantified using three resistances in series; the resistance of the boundary layer at the surface of the lung, the resistance of lung wall and the resistance of the plasma and erythrocyte. The boundary layer resistance can be quantified as:

$$R_{Lbl} = \textbf{thickness of the boundary layer}/K_{bl} \times \textbf{lung surface area}$$

where K_{bl} is the diffusion coefficient of 76.5 and 59.9 nmol/s cm kPa for O_2 and CO_2 respectively.

Another function for the lung septa suggested by their anatomy, having both supportive and contractile elements, is to "move" and disrupt pulmonary boundary layers, though there are no data supporting this function. The lungs also have smooth muscle that could contract and create pressure waves that might disrupt boundary layers. Couple this to pressure waves transmitted through the pleuroperitoneal cavity by buccal contractions and there exist a variety of potential mechanisms to create convective currents within the lung that would decrease boundary layers and promote gas mixing. There are no actual measurements of the boundary layer thickness on the lung surface, but a layer thickness of 600 μm has been used in the calculations. Estimating the boundary layer resistance, using measured surface area data of Czopek and his colleagues (see Czopek 1965) and Krogh diffusion coefficients for air, give values between 9×10^{-6} and 2×10^{-4} s kPa/nmol for O_2 and 28% higher for CO_2 in amphibians (see below).

Another element that can contribute to the overall resistance of the boundary layer is the filtration of plasma (formation of lymph) from the pulmonary capillaries that might accumulate within the alveoli. This would then add an additional liquid boundary layer with 10^4 the resistance of an air boundary layer of equivalent thickness. Amphibian lung capillaries have high filtration coefficients (Smits *et al.* 1986), see the same pressure as the systemic circuit (Langille and Jones 1977) and have low plasma colloid pressures (Hillman *et al.* 1987). Consequently, amphibian lungs have high pulmonary filtration rates even when normotensive and ten-fold greater rates when hypertensive (Smits *et al.* 1986). Despite these characteristics that should lead to the accumulation of fluid on the pulmonary surface, edema is apparently prevented by a well-developed pulmonary lymphatic system (Smits 1989, 1994, 1998). Little is known about the pulmonary lymphatic system, but lymph moves unidirectionally, through one-way valves in the microvessels and the driving force is contraction of the lymph vessels (Koyama and Horimoto 1988). What may prevent long-term pulmonary filtration problems is a well-developed atrial natriuretic protein system in the lungs (Glass *et al.* 1996).

The surface of the lung and its septa is a simple squamous epithelium facing the lung lumen (31% of total thickness), an intercellular matrix of connective tissue (42%) and an endothelial cell forming the pulmonary capillary (27%) (Meban 1980). The resistance to gas flux across each of these barriers is proportional to their individual thicknesses, and the total resistance is equal to the sum of each of the three barriers. Macrophages working on the lung's surface attack foreign material and invading organisms, that would otherwise increase the diffusion distance and damage the epithelial layers (Welsch 1983).

$$\mathbf{R_{Lwall} = thickness\ of\ the\ three\ series\ anatomic\ elements/K_{tissue} \times lung\ surface\ area}$$

where K_{tissue} is 1.04×10^{-4} and 3.9×10^{-3} nmol/s cm kPa for O_2 and CO_2, respectively.

The mean thickness of the three layers seems somewhat conservative at an ordinal level with anurans about 1.9 μm, caecilians 2.35 μm and urodeles about 2.6 μm (Meban 1980; Maina and Maloiy 1988). This three-layer series design is a conserved anatomic arrangement in all vertebrate lungs from fish to mammals (see review of Maina and West 2005). The actual appearance of the luminal surface can be assessed by the ratio of the mean thickness to harmonic mean thickness around pulmonary capillaries (Meban 1980). In amphibians this ratio is about 1.3:1 the lowest tetrapod ratio known, the largest is found in birds at 8:1 (Meban 1980). This indicates a rather uniform thickness across the inner surface of amphibian lungs and septa in comparison to other tetrapods. Estimated resistances for amphibian lungs from the data of Czopek and his colleagues for surface area and thickness data of Meban (1980) give values that range from 5×10^{-2} to 25 s kPa/nmol for amphibians, much of the variation is related to body mass (see below). In comparison to these values it is clear the lung boundary layer resistances are less than 1% of the lung anatomic resistances.

The third element in the series is the resistance of the plasma and erythrocytes to diffusional flux. Lung capillary resistance has been quantified by Roughton and Forster (1957) as:

$$\mathbf{R_{Lcap} = (1/volume\ of\ blood\ in\ lung\ capillaries) \times \theta}$$

where θ = the ml of O_2 taken up by erythrocytes in 1 ml of blood/s kPa

As the hemoglobin concentration of blood or its volume flux through the lung increases, the resistance to O_2 flux decreases. There are no data for the actual volumes of blood contained in amphibian lungs, but it should be proportional to the number of capillary meshes/surface area of the lungs. Hence we can estimate the effect of interspecific variation of vascularity on resistance by weighting the anatomic resistance by the capillary mesh density in the following manner: $R_{lung} \times$ mean lung capillary meshes/actual capillary meshes. This has the effect of multiplying R_{lung} by 0.67 to 4 dependent upon a species' lung capillarity. Recognize this will tend to overestimate the actual resistance since it is only a way to standardize within a given surface region and does not correct for the absolute volume or the effects of hemoglobin. This will be most dramatic for the pulmonary surface as it has a capillary density five times greater than skin or buccal regions (Fig. 2.24). It is generally assumed that amphibian lungs are most probably perfusion limited, hence blood flow would be an important determinant of overall resistance. There are no definitive data on the effect of perfusion on lung O_2 or CO_2 resistances at this time.

Amphibian lungs vary in volume (Tenney and Tenney 1970), thickness (Meban 1980), surface area (Czopek 1965), degree of septation (Czopek 1965; Tenney and Tenney 1970), and the presence of arterial–venous shunts in some species (Smith and Rapson 1977). Unfortunately the current literature attributes little functional consequence to the interspecific variation in these data, which we will attempt to correct in Chapter 3.

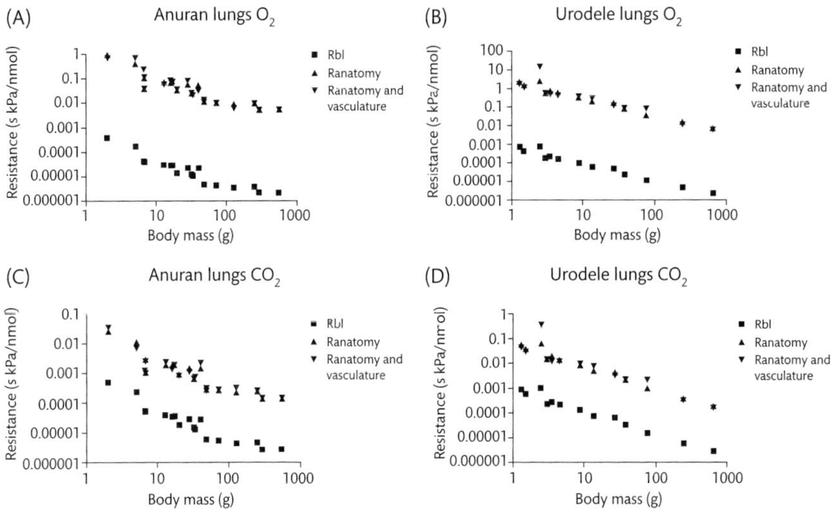

Fig. 2.24 A summary of calculated resistances for O_2 (A and B) and CO_2 (C and D) of pulmonary boundary layer resistance, anatomic resistance and pulmonary resistances weighted for vascularity in anurans (A and C) and urodeles (B and D) with body mass.

Ventilation

Lungs are different from the skin as a respiratory exchange surface in that the P_{O_2} at the surface is determined not only by environmental P_{O_2} but also ventilation and metabolism. The convective ventilatory process of moving air from the environment to the pulmonary surface is described as effective ventilation (V_{eff}; see Jackson 1978) and is described as:

$$V_{eff} = f_R \times (V_T - V_{DS})$$

where f_R is the frequency of ventilation, V_T is the volume/breath, and V_{DS} is the anatomic and physiological dead space.

Amphibian lungs have trachea and short primary bronchi that vary in length. Caecilians, being elongate, a tend to have a long trachea, and anurans a short trachea. Hence anatomic dead space will vary. Amphibians appear to have a dead space of 30–40% of tidal volume as a consequence of, principally, buccal mixing of inspired and expired gas (Fernandes *et al.* 2005). The f_R is somewhat more complicated in amphibians than other vertebrates because only a small fraction (20%) of buccal oscillations are actually involved in ventilation; most buccal oscillations simply exchange air in the buccal cavity and appear to be most important in olfaction (see Jorgensen 2000). The actual product of f_R and V_T known as V_I for inspiratory minute volume is 70–120 ml/kg min at rest and can reach 500–1100 ml/kg min when stimulated by hypercapnia or activity at 25°C (Withers and Hillman 1983; Kruhoffer *et al.* 1987; Branco *et al.* 1992, 1993).

Amphibians are unique as far as tetrapods are concerned in how they mechanically ventilate their lungs. Air is brought into the buccal cavity by lowering the floor of the buccal chamber with the nostrils open. Contraction of buccal floor intermandibularis muscles will force the air out of the buccal cavity through the nostrils if the glottis opening to the trachea is closed. If the lungs are going to be ventilated, the nostrils are partially closed (Jones 1982) the glottis opened and the intermandibularis muscles in the floor of the buccal cavity contract to raise buccal pressure and move air from the buccal cavity into the lung (Gans *et al.* 1969). The glottis is then closed, the nostrils fully opened and the pattern can repeat over 2–20 cycles during a lung inflation cycle. The number of buccal cycles used during inspiration seems to be related to head width, lung volume and the eventual pressure developed in the lung (see Brainerd 1999), consistent with the arguments of Ruben and Boucot (1989) concerning limitations presented by cephalic narrowness. Apodans have 10–20 cycles per lung ventilatory event (Bennett *et al.* 1999; Carrier and Wake 1995; KC *et al.* 2000; Gardner *et al.* 2000). The net result is that the filled lung is at a positive pressure relative to the environment (0.2–0.6 kPa in anurans and urodeles, 1.5 kPa in caecilians). Expiration is a consequence of elastic recoil of the lung and body and can entail the use of hypaxial trunk muscles (Polos and Hedrick 2006). This is why amphibians are described as positive pressure or buccal pump ventilators, whereas other terrestrial vertebrates have intrapulmonary pressures equal to atmospheric pressure during gas exchange. The intrapulmonary pressure is determined by the number of pulmonary ventilatory cycles that occur during lung filling, the volume exchange per buccal pump and the compliance of the lungs and body wall (see Hughes and Vergara 1978; Dupre *et al.* 1985). Mammals, birds and reptiles are aspirational ventilators since they create a lower pressure in their lungs than the environment. This low pressure is achieved by expanding the pleural space around the lungs, which allows gas to move from the environment to the lungs through open nostrils and glottis to equalize the pressure difference. Recently several lizard species have been demonstrated to use buccal (gular) pumping in concert with aspirational breathing to augment inspiration during exercise (Owerkowicz *et al.* 1999). A third type of ventilatory event described in aquatic salamanders is open mouth gulping (Simons *et al.* 2000). This is an occasional type of ventilation in aquatic species breathing at the surface. Muscular work generates the pressure difference necessary for convective gas movement whether it is an aspirational or buccal ventilator, the difference is the embryonic origin of the skeletal muscles participating in creating the pressure difference.

The diffusional exchange of oxygen across the lungs is driven by a partial pressure difference from the lung to pulmonary capillary. The P_{O_2} within the lung at the alveolus is determined by the following relationship:

$$PL_{O_2} = PI_{O_2} - (V_{O_2}/V_{eff})\,(P_B - P_{wv})$$

where PL_{O_2} is the lung O_2 pressure, PI_{O_2} is the inspired O_2 pressure, V_{O_2} is the rate of metabolism, V_{eff} is the effective alveolar ventilation rate (V_A), P_B is the barometric pressure and P_{wv} is the pressure of water vapor at saturation for the temperature of the organism. At sea level (PI_{O_2} = 153 mm Hg = 20.4 kPa) and during maximal activity PA_{O_2} is about 130 mm Hg (17.3 kPa), and V_A was 19 and 32 ml/min for *R. pipiens* and *B. cognatus* respectively (Withers and Hillman 1983). We know from Chapter 1 that PI_{O_2} will decrease with altitude.

The P_{CO_2} within the alveolus is determined by the following relationship:

$$\mathbf{PL_{CO_2} = PI_{CO_2} + (V_{CO_2}/V_{eff})\,(P_b - P_{wv})}$$

where PL_{CO_2} is the lung CO_2 pressure, PI_{CO_2} is the inspired CO_2 pressure, V_{CO_2} is the rate of CO_2 production, V_{eff} is the effective rate of alveolar ventilation. Frequently PI_{O_2} is close to zero, as CO_2 comprises only 0.03% of the gas in the atmosphere or 0.03 kPa at sea level.

2.4.3 Skin

Exchange of oxygen across the skin entails O_2 diffusing in a series fashion across a boundary layer, across the epidermis and capillary endothelial cell layers, and finally through the plasma and into the erythrocytes. Exchange of CO_2 follows the same series pathway in reverse order. The formulas for calculating the resistances are the same as those for the lungs delineated above, except of course thicknesses and surface areas refer to cutaneous values. The variables involved in determining the boundary layer thickness were dealt with in Chapter 1 and might vary from 10 to 1000 μm in air and 10–700 μm in water (see Pinder and Feder 1990). There are behaviors, like activity and rocking in hellbenders, that will minimize boundary layers and increase diffusional fluxes (Harlan and Wilkinson 1981). The thickness of the epidermis and vascularity varies with region of the body (Czopek 1965), as does the skin regional rate of gas exchange (Talbot 1992).

Species will vary in their surface area to mass ratios based on body shape; long thin caecilian and salamander shapes will have more cutaneous surface area than the more globose anuran body shape. Species can also have elaborations of surface area like the 1 to 2-cm-long, hair-like papillae on the dorsolateral surfaces of the body and thighs in male hairy frogs (*Trichobatrachus*, Noble 1925; Jones 1971) or the lateral skin folds of hellbenders (*Cryptobranchus*) and the Lake Titicaca frog (*Telematobius*). Adaptations that increase the skin resistance to water loss also appear to increase the skin resistance to gas flux (Stinner and Shoemaker 1987).

Resistances calculated for cutaneous boundary layers in air are similar to lung boundary layer resistances, that is $9 \times 10^{-5} - 1.2 \times 10^{-4}$ s kPa/nmol. Aquatic boundary layers are much more resistive for O_2 flux than aerial boundary layers because of the much lower diffusion coefficients and capacitances; $K_{blwater}$ is

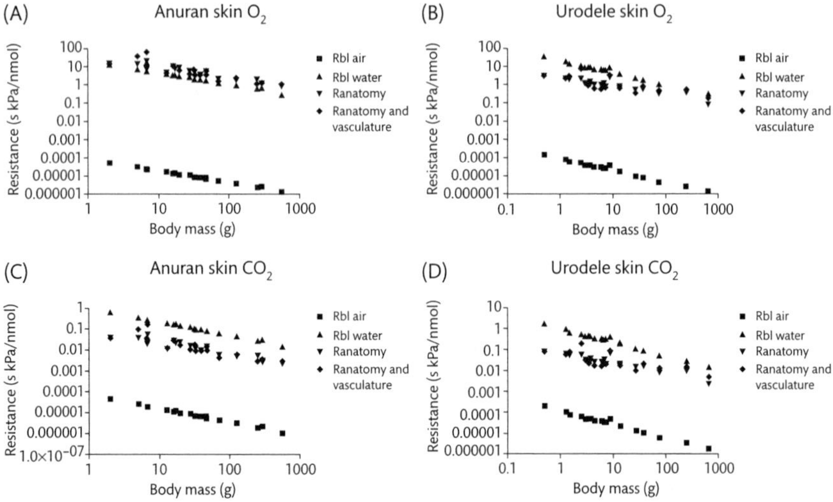

Fig. 2.25 A summary of calculated resistances for O_2 (A and B) and CO_2 (C and D) of cutaneous boundary layer resistance, anatomic resistance and cutaneous resistances weighted for vascularity in anurans (A and C) and urodeles (B and D) with body mass.

3.3×10^{-4} and 6.8×10^{-3} for O_2 and CO_2 respectively. The calculated cutaneous boundary layer resistances in water range from 0.1 to 5.7 s kPa/nmol for O_2 and 3×10^{-2} to 1.8 s kPa/nmol for CO_2, again mainly dependent upon thickness and body mass. The R_{skin} calculated from the morphologic data of Czopek *et al.* (1965) range from 1 to 23 s kPa/nmol (Fig. 2.25). The R_{bl} is a small fraction of this R_{skin} in air but approximately 50–60% of the R_{skin} in water. This analysis agrees with both models and experimental measurements of the importance of boundary layers in water for cutaneous O_2 exchange (Pinder and Feder 1990; Booth 1994). Amphibian skin is presumed to be diffusion limited (Piiper and Scheid 1977; Feder and Burggren 1985), though some data suggests an important role of perfusion (Burggren and Moali 1984; Burggren and Feder 1986).

2.4.4 Buccal

Gas exchange across the buccal region is possible and must be considered given the frequency of buccal ventilatory oscillations (see Jorgensen 2000), little dead space, and presumably thin boundary layers as a consequence of the highly convective environment. The reality of a small buccal surface area, large diffusion distances and relatively poor vascularization work in opposition to the buccal ventilatory advantages and the overall resistance to O_2 diffusion is very large in comparison to both skin and lungs as can be seen in Fig. 2.26 in the analysis of the available data

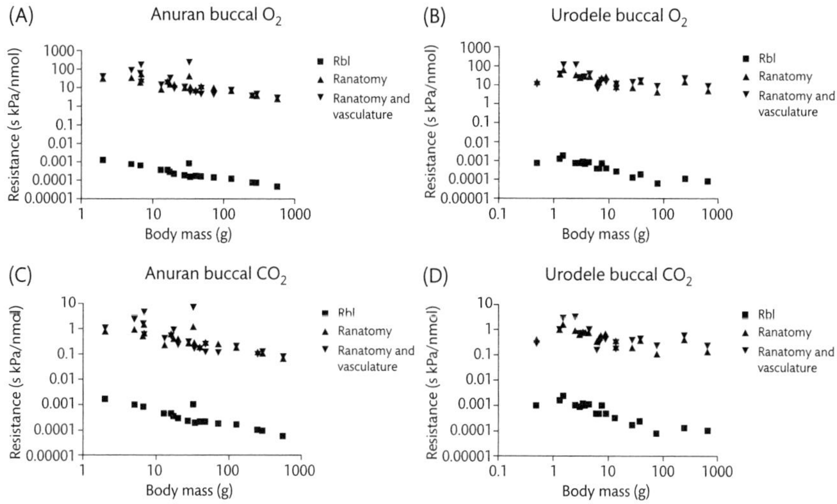

Fig. 2.26 A summary of calculated resistances for O_2 (A and B) and CO_2 (C and D) of buccal boundary layer resistance, anatomic resistance and buccal resistances weighted for vascularity in anurans (A and C) and urodeles (B and D) with body mass.

(see Czopek 1965). The buccal resistance decreases with increasing body mass, principally as a consequence of an increase in buccal surface area with increasing mass. The contribution of a 600 μm air boundary layer provides essentially nothing to the overall resistance.

2.4.5 Gills

Gills are present in larval amphibians, and the adults of some apodans and urodeles. Anuran larval gills are generally internal while the gills of apodans and urodeles are external. The convective movement of water across external gills is achieved by forcefully fluttering or waving the external gills in water. Ventilation of internal gills involves moving the water across the gills from buccal to opercular cavities based on expansion and contraction of the buccal cavity by skeletal muscles.

Gills seem to play a far more significant role in larval amphibians than adults (see Szarski 1964). During normal larval development the number and length of branchial filaments increase (Czopek 1958, 1959; Strawinski 1956), increasing the ratio of gill capillaries to body mass. When aquatic salamander larvae and anuran larvae are raised in hypoxic water conditions, there is an increase in the number and size of gill filaments increasing the surface area for gas exchange (Bond 1960; Burggren and Mwalukoma 1983). The treefrog, *Agalychnis callidryas*, lays its eggs in terrestrial situations, which then hatch and fall into water to complete metamorphosis. The terrestrial-phase embryos have external gills which characteristically regress with hatching and the occupation of their aquatic larval phase (Warkentin 2000). If the eggs are exposed to hyperoxic environments or their surface area exposure to air is increased the gills begin to regress before hatching (Warkentin 2000, 2002). If the hatched larvae are exposed to hypoxic water conditions the gill regression is inhibited (Warkentin 2000). These data suggest the gills are important in maintaining gas exchange during both the embryonic and larval phases.

The importance of gill gas exchange developmentally will obviously change as lungs are developed. Tadpoles of *Rana catesbeiana* without lungs exchange about 40% of their O_2 across their gills, the remaining 60% across their skin (Burggren and West 1982). This fraction decreases as lungs are developed. In tadpoles with lungs and gills there is a shift to greater reliance on lungs with a decreased ventilation of gills when exposed to aquatic hypoxic conditions (West and Burggren 1982, 1983). This would prevent diffusional loss of O_2 to the environment across the gills. Aquatic P_{O_2} is sensed by O_2 chemoreceptors in the first gill arch to control this pulmonary gill balance of O_2 exchange (Jia and Burggren 1997; Straus *et al.* 2001; Saltys *et al.* 2006). There are also vascular shunts between the afferent and efferent gill arteries which would prevent perfusion of gill capillaries and consequential loss of O_2 across the gills in hypoxic aquatic environments in both urodele larvae (Boas 1882; Figge 1936) and anuran larvae (Saint-Aubain 1981). There are no data on the control of these vascular gill filament shunts.

2.4.6 Integration of Exchange Pathways

The multiplicity of parallel gas exchange surfaces makes it somewhat difficult to attribute the relative significance of each pathway to the total gas flux. Since the exchange surfaces are in a parallel arrangement, the use of conductances, rather than resistances, make the math somewhat easier as the total conductance will be the sum of the individual conductances. As we have the individual resistances of each pathway we only need to take the reciprocal of these to calculate the conductances, which are presented in Fig. 2.27.

There are four obvious conclusions to be reached from Fig. 2.27: (1) conductances for the three pathways increase with body mass, principally as a consequence of increased respiratory surface area; (2) the allometry of total conductance is proportional to body mass$^{0.91}$; (3) the lung conductances are the highest of the three parallel pathways; and (4) anurans have greater total conductances than urodeles. The pattern for anurans is that lungs represent an order of magnitude greater conductance for O_2 and CO_2 than skin except at very small masses. This is due principally to the shorter diffusion distances in the lung. This difference would actually be greater when the enhancement of conductance by the circulation in lungs is considered.

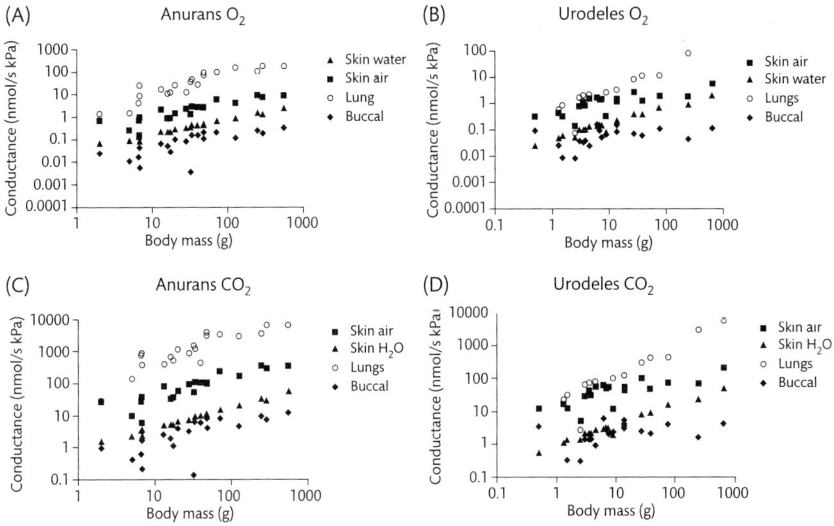

Fig. 2.27 A summary of calculated conductances for O_2 (A and B) and CO_2 (C and D) of skin (air and water), lungs and buccal cavity in anurans (A and C) and urodeles (B and D) with body mass.

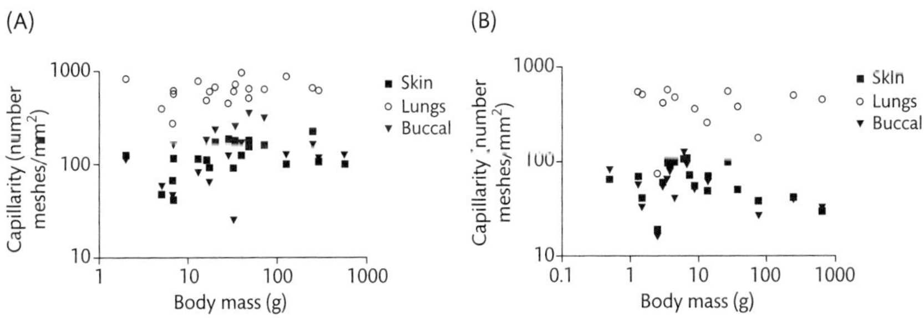

Fig. 2.28 The degree of capillarity of skin, lungs and buccal cavity for anurans (A) and urodeles (B) with body mass.

Figure 2.28 demonstrates that the capillary density is five to six times greater in lungs than skin or buccal regions. This difference has not been accounted for in the conductance calculations. The greater lung conductance for CO_2 than skin only strengthens the enigma of greater resting CO_2 flux across the skin compared to the lung in amphibians. The conductances of O_2 and CO_2 for the buccal region are very low in comparison to the lungs and skin in both anurans and urodeles.

Salamanders generally have lower pulmonary conductances for O_2 and CO_2 than anurans, and at small masses the skin and lungs have almost the same conductances. You would predict greater rates of pulmonary O_2 flux as a consequence of the five-fold greater capillary density.

Conductances have also been experimentally determined by estimating or measuring both a gas flux and its driving force. A cutaneous conductance for CO_2 in water determined for *R. catesbeiana* at 20°C of 0.09 nmol/s kPa (Mackenzie and Jackson 1978) is within 10% and the experimental error of our value for all anurans calculated only from the morphological data. The mean cutaneous conductance determined for O_2 from the morphological data (0.021–0.032 nmol/s kPa cm^2) compares favorably to a range of experimentally determined conductances 0.006–0.071 nmol/s kPa cm^2 (Pinder *et al.* 1991; MacKenzie and Jackson 1978; Moalli *et al.* 1980; Pinder 1987; Malvin and Hlastala 1989; Piiper *et al.* 1976). The pulmonary conductance calculated from the morphological data for *R. catesbeiana* was 179 nmol/kg s kPa, which is comparable to a measured CO_2 conductance in *R. catesbeiana* of 120 nmol/kg s kPa (Glass *et al.* 1981). The morphologic data alone, therefore, seem to provide estimates of gas conductances consistent with estimates derived from experimental flux measurements. This strengthens the approach of trying to quantitatively delineate the physical variables involved.

How well does conductance predict actual fluxes? Using conductances alone one would predict that most exchange of both CO_2 and O_2 should occur across the lung (1.5:1 for O_2 and 16:1 for CO_2), and CO_2 fluxes should be 40–400 times greater than O_2 fluxes, yet experimental data show O_2 fluxes are greater than CO_2 fluxes in the lung (2:1), CO_2 fluxes exceed O_2 fluxes in the skin (2:1) and total fluxes across the skin and lungs are about the same during standard metabolic rates (Fig. 2.21 above). We need to look at the product of conductance and the partial pressure difference of O_2 and CO_2 in order to understand the measured fluxes. The average partial pressure differences for CO_2 and O_2 across the skin and lungs are difficult to assess in part because: (1) the skin is perfused by both systemic and pulmocutaneous flows; (2) the fraction of pulmocutaneous blood going to the lung or skin can vary; (3) the presence of cardiac shunts can vary the volume of blood going to the skin and lungs versus the systemic circuit over time; and finally (4) intracardiac mixing can vary the partial pressures in the pulmocutaneous and systemic circuits over time. These concerns are not going to be significant variables at rest, since there is very little difference in the partial pressures between arterial and venous blood (see Fig. 2.35). Hence mixing and shunting will have very little effect on the partial pressure differences involved in the diffusion of gases across the respiratory surfaces. These potential variables will take on more importance during activity, especially at the limits of O_2 and CO_2 fluxes. The atmospheric pressures are 20.4 and 0.03 kPa for O_2 and CO_2 respectively. Pulmonary pressures at rest are 17.1 and 1.48 kPa for O_2 and CO_2 respectively (Fig. 2.29) (Wang *et al.* 1998; Fernandes *et al.* 2005; Withers and Hillman 1983). The arterial and venous pressures for CO_2 at rest are indistinguishable from the P_{LCO_2}, hence there should be little efflux of CO_2 into the lung

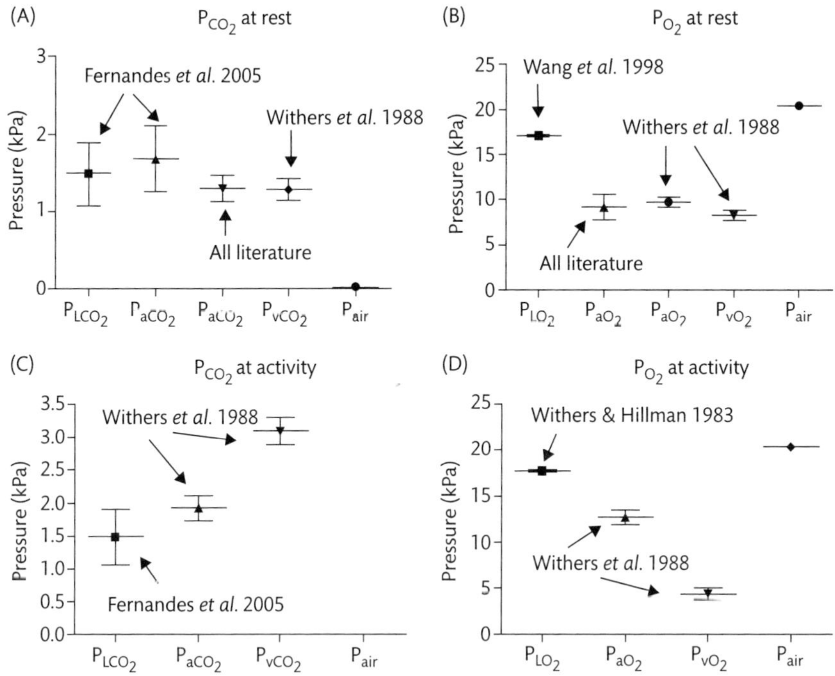

Fig. 2.29 The partial pressures at various points, during normoxia, for the respiratory exchange of O_2 (B and D) and CO_2 (A and C) during rest (A and B) and activity (C and D) in the genus *Bufo*. Values are the means and 95% confidence interval of the mean. All literature comprises values from the literature cited in the figures and values from Branco *et al.* 1992; Kruhoffer *et al.* 1987; Boutilier and Toews 1977; Boutilier *et al.* 1979a, b, c; Boutilier and Heisler 1988; Portner *et al.* 1991; Andersen *et al.* 2001; Branco and Glass 1995.

(Fig. 2.29). The arterial and venous pressures for O_2 at rest are 9.16 and 8.24 kPa respectively, hence the driving force for O_2 influx is about 8–9 kPa across the lungs and 11–12 kPa across the skin (Fig. 2.29). This means that at rest there should be little or no pulmonary CO_2 efflux because there is little or no driving force, and flux of O_2 should predominate across the lungs because the lungs have a greater O_2 conductance and only a slightly lower driving force than across the skin. This is exactly what has been documented for measured metabolic gas fluxes at rest. During activity, P_{vO_2} decreases and P_{vCO_2} increases (Withers *et al.* 1988), which should dramatically increase CO_2 efflux and O_2 influx across the lungs and increase O_2 influx and CO_2 efflux across the skin only marginally because most of the right side output (80%) goes to the lungs and the lungs have greater conductances. As P_{aO_2} increases with activity, the driving force for O_2 influx from the environment decreases. As much of the skin is perfused by this systemic flow, the cutaneous O_2 influx would decrease. The experimental data for O_2 and CO_2 exchange support a greater reliance on pulmonary gas fluxes when metabolism is increased by increasing body temperature (Hutchison *et al.* 1968; MacKenzie and Jackson 1978) or activity (Gottlieb and Jackson 1976; Withers and Hillman 1983; Pinder and Burggren 1986). It is interesting that the data for the aerial breathing *B. marinus* portray a respiratory system that is essentially hyperventilated at rest. Increased pulmonary metabolic fluxes are driven by changes in the driving force for gas flux (increase in P_{vCO_2} and decrease in P_{vO_2}) rather than conductance changes. Fluxes can be accommodated during apneic periods via the store of O_2, and sink for CO_2, present in the large amphibian lung volumes in comparison to other vertebrates (Tenney and Tenney 1970). These are essentially the same conclusions reached for *X. laevis* during voluntary diving episodes (Boutilier and Shelton 1986). Hence in the short term, ventilation would not have to be tightly coupled to either resting or increased metabolic demand associated with activity or temperature via their chemoreceptor signals. There is experimental evidence supporting episodic breathing initiated by the central pattern generator without changes in blood gasses or pH (Smatresk and Smits 1991; Kinkead and Milsom 1994).

2.4.7 Physiological Control Systems of Respiration

The potential physiological control that will match long-term respiratory exchange to rates of metabolism can be broken down into the control of ventilation (convection) and the control of perfusion of the exchange surface (diffusional exchange). The normal amphibian response of systemic vascular beds to either a decline in tissue P_{O_2} or increase in P_{CO_2} is a dilation of precapillary sphincters, decreasing resistance and increasing flow to that tissue (Carrier 1965; Becker *et al.*1970; Gentry and Johnson 1972). This local vascular response will compensate for conditions that would prevail when O_2 demand exceeds blood flow at the tissue level. Pulmonary vascular beds generally show a vasoconstriction in response to a decrease in P_{O_2} or an increase in P_{CO_2} (Koyama and Horimoto 1982, 1983, 1988). Pulmonary hypoxic vasoconstriction shunts blood to lung regions where there is a greater pressure difference for diffusive gas exchange, and contribute to matching ventilation and perfusion (see Moudgil *et al.* 2005). Since skin is a significant conductance site for gas exchange, the obvious question is how does skin respond, from a vascular perspective, to a decrease in environmental P_{O_2}? The actual data on skin blood flow can be grouped into two categories: (1) vasodilation to enhance conductance (Burggren and Moalli 1984; Burggren and Feder 1986); or (2) vasoconstriction to shunt blood to more favorable gas exchange sites and even retard O_2 loss in hypoxic aquatic environments (Malvin 1988, 1993; Malvin and Hlastala 1986, 1989). The preponderance of data and experimental techniques used for skin flow measurements seem to favor a skin response where a decrease in blood flow accompanies a decrease in P_{O_2} and increase in P_{CO_2}, just like in lungs.

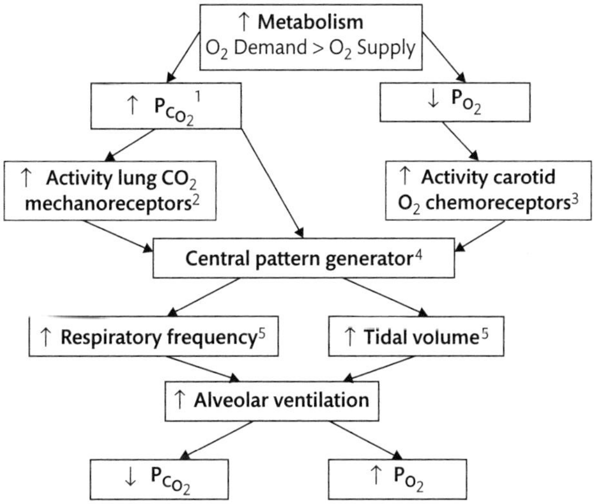

Fig. 2.30 The negative feedback control of ventilation: [1]Smatresk and Smits 1991, Branco *et al.* 1992, 1993; [2]Kuhlmann and Fedde 1979, Milsom and Jones 1977; [3]Shelton *et al.* 1986; [4]Sakakibara 1984, Kogo *et al.* 1994, Wang *et al.* 1999; [5]Smyth 1939.

The control of ventilation is summarized in Fig. 2.30. It principally entails a central respiratory pattern generator whose activity can be modulated by central and peripheral sensors of P_{O_2} and P_{CO_2} to vary both the rate and depth of ventilation.

2.5 Erythrocytes Oxygen and Carbon Dioxide Exchange

The next step in the flux of O_2 and CO_2 is the interaction between the erythrocytes (red blood cells, RBCs) and CO_2 and O_2. In larval amphibians erythrocytes are produced in the kidney and liver and in adults anurans by the liver, spleen and bone marrow. In larval and adult urodeles and caecilians erythrocytes are produced in the spleen (see Turner 1988). Erythrocyte production is hormonally stimulated by erythropoietin, apparently produced in the kidney (Frangioni and Borgioli 1980). The stimulus for erythropoiten production appears to be an increase in lactate concentration that would normally accompany hypoxia associated with too little O_2 delivery to tissues (Frangioni and Borgioli 1978). Once RBCs are produced, they have an estimated life span of 2–4 years (Altland and Brace 1962). Amphibian erythrocytes are generally nucleated and large (urodeles and caecilians) in comparison to other vertebrate (Fig. 2.31). Their appearance is of a biconvex oval, the convexity comprising the nucleus. The cells vary in size

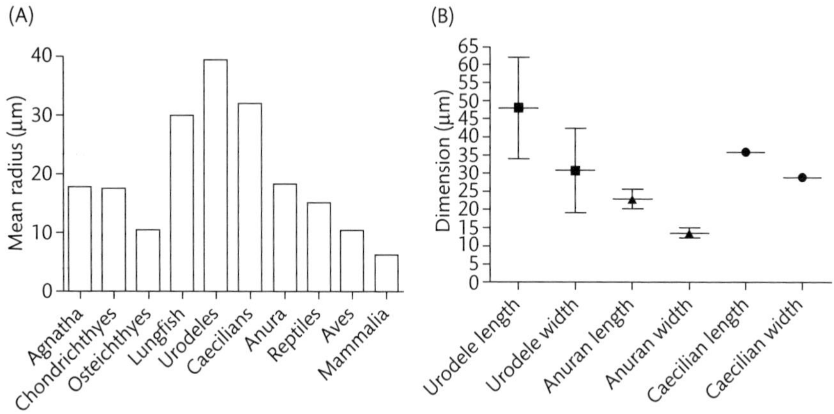

Fig. 2.31 (A) The mean radius of vertebrate erythrocytes and (B) a summary of the length-width relationships from the three amphibian orders; data from Snyder and Sheafor (1999).

(Fig. 2.31) with anuran RBCs smaller than urodele RBCs in length and width. The nucleus size is correlated with genome size (Olmo and Morescalchi 1975, 1978) and is proportional to RBC size (Kuramoto 1981). It has been suggested that cell size might be a useful predictor of genome size (Kuramoto 1981). The current hypothesis for the variation in RBC size is that it reflects variation in capillary dimensions (Snyder and Sheafor 1999). Larger capillaries have lower resistance to fluid flow through them. RBC width is about 25% larger than capillary diameter in vertebrates, which means the RBC must deform to pass through capillaries, enhancing diffusional exchange. It has been proposed that the development of an additional vascular bed and its additional vascular resistance with the evolution of the lungs in urodeles and lungfish, drove large capillaries (and RBC dimensions) in urodeles to lower overall vascular resistance and decrease the work of the heart (Snyder and Sheafor 1999). This hypothesis clearly must be tested further.

Erythrocytes contain both hemoglobin, which reversibly carries O_2, and the enzyme carbonic anhydrase (see Toews *et al.* 1978), which catalyzes the reaction of water and CO_2 to form H_2CO_3 (see below). Carbonic anhydrase is also present in many other amphibian tissues including the lung endothelium, skin epithelium, kidney, muscle, gill and urinary bladder epithelium (Fain and Rosen 1973; Rosen *et al.* 1974; Rosen and Friedley 1973; Toews *et al.* 1978). Hemoglobin functionally increases the O_2 capacitance coefficient about 10–20 fold compared to plasma alone, hence considerably more O_2 is carried in the blood with RBCs present. This functionally means that less blood has to be transported by the heart to accommodate metabolism. The interaction of hemoglobin with O_2 is determined by: (1) its concentration, (2) its structure and (3) the concentration

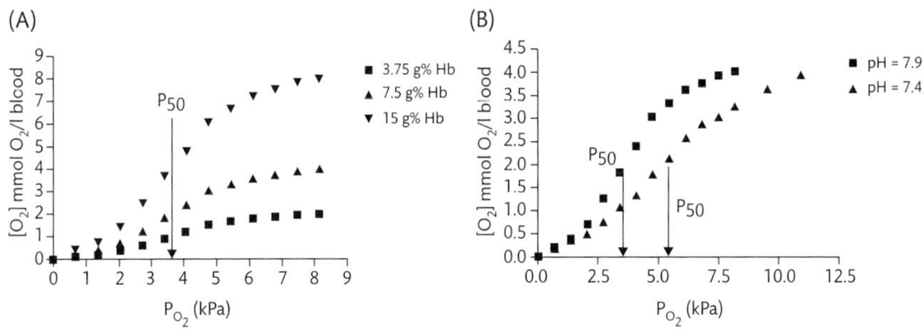

Fig. 2.32 The effects of (A) hemoglobin concentration and (B) pH on the amounts of oxygen carried by the hemoglobin at different oxygen partial pressures.

of intracellular modulators of O_2 binding. Figure 2.32 first depicts the effect of P_{O_2} on the O_2 concentration in blood, and by inference the effect of hemoglobin concentration on O_2 capacitance. The first thing to note is that the relationship is not linear, but sigmoidal. As hemoglobin is a tetrameric molecule, comprised of two α chains and two β chains, the sigmoidal O_2 loading curve means that not all of the four O_2 molecules that bind to a tetrameric hemoglobin molecule load to the hemoglobin with equal ease, hence O_2 loading is described as cooperative. The degree of cooperativity is quantified by the Hill coefficient (Table 2.3), which is the slope of the relationship between log (% saturation/(100– % saturation) and log P_{O_2}, at the partial pressure at which the hemoglobin is half saturated (P_{50}). This method for quantifying cooperativity was developed for human blood where the slope is essentially constant for all levels of hemoglobin saturation. Amphibian hemoglobin presents a difficulty with this approach as the Hill coefficient varies over various levels of saturation from 1.7 at low saturation levels to over 4 at high saturation levels in some species (Lykkeboe and Johansen 1978; Maginniss *et al.* 1980). This may in part result from different isoforms of hemoglobin being present in amphibian erythrocytes that have intrinsically different O_2 affinities (Maginniss *et al.* 1981). This also suggests a "supercooperativity" of O_2 binding dependent upon reversible deoxygenation-dependent tetramer–tetramer associations creating low O_2 affinities (see Riggs 1998). High Hill coefficients favor unloading of O_2 at the tissue level.

The P_{50} can be modified by pH, as can be seen in Fig. 2.32B. The mechanism for right-shifting the loading curve can be attributed to the fact that the binding of O_2 to hemoglobin releases a H^+ as in the following reaction:

$$HHb + O_2 \leftrightarrow HbO_2 + H^+$$

Hence, by mass action, increasing the $[H^+]$, that is lowering the pH, will drive the above reaction to the left, and drive O_2 off the hemoglobin. This pH mediated

Table 2.3 Blood loading characteristics 20–25°C

Species	P_{50} torr	Bohr effect $\Delta \log P_{50}/\Delta pH$	Hill coefficient	Reference
Anurans				
X. laevis	30	−0.37	–	Boutilier and Shelton 1986
X. laevis	27	−0.4	2.1	Jokumsen and Weber 1980
R. catesbeiana				
larval	5.5	−0.18	1.8	Johansen and Lenfant 1972
larval	13	–	–	Hazard and Hutchison 1978
larval	9.4	−0.60	2.5	Pinder and Burggren 1983
adult	33	−0.65	2.4	Pinder and Burggren 1983
adult	50	−0.29		Lenfant and Johansen 1967
adult	–	−0.18	1.95	Johansen and Lenfant 1972
adult	42	−0.27	2.6	Tazawa *et al.* 1979
adult	37	–	–	Hazard and Hutchison 1978
adult	37	–	–	Maginniss *et al.* 1980
R. brevipoda	52	−0.18	2.5	Tazawa *et al.* 1979
R. esculenta	40	−0.26	–	Gahlenbeck and Bartels 1968
B. marinus	53	−0.23	2.78	Boutilier and Toews 1981
T. culeus 10°C	16	−0.30	2.78	Hutchison *et al.* 1976
Urodeles				
C. alleganiensis	24	−0.24	2.9	Boutilier and Toews 1981
A. tridactylum	27	−0.20	1.95	Lenfant and Johansen 1967
N. maculosus	15	−0.13	1.4	Lenfant and Johansen 1967
D. ensatus				
larval	22	−0.13	–	Wood 1971
adult	32	−0.14	–	Wood 1971
A. mexicanum				
larval	22	−0.13	–	Gahlenbeck and Bartels 1970
adult	32	−0.14	–	Gahlenbeck and Bartels 1970
T. granulosa	37	−0.12	–	Wood 1970
S. salamandra	38	−0.22	–	Gahlenbeck and Bartels 1970
D. fuscus	27	–	2.64	Gatz *et al.* 1974
Caecilians				
T. compressicauda				
larval	9.3	–	–	Toews and Macintyre 1977
adult	22	−0.08	2.15	Toews and Macintyre 1977
B. taitanus	27	−0.21	1.79	Wood *et al.* 1975

right shift of the loading curve is known as the "Bohr effect". When we consider CO_2 at the same time, the $[CO_2]$ should be high in capillary beds in the tissue where it is produced and low in capillary beds at the respiratory exchange surface where it is lost to the environment. CO_2 is related to $[H^+]$ by the carbonic anhydrase catalyzed reaction:

$$CO_2 + H_2O \leftrightarrow H_2CO_3 \leftrightarrow H^+ + HCO_3^-$$

There are two processes involving both H^+ and HCO_3^- that prevent the accumulation of these products from slowing this reaction by mass action when cells see high P_{CO_2} in the tissue. First, a HCO_3^- is exchanged across the erythrocyte cell membrane for a Cl^- from the plasma, in a process known as the "chloride shift" (Boutilier and Toews 1981). Second the H^+ is bound to the hemoglobin favoring the unloading of O_2, or it is buffered by binding to the histidine residues of the hemoglobin molecule (see Nikinmaa 1990, 1997), or it is exchanged for Na^+ across the red cell membrane. Hence RBCs in respiratory capillaries (gills, skin and lung) will see low P_{CO_2} (lower $[H^+]$, higher pH) and the loading curve will be left shifted, favoring O_2 loading, and tissue capillaries will see higher P_{CO_2} (higher $[H^+]$, lower pH) favoring O_2 unloading from the hemoglobin. The magnitude of the Bohr effect is generally quantified as $\Delta \log P_{50}/\Delta pH$ at P_{50}. Like Hill coefficients, the Bohr coefficient in amphibians is dependent upon the level of O_2 saturation of the hemoglobin, tending to be low at low saturations and high at high saturations (Maginniss *et al.* 1980). Values for Bohr coefficients vary from -0.08 to -0.4 in amphibians and tend to be greater in anurans than other amphibians (Fig. 2.33, Table 2.3). There is a reported value greater than -0.6 for *Rana catesbeiana* larvae and adults (Pinder and Burggren 1983) which appears

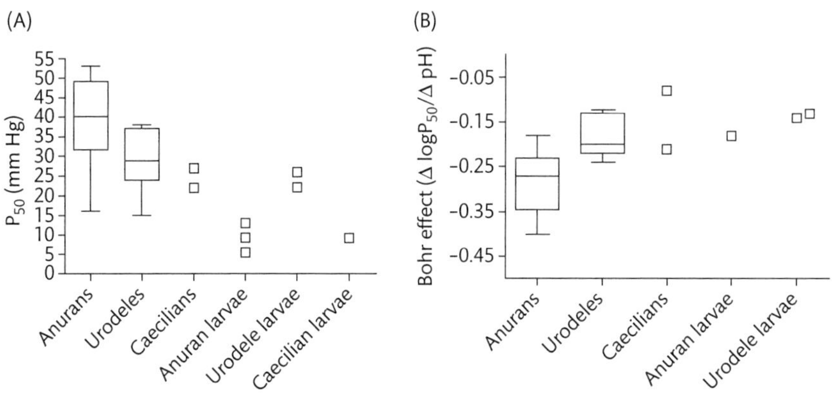

Fig. 2.33 A summary of (A) the partial pressures of O_2 needed for half saturation of the hemoglobin and (B) the magnitude of the effect of pH on this parameter for amphibian blood.

to be an outlier in comparison to other measurements on this species. It is not clear what selective advantage of having hemoglobin more sensitive to pH is provided for anurans, but a fascinating account in other vertebrate species can be found in a review by Giardina *et al.* (2004). Modulators, other than H^+, that can right shift the hemoglobin O_2 loading curve are an increase in temperature (Jokumsen and Weber 1980; Boutilier *et al.* 1987), an increase in intracellular Cl^- concentration and an increase in intracellular organic phosphate concentration. In amphibians, the principal intracellular phosphate compounds are ATP and 2,3-diphosphoglycerate (Hazard and Hutchison 1982). The molar ratio of these phosphate compounds to hemoglobin can vary developmentally, with temperature and interspecifically (Wood 1971; Hazard and Hutchison 1978; Hutchison and Hazard 1984).

The actual structure of hemoglobin can vary developmentally and interspecifically. Most anuran species produce a form of hemoglobin, as aquatic larvae, different from the form produced as terrestrial adults (see Frieden and Just 1970; Just and Atkinson 1972; Pinder and Burggren 1983). The larval hemoglobin characteristically is very much left-shifted with P_{50} values in the range of 0.7–1.8 kPa, while the adults of these larval forms are in the range of 3–6.8 kPa. The larval anuran red blood cells are removed by the spleen after metamorphosis and replaced with adult cells (Tamori and Wakahara 2000). Adult anurans tend to have a greater P_{50} than other amphibians (Fig. 2.33, Table 2.3). The lowest anuran P_{50} values are for the aquatic *Xenopus laevis* and *Telematobius culeus*. An interesting example of an aquatic amphibian is *T. peruvianus* from the Andes of South America. Not only will it normally experience lower environmental P_{O_2} as a consequence of being aquatic, but it ranges up to 3800 m in altitude, which would exacerbate the hypoxic aquatic environmental stress. The hemoglobin of *T. peruvianus* has a low P_{50} that is insensitive to Cl^- (Weber *et al.* 2002). Urodele larvae do not appear to shift the type of hemoglobin produced but rather modify its characteristics via a change in the concentration of intracellular organic phosphates during metamorphosis (Wood 1971; Woody and Justus 1974). This left-shifted larval hemoglobin-loading curve is thought to be an adaptation to load hemoglobin at low P_{O_2} in aquatic environments. As we saw in Chapter 1, aquatic compared to terrestrial environments are much more likely to see low P_{O_2} as a consequence of the total metabolism of aquatic organisms depleting the O_2 in solution.

The larvae of the viviparous caecilian *Typhlonectes compressicauda* also have significantly left-shifted loading curves in comparison to the female carrying the larvae in her uteri, an adaptation presumed to allow O_2 transfer from the mother to offspring (Toews and Macintyre 1977). The mechanism for this difference appears to be entirely mediated by differences in organic phosphate concentrations, the same mechanism as found in urodeles (Garlick *et al.* 1979).

The concentration of hemoglobin determines the absolute volume of O_2 that can be carried in the blood. The concentration of hemoglobin can be directly determined or inferred from associated metrics. The first is the direct determination of hemoglobin concentration and is usually presented in the literature as g/100 ml (g%), this strange unit is a carry-over from the clinical literature, and we will use g Hb/ml blood. A gram of amphibian hemoglobin will bind 1.3 ml of O_2 or 58 μmol of O_2 (Hillman 1976). An indirect metric of hemoglobin concentration is the cellular fraction by volume that erythrocytes represent in the blood or hematocrit (% by volume that erythrocytes represent in blood). A third common concentration metric is mean corpuscular hemoglobin concentration (MCHC), which is the [Hb]/volume of hematocrit (g Hb/ml erythrocytes). This provides an estimate of the density of hemoglobin/volume of RBC. If hematocrit is a useful predictor of [hemoglobin] in amphibians, then we would expect MCHC to not differ between different groups which is what we see, MCHC is remarkably constant at 0.255 ± 0.008 for metamorphosed amphibians. This means variation in hematocrit reflects variation in hemoglobin concentration and both measures reflect variation in blood oxygen carrying capacity.

Hematocrit and hemoglobin concentration are quite variable intraspecifically as can be seen in the following examples of extensive field sampling for four species (Fig. 2.34). This variation can reflect differences in either plasma volume or red blood cell volume. Hematocrit is known to vary on a short temporal basis with exercise (Boutilier *et al.* 1986; Tufts *et al.* 1987; Hedrick *et al.* 1999), hypoxia (Wood and Malvin 1991; Portner *et al.* 1991), catecholamine administration (Herman 1977; Mbangkollo and de Roos 1983) and handling stress (Mbangkollo and de Roos 1983), all phenomena known to increase blood pressure. The hemoconcentration mechanism is probably a consequence of an elevated capillary pressure induced ultrafiltration of plasma into the interstitial space (Hillman *et al.* 2004); osmotic fluid uptake by active cells (Boutilier *et al.* 1986); and inhibition

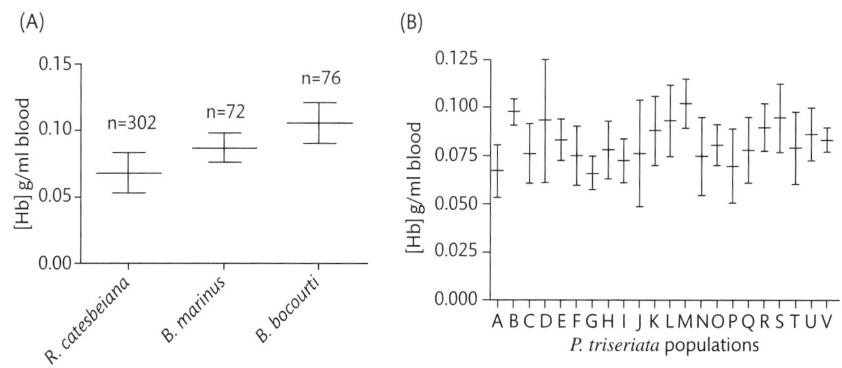

Fig. 2.34 (A) Interspecific variability in the mean hemoglobin concentration in different species of anurans (Stuart 1951; Coppo *et al.* 2005) and (B) along an elevational distribution in populations of *Pseudacris triseriata* (Packard and Stiverson 1976).

of lymphatic return (Malvin *et al.* 1995). Given the temperature at which these experiments were conducted, it seems unlikely that the liver plays a role in these responses (Frangioni and Borgioli 1989, 1991, 1993). Handling stress that elevates blood pressure before and during blood sampling can alter [hemoglobin] by 80% (Hedrick *et al.* 1999), which certainly will contribute to the variation in hemoglobin concentration determinations. In urodeles, exposure to hypoxia is known to elevate hematocrit, which appears to be mediated by release of erythrocytes from the spleen (Frangioni and Borgioli 1991, 1993), though ultrafiltration as a mechanism has never been tested in urodeles. We are unaware of any data on caecilians.

What this means functionally is that one has to exercise care in demonstrating and interpreting interspecific or experimental differences in hematocrit or [hemoglobin] because of the dynamic nature of plasma volume and its exchange with the interstitial space (see also lymphatic system). Figure 2.34 shows the mean and standard deviation for hemoglobin concentrations of four anuran species with extensive sampling. The standard deviation is large in comparison to the mean in all these studies, emphasizing the large intraspecific variation. The annual sampling of farmed *R. catesbeiana* represents over 300 individuals (half are males) sampled over a size range of 50–350 g. The range in concentration was 0.05–0.11 g Hb/ml blood. The 22 *Pseudacris* populations (Packard and Stiverson 1976) were sampled along a transect of increasing elevation. As can be seen there is no change in hemoglobin concentration along the elevational gradient, but there is large variation between populations leading to large intraspecific variation. Body mass, season, sex and temperature (Krishnamoorthy and Shakunthala 1974) have all been correlated to differences in hemoglobin concentration and/or hematocrit, as well as variables that affect blood pressure.

Hemoglobin concentration determines the absolute volume of O_2 that can be carried in the blood, but the P_{O_2} in relation to the loading curve determines the actual volume of O_2 in the blood. The figure below integrates measured P_{O_2} with O_2 loading curves to calculate the degree of hemoglobin saturation during those activities for various species for which data are available. Two features are notable. Normally the systemic arterial hemoglobin is about 90% saturated with O_2 at rest, except in obligate and predominant skin breathers where it is only 60–70% saturated (Fig. 2.35). This means that the loading curve has a lower affinity for O_2 than necessary to allow near saturation at the partial pressure of O_2 present in the artery. This may be a consequence of a limited O_2 respiratory conductance or the admixture of cutaneous oxygenated blood with systemic venous blood. At rest the systemic venous saturations are also high 70–80%. Activity widens the difference in arterial–venous saturations, primarily by lowering venous saturations to 35% (Fig. 2.35). During diving both arterial and venous saturations decline as O_2 is consumed from the lung (Fig. 2.35). The same thing occurs if gilled forms are removed from the water (Fig. 2.35). The hemoglobin always appears

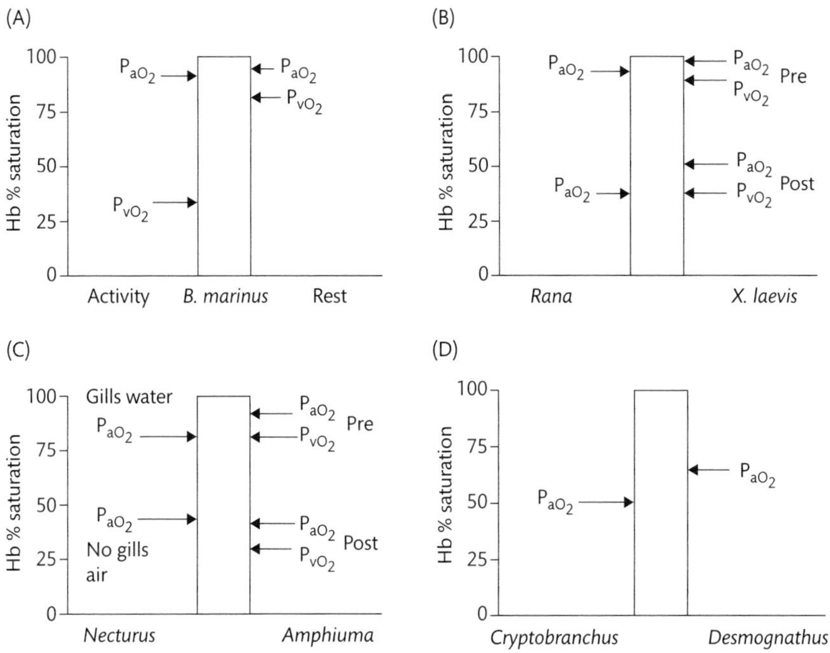

Fig. 2.35 The percent of full hemoglobin saturation for arterial and venous blood during different activities and morphological states: (A) rest-activity, data from Withers *et al.* 1988, Hedrick *et al.* 1999; (B) diving, data from Emilio 1974, Tazawa *et al.* 1979a, Pinder and Burggren 1983, Emilio and Shelton 1974, Jokumsen and Weber 1980, Boutilier and Shelton 1986; (C) dive-emersion, data from Lenfant and Johansen 1967, Toews *et al.* 1971; (D) skin breathers, data from Gatz *et al.* 1974, Boutilier and Toews 1981.

to retain some O_2 as the % saturation does not fall below 25%. This is probably an inflated value since venous blood also has oxygenated blood returning from the skin. When *Bufo marinus* were exercised in an environment where the skin was exposed to N_2, eliminating oxygenated cutaneous venous blood, the level of venous saturation fell to 14% (Hedrick *et al.* 1999).

Carbon dioxide is carried in the blood as CO_2 dissolved in solution, as HCO_3^-, and bound loosely to protein as carbamino-CO_2. Since hemoglobin is the most abundant protein in the RBC, hemoglobin carries most of the CO_2 present as carbamino-CO_2. Most of the CO_2 (60–85%) in blood is carried as HCO_3^- in the plasma (Fig. 2.36 Emilio and Shelton 1980; Boutilier and Toews 1981; Stinner *et al.* 1994). A CO_2 loading curve (Fig. 2.36) shows that oxygenation decreases the CO_2 concentration of the blood, while deoxygenated blood carries more CO_2. This right shift of the CO_2 loading curve with oxygenation is known as the Haldane effect. The Haldane effect favors CO_2 loading in the tissues where it is

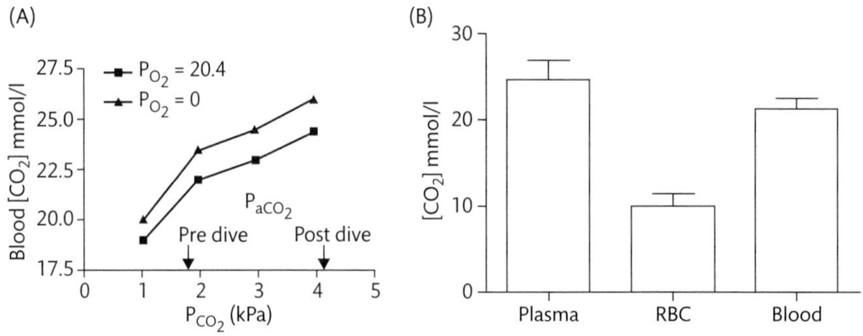

Fig. 2.36 (A) The effect of the partial pressures of carbon dioxide and oxygen on the amount of carbon dioxide carried in the blood of *X. laevis* (Boutilier and Shelton 1986; Emilio and Shelton 1980) and (B) how carbon dioxide is partitioned between the plasma and erythrocytes (RBC) in *Bufo marinus* (Stinner *et al.* 1994).

produced as O_2 is unloaded from the hemoglobin and promotes CO_2 unloading at the respiratory surface as the O_2 is loaded on the hemoglobin. This dance between CO_2 and O_2 and the hemoglobin molecules is essentially mediated via changes in the H^+ concentration and the partial pressures of CO_2 and O_2 in various regions. The magnitudes of the Bohr and Haldane effects for blood of *X. laevis* appear to be correlated (Boutilier and Shelton 1986).

Another function for erythrocytes, independent of increasing O_2 capacitance of the blood, may be as a sensor of oxygenation state and eventual affector of vascular tone, helping to match flow and the amount of O_2 carried in the blood (Ellsworth *et al.* 1995). This response appears to entail the release of ATP from RBCs in response to deoxygenated hemoglobin. The ATP released binds to vascular endothelial purine receptors leading to vasodilation (Jagger *et al.* 2001). This appears to be mediated via interactions of the hemoglobin with band 3 membrane proteins of the red cell membrane (Weber *et al.* 2004). This function has established importance in mammals, but has not been investigated in amphibians, and awaits a comparative analysis for its significance in vertebrates.

2.6 Cardiovascular Oxygen and Carbon Dioxide Exchange

One of the exciting aspects of the amphibian circulatory system is that the pulmonary and systemic circulations are in parallel and not in series. Only reptiles share with amphibians the possibility of parallel systemic and pulmonary vascular circuits. Functionally, this means blood flow rates can be different in the two circulatory beds, whereas flow obviously has to be equal in a series circuit. This differential flow capacity is achieved by shunting blood from one circuit to the other. Assuming a default of a series circuit where pulmonary and systemic flows are equal, a right to left shunt entails shunting blood destined for the pulmonary circuit to the systemic circuit, while a left to right shunt means that blood destined for the systemic circuit is shunted to the pulmonary circuit. Shunts are possible based on the anatomy of amphibian hearts, in particular a single ventricle and conus arteriosus. The ability to shunt blood has the disadvantage of oxygenated and deoxygenated blood mixing in the common ventricle and conus arteriosus. This in turn lowers systemic P_{O_2} as well as pulmonary and cutaneous P_{CO_2} and raises pulmonary P_{O_2} as well as systemic P_{CO_2}, thereby decreasing the efficiency of diffusional gas exchange in lungs, skin and tissues. This leads to less O_2 and more CO_2 being carried in the systemic arterial blood, hence increasing the amount of blood that has to be convectively pumped to achieve the same gas flux to support metabolism. The single pressure sources (ventricle and conus arteriosus) dictate that the arterial pressure of the systemic and pulmocutaneous circuits will be the same in this parallel circuit.

2.6.1 Heart Anatomy

The amphibian heart is contained within a pericardial sac that presumably sets limits to the degree of heart expansion. The pericardial sac contains pericardial fluid of unknown origin and fate. The pericardial fluid helps lubricate the surface of the heart and minimize frictional damage during the dynamic filling and emptying of the heart. Ultrastructural evidence from the pericardium suggests a lymphatic origin and disposition of pericardial fluid in frog hearts (Cerra *et al.* 2003), though this has not been validated experimentally. Disruption of the integrity of the pericardial sac has been implicated in functional compromise of the degree of separation of blood flow in the heart (Tazawa *et al.* 1979).

The anuran heart is composed of a sinus venosus, two atria (auricles), a single ventricle, emptying via a conus arteriosus (bulbus cordis) that contains a spiral valve (Sharma 1957, 1961; Langille and Jones 1977). The spiral valve contributes to the separation of blood into a pulmocutaneous circuit and systemic circuit (Fig. 2.37). Urodele hearts follow the same general plan except they are modified in species with a decreased reliance on lungs for gas exchange. Lungless salamanders have no atrial septum, a rudimentary left atrium and no spiral valve in the conus in the absence of pulmonary venous return. Other salamanders with reduced lungs show a fenestrated atrial septum, allowing exchange of blood between the two atria, a reduction in the size of the left atrium and a reduction or loss of the spiral valve (Noble 1925). The caecilian heart is characterized by a large sinus venosus partitioned for venous drainage from the cranial and caudal regions (Lawson 1966). The right atrium is much larger than the left and the septum may be incomplete. The ventricle is trabeculated, with a central trabeculum that runs the length of the ventricle. This central trabeculum has been interpreted as

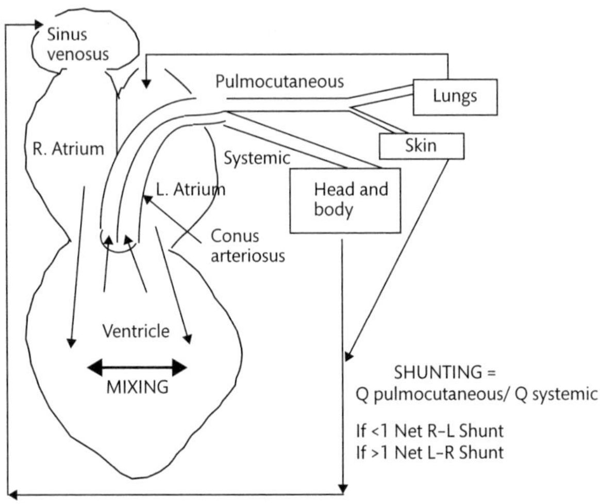

Fig. 2.37 A diagrammatic representation of the anatomy and pattern of blood flow in an anuran heart.

either the remnants or start of an interventricular septum (Marcus 1935; Schilling 1935), though the most recent interpretation is that it cannot effectively partition the blood in the ventricle (Lawson 1966). There may or may not be the equivalent of the spiral valve in the conus (Ramaswami 1944; Lawson 1966). There is tremendous interspecific variation in apodan hearts. There are no functional data to interpret this variation. Overall, there is no typical "amphibian" heart. The extant array of cardiac anatomic variation awaits critical testing of its functional consequences.

The walls of the sinus venosus, atria, ventricle and conus arteriosus all contain cardiac striated muscle that is responsible for contracting and decreasing the volume of the respective chambers. The pressure generated from the contraction of these muscles is responsible, in part, for filling the next cardiac chamber in series. The mass of the heart is larger than that of fish of equivalent mass (Poupa and Ostadal 1969). The energetics of shortening in amphibian cardiac muscle resembles mammalian cardiac muscle (Holroyd and Gibbs 1993). There are passive one-way valves between the chambers and between the cardiac chambers and vessels that prevent retrograde flow and only allow a unidirectional movement of blood. The pattern of contraction of the heart starts with a spontaneous depolarization initiating an action potential in the sinus venosus. The action potential leads to an increase in intracellular calcium concentration, in part the result of sarcoplasmic reticulum release of calcium (Ju and Allen 1999) and in part influx of calcium from the extracellular space (Morad and Goldman 1973; Cleeman *et al.* 1984). The calcium binds troponin and starts the interaction of actin and myosin leading to shortening of the small amount of cardiac muscle in the sinus venosus. The spontaneous action potential generation in the sinus has led to its use as a model for sinoatrial node pacemaker activity in mammalian hearts (Irisawa 1978; Hartzell 1979; Shibata and Giles 1985). The action potential also spreads to and through the branching cardiac muscle fibers of the atria, initiating contraction via an increase in intracellular calcium concentration of atrial muscle fibers from the top of the atria to their base forcing blood from the atria into the ventricle. The action potential is prevented from propagating directly to ventricular fibers by connective tissue separating the atria and ventricle except at some atrioventricular myocardial junction (Sedmera *et al.* 2003). There is a temporal delay and the action potential and initiation of muscle contraction then appears at the apex and moves dorsally to the base, forcing blood movement from the bottom up, just like in mammalian hearts. This parallel behavior is not predicted from the anatomy. Mammalian hearts have specialized nodal tissue, the atrioventricular node, the Bundle of His and Purkinje fibers, which coordinate this pattern of excitation. These nodal tissues are absent in amphibians. The only structures that run from base (top) to apex (bottom) in amphibian hearts are the trabeculae and are inferred to be the most likely equivalent pathway for action potential propogation in the ventricle (Sedmera *et al.* 2003). The action potential upon reaching the base of the ventricle is then transmitted to the conus arteriosus, eventually leading to its contraction. There are time delays between the arrival of the action potential in the atria, ventricles and conus and the action potential in the adjacent chamber allowing time for blood movement and filling from one chamber to the next before contraction occurs in the next chamber (Sedmera *et al.* 2003).

The sinus venosus containing mixed venous return (systemic and cutaneous) empties into the right atrium while the pulmonary venous return from the lung empties into the left atrium. The anatomical division of the atrium into two chambers maintains the division of systemic and pulmonary venous returns. The presence of two atria is a morphologic characteristic only found in some of the current lungfish (Burggren and Johansen 1986), none of the other extant fishes and all other tetrapods. Partitioning of systemic and pulmonary blood in the ventricle and conus is where our understanding of the mechanisms for flow separation begins to get cloudy. All other tetrapods have a partial or a complete septum anatomically dividing the ventricle. Using only anatomic features, the design looks as though right and left atrial blood should mix in the ventricle, given the lack of physical separation and equivalent systolic pressures in the systemic and pulmonary circuits over time (De Graaf 1957; Johansen 1963; Shelton and Jones 1968; Jones and Shelton 1972; West and Smits 1994). Functional data assessing the degree of mixing, using a variety of visual, radiographic, volumetric and P_{O_2} measurements, runs from complete mixing to only 15% mixing (85% separation) (Ozorio de Almeida 1923; Noble 1925; Foxon 1951; De Graaf 1957; Simons 1959; Delong 1962; Johansen 1962, 1963; Johansen and Ditadi 1966; Haberich 1965; Shelton and Jones 1965, 1968; Tazawa and Mochizuki 1977; Tazawa *et al.* 1979; Withers *et al.* 1988; Hedrick *et al.* 1999). The degree of mixing can vary over time, and the variables determining mixing have yet to be completely understood but of paramount importance is the degree of ventricular filling (Haberich 1965), both low and high end diastolic volumes decreased the efficiency of artificially perfused frog hearts (Haberich 1965). This is consistent with *in vivo* data, where hearts were least efficient with higher mixing at low and high heart rates, an indirect metric for filling time (Hedrick *et al.* 1999). The mechanisms for the continued separation of systemic and pulmonary venous blood in the non-septate ventricle are not understood. The presence of trabeculae (clefts) within the walls of the ventricle is consistently implicated in helping maintain the separation of blood in the ventricle (see Noble 1925; Angell *et al.* 1966; Haberich 1965; Johansen and Hanson 1968). The blood then moves from the ventricle into the conus arteriosus.

The anuran conus arteriosus is contractile and incompletely partitioned by a spiral valve (Morris 1974; Liberthson *et al.* 1975; Langille and Jones 1977). The presence of a contractile conus arteriosus with a coronary artery and spiral valve has attracted significant attention and speculation as to its functional significance. Morris (1974) proposed the spiral valve acts as a flap, physically occluding the flow to the pulmocutaneous circuit with contraction of the conus arteriousus, thereby helping to partition systemic and pulmocutaneous flows via a valvular mechanism within the conus. Subsequent work monitoring the time course of the pressure trace in the pulmocutaneous artery and ventricle found no evidence for a valvular mechanism unless conus contractions are artificially paced without allowing for the normal delay following ventricular contraction and subsequent conus filling (Langille and Jones 1977). The presence of a spiral valve appears to contribute to the separation of right and left ventricular blood by streaming flows from right and left sides of the ventricle to the pulmocutaneous and systemic circuits (Langille and Jones 1977). The significant role of the spiral valve in maintaining independent systemic and pulmocutaneous flows is anatomically reinforced in salamanders with reduced lungs having a reduction in length in the spiral valve, and the complete loss of the spiral valve in species with fenestrated interatrial septa and/or lungless species with only a right atrium (Noble 1925). The contraction of the conus does not normally contribute substantially to cardiac output, as its volume is low in comparison to ventricular volume (Shelton and Jones 1965). There is evidence that during diving and/or hemorrhage the role of conus contraction in moving blood may be more significant in adult anurans (Liberthson *et al.* 1975). Interestingly, the conus arteriosus in young larval bullfrogs appears to be the major cardiac pump transcending the normal role of the ventricle (Pelster and Burggren 1991).

2.6.2 Vascular Anatomy

The anuran conus arteriosus has three parallel vascular tracks arising from both the right and left sides: a carotid arterial branch, a systemic arch branch and a pulmocutaneous artery branch. The right and left systemic arches merge posteriorly to form the dorsal aorta. The carotid artery and systemic arch carry blood returned from the lung (high P_{O_2}, low P_{CO_2}) via the left atrium, while the pulmocutaneous artery normally carries blood returning from the systemic veins (low P_{O_2}, high P_{CO_2}) as well as venous return from the skin (high P_{O_2}, low P_{CO_2}) via the right atrium. Without any mixing and/or shunting there is already a functional compromise, the venous return from the skin raises the P_{O_2} and lowers the P_{CO_2} of the blood that is returned from the systemic venous return, that is then sent to the lungs and skin. This will decrease the diffusional driving force for gas exchange across both of these respiratory surfaces. There are no data to establish how great a compromise this may represent, as there are no measurements of P_{O_2} and P_{CO_2} from both cutaneous and pulmonary venous returns. Given the small driving force for CO_2 across the lungs at rest in conjunction with the greater pulmonary conductance compared to skin (see lung section), it is difficult to project the consequences. As we saw in the blood section, at least at rest, it does not prevent systemic arterial blood from being 85–95% saturated with O_2.

Mechanically, the anuran arteries are highly distensible, resilient and non-linearly elastic (Gibbons and Shadwick 1991a). The transition from high compliance to low compliance occurs above the physiological pressure range (Gibbons and Shadwick 1991a). The transit time for the pressure pulse through the arterial tree is only about 3% of the cardiac cycle and does not differ in profile along its length (Gibbons and Shadwick 1991b). These characteristics have led to the conclusion that the arterial system serves as a simple Windkessel, expanding during systole and recoiling elastically during diastole, thereby storing then releasing energy and so reducing and smoothing the pressure pulse and protecting the microvasculature from extreme pressures (Gibbons and Shadwick 1991a, b).

The urodele heart has a different vascular anatomy leading from the heart. There is a pulmonary branch alone, with no cutaneous connection. The skin, therefore, is being perfused by systemic arterial blood, leading to lower driving forces for O_2 across the skin. Apodans show more asymmetry of their vasculature than the other groups. There may be a single pulmonary artery from the conus in some species, rather than the usual pairing (Ramaswami 1944).

2.6.3 Physiology

The contraction and shortening of the heart's cardiac muscle creates a pressure over and above the mean circulatory filling pressure (MCFP). The MCFP is the pressure within the vascular system when the heart is stopped and is determined by the volume of blood present and the vascular compliance. The vascular compliance (Δ volume/Δ pressure) is principally determined by the compliance of the venous circuit as this is where 60–70% of the total blood volume is located. Any variable that will decrease the volume of the venous vasculature will decrease the vascular compliance and increase MCFP. Increased MCFP will increase cardiac heart filling, which will increase heart output. The pressure above MCFP moves fluid throughout the various vascular beds and this blood flow can be described as:

Cardiac Output = Driving Force/Resistance = $(P_a - P_v)/R$

where P_a is mean arterial pressure, P_v is central venous pressure and R is the total peripheral resistance contributed by the frictional loss of energy as blood moves through all the vasculature. P_a is calculated as $P_{diastole} + 1/3\ (P_{systole} - P_{diastole})$. The $P_{systolic}$ is the same for pulmocutaneous, systemic and carotid circuits. $P_{diastolic}$ tends to decrease to lower levels in the pulmocutaneous circuits (Langille and Jones 1977), principally as a consequence of lower resistance in the pulmonary circuit (West and Smits 1994). Consequently, mean P_a tends to be slightly less in the pulmocutaneous circuit.

A second indirect way to measure systemic output is utilization of the Fick Principle or conservation of mass where:

V_{O_2} = Systemic blood flow rate × $(C_{aO_2} - C_{vO_2})$

where V_{O_2} is the rate of O_2 flux, C_{aO_2} is the arterial O_2 concentration, and C_{vO_2} is the venous O_2 concentration. The Fick principle can be a difficult application in amphibians because of the gas exchange across the skin, adding O_2 and removing CO_2. Consequently, the concentrations of O_2 and CO_2 in the venous return reflect more than O_2 consumption and CO_2 production.

Mechanically, cardiac output is:

Cardiac Output = Stroke Volume × Heart Rate

where stroke volume is the volume ejected per beat (V_s) and heart rate is the number of beats per min (f_h). Stroke volume can vary with time, principally under the control of filling volume and the degree of ventricular emptying or ejection fraction. The basic Frank–Starling law of the heart, "the heart will eject the volume returned", was initially described in amphibians (Frank, translated by Chapman and Waterman 1960; McKean *et al.* 1997). The mechanism for the Frank–Starling law is inherent in the sensitivity of troponin binding sites for calcium—they have greater affinity for calcium when stretched (Ruegg 1987). Mediators that increase the amount of intracellular calcium, like epinephrine, or increase the affinity of the troponin site for calcium, like stretch, will increase the ejection fraction for the same amount of calcium and increase stroke volume. Variables that change the mean circulatory filling pressure will change the degree of stretch and change stroke volume. An increase in blood volume (osmotic water uptake or enhanced lymphatic return) or constriction of the venous circuit increasing MCFP and mobilizing venous return will increase stroke volume while a decrease in blood volume (decreased lymphatic return, hemorrhage or dehydration) will decrease stroke volume (Hillman and Withers 1988). Variables that decrease the concentration of intracellular calcium or lower the troponin site affinity for calcium (Ruegg 1987) will also decrease stroke volume. Heart rate control is based on the interaction of sympathetic (increase rate) and parasympathetic (slows rate) nervous stimulation of the heart (Morris and Nilsson 1994).

The total peripheral resistance is a little more complicated and the variables involved are inferred from the Poiseuille–Hagen flow formula for a Newtonian fluid as:

$$R_{peripheral} = 8\eta l/\pi r^4$$

where η is blood viscosity, l is the length of the vasculature, and r is the radius of the vasculature. Recognize that blood is a non-Newtonian fluid, but the applicability of this formulation in a general sense is useful. Since blood vessels contain smooth muscle they can vary their radius and consequently vary their resistance in the short term. Variation in vessel diameter can occur in all the vessels of each vascular bed, but the greatest range in vessel diameter is most probably the arterioles as they contain the most smooth muscle and are most responsible for shunting blood. There are a variety of factors known to influence the degree of smooth muscle tone in the amphibian vasculature (see Figs 2.44 and 2.45). The length and mean radius of the vasculature will change with growth, hence it will be related to size and development. As both the length (increase resistance) and cross sectional area (decrease in resistance) change with growth, the effects of growth are not predictable. A good rule of thumb is that it takes a about a four-fold increase in cross-sectional area of vessels in parallel to compensate for the increased frictional loss of those vessels in length. Vascular resistance decreases during developmental growth in tadpoles, indicating the effect of growth of parallel vascular beds, increasing the mean cross-sectional area, is a more important variable than the increase in length of the vasculature during this phase of growth (Pelster and Burggren 1991; Hou and Burggren 1995).

The principal variable in determining blood viscosity is hematocrit; cell size does not appear to play a role in viscosity except with yield shear stress (Hedrick 1985). The viscosity of amphibian blood is exponentially related to hematocrit (Weathers 1976; Hillman *et al.* 1984; Hedrick 1985; Withers *et al.* 1988). A typical relationship for *B. marinus* is presented in Fig. 2.38. Viscosity of blood is also related to temperature, that is it is more viscous at the same hematocrit at lower temperatures (Weathers 1976). Since blood O_2 carrying capacity is linearly related to hematocrit, there should be an optimum hematocrit where the product of blood flow rate and O_2 carrying capacity (systemic O_2 transport capacity) is maximized. At higher than optimal hematocrits, systemic O_2 transport capacity will decrease because the greater viscosity decreases blood flow, canceling out the positive effect of increased O_2 carrying capacity. At lower than optimal hematocrits, O_2 carrying capacity decreases faster than blood flow increases. Weathers

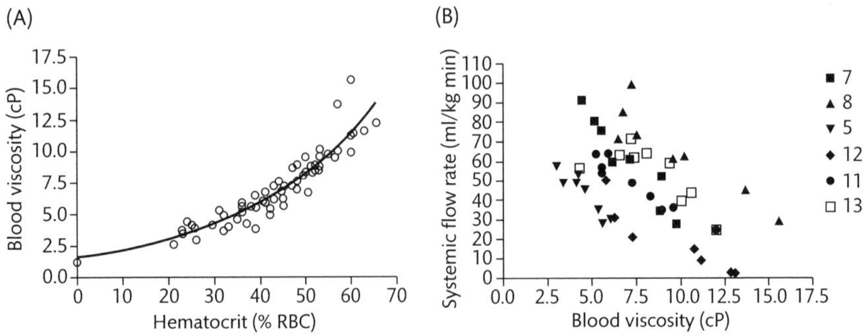

Fig. 2.38 (A) The effects of hematocrit on blood viscosity and (B) the effects of blood viscosity on maximal systemic blood flow rates of *B. marinus* (Hillman *et al.* 1985).

(1976) tested the concept of optimal hematocrit in amphibians, using *in vitro* viscosity and blood flow measurements, to determine the optimal hematocrit for systemic O_2 transport for bullfrog blood over a variety of temperatures. The *in vivo* hematocrit increased as expected with acclimation to warmer temperatures, since viscosity decreases with increasing temperature allowing the animal to increase hematocrit to achieve the same viscosity. This hematocrit response was present in winter bullfrogs but not summer animals. Viscosity measured *in vitro* actually reflects patterns of blood flow measured *in vivo*, at least under maximal cardiac output conditions. Systemic blood flow rates decrease in an inverse relationship to viscosity (Fig. 2.38, Hillman *et al.* 1984; Withers *et al.* 1988). The concept of an optimal systemic O_2 transport is a useful concept to keep in mind, but given the marked individual variation in hematocrit induced by variation in P_a, changes in lymphatic function and release of cells by the liver and spleen discussed earlier, it is an extremely difficult concept to test in amphibians. Suffice to say, increased blood viscosity resulting from increased hematocrit will decrease blood flow *in vivo*.

There are three parallel vascular beds that contribute to the total peripheral resistance: the pulmocutaneous vasculature, the carotid vasculature and the systemic vasculature. Because the vascular beds are in a parallel arrangement: $1/R_{total} = 1/R_{pc} + 1/R_{sys} + 1/R_c$. The results of West and Smits (1994) in Fig. 2.39 show the blood flow rates in and vascular resistances of these beds as well as total cardiac out and peripheral resistance in the toad *Bufo marinus*. Note the equivalent flows and resistances of the pulmocutaneous and systemic circuits. About 80% of the flow in the pulmocutaneous circuit goes to the lungs and 20% to skin (West and Burggren 1984; Boutilier *et al.* 1986). The carotid circuit has a high resistance and low flow, a consequence of the lower mean cross-sectional area.

Since the vessels comprising the vasculature are both in series and parallel, it is useful to understand where the predominance of the vascular resistance lies. The venous side of the systemic circuit has about one-fourth the resistance of

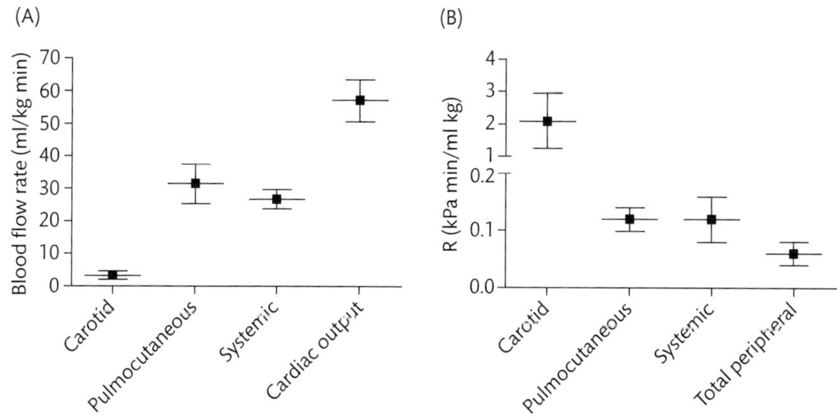

Fig. 2.39 A summary of (A) blood flow rates and (B) the resistances of pulmocutaneous, carotid and systemic circuits for *Bufo marinus* (West and Smits 1994).

the arterial–arteriole side of the capillary bed, hence the R_v/R_a is 0.23 (Hancock *et al.* 2000). This is important in the calculation of capillary pressures from the Pappenheimer–Soto-Rivera equation:

$$P_{cap} = [P_v + R_v/R_a\, P_a]/[1 + (R_v/R_a)]$$

Capillary pressure is important in filtration across the capillaries into the interstitial space forming lymph, as we will see later.

What kind of values do we see for these cardiovascular parameters? For some of the cardiovascular parameters, like arterial pressures, there is an extensive database. Other parameters that are technically and surgically more challenging, like blood flow measurements, have a limited database. Because of the technical difficulty in general, as well as the propensity of these measurements to vary with time, many of these cardiovascular measurements were made on anesthetized animals. Our most complete data for unanesthetized animals are for *Bufo marinus* because of its large size. Additionally, they handle surgery well and are generally available from commercial suppliers. Recognize size, availability and tenacity are appropriate experimental rationales to collect basic data on anuran cardiovascular function. The data are too phylogenetically narrow at this time to evaluate the potential variability present in these cardiovascular parameters and allow evolutionary interpretations.

2.6.4 Shunting

There is either no net shunt or a net right to left (R–L) shunt at rest (Shelton 1970; Emilio and Shelton 1972; Meyers *et al.* 1979; Tazawa and Johansen 1987; West and Smits 1994; Hedrick *et al.* 1999; Andersen *et al.* 2003). This R–L shunt is most obvious during apnea, diving and at low temperatures (Emilio and Shelton 1972; Langille and Jones 1977; Meyers *et al.* 1979; Gamperl *et al.* 1999; Hedrick *et al.* 1999; Andersen *et al.* 2003). A R–L shunt will decrease the amount of gas exchanged across the pulmonary surfaces, consistent with the significance of cutaneous exchange at rest. The R–L shunt is also associated with the periods of lowest cardiac outputs. Given the capacity for cutaneous exchange in amphibians, minimizing pulmonary flow would decrease cardiac work. The possibility of a second intrapulmonary shunt exists based on the description of intrapulmonary arteriovenous connections for the clawed frog, *X. laevis* (Smith and Rapson 1977). These would allow blood to bypass the pulmonary exchange surfaces, which could be a selective advantage during long apneic bouts where intrapulmonary gas pressures might actually favor O_2 loss and CO_2 gain if venous blood, enriched by cutaneous exchange, were to perfuse the lung. The clawed frog is a totally aquatic anuran where long apneic periods occur (Boutilier 1984). There are no data on the functional consequences of these intrapulmonary anatomical shunts. There is a left to right shunt following ventilation, exposure to hypoxic and/or hypercapnic environment and during activity (Emilio and Shelton 1972; Langille and Jones 1977; West and Smits 1994; Hedrick *et al.* 1999; Andersen *et al.* 2003). Examples of the effects of temperature, activity and hypoxia on shunts are presented in Fig. 2.40. The left to right shunt with activity would result from an increased venous return based on a sympathetically mediated decrease in venous compliance coupled to an actual shunt of blood from the now enhanced left heart return from the pulmonary vein.

2.6.5 Mixing

The most recent summary for mixing data can be found in Fig. 2.40, derived from Hedrick *et al.* (1999), who organized the degree of mixing data around f_h. There is almost complete mixing at low heart rates (at rest in low temperatures). The heart does not maintain separation of pulmocutaneous and systemic flows, compromising diffusional gas exchange in both circuits. Gas exchanges under these resting conditions are low enough that no limitation to flux is presented. As heart rate increases to supply increased gas fluxes associated with increasing temperature and activity, the degree of mixing declines precipitously, increasing the efficiency of diffusional gas exchange across the skin and lungs and delivery to the tissues. At the highest heart rates associated with activity at elevated temperature (30°C) mixing increases, decreasing respiratory gas exchange efficiency, which could contribute to a lower Q_{10} for activity metabolism at higher temperatures than expected (Seymour 1973; Carey 1979). The greatest efficiency for diffusional exchange across the pulmonary and systemic circuits (least mixing) is at the heart rates found with resting metabolism and moderate activity at preferred body temperatures (Hedrick *et al.* 1999).

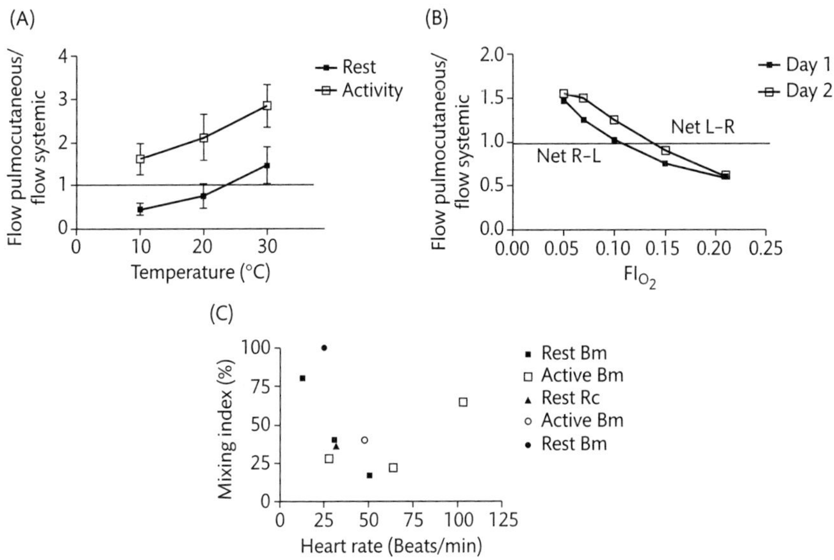

Fig. 2.40 The degree of shunt (relative blood flow rates of the pulmocutaneous to systemic circuit) (A) at different temperatures, with activity and (B) at different oxygen partial pressures for *Bufo marinus* (Hedrick *et al.* 1999; Andersen *et al.* 2003) and (C) the degree of intracardiac mixing at different heart rates in anurans (Hedrick *et al.* 1999).

2.6.6 Arterial Pressure

The most recent summary of systolic arterial pressures in amphibians can be found in Weathers (1975). A good estimate for anuran mean arterial pressure under resting conditions is around 3.4 kPa (Shelton and Jones 1968). There are far fewer measurements of venous pressures but centrally they range from 0.2–0.6 kPa (Weathers 1976; West and Burggren 1984) and 0.7–1 kPa on the upstream side of the hepatic and renal portal systems (Withers *et al.* 1988). Arterial pressure increases during development (Pelster and Burggren 1991; Burggren *et al.* 1992; Hou and Burggren 1995), and in the clawed frog, *Xenopus laevis*, body mass through development to adulthood predicts half the variation in mean arterial pressure (Hou and Burggren 1995). Arterial pressure increases with activity (Withers *et al.* 1988), even though peripheral resistance decreases with activity (see Fig. 2.41), indicating cardiac output increases more than peripheral resistance decreases during activity. Season has also been shown to influence arterial pressures (Segura and D'Agostino 1964; Pelster and Burggren 1991).

2.6.7 Peripheral Resistance

Peripheral resistance in adult anurans varies in the different circuits as we saw above. Systemic circuit resistance will decrease with activity indicating the opening of vascular beds (Fig. 2.41). There is a suggestion it may vary interspecifically as evidenced by the *Bufo marinus* and *Rana catesbeiana* systemic resistance comparison. Pulmonary resistance will decrease when fresh air is ventilated and during activity, but increases if the pulmonary surface is exposed to low O_2 or high CO_2 air and is most sensitive to the combination of hypoxia and hypercarbia (Fig. 2.41; West and Burggren 1984; Hedrick *et al.* 1999). These pulmonary resistance changes are mediated by both a central mechanism via vagal innervation of the pulmonary artery (St Aubain and Wingstrand 1979) and local hypoxic vasoconstrictor responses (Koyama and Horimoto 1982, 1983, 1988). Overall peripheral resistance declines during developmental growth with the addition of parallel vascular beds (Fig. 2.41; Hou and Burggren 1995; Fritsche 1997).

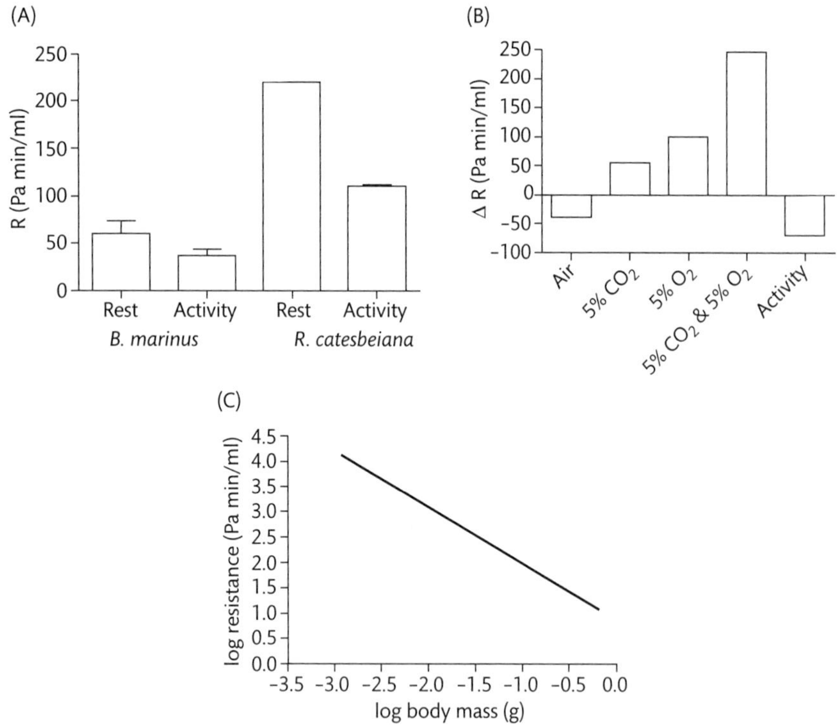

Fig. 2.41 (A) The effects of activity on systemic and pulmonary vascular resistances; (B) the effects of various gas mixtures on pulmonary resistances; and (C) the effect of body mass on total peripheral resistance in *X. laevis* tadpoles (Weathers 1976; West and Burggren 1984; Hillman 1987; Hillman and Withers 1988; Withers *et al.* 1988; West and Smits 1994; Hou and Burggren 1995; Hedrick *et al.* 1999).

2.6.8 Stroke Volume

The stroke volumes for the few amphibians measured to date run from 1 to 3 ml/kg at rest (Fig. 2.42; Johansen 1963; Hoyt *et al.* 1994; Gatz *et al.* 1975; Hou and Burggren 1995; Jacobsen and Fritsche 1999; Shelton and Jones 1965; Tazawa *et al.* 1979; West and Smits 1994; Yano *et al.* 1995; Hedrick *et al.* 1999; Gamperl *et al.* 1999; Andersen *et al.* 2003). Stroke volume appears to vary interspecifically (Fig. 2.42) and increases with activity (Hedrick *et al.* 1999). Anurans and urodeles demonstrate substantial variation (about three-fold) in ventricle mass (Hillman 1976, 1980; Hillman *et al.* 2000; Withers and Hillman 2001), corroborating the measured interspecific variation of stroke volume. Stroke volume scales with body mass ($mass^{1.4}$) during development in *X. laevis* (Hou and Burggren 1995; Jacobsen and Fritsche 1999).

2.6.9 Heart Rates

The variables known to influence heart rate in amphibians include: temperature (Weathers 1975, 1976; Carey 1979; Hedrick *et al.* 1999); body mass (Burggren and Doyle 1986; Burggren *et al.* 1990; Hou and Burggren 1995); season (Segura and D'Agostino 1964; Weathers 1975); developmental stage (Burrgren and Doyle 1986a, b; Burggren *et al.* 1990; Burggren and Pinder 1991; Pelster and Burggren 1991; Burggren *et al.* 1992; Hou and Burggren 1995; Jacobsen and Fritsche 1999); and activity (Hillman 1976; Carey 1979; Burggren and Doyle 1986; Hedrick

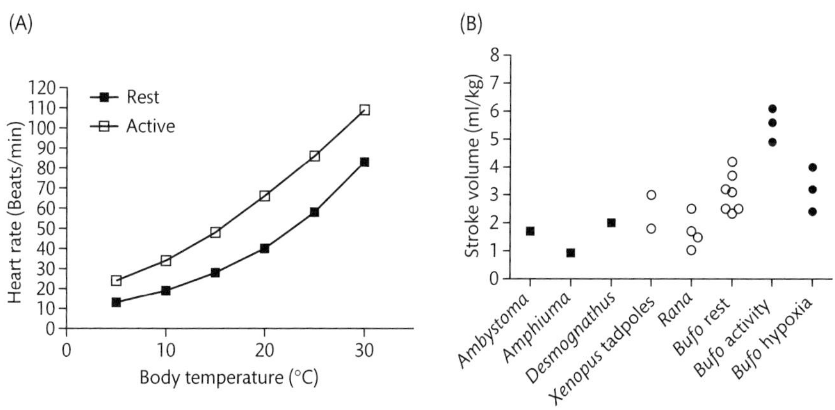

Fig. 2.42 (A) The effects of temperature on resting and maximal heart rates of *B. boreas* (Carey 1979) and reported stroke volumes for different amphibian species (B) at rest (Johansen 1963; Hoyt *et al.* 1994; Gatz *et al.* 1975; Yano *et al.* 1995; Hou and Burggren 1995; Jacobsen and Fritsche 1999; Shelton and Jones 1965; Tazawa *et al.* 1979; West and Smits 1994; Hedrick *et al.* 1999; Gamperl *et al.* 1999; Andersen *et al.* 2003).

et al. 1999). Heart rates at rest and during activity define a scope (maximum–minimum) for heart rate possibilities in a species. A representative example of the resting and active heart rates for the boreal toad *B. boreas* at different body temperatures (Carey 1979) is presented in Fig. 2.42. Anuran maximal heart rates do not seem to vary interspecifically (Hillman 1976), hence interspecific differences in cardiac output have never been attributed to differences in maximal heart rate, only stroke volume.

2.6.10 Cardiac Outputs

A difficulty with trying to interpret the available data on interspecific variation in cardiac outputs is that heart rates differ and are a significant source of the variation in blood flow rates. For instance, Weathers (1976) presents "resting" flow measurements for *R. catesbeiana*, that have heart rates equal to bullfrogs during "maximal" activity (Withers *et al.* 1988), an inconsistent proposition with "resting". The lack of overt anatomical features to evaluate stress in amphibians makes it impossible to assure a resting state through any visual cue. One way to compensate for the state of arousal of the animal is to normalize flow measurements to individual heartbeats, measure heart rate, and interpret those rates in light of the scope for heart rate. A heart rate approaching maximal, at a specific temperature, precludes the use of the term resting to describe the behavioral state of the animal. An excellent example of the perils that a lack of understanding about resting cardiovascular variables can introduce in the interpretation of cardiovascular results is obvious in the work of Cabanac (1999). The absence of an "emotional tachycardia" upon physical handling in bullfrogs, and its presence in a reptile led Cabanac (1999) to conclude that "emotion" emerged in the evolutionary lineage between amphibians and reptiles. The difficulty with this interpretation is that the measured resting heart rates for the bullfrogs in his experiments, are equal to maximal heart rates during activity for this species as measured in other studies (Hillman and Withers 1988; Hillman *et al.* 1987; Withers *et al.* 1988). The bullfrogs were already demonstrating an emotional tachycardia in the resting experimental environment and the investigator did not recognize the "resting" heart rates as maximal. Hence there was no further heart rate scope to allow a tachycardia upon being physically handled, not because of the lack of an "emotional tachycardia".

A summary of the stroke flows in various vascular beds for *Bufo marinus*, a species where the most data exist, is presented in Fig. 2.43 (Hillman *et al.* 1985; Hillman *et al.* 1987; Withers *et al.* 1988; Hillman and Withers 1988; West and Smits 1994; Hedrick *et al.* 1999; Gamperl *et al.* 1999; Andersen *et al.* 2003). At rest, most of the stroke blood flow is going into the systemic circuit. During activity, pulmocutaneous stroke blood flow increases three-fold to exceed systemic stroke flow, which changes little during maximal output conditions. These data suggest that the left to right shunt is important for gas exchange and resting

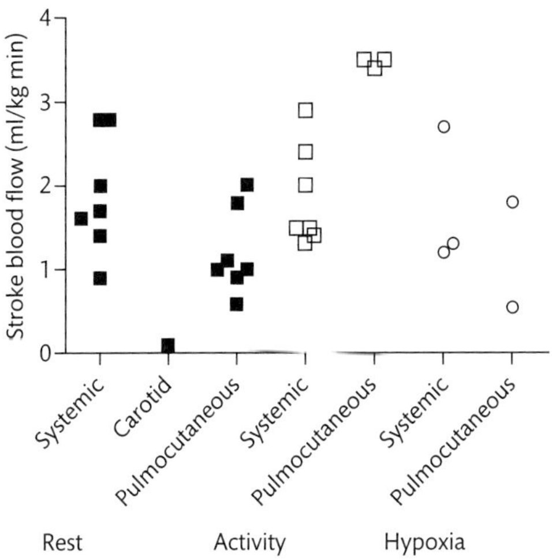

Fig. 2.43 A summary of blood flow rates in different vascular beds of *Bufo marinus* at rest, with activity and during hypoxia (Hillman *et al.* 1985; Hillman *et al.* 1987; Withers *et al.* 1988; Hillman and Withers 1988; West and Smits 1994; Hedrick *et al.* 1999; Gamperl *et al.* 1999; Andersen *et al.* 2003).

systemic flows are basically maintained and sufficient to handle gas exchange in the tissues during activity. Augmentation of O_2 delivery to, or CO_2 removal, from tissue is achieved by increasing heart rate and/or widening the arteriovenous difference. A corollary perspective is that systemic stroke flow rates at rest are much greater than necessary to handle standard levels of aerobic metabolism when flow measurements are viewed in light of the high degree of venous saturation at rest. This suggests the L–R shunt during activity is necessary to handle pulmonary gas flux, not systemic gas transport. Whether the lungs and their ventilation are under-designed and hence provide a significant resistance to the elevated pulmonary gas flux associated during activity is a question that will come up under aerobic metabolism in Chapter 3.

2.6.11 Integration with Blood Data

The actual systemic O_2 transport for an amphibian (*B. marinus*) at rest and during activity can now be calculated by combining the flow data from Section 2.6 with the O_2 content data from Section 2.5. At rest the arterial blood is 95% saturated, with a full saturation content of 0.1 ml O_2/ml blood. During activity the blood is about 90% saturated but as a consequence of activity induced hemocontration has a 20% greater hemoglobin concentration, hence a greater full saturation content of 0.11 ml O_2/ml blood. The systemic cardiac output at rest is 56 ml/min kg (28 beats/min × 2 ml/kg beat) and during activity 112 ml/min kg (56 beats/min × 2 ml/kg beat). The overall O_2 delivery to tissues is 5.6 ml O_2/kg min (250 mmol/kg min) at rest and 12.3 ml O_2/kg min (550 mmol/kg min) during activity. Standard metabolic rates are about 15 mmol/kg min (Bentley and Shield 1973) while activity rates are 520 mmol/kg min (Hillman 1987). This means arteriovenous differences should be low at rest and large during activity, exactly what has been demonstrated (see Fig. 2.35). It also reinforces the concept that resting tissue is being convectively supplied with O_2 at rates far greater than metabolic demand dictates. Recognize the greater rate of O_2 delivery than metabolism demand at rest provides a buffer or safety factor to maintain aerobic metabolism during short activity bursts without an increase in cardiac output. This precludes the need for physiological control systems that would immediately couple increases in cardiac output to increases in metabolism.

2.6.12 Cardiovascular Control under Volume Stress

The negative feedback control loops involved in the control of cardiovascular function when blood volume increases and decreases are presented in Figs 2.44 and 2.45. Note that there are hormonal, mechanical, physical and neural elements

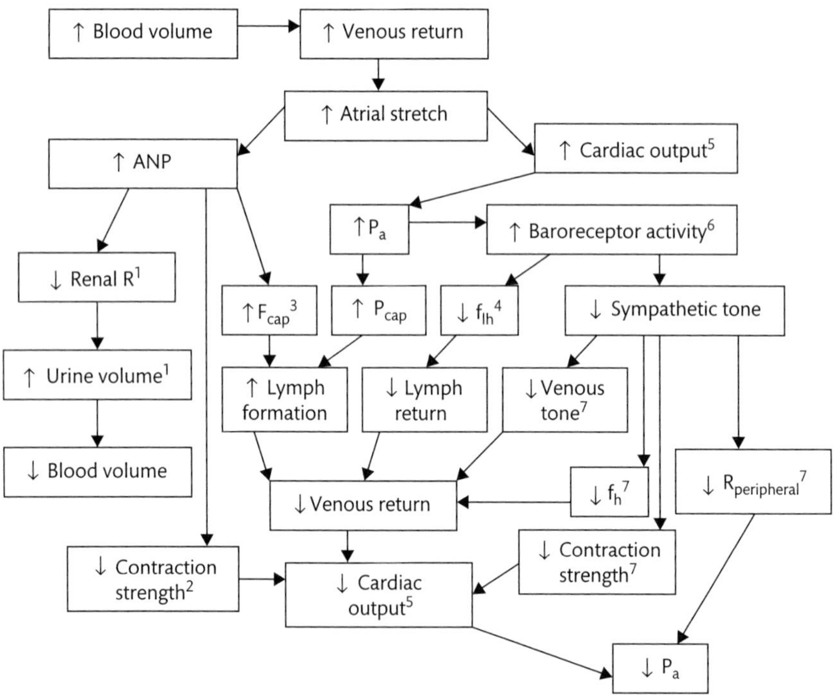

Fig. 2.44 The control of cardiovascular function in amphibians following a blood volume expansion. [1]Donald *et al.* 2002; [2]Uchiyama *et al.* 1998, Cerra *et al.* 2003; [3]Meyer *et al* 1992; [4]Crossley and Hillman 1999; [5]Frank translated by Chapman and Waterman 1960; [6]West and Van Vliet 1994; [7]Morris and Nilsson 1994.

involved in controlling the renal, lymphatic and cardiovascular reflexes that return blood volume to its resting state after it is expanded or depleted.

2.7 Tissue Oxygen and Carbon Dioxide Exchange

The same variables we discussed in the respiratory diffusion section apply here. The flux from the tissue capillaries to the surrounding cells, and eventually to the intracellular mitochondria, for O_2 and the reverse for CO_2, is driven by the partial pressure differences from blood in the capillaries to the mitochondria for O_2 and CO_2. The mean diffusional distance from the capillary to the mitochondria and the tissue diffusion coefficient are the two major anatomical variables determining the resistance to gas flux. Diffusion distance is principally determined by the degree of capillarity in a given organ. A greater capillary density shortens the

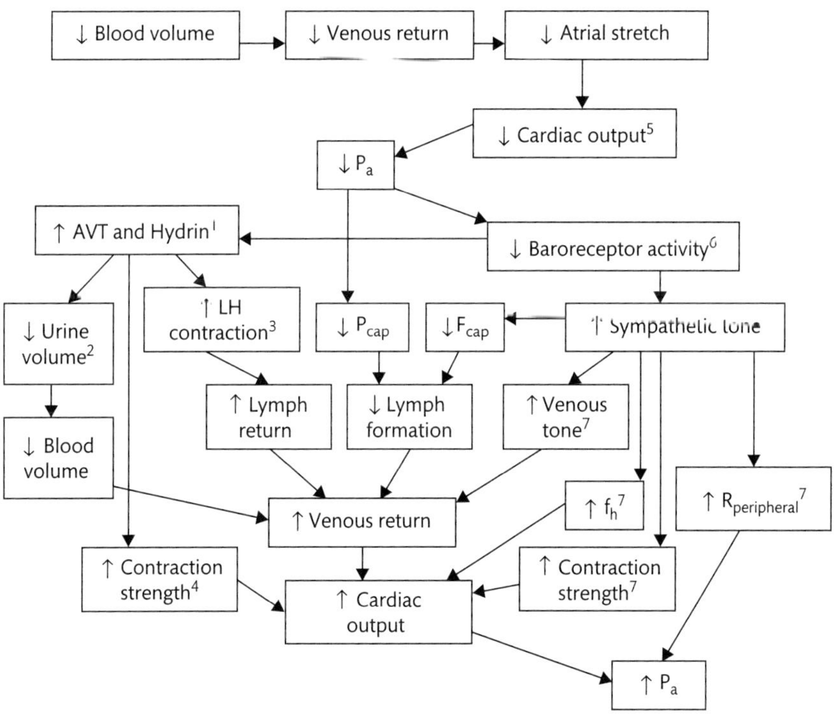

Fig. 2.45 The control of cardiovascular function following a decrease in blood volume: [1]Nouwen and Kuhn 1985, Kono *et al.* 2005; [2]Sawyer 1957, Henderson *et al.* 1972, Gallardo *et al.* 1980, Pang *et al.* 1980; [3]DeGrauw and Hillman 2004; [4]Sham *et al.* 1989; [5]Frank translated by Chapman and Waterman 1960; [6]West and Van Vliet 1992; [7]Morris and Nilsson 1994.

mean diffusion distance and lowers the resistance to gas flux in that region. Since not all capillaries are perfused at rest, as more capillaries are perfused with activity the mean diffusion distance decreases (Krogh 1919).

The major variable modifying tissue diffusion coefficients besides capillary density is the presence of intracellular myoglobin (Wittenberg 1970; Scholander 1960). Myoglobin is a monomeric hemoprotein and essentially represents one of the monomers of the hemoglobin tetramer. It contains a single O_2 binding site and normally demonstrates hyberbolic saturation kinetics rather than the sigmoid cooperative binding kinetics of hemoglobin. Recent evidence suggests that under conditions that mimic increased metabolism, the myoglobin curve shifts right and becomes more sigmoidal, enhancing O_2 exchange to the mitochondrial cytochromes (Seiyama 2006). The presence of myoglobin has two independent effects on the diffusional flux of O_2 in cells. First, it increases the amount of O_2 dissolved in the cytoplasm at any given P_{O_2}. Functionally this slows the rate of change in P_{O_2} as it is used during aerobic metabolism of the mitochondria. This maintains a greater driving force for diffusional O_2 flux if myoglobin is present, acting as a P_{O_2} buffer. The second effect of intracellular myoglobin is that it increases the diffusion coefficient for O_2 in the cytoplasm. Hence for the same P_{O_2} difference, diffusional O_2 flux will be greater with higher myoglobin in the cytoplasm (Kreuzer 1970; Wittenberg 1970). This is known as facilitated O_2 diffusion.

There are some key elements to understand about the driving force for O_2 uptake and CO_2 removal in tissues. At any given time, only a fraction of an organism's capillaries are being perfused with blood. There is not enough blood volume to fill all the vasculature of every organ. This partitioning is, in part, centrally mediated at the level of the arterioles (Tyml and Groom 1980), determining what proportion of the systemic output is channeled to a specific organ. At the level of capillary beds, not only is the pressure from the arteriole important (Tyml 1987) but also the tonus of the capillary muscles is under the control of the local environment of the tissue (Carrier 1965; Becker *et al.* 1970; Gentry and Johnson 1972). Tissues conditions characteristic of insufficient perfusion relative to tissue metabolism (low P_{O_2}, high P_{CO_2}, low pH, elevated K^+ concentration and elevated osmotic concentration) will lead to relaxation of the smooth muscle of the precapillary sphincter. This relaxation will lead to increased blood flow in this region and the return of tissue conditions to perfusion matching demand (high P_{O_2}, low P_{CO_2}, higher pH, lower K^+ and osmotic concentrations) and the eventual increased tonus of the precapillary sphincter. The increase in blood perfusion with muscle contraction results from an increase in red blood cell velocity, an increase in the number of red blood cells per unit capillary length and the density of perfused capillaries (Tyml 1987), and is mediated by both arteriolar and local effects (Tyml and Groom 1980). This leads to a periodicity of blood flow locally and a consequential temporal variation in local P_{O_2} and P_{CO_2}. Examples of this temporal variation in amphibian skeletal muscle are presented in Fig. 2.46 in two ways; first, how it varies with depth through a skeletal

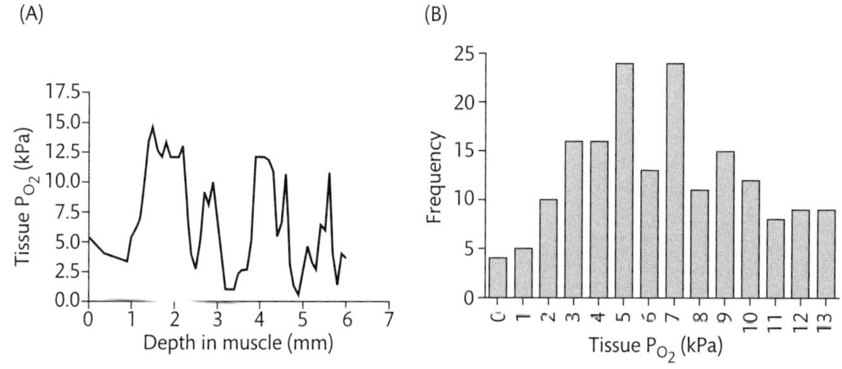

Fig. 2.46 (A) The spatial variation and (B) a histogram of measured tissue P_{O_2} in *Rana* skeletal muscle (from Becker *et al.* 1970). Note the wide array of variation reflecting the characteristic of periodic blood perfusion.

muscle and, second, as a histogram of measured skeletal muscle tissue P_{O_2} (Becker *et al.* 1970). There is tremendous variation around the mean of 6.6 kPa, a consequence of temporal variation of both blood flow and tissue metabolism. The resting rate of aerobic metabolism in this skeletal muscle (2.4 μmol/g h) was not slowed until the tissue P_{O_2} reached 0.8–1.1 kPa (Becker *et al.* 1970). Obviously, the local active hyperemia prevented this from occurring in 99% of the tissue P_{O_2} measurements. The other advantage of knowing the critical tissue P_{O_2} that supports aerobic metabolism is that it allows us to calculate an average minimal tissue resistance, $R_{tiss} = \Delta P_{O_2}/V_{O_2}$ where the ΔP_{O_2} is the difference from tissue to mitochondrion. Amphibian mitochondrial V_{O_2} are generally not limited until the solution falls below 0.01 kPa for state 4 and 0.1 kPa for state 3 (St Pierre *et al.* 2000), hence the ΔP_{O_2} is about 0.8 kPa and the V_{O_2} is 0.7 mol/g s, giving a mean tissue resistance of 1.1 s g kPa/nmol or a conductance of 0.91 nmol/s g kPa for the skeletal muscle at rest. This value is about equivalent to the conductance/g of anuran lungs. It approximates the skeletal muscle conductance of rats (0.5–0.7 nmol/s g kPa) (Henderson *et al.* 2002) and dogs (1.6–2 nmol/s g kPa) (Hepple 2000). The only other available estimate for tissue conductance in amphibians is 9.2 nmol/s g kPa for active animals when more blood is flowing to the skeletal muscle (Withers and Hillman 1988). The tissue conductance for CO_2 can be estimated as the tissue conductance for $O_2 \times K_{CO_2}/K_{O_2}$, or 0.91 (500/14) or 32 nmol/s g kPa.

2.8 Blood Volume Control

A principal variable in the control of stroke volume and its consequential effect on cardiac output is the amount of blood that returns to the heart via the venous

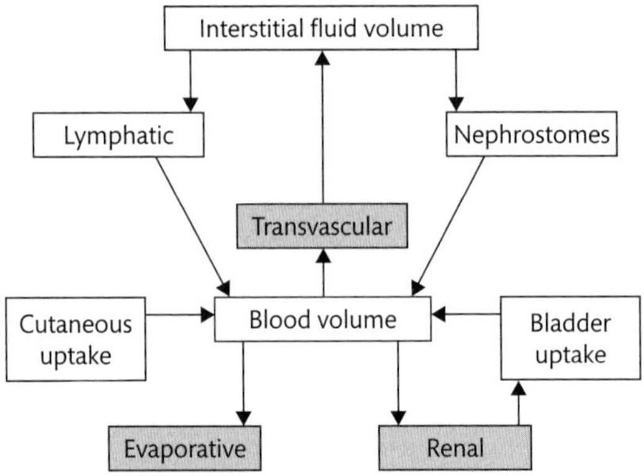

Fig. 2.47 A blood volume budget for amphibians.

system (Frank–Starling law). An important variable in determining the mean circulatory filling pressure and venous pressure is the volume of blood contained within the circulatory system. Let us first delineate the routes of fluid entry and loss in order to develop a blood volume budget for amphibians (Fig. 2.47). As can be seen from this diagram, fluid is lost from the blood via evaporation, renal and transvascular filtration and gained from the capillaries of the skin and bladder as well as carried directly to the venous system by the lymphatic system and nephrostomes. There is a much more varied potential series of exchanges possible in the regulation of blood volume in anurans than simply a transcapillary exchange.

A blood volume budget for a toad in aerial and aquatic environments, during control and maximal flux conditions appropriate to the environment, is shown in Fig. 2.48.

2.8.1 Skin and Bladder Influx

The assumption used in calculating the cutaneous and bladder influx rates was that the water enters the capillaries and not the lymphatic/interstitial space. This assumption holds for skin (Word and Hillman 2005), but has never been tested for bladder exchange. Bufonid cutaneous uptake rates range from 0.5 to 0.6 ml/kg min in hydrated states and reach 2.4–2.7 ml/kg min following salt load or dehydration (Shoemaker 1965; Shoemaker and Waring 1968; Jorgensen *et al.* 1993). Bufonid bladder fluxes range from hydrated fluxes of 0.1 ml/kg min to 0.4–1.1 ml/kg min following hemorrhage or salt load (Middler *et al.* 1968a). Ranid cutaneous uptake rates are characteristically lower, hydrated around 0.1–0.3 ml/kg min to 0.9–1.2 ml/kg min in stimulated states (Parsons *et al.* 1990; Parsons and Mobin 1991). Ranid bladder uptake rates are also lower than those of bufonid, 0.05 to 0.7 ml/kg min for control and stimulated rates (Sinsch 1991).

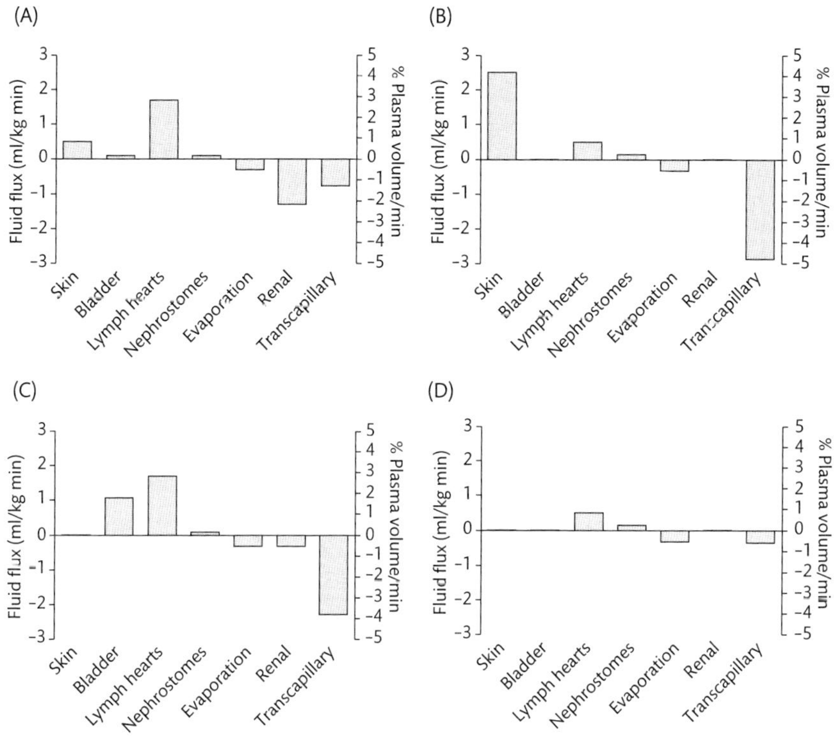

Fig. 2.48 Blood volume budgets for *Bufo marinus* in different environmental conditions: (A) hydrated in water; (B) dehydrated in water; (C) hydrated in air; (D) dehydrated in air.

2.8.2 Nephrostome Influx

The nephrostomes or peritoneal funnels of adult amphibians are ciliated funnels located on the ventral surface of the kidney (see Chapter 3). The funnels open to the peritoneal space and dump directly renal veins or nephron tubules. Rugh (1938) demonstrated rapid uptake (5 min) of ink injected intraperitoneally via this route. Morris (1981) has actually measured the flux of fluid moving through the peritoneal funnels into the renal veins *in vitro* using isolated *B. marinus* kidneys. Her estimates for volumes exchanged range from 0.08 to 0.16 ml/kg min in control and AVT stimulated states.

2.8.3 Renal Loss

The actual volume lost from the blood volume is a composite of the glomerular filtration rate (GFR) minus the volume of that filtrate which is reabsorbed by the kidney tubules (see Section 3.7). The volume that is not reabsorbed and remains in the tubules is either stored in the bladder or lost to the environment with urination. The principal control of volume efflux is glomerular, though tubular components can be significant (Yokota *et al.* 1985; Henderson *et al.* 1972; Shoemaker and Bickler 1979). Renal losses of plasma are essentially zero when water loss exceeds gain (Pang *et al.* 1980; Henderson *et al.* 1972). Renal losses increase with hypervolemia, and the magnitude is dependant upon the osmotic concentration of the expansion fluid (Hillman and Schimpf 1994). Hyperosmolality overrides the effect of expansion in both *Bufo* and *Rana*, while hypo-osmolality enhances the response to hypovolemia (Hillman and Schimpf 1994). The actual fluxes from control and hypo-osmotically expanded states for *Bufo* are 0.3 to 1.3 ml/kg min and for *Rana* are 0.2 to 0.6 ml/kg min.

2.8.4 Lymphatic Fluxes

Lymphatic return accounts for about 90% of the known return of interstitial fluid in anurans. Understanding the control of lymphatic flux is crucial for understanding the control of their blood volume. This is also a conclusion of Baldwin *et al.* (1993) and Malvin *et al.* (1995). What is clear from these blood volume budget data are that the major loss of fluid from the vasculature is transvascular filtration to the body and the major route of influx is lymphatic return via lymph hearts. Consequently, any understanding of the control of blood volume, and its consequential effect on cardiovascular function then pivots on an understanding of the mechanisms of lymph formation and the processes that determine its return to the circulation.

The presence of a circulatory system driven by hydrostatic pressure necessarily entails the potential of loss of fluid via capillary leaks through gaps between the capillary endothelial cells into the interstitial fluid space. The fluid that escapes is known as lymph. Lymph generally is just like plasma but with a lower protein concentration. That means the blood cells and some proteins normally do not pass through the gaps in the capillary walls but the other solutes are permeable. The rate of lymph formation is generally interpreted from the Starling fluid flux equation:

$$\textbf{Capillary Lymph Formation} = F_C\,((P_{cap} - P_{ist}) - \sigma\,(\pi_{cap} - \pi_{ist}))$$

where, F_C = filtration coefficient of the capillaries; P_{cap} = capillary hydrostatic pressure; P_{ist} = interstitial hydrostatic pressure; σ = reflection coefficient for plasma proteins; π_{cap} = colloid osmotic pressure of the plasma; π_{ist} = colloid osmotic pressure of the interstitial fluid.

Normally, the balance of hydrostatic forces favors efflux of fluid from the capillary, while the colloid forces favor influx of fluid. On average, the hydrostatic forces dominate in the circulation and there is a consequential efflux of fluid from the vasculature across the capillaries into the interstitium. If fluid is added to the plasma space, P_{cap} will increase and π_{cap} will decease, both will lead to greater efflux and a return to the original plasma volume. If fluid is lost from the plasma space, P_{cap} will decrease and π_{cap} will increase, both favoring increased absorption of fluid from the interstitium and a return to the original plasma volume. Consequently, there are physical mechanisms inherent in this formulation that tends toward homeostatic redistribution of blood and interstitial volumes independent of a regulated negative feedback loop. The balance of hydrostatic and oncotic osmotic forces favors ultrafiltration in anurans. This conclusion is based on: (1) an in-depth analysis of the balance of hydrostatic and oncotic osmotic forces during dehydration (Hillman *et al.* 1987); (2) plasma is lost and gained with variation in lymph heart function in control states (Baldwin *et al.* 1993); and (3) that after the lymphatic hearts are cauterized death quickly ensues as a consequence of loss of plasma volume and increase in hematocrit (hemoconcentration), even in normal hydrational states (Zwemer and Foglia 1943). Therefore, oncotic osmotic forces do not mobilize interstitial fluids in amphibians, and fluid lost across body capillaries must re-enter the vascular space via the nephrostomes or lymphatic pathways. This is fundamentally different than the mammalian paradigm where transcapillary uptake is a significant route of fluid return during low pressure, hypovolemic conditions.

Capillaries vary in their individual filtration coefficients (arteriolar, true, venular: see Bundgaard and Frokjaer-Jensen 1982), in whether they are being perfused, and in their number and filtration coefficients between different organs (Michel 1984; Taylor and Townsley 1987). The literature is rich in describing the characteristics and control of filtration across individual frog mesenteric capillaries (Meyer and Huxley 1990; Meyer *et al.* 1992). This model has proved invaluable in delineating control of the microvasculature, but is difficult to extrapolate to systemic circulatory filtration in general. The unknowns in such integrations are many; numbers, relative proportion of mass that various organs comprise, surface area, proportion of perfused capillaries and how representative are mesenteric capillaries of an average capillary. The use of a capillary-focused equation to describe blood volume regulation and lymph formation would require an integration of both the blood flow and Starling variables for each and every organ in the body, an impractical approach; hence it has limited ability to interpret redistribution of fluid at an organismal level. A more useful model at an organismal level for understanding the short-term effects of volume loading and hemorrhage on blood volume and lymph formation is a derivation of the Starling fluid flux equation, emphasizing whole-body characteristics and incorporating the interaction of volume flux on pressures. This approach entails incorporating the compliances (capacitances) of the vascular and interstitial spaces as well as delineating a whole-body filtration coefficient. It essentially looks at the kinetics of fluid flux and is described by the following equation (Tanaka 1979):

$$\textbf{Whole Body Lymph Formation} = F_C\,(V_{vas}/C_{vas} - V_{ist}/C_{ist})$$

where, F_C = whole-body filtration coefficient; C_{vas} = compliance of the vasculature (Δ volume/Δ pressure); C_{ist} = compliance of the interstitium (Δ volume/Δ pressure); V_{vas} = volume of fluid in the vasculature; V_{ist} = volume of fluid in the interstitium.

Fluid shifts in this approach can be analyzed without knowing absolute capillary hydrostatic pressure, interstitial fluid pressure, or plasma or interstitial colloid osmotic pressure of the whole body and flux is based on relative changes in pressure related to the compliances of the vasculature and interstitium. In this approach, the hydrostatic pressure and oncotic pressure terms of the Starling balance are embedded into net vascular pressure ($P_{vas} - \pi_{cap} = V_{vas}/C_{vas}$) and interstitial pressure ($P_{ist} - \pi_{ist} = V_{ist}/C_{ist}$). Anuran C_{vas} falls within a fairly standard vertebrate range of 28–47 ml/kg kPa (Hillman *et al.* 2004). Anuran C_{int} is 2000 ml/kg kPa, a value 25 to 50 times greater than reported in any other vertebrate (Hillman *et al.* 2004). Couple the C_{ist} to a high F_C (Hancock *et al.* 2000; Hillman *et al.* 2004) and anuran amphibians should have little counter balance to any fluid added to the blood favoring filtration and a high turnover of plasma. We see this ease of filtration exhibited when pressure increases with exercise and stress (i.e. exercise and stress induced hemoconcentration) and in high rates of plasma turnover (DeGrauw 1998).

2.8.5 Lymphatic System Anatomy and Physiology

Excellent reviews of anuran lymphatic structure and function exist (Toews and Wentzell 1995; Hillman *et al.* 2004, 2005). In anurans, lymph is generally returned to the venous circulation by two pairs of lymph hearts, the anterior pair lateral to the third vertebra under the suprascapular cartilage and the posterior pair located lateral to the urostyle at the nexus of the subvertebral, lateral, iliac and pubic lymph sacs. Some anurans, especially some ranid frogs, have more than one pair of posterior lymph hearts (Kampmeier 1969). The hearts contract rhythmically, driven by spinal motor center stimulation of the lymph heart muscle fibers (Flindt 1966) via cholinergic synapses (Greber and Schipp 1986). The fibers seem to be modified skeletal muscle fibers based on their embryogenesis (Greber and Schipp 1990), polynucleation and absence of intercalated discs (Schipp and Flindt 1968), and the occurrence of satellite like cells (Rumyantsev and Shmantzar 1967).

Lymphatic heart output can be rapidly modulated by both rate and stroke volume changes (see Jones *et al.* 1997). Lymph heart rate is modified by spinal nerve efferent activity (Priestley 1878). A negative feedback control loop between arterial baroreceptors and lymphatic heart rate has been described (Yamane 1990; Crossley and Hillman 1999). Natural variation in arterial pressure was inversely related to lymph heart rate, this relation was lost following denervation of the baroreceptor and stimulation of the baroreceptor afferent nerves shut off the lymph hearts without a change in blood pressure (Crossley and Hillman 1999). Little is known about the effect of physical factors, such as interstitial fluid pressure or venous pressure on lymphatic heart stroke volume, even though stroke flow shows a wide range of variation (Jones *et al.* 1997). Clearly lymph heart stroke volume (like cardiac heart stroke volume) is in part determined by lymph input; the lymph heart cannot pump what it does not have. This limitation is apparent from the following three findings: (1) lymph heart stroke volume declines with dehydrational fluid loss, despite no change in lymph heart systolic pressure (Jones *et al.* 1997); (2) postural shifts in anesthetized animals that favor pooling of lymph around the posterior hearts caused a dramatic rise in stroke volume (Jones *et al.* 1997); and (3) lymph heart output in general has been reported to correlate with the rate of lymph formation (Baldwin *et al.* 1993).

The anuran lymphatic system, outside of the hearts, consists of interconnected subcutaneous sacs separated by connective tissue walls that have one-way valves. The valves appear to be controllable rather than being simply passive flaps (Jolly 1946). The various lymph sacs have been generally described and can vary interspecifically (Carter 1979). The mechanisms involved in moving this lymph are a combination of differential segment compliance creating a tissue compliance pump with lymph formation (Hillman *et al.* 2005), skeletal muscle pumps varying lymph sac compliance and hence lymph sac pressure (Hillman *et al.* 2004; Drewes *et al.* 2007), pulmonary ventilation moving lymph to axillary regions and creating a negative pressure upon expiration (Hedrick *et al.* 2007) and finally lymph hearts composed of an uncertain skeletal muscle type responsible for generating sufficient pressure to move lymph into the venous circulation through a one-way valve. There are interspecific differences in the capacity to mobilize lymph to buffer short-term variation in blood volume due to hemorrhage and dehydration (Hillman *et al.* 1987; Hillman and Withers 1988), giving an interesting and unique comparative model for evaluating the efficiency and capacities of these lymph movements.

Caecilians and urodeles have a very different lymphatic anatomy than anurans. There are no subcutaneous lymphatic sacs, the dermis of the skin is tightly bound to the connective tissue sheaths of the underlying skeletal muscle. Consequently there should be low C_{ist} in these groups compared to anurans, though C_{ist} has never been measured in these groups. There is also a pair of lymph hearts per body segment (Kampmeier 1969), not just anterior and posterior pairs as in anurans. Functionally the difference in C_{ist} would favor storage of lymph in anurans (high C_{ist}) and the development of high interstitial pressures with any capillary filtration (low C_{ist}) in urodeles and caecilians. This would slow plasma turnovers and provide less residual lymph to mobilize during periods of low blood volume stress in urodeles and caecilians. The high C_{ist} and ease of lymph storage favors anurans during dehydration, where the lymph can be used to stabilize plasma volume (see Section 3.3 Dehydration Tolerance).

Lymphatic volume flux ranges from 0.7 to 1.6 ml/kg min in control and post-hemorrhage for *Bufo marinus* (Baustian 1988). Toews and colleagues report fractional outputs from 0.25 to 0.64 ml/kg min using cannulated lymph heart outputs, but direct outputs of 1.7 ml/kg min using implanted Doppler flow probes for *B. marinus* (Jones *et al.* 1997), essentially the same as estimates derived from Baustian's (1988) data. Malvin *et al.* (1995) report lymph heart outputs using cannulated hearts of 16 ml/kg min for *B. woodhousei*, an output that appears out of context with all other data. Ranid values range from 0.7 to 0.9 ml/kg min (Baldwin *et al.* 1993).

Blocking the lymphatic pathway by thermocautery of the lymph hearts leads to death of toads within a couple of days following a shift of fluid from the blood to the interstitial space (Zwemer and Foglia 1943). Middler *et al.* (1968) present conflicting data on the significance of the lymph hearts in mobilizing interstitial fluid following hemorrhage. More recently Baustian (1988) has shown that animals with no bladder reserves and cauterized lymph hearts are unable to replace fluids lost through hemorrhage.

3

Specialized Physiological Characteristics of Amphibia

3.1 Skin and Urinary Bladder as Models for Water and Ion Exchange

In the present section, current models for salt and water transport across amphibian skin and urinary bladder are described. In Chapter 4, these mechanisms will be related to ionic and osmotic balance in species that inhabit terrestrial, aquatic and brackish environments. In Chapter 5, the hypotheses and experimental methods leading to the current models will be discussed from an historical perspective.

3.1.1 Anatomy and transport mechanisms of the skin

Farquhar and Palade (1964) conducted a detailed microscopic study of the frog epidermis (Fig. 3.1A) (see Chapter 4). The outer or cornified layer (stratum corneum) is made of dead cells that, with the exception of some cocoon forming species, do not constitute a permeability barrier for water or salt transport. The outer most living cell layer is the s. granulosum whose cells are coupled by tight junctions (zonulae occludens) which separate the apical and basolateral membranes so that the apical membrane is the limiting barrier for salt and water transport across the epithelium. There is a regular pattern of cell division in the s. germinativum and transition to the s. granulosum which culminates in shedding of the s. corneum. Cells of the s. granulosum, s. spinosum and s. germinativum are termed principal cells. They are structurally connected by desmosomes in highly convoluted membrane contacts and form an electrical syncytium via gap junctions (Rick *et al.* 1978). Mitochondria-rich (MR) cells are flask shaped with the narrow end protruding to the outer surface and the wider end projecting into the s. granulosum. MR cells are interspersed among the principal cells and their apical membranes are attached to adjacent principal cells by tight junctions (Fig. 3.1B). MR cells are not coupled electrically with the principal cells. In toad skin they constitute less than 0.2% of the epidermal volume and their apical membranes make up less than 1% of the total skin area. The s. corneum is not an appreciable

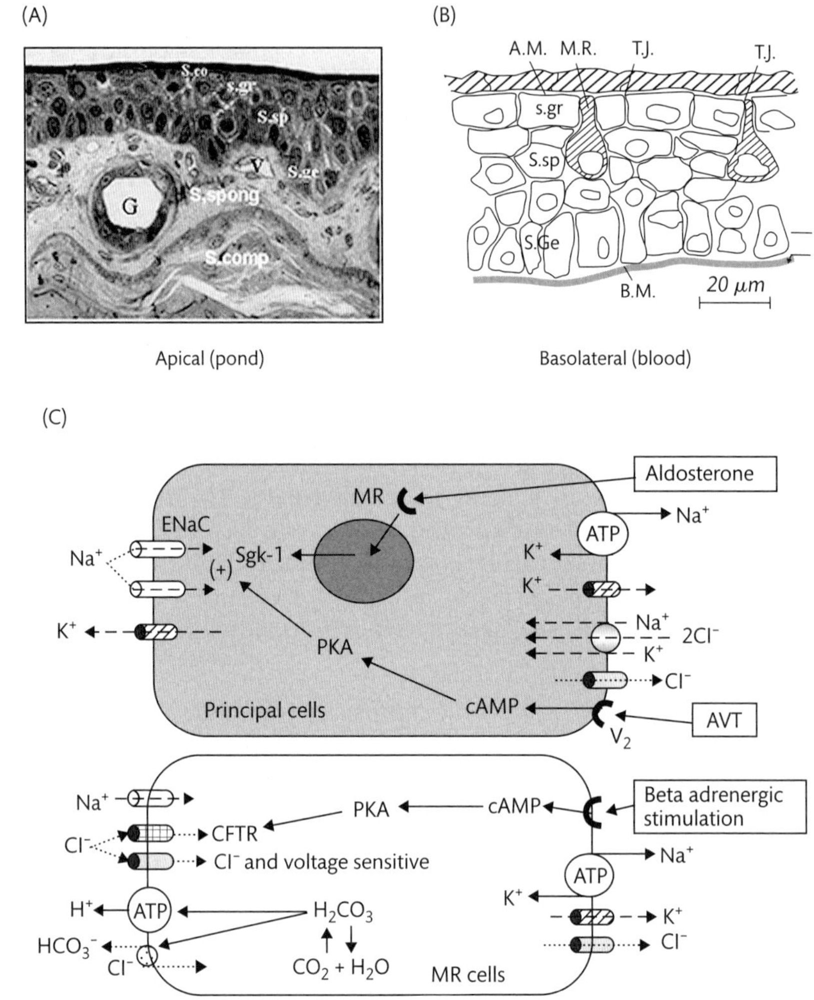

Fig. 3.1 (A) Cross section of frog skin showing epithelial cell layers that include principal cells: stratum corneum (SC), stratum granulosum (SG), stratum spinosum (SS) and stratum germinativum (SGe) and mitochondria-rich cells (M). The underlying dermal layers include the stratum spongiousum (SSp) that contains blood vessels and skin glands and the s. compactum (SCo) that contains a network of collagen fibers (from Farquhar and Palade, 1964). (B) Drawing of epidermis showing principal versus MR cells (from Larsen, 1991). (C) Transport pathways across the functional syncytium of principal cells and MR cells that are not coupled to the principal cells.

barrier except in a few species that accumulate layers of s. corneum when burrowed underground (see Chapter 4).

Models for Na^+ and Cl^- transport across the skin are based on studies of Ussing and Zerahn (1951) in which isolated skin is placed as a diaphragm between two chambers that contain an identical Ringer's solution. Ion transport can be measured with isotopes or electrically as the short-circuit current that is required to maintain a zero potential gradient across the skin. The technique is described in detail in Chapter 5. Figure 3.1C (adapted from Larsen 1991) provides a schematic drawing that depicts the principal cell syncytium as a single compartment and describes transport pathways in the apical and basolateral membrane of both cell types. Na^+ entry across the apical membrane of the s. granulosum is regulated by epithelial Na^+ channels (ENaCs) which are highly selective for this ion (Kellenberger and Schild 2002). ENaCs are reversibly blocked by the diuretic, amiloride, while basolateral extrusion is mediated by Na^+–K^+ ATPase that is inhibited by the cardiac glycoside, ouabain. Mills *et al.* (1977) showed cells in the s. spinosum and s. germanitivum to be non-polarized, with the Na^+–K^+ pump located in membranes that face the extracellular spaces of the epidermis and collectively form the basolateral membrane of the principal cell syncytium. The basolateral membrane is permeable to K^+ which leaks out to establish a negative potential inside the cells; this has been measured to be as great as −100 mV (Willumsen and Larsen 1986) and is near the equilibrium potential predicted by the concentration difference for this ion across the basolateral membrane. The basolateral membrane also has the Na^+–K^+–$2Cl^-$ cotransport protein (NKCC) that is proposed to be activated when cells shrink during exposure to hypertonic media and, by drawing electrolytes into the cell, mediates a regulatory volume increase (Larsen 1991). A K^+-selective channel has been described in the apical membrane in the skin of frogs exposed to KCl in their water source (Zeiske and Van Driessche 1980). This is consistent with observations by Frazier and Vanatta (1981) that intact frogs loaded with KCl will secrete K^+ across their skin.

MR cells have a similar distribution of ENaCs and the Na^+–K^+ ATPase in the apical and basolateral membranes, respectively. In addition, MR cells are proposed to be the primary pathway for Cl^- transport (Larsen 1991). Several mechanisms have been demonstrated for Cl^- transport. Absorption from very dilute solutions (c. 10^{-5}–10^{-4}M) requires that the apical membrane be hyperpolarized by H^+ ATPase (V-ATPase), located only in the apical membrane of the MR cells (Klein *et al.* 1997). This drives Cl^- entry by way of a Cl^-–HCO_3^- anion exchange protein (AE) also in the apical membrane. Cellular Cl^- diffuses out of the cell via basolateral Cl^- channels. In dilute solutions, the H^+ ATPase will also reverse the transepithelial potential to draw Na^+ across the skin via either the MR or principal cells. With NaCl concentrations comparable to most freshwater environments (c. 0.1–2–3 mmol/l) the H^+ ATPase contributes to Cl^- uptake. However, the apical membrane potential of principal cells, established by the Na^+–K^+ pump and basolateral K^+ efflux, is sufficient for Na^+ absorption. Finally, with NaCl concentrations above c. 3 mmol/l, Cl^- uptake is passive, that is positive charge mediated by active Na^+ transport via the Na^+–K^+ pump fuels transport of both ions. Passive Cl^- flux is mediated by apical and basolateral channels in MR cells and also by paracellular transport through tight junctions whose permeability increases with increasing osmolality of the apical bathing solution (Ussing and Windhager 1964). Recently, Willumsen *et al.* (2002) demonstrated the presence of the cystic fibrosis transport regulator protein (CFTR) in MR cells as an alternative pathway for passive Cl^- entry across the apical membrane. This pathway is activated by β adrenergic stimulation. The function of the voltage-gated pathway versus CFTR remains to be determined.

MR cells have also been associated with acid–base balance. Living frogs cutaneously secrete H^+ when acidotic and $HCO3^-$ when alkalotic (Vanatta and Frazier, 1981). During acidosis the H^+ pump acidifies the external bathing solution while alkalosis stimulates net $HCO3^-$ secretion. These functions have established models that require differential expression of H^+ ATPase and AE proteins in the apical and basolateral membranes of the MR cells. Acid secreting cells with apical expression of these proteins are termed a or α cells while bicarbonate secreting cells, termed b or β cells express them in the apical membrane. The model presented in figure 3.1 C has apical expression of these transport proteins in adddition to apical and basolateral channels that permit passive (rheogenic) Cl^- transport driven by active Na^+ update. Larsen (1991) has termed cells with this capacity to be γ type MR cells.

3.1.2 Water Absorption

The model for salt transport (Fig. 3.1) is generally applicable to water absorption. The apical membrane is the limiting barrier, and the basolateral membrane is formed by intracellular facing membranes of all of the living cell layers. Because they constitute a small fraction of the total skin area, MR cells are not believed to be a significant factor in water absorption. Among terrestrial and arboreal anurans, a specialized area of skin in the posteroventral region, termed the seat patch or pelvic patch, is more permeable, highly vascularized and accounts for most water absorption across the skin (Christensen 1974; Bentley and Main 1972; McClanahan and Baldwin 1969; Marrero and Hillyard 1985: Fig. 3.2). Permeability of the seat patch can be greatly increased by the amphibian antidiuretic hormone, arginine vasotocin (AVT), a phenomenon termed the hydro-osmotic response (Heller 1974), while the skin of aquatic species such as *Xenopus laevis* lack specialized regions, have lower rates of water absorption and a minimal or absent hydro-osmotic response.

The discovery of aquaporins as a family of water conducting channels (Preston and Agre 1991) has provided a mechanism for the rapid water movement across membranes of plant and animal tissues (see Chapter 5). For absorbing epithelia, such as the amphibian skin, water must first enter the cells across the apical membrane and then leave across the basolateral membrane and be taken up by capillaries that underlie the epidermis. The regulated step for water absorption is mediated by aquaporins, residing in subapical vesicles that become inserted into the apical membrane when the tissue is stimulated by AVT. Water then diffuses out of the principal cell syncytium via a constitutively expressed aquaporin

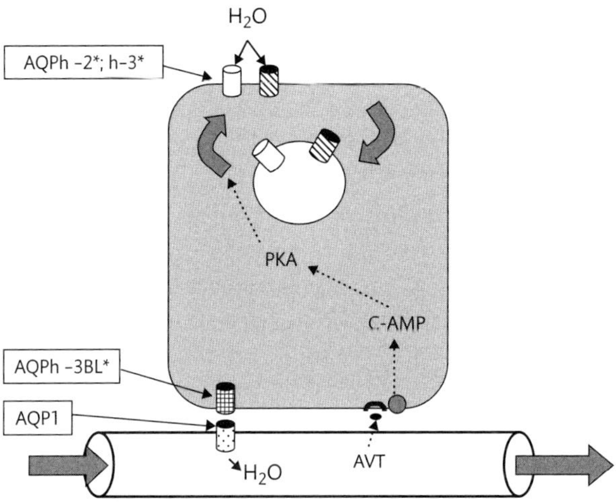

Fig. 3.2 The pathway for water absorption across the seat patch skin. Apical entry is regulated by arginine vasotocin (AVT) binding to V2 receptors and stimulating the formation of cyclic AMP and protein kinase A (PKA) that stimulates the insertion of vesicles containing aquaporins h2 and h3. Water flow across the basolateral membrane is mediated by aquaporin h3BL and continues into cutaneous capillaries via aquaporin h-1.

isoform in the basolateral membranes and taken up by the circulation (Fig. 3.2). The classification of AVT-stimulated versus constitutive aquaporins in amphibians is complicated because of different notations used by different laboratories studying different species and is largely based on mammalian isoforms that have been identified by cloning and sequencing of cDNAs. A complete phylogeny of amphibian aquaporins has been compiled by Suzuki (2007). Table 3.1 summarizes relevant amphibian isoforms and their distribution relative to body fluid homeostasis.

Comparison of AQP distribution in the skin of terrestrial, scansorial, semi-aquatic and aquatic species has shown AQP-h3 and homologous isoforms to be uniformly present in the pelvic skin of all three groups (Table 3.1). Isoforms homologous to AQP-h2 are found in the bladder of all four groups but only in the pelvic skin of terrestrial and scansorial species that have a specialized seat patch region in the pelvic skin. Both AQP-h2 and AQP-h3 in the skin of *Hyla japonica* translocate from vesicles in the cytosol to the apical membrane of granular cells following treatment with AVT (Tanii *et al.* 2002; Hasegawa *et al.* 2003). Thus, the greater permeability of the seat patch is related to expression of a urinary bladder-like AQP-h2 in addition to the more widely distributed ventral-like AQP-h3. Water flow continues across the basolateral membrane via aquaporins homologous to AQP-h3BL and into the circulation via AQP-1 that Willumsen *et al.* (2007) have demonstrated in the endothelial cells of capillaries in the seat patch.

Table 3.1 Classification of aquaporins in skin, kidney and bladder

Aquaporin Types	Mammalian Isoforms*	Amphibian Isoforms	Location and AVT sensitivity
Type 1	AQP1 Mouse, Human AQP1	*Xenopus tropicalis* (AQP1) *Bufo marinus* (AQP-t1) *Rana esculenta* (FA-CHIP) *Hyla japonica* (AQP-h1) *Hyla chrysoscelis* (HC-1)	Constitutively expressed in capillary endothelium of skin kidney and bladder
Type a2	AQPa2 Anuran specific*	*Hyla japonica* (AQP-h2) *Bufo marinus* (AQP-t2) *Hyla japonica* (AQP-h3) *Bufo marinus* (AQP-t3)	AVT stimulates insertion of AQP–t2 and AQP–h2 into apical membrane of the urinary bladder. In the skin AVT stimulates insertion of both the –h2, -t2 and –h3, -t3 isoforms into the apical membrane
Type 2	AQP2 Mouse, Human AQP2	*Hyla japonica* (AQP-h2K) *Hyla chrysoscelis* (HC-2)	Primarily in the kidney, homologous to ADH-stimulated isoforms in mammalian collecting duct
Type 3	AQP3 Mouse, Human AQP3	*Hyla japonica* (AQP-h3BL) *Hyla chrysoscelis* (HC-3)	Found in basolateral membranes of skin kidney and bladder

Amphibian and mammalian AQP-1 isoforms are constitutively expressed in a number of tissues, including capillary endothelium of the skin. Note that AVT stimulated AQPs in the skin and bladder are in an anuran-specific group (Type a2) and that AQPs -h3 and -t3 are classified in this group. Amphibian homologues to mammalian AQP2 isoforms are located primarily in the kidney. Basolateral AQPs in amphibians are homologous to mammalian AQP-3 and expressed in skin, kidney and bladder. Note that the designation for *H. japonica* specifies AQP-h3BL as the basolateral isoform as opposed to AQP-h3 in the apical membrane.

Circulation–Perfusion versus Diffusion

Once water has been absorbed across the epithelium of a dehydrated amphibian, it must be taken into the plasma and lymphatic fluid to restore osmotic balance. Blood vessels in the s. spongiosum are closely associated with the basal cell layer of the epidermis and form a dense network, especially in the seat patch region of terrestrial anurans. Overton (1904) observed an increase in circulation in the skin when European treefrogs, *Hyla arborea*, rehydrated. Christensen (1974a) found the elevation of water uptake across the isolated seat patch region of *Bufo bufo* by AVT was only observed when the vasculature was perfused with a Ringer solution. In living frogs, water uptake across the seat patch region of dehydrated *Rana catesbeiana* was reduced when the heart was stopped (Parsons and Mobin 1991; Parsons and Schwartz 1991). In both of these studies the tissue concentration of Na^+ was depleted in the absence of perfusion and it was proposed that cutaneous circulation is required to maintain the osmotic gradient across the skin as water is absorbed. Other studies have shown varying degrees of dependence upon cutaneous circulation for sustaining elevated water uptake by dehydrated animals or in isolated skin. The rate of water uptake across the isolated seat patch region of *Bufo punctatus* that was not perfused averaged 208 mg/cm^2 h compared to 423 mg/cm^2 min for the seat patch of living toads with an intact circulation, that is perfusion accounted for about half of the AVT-stimulated rate for water absorption.

Viborg and Rosenkilde (2004) developed a technique for concurrently measuring water absorption and blood flow in the seat patch of conscious, unrestrained *Bufo bufo*. Toads were placed over a chamber that contained water and a probe for a laser – Doppler flowcytometer that gave a voltage output, termed blood cell flux (BCF), calculated from the density and velocity of red cells in the skin. It was found that dehydration alone has little effect on BCF and that water potential receptors in the skin appear to initiate a reflex stimulation of BCF (Fig. 3.3A, B). Water absorption and BCF were significantly elevated when dehydrated toads were placed on water and both declined as rehydration occurred over a 120-minute period (Fig. 3.3C). Isoproterenol produced a similar increase in water uptake and BCF by hydrated toads suggesting the reflex is mediated by the sympathetic nervous system. In contrast, AVT stimulated water absorption by hydrated toads without an increase in BCF, although the increase was smaller and shorter lived than that seen in dehydrated toads (Fig. 3.3D). Viborg and Hillyard (2005) and Viborg *et al.* (2006) have subsequently used this technique to measure the initial rate of water absorption and BCF in other toads, *Bufo woodhousei, B. punctatus, B. marinus and B. alvarius.* While both parameters increase, they are not linearly correlated as would be expected if blood flow was the limiting factor. The authors suggested that dehydrated toads, upon detecting available water, constitutively increase BCF to maintain an osmotic gradient (perfusion) and that water permeability, mediated by AVT activation of aquaporins, stimulates water transport across the epithelium (diffusion). The integration of these processes and the animals' behavior is an important area for future study.

Behavioral Considerations

Townson (1795) observed dehydrated frogs to press their ventral skin to a moistened surface "...until they looked plump..". Many years later, Stille (1958) described this same behavior in seven anuran species and termed it the water absorption response (WR), which has more recently been used to study the sensory properties of the skin and hormonal regulation of the behavior (reviewed by Hillyard *et al.* 1998). This behavior is most pronounced in terrestrial and scansorial species with a specialized seat patch (Fig. 3.3A) and allows animals to obtain water from moist surfaces where standing water is absent.

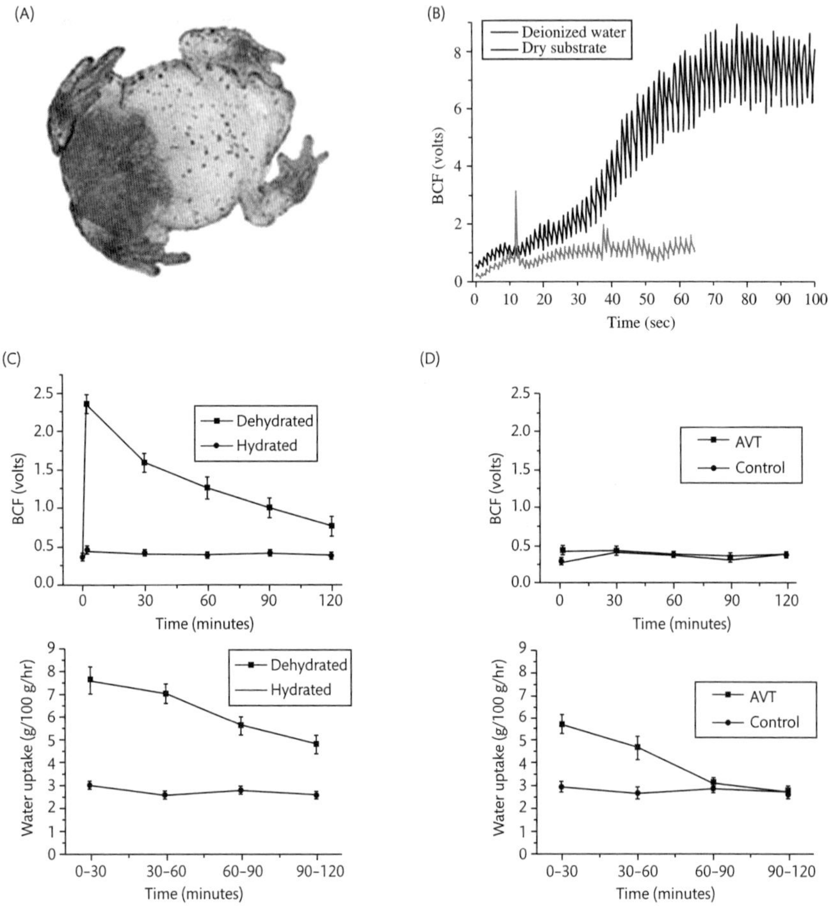

Fig. 3.3 (A) *Bufo punctatus* displaying the water absorption response with extensive capillary perfusion of the seat patch region. (B) Output of a Laser-Doppler flowmeter recording from the seat patch of a dehydrated *B. punctatus*. Toads were placed on a wet or dry surface in a rehydration chamber that contained the flow probe. Blood cell flow (BCF) is a function of the number and velocity of red blood cells detected by the probe. (C) BCF and water absorption were measured concurrently as dehydrated *B. bufo* rehydrated. Both were significantly greater than observed for hydrated toads given the same rehydration conditions. (D) Arginine vasotocin (AVT) injection increased water absorption without a change in BCF.

3.1.3 Larval Skin

Most experiments on transport mechanisms in the skin of larval anurans have been done with *Rana catesbeiana* because of their large size. Anatomically, the larval skin consists of two cell layers (Weed 1933): an outer or apical cell layer, which forms the high resistance barrier, and a basal cell layer, which is not electrically coupled to the apical cells. The skin is interspersed with skein cells that contain bundles of intermediate filaments. The larval frog skin contributes relatively little to total ion and water exchange, most being across the gills (Alvarado and Moody 1970). With regard to the low water permeability, Willumsen *et al.* (2003) were unable to obtain RNA that indicated expression of aquaporins in the ventral skin. Early studies in Ussing chamber preparations revealed the presence of an apical channel that is poorly selective for the alkali metal cations (Hillyard *et al.* 1982). This channel can be transiently stimulated by amiloride, ATP and acetylcholine. Takada *et al.* (2006) have recently shown that all three subunits of the Na^+ selective ENaC are expressed in the larval skin and have the same amino acid sequence as in the adult skin. However, they remain inactive. The relationship between the larval and adult channels remains obscure. Factors regulating the development of the adult pattern of transport are described in Section 3.8.

The apical cell layer is retained during larval stages but during metamorphosis it is shed and replaced by principal cells that are derived from the basal cell layer (Robinson and Heinzelman 1987). Hillyard *et al.* (1982) showed that the increase in active Na^+ transport during the climax stages of metamorphosis corresponded with an increase in functional ENaC expression as the adult pattern of epithelial cells developed. There is also a shift in the distribution of Na^+–K^+ ATPase during metamorphosis that can be demonstrated by the binding of radiolabeled ouabain, an inhibitor of the enzyme (Robinson and Mills 1988a, b). Prior to metamorphosis the enzyme is located primarily in the outer third of the skin, corresponding to the basolateral membrane of the apical cells. During metamorphosis, the enzyme becomes located in the inner two-thirds of the skin, corresponding to the basolateral membranes of the spinous and germanitive cell layers. Robinson and Heinzelman (1987) did not observe MR cells in the bullfrog skin prior to metamorphosis although Fox (1983) reported they have been recognized in the skin of larval *Rana temporaria*.

3.1.4 Urodele and Apodan Skin

Skin of aquatic urodeles shows a wide range of water permeability and Na^+ transport that can be related to phylogeny and habitat. *Necturus* (Proteiidae) is a neotenic, aquatic urodele that has low water permeability and no measurable, active Na^+ transport (Bentley and Yorio 1977). Like the larval frog the gills appear to be the primary site of salt and water exchange. Other aquatic urodeles, *Siren lacertina* (Sirenidae) and *Amphiuma means* (Amphiumidae), have amiloride-sensitive Na^+ transport across the skin but no AVT-stimulated increase in water permeability (Bentley 1975). Larval *Ambystoma tigrinum* (Ambystomatidae) are entirely aquatic while the adults may be found in highly terrestrial environments (Delson and Whitford 1973). The larvae have gills and the skin transports Na^+ at a rate one-tenth that of the adult. AVT stimulates Na^+ transport in the adult but not the larvae. Water permeability of the skin is comparably low in both adult and larval skin and not stimulated by AVT in either (Bentley and Baldwin 1980). However, Spight (1967) did observe *Ambystoma opacum* that were dehydrated by 20% to rehydrate at a greater rate than predicted from the osmotic gradient and suggested other mechanisms for increased water permeability than AVT might be involved.

As noted earlier, newts (Salamandridae) initially metamorphose into a terrestrial eft stage and return to the water in an aquatic phase but AVT-stimulation of cutaneous water permeability is very large in both phases and comparable to that of terrestrial anurans. Similar results have been reported for *Notophthalmus viridescens* (Brown and Brown 1977), *Taricha torosa* (Brown and Brown 1980) and *Triturus vittatus* (Warburg and Goldenberg 1978). In contrast, water permeability of *Salamandra maculosa* skin is not stimulated by AVT. Rather, water retention is the result of AVT-stimulation of water reabsorption across the urinary bladder (Bentley and Heller 1965). There is a similar lack of AVT-stimulated water permeability across the skin of *Triturus alpestris* and *T. cristatus* (Bentley and Heller 1964). Unlike *S. salamadra*, water retention is due to a reduction in urine formation rather than reabsorption from the urinary bladder. Both species do show AVT-stimulation of Na^+ transport across the skin as does the skin of *Taricha granulosa* (Brown *et al.* 1988), indicating this to be a common feature of the Salamandridae.

Lungless salamanders (Plethodontidae) have a slender body and high surface area for water absorption. In fact, Spight (1967) found dehydrated *Plethodon jordani* to rehydrate at a rate similar to a value reported by Cohen (1952) for *Aneides lugubris*, at that time the greatest rehydration rate reported in the literature. Hillman (1974) later showed the rate of water absorption by *A. lugubris* was stimulated by AVT, as was Na^+ transport across the skin.

3.1.5 Anatomy and transport mechanisms in the urinary bladder

In contrast to mammals that conserve water by forming a small volume of concentrated urine, frogs (*Rana pipiens*) cease to form urine when water is deprived (Adolph 1927). When water is available the amphibian kidney forms a large volume of dilute urine that is stored in a large urinary bladder, which is a diverticulum of the cloaca. Adolph (1927) also showed that the amount of water stored in the urinary bladder corresponds to the weight gain when frogs are immersed in water, that is urine formation corresponds closely with cutaneous absorption. Steen (1929) repeated these experiments and further showed that water loss during dehydration corresponds with depletion of stored urine.

The use of the urinary bladder as a water storage organ is seen in all three amphibian orders and is most important in fossorial and scansorial species. A possible exception is the aquatic apodan, *Typhlonectes compressicauda*, which is reported to have a large urinary bladder that Wake (1970) suggested "... may be used to store larger quantities of urine than required by non-aquatic caecilians." Values reported in the literature for anurans and urodeles (Table 3.2) are often averages obtained from animals in the laboratory with water available, *ad lib* (Bentley 1966; Ruibal 1962; Shoemaker 1964; Shoemaker *et al.* 1969; Hillman 1974; Brown and Brown 1977; Stiffler *et al.* 1990). In other cases, the largest value recorded is taken as a representative maximal value. For example, Ruibal (1962) found *Bufo cognatus* to store 19–31% of their hydrated weight as bladder water although Shoemaker *et al.* (1969) report excavating a burrowed *B. cognatus* with bladder water storage in excess (103%) of the hydrated weight. van Beurden (1984) similarly found water storage by *Cyclorana platycephalus* to be 130% of the hydrated weight. As we suggested in Chapter 2, measurement of bladder compliance may be a more objective measure for interspecific comparisons.

Because of their large size, the urinary bladders of bullfrogs (*Rana catesbeiana*) and the marine toad (*Bufo marinus*) became popular tissues for experimentation and subjects of detailed electron microscopic examination (*R. catesbeiana*, Strum and Danon 1974; *B. marinus*, Peachy and Rasmussen 1962). In both species,

Table 3.2 Representative bladder capacities of amphibians is greatest in fossorial and arboreal species

Species	% Body weight	
Anura		
Cyclorana platycephalus	57 (130)	Arid, fossorial, Australia
Neobatrachus wilsmorei	50	Arid, fossorial , Australia
Notaden nicholsi	50	Semi arid, fossorial , Australia
Bufo cognatus	31 (103)	Arid-semi arid, fossorial, N. America
Bufo marinus	25	Tropical, terrestrial, Central America
Scaphiopus couchi	33	Arid, fossorial, N. America
Hyla moorei	30	Wet, arboreal, Australia
Xenopus laevis	1	Aquatic, Africa
Urodela		
Aneides lugubris	50	Temperate, arboreal, N. america
Salamandra maculosa	34	Temperate, terrestrial, S. Europe
Ambystoma tigrinum	20	Temperate, fossorial, N. America
Notopthalmus viridescens	15–20	Temperate, terrestrial (eft), N. America
Triturus cristatus	2	Temperate, mostly aquatic, Europe
Necturus maculosus	5	Temperate, aquatic, N. America

Values in parentheses indicate extremely high bladder water reserves recorded from animals in the field.

the most abundant cells are squamous epithelial cells, called granular cells, that form the limiting layer for transport of salts and water. Less abundant are mitochondrial-rich cells. Peachy and Rasmussen (1962) observed mitochondria-rich cells and noted they were not abundant in *B. marinus* bladder although Scott *et al.* (1974) showed them to consist of 15% of the epithelial cells (Fig. 3.4). In the bullfrog, MR cells were observed to be about 25% of the epithelial cells. As with the skin, it is likely that the number of MR cells varies with the ionic demands of the animal.

The effects of dehydration on urine reabsorption *in vivo* could be duplicated by posterior pituitary extracts (Ewer 1952) and Bentley (1958) showed a similar effect with isolated bladders. As noted for the skin (Fig. 3.2), the urinary bladder contains an AVT-stimulated aquaporin in the apical membrane, a basolateral–specific form in that membrane and aquaporin 1 in the associated blood capillaries. The reabsorption of bladder water during dehydration is controlled by regulation

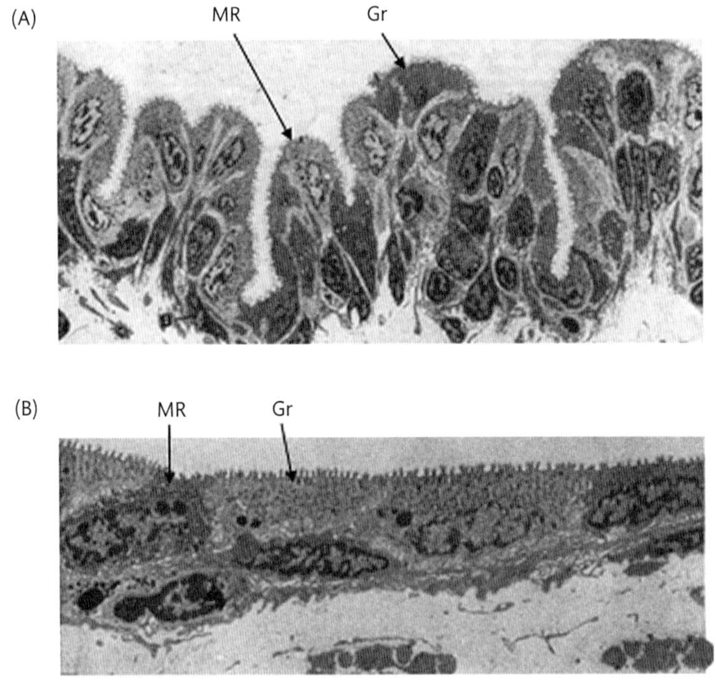

Fig. 3.4 (A) Folds in bullfrog (*Rana catesbeiana*) urinary bladder with granular (GR) and mitochondria-rich (MR) cells forming the interface with the bladder lumen. Underneath these cell layers are basal cells. (B) Distended bladder to illustrate the high compliance of the tissue.

of the AQP-2 type isoform in the apical membrane as is the case for the skin. The hydro-osmotic effect of AVT on the isolated bladder is much greater than that on the isolated skin, probably because of the large surface area in the distended state and a high density of AVT-stimulated aquaporins (Suzuki *et al.* 2007).

More recently, an AVT stimulated aquaporin 2 class protein has been identified in the collecting ducts of kidneys in *Bufo marinus* (Uchiyama and Konno 2006). It is known that dehydration and AVT reduce glomerular filtration so urine formation is reduced (Schmidt-Nielsen and Forster 1954; Shoemaker and Bickler 1979). Presumably, the increase in water reabsorption would allow urine that is formed during dehydration to equilibrate with the body fluids and, during periods of rehydration and increased urine formation, would allow water to be reabsorbed directly into the renal portal circulation (Section 3.7). As with the skin and urinary bladder aquaporin 1 is localized in the peritubular capillaries of the kidney.

At about the same time the bladder was becoming a model tissue for studying water transport, it was discovered that the isolated toad bladder transported Na^+ from the lumen to the serosa via an amiloride-sensitive pathway and Na^+ transport was stimulated by antidiuretic hormone (Sawyer and Schisgall 1956; Leaf *et al.* 1958). In living toads, Middler *et al.* (1968) demonstrated that Na^+ transport in the bladder mediates reabsorption from the urine in a manner analogous to the cortical collecting duct of mammals. Like the skin (Fig. 3.1C), the urinary bladder became a model for the study of ion and water transport mechanisms across the mammalian collecting duct and produced a large literature that can be applied to its role in salt and water metabolism in amphibians.

3.1.6 Regulation of salt and water transport

The amphibian skin and urinary bladder have been used as model tissues to study the effects of several hormones on salt and water transport by the mammalian kidney. Historically, the effects of the hormones are often obtained at pharmacologic doses so extrapolation to the whole animal in an environmental context must be done with caution. Given this caveat, plasma levels of some hormones can be related to their physiological effects, in the context of the natural history of particular species as can agonists and antagonists of the sympathetic nervous system.

Antidiuretic Hormone and the Water Balance Response

Brunn (1921) showed injection of extracts from the posterior lobe of mammalian pituitary glands stimulated the rate of water absorption by living frogs. This became known as the "Brunn" or water balance effect (Heller 1974). Pituitary extracts were also shown to stimulate water permeability and active Na^+ transport across isolated frog skin (Sawyer 1957; Ussing and Furhman 1951). The stimulation of water permeability by pituitary extracts is termed the "hydro-osmotic response" while the stimulation of Na^+ transport is the "natriferic response" (Heller 1974). The posterior pituitary hormone that mediates the hydro-osmotic and natriferic responses was shown to be arginine vasotocin (AVT = arginine 8 oxytocin: Pickering and Heller 1959; Sawyer *et al.* 1959; Katsoyannis and Du Vigneud 1959). As noted above, the hydro-osmotic effect of AVT is mediated by the insertion of vesicles containing aquaporins in the apical membrane of granular cells of the skin and urinary bladder and principal cells of the renal collecting ducts, in addition to constriction of smooth muscle regulating glomerular filtration.

The combined effect of AVT on these three tissues has been termed the antidiuretic response or the water balance response (Heller 1974) and is most prominent in terrestrial species that routinely venture away from water (Bentley 1966). As described above for the individual tissues, the response conserves fluid volume by reducing urine formation, increasing reabsorption of bladder water and increasing cutaneous water permeability for more rapid rehydration when the animals return to water. This might appear to be disadvantageous because of the potential for increased evaporative water loss while away from water. It will be seen in Chapter 5 that diffusional water loss is relatively insensitive to the cross sectional area of aquaporins while osmotic water gain when water is available is very sensitive to the radius of the conducting pore in the aquaporin water channel. The water balance response is initiated by dehydration that will produce an increase in plasma osmolality along with a decrease in plasma volume. Experimentally, it has been shown that injection of hyperosmotic NaCl solutions with no decrease in plasma volume or depletion of plasma volume by hemorrhage, with no change in osmolality, will both evoke the water balance response in *B. marinus* (Shoemaker 1965).

Elevated AVT levels have been measured in anuran species subject to either hyperosmotic or hypovolemic stimuli. Nouwen and Kuhn (1983, 1985) found that dehydration and iso-osmotic hypovolemia due to blood loss resulted in elevated plasma levels of AVT in frogs (*Rana ridibunda*) while Konno *et al.* (2005) found elevated levels in toads (*B. marinus*) subject to dehydration or immersion in hyperosmotic NaCl solutions. In the latter study, a control group of toads maintained with water *ad lib* had plasma osmotic concentration and reduced AVT levels that were similar to those of toads maintained in tap water which supports the concept of "anticipatory drinking" proposed by Jorgensen (1994), that is with water available *ad lib*, toads will retain a hydrated state in the absence of elevated AVT levels.

The natriferic effect of AVT in frog (*Rana pipiens*) and toad skin (*Bufo woodhouseii*) results from an increase in the number of ENaCs per unit of membrane area (Baker and Hillyard 1992; Helman *et al.* 1983). Baker and Hillyard (1985) observed a fivefold increase in amiloride-sensitive Isc (I_{Na}) following AVT stimulation of isolated toad skin, which correlated significantly with the density of active Na^+ channels. Apical membrane area, evaluated from the change in capacitance, increased significantly by a factor of about 18%. These results might be explained by the insertion of vesicles with a high density of channels (see review by Garty and Palmer 1997), a combination of insertion and activation.

The natriferic and hydro-osmotic effects of AVT can be duplicated by the addition of cyclic adenosine monophosphate (cAMP) and the tissue level of cAMP is elevated following AVT treatment of the isolated skin (Johnsen and Nielsen 1984; Marrero and Hillyard 1985). The current model is that AVT binds to a Gs-protein coupled receptor (V2) to stimulate adenylate cyclase. The resulting increase in cAMP activates protein kinase A, which phosphorylates putative proteins associated with the activation of ENaCs in the apical membrane (Fig. 3.1C). It remains to be seen if the phosphorylated protein is an ENaC subunit resident in the membrane or if phosphorylation of cytoskeletal proteins promote insertion or inhibit withdrawal of vesicles containing ENaCs, as is seen with aquaporins. In contrast, the reduction of glomerular filtration rate (GFR) during dehydration is mediated by AVT binding to a Gq-protein coupled receptor (V1) that activates phospholipase C to form diacylglycerol (DAG) and inositol triphosphate (IP3) (Ammar *et al.* 1994).

Hydrins

A second group of neurohypopyseal peptides, termed hydrins, have been isolated from anurans (Acher *et al.* 1997) but appear absent in urodeles (Chauvet *et al.* 1991). Two hydrins, 1 and 2, are produced by less complete processing of provasotocin:

AVT: Cys-Tyr-Ile-Gln-Asn-Cys-Pro-Arg-Gly
Hydrin 1: Cys-Tyr-Ile-Gln-Asn-Cys-Pro-Arg-Gly-Gly
Hydrin 2: Cys-Tyr-Ile-Gln-Asn-Cys-Pro-Arg-Gly-Gly-Lys-Arg

Hydrin 1 in addition to hydrin 2 has been identified in *Xenopus laevis*. Hydrins differ from AVT in that they stimulate water permeability of the skin and bladder with little renal effect. Acher *et al.* (1997) propose that at physiological concentrations the primary effect of AVT is antidiuresis while hydrins have a higher affinity for V2 receptors in the skin and bladder to stimulate rehydration and utilization of bladder water. Rehydrating toads are able to rapidly restore bladder water storage (Steen 1929), which suggests a diuretic condition in concert with elevated cutaneous water permeability. However, the relative importance of hydrins versus AVT in amphibian water balance remains to be determined.

Renin–Angiotensin–Aldosterone

Aldosterone stimulates Na^+ transport across isolated preparations of the skin and urinary bladder as well as the colon (reviewed by Uchiyama and Konno 2006) and is proposed to maintain plasma Na^+ concentration when amphibians are subject to osmodilution in hypo-osmotic media (Chapter 4). In the current model, aldosterone binds to a mineralocorticoid receptor (MR) that serves as a transcription factor to stimulate the synthesis of proteins that result in a greater number of conducting ENaCs in the apical membranes of absorbing epithelia and also greater activity of Na^+–K^+ ATPase in the basolateral membranes (Pascual-Le Tallec and Lombes 2005; Connell and Davies 2005). One of the proteins identified is a kinase, Sgk1, that increases ENaC density by inhibiting their removal from the apical membrane (Fig. 3.1C).

It is instructive to interpret literature values in the context of the experimental procedures used to manipulate plasma aldosterone levels. This is evident in two studies with *B. marinus*: Garland and Henderson (1975) maintained toads unfed for 1 week prior to experiment and then for 3 weeks in tanks containing 2.5 cm deep solutions of 9% NaCl (285 mOsm/l) or deionized water (DI, 0 mOsm/l). Toads in DI had significantly lower plasma osmolality, Na^+ concentration and elevated aldosterone levels as predicted from the known effect on isolated tissues. In contrast, Konno *et al.* (2005) fed toads up to the experiment and kept them immersed in either 300 mOsm/l NaCl or tap water (Na^+ concentration not stated) for only 7 days. Other toads were dehydrated by water deprivation or, as a control, allowed to remain in maintenance terraria with water available *ad lib.* As with Garland and Henderson (1975), plasma osmolality and Na^+ concentration were reduced in hypo-osmotic tap water relative to the other treatments. However, aldosterone levels were significantly lower in the tap-water-exposed toads than either the control or dehydrated toads and the levels in the dehydrated toads were many times greater than any of the other groups. Their conclusion was that aldosterone is released in response to decreased plasma volume in addition to decreased N^+ concentration. The main difference between the experiments is the nutritional status and time of exposure to the dilute solution. It would appear that aldosterone levels are elevated more rapidly in response to plasma volume depletion than to reduced Na^+ concentration. This is consistent with observations that aldosterone levels of *Bufo japonicus* in the field were highest during the summer period when the toads ventured away from water to their terrestrial habitats and presumably would encounter dehydrating conditions (Jolivet-Jaudet *et al.* 1983).

The regulation of plasma aldosterone has been assumed to resemble that of mammals where a decrease in plasma volume is detected by the kidneys, renin is released and catalyzes the formation of angiotensin I from angiotensinogen (Grill *et al.* 1972). Angiotensin I is converted to angiotensin II (AII) by a converting enzyme (Hasegawa *et al.* 1983) and AII stimulates aldosterone release from the amphibian adrenocortical (inter-renal) tissue. In support of this hypothesis, Konno *et al.* (2005) found parallel increases in AII and aldosterone in dehydrated *B. marinus*. It should be noted that De Ruyter and Stiffler (1985) found that adrenocoticotrophic hormone (ACTH) was able to stimulate aldosterone release and restore plasma Na^+ concentration in larval salamanders, *Ambystoma tigrinum*. These authors cite references that suggest this mechanism is operative in other urodeles.

Bishop *et al.* (1961) found that reduced Na^+ transport across the skin of frogs (*Rana pipiens*) whose inter-renal tissues had been removed could be partially restored by either hydrocortisone or aldosterone, suggesting similar effects of both glucocorticoids and mineralocorticoids. In fact, Schmidt *et al.* (1993) found that stimulation of Na^+ transport in cultured cells from the amphibian kidney (A6, see Chapter 5) by aldosterone was mediated by a glucocorticoid receptor (GR) rather than a MR. In this regard, Jungreis (1979) hypothesized that aldosterone also has glucocorticoid activity and stimulated gluconeogenesis and urea synthesis of *R. pipiens* subject to saline dehydration in the summer but not winter months. This is consistent with the observation by Konno *et al.* (2005) that urea levels were elevated in dehydrated toads that displayed elevated plasma aldosterone concentrations and the field observations of Jolivet-Jaudet *et al.* (1983). The issue of mineralocorticoid versus glucocorticoid function for aldosterone and seasonal variation in hormone responsiveness remain confounding issues that should be considered when examining literature citations or designing experiments.

In addition to stimulating aldosterone secretion, AII is a potent vasoconstrictor that can maintain blood pressure during hypovolemia caused by either dehydration of blood loss (Guyton and Hall 2006). AII also stimulates thirst and oral drinking in mammals, birds and reptiles (Fitzsimons 1998) to restore plasma volume. Experiments with exogenously administered AII have demonstrated an increase in WR behavior and water retention in several anuran species (Hoff and Hillyard 1991; Propper and Johnson 1994; Viborg and Rosenkilde 2004), suggesting analogy between the mechanisms that control oral and cutaneous drinking. These results are consistent with those of Konno *et al.* (2005) showing elevated AII levels in dehydrated toads. Injection of AII into the lymphatic spaces or within the cerebral ventricles (Propper *et al.* 1995) both stimulate WR indicating thirst is regulated by centers within and external to the blood–brain barrier. Stimulation of the WR by AII is sensitive to environmental factors, including changes in barometric pressure (Hoff and Hillyard 1993). Also, dehydrated *Scaphiopus couchi* will display the WR without measurable increases in plasma or brain levels of AII, suggesting other factors may stimulate cutaneous drinking (Johnson and Propper 2000). Thirst and water seeking remains an interesting subject for research and is particularly important in regions where climate change may result in reduced rainfall and water availability.

Prolactin

Water-seeking behavior has long been observed in newts (Salamadridae). The larvae metamorphose and emerge in a terrestrial eft stage with a dry skin. During the breeding season they return to water and develop an aquatic form with moist skin. Exogenous prolactin prolongs the larval stage while thyroxine stimulates metamorphosis. Conversely, prolactin stimulates development of the aquatic form and the return to water, termed the "water drive". Despite the difference in skin texture, rehydration rates and the response to AVT are comparable between the aquatic and terrestrial phases of *Notophthalmus viridescens* (Brown and Brown 1977). However, more recent studies with terrestrial phase *Taricha torosa* show prolactin to reduce cutaneous water permeability produced by AVT which is consistent with reducing water uptake when the animals might return to water (Brown and Brown 1982). Prolactin has been shown to reduce or prevent the decrease in Na^+ concentration following hypophysectomy in larval *Ambystoma mexicanum* (Whittouck 1972a, b) and in the aquatic salamander, *Necturus maculosus* (Pang and Sawyer 1974). Prolactin did not restore plasma Na^+ in hypophysectomized newts (*N. viridescens*; Brown and Brown 1973 and *T. torosa*; Crim 1972). This effect seems to be one of reducing efflux of Na^+ into dilute media by more aquatic species as is the case for the effect of prolactin on gills of euryhaline fish transferred to freshwater (Maetz *et al.* 1967).

In anurans, exogenous prolactin similarly has been shown to prolong the larval stage (Etkin and Gona 1967; Brown and Frye 1969). More recent studies by White and Nicoll (1979) show that prolactin receptors are present in the gills during the larval stage and increase rapidly in the kidneys at metamorphosis, suggesting the osmoregulatory role of prolactin shifts from the gills to the kidneys during metamorphosis. Receptor density in the kidneys is stimulated by thyroxine, which is also increasing during metamorphosis.

Early experiments showing delay of metamorphosis were done with ovine prolactin. It should be noted that Huang and Brown (2000) created transgenic *Xenopus laevis* with over expression of genes for both ovine and *Xenopus* prolactin. The only effect of this overexpression on development was the retention of tail rudiments in newly metamorphosed froglets; developmental time was not affected. These authors questioned a role for prolactin as a "juvenile hormone", at least in *Xenopus*.

The mechanism for prolactin stimulation of Na^+ transport has been studied in isolated skin from premetamorphic larvae and newly metamorphosed bullfrogs. With larval skin, treatment with either aldosterone or glucocorticoids induces the appearance of functional ENaCs and amiloride-sensitive Na^+ transport (Takada *et al.* 1995). Prolactin inhibits these effects and promotes the continued expression of a larval-type cation channel, referred to earlier, that is activated by amiloride and ATP (Takada *et al.* 1996). In the skin of newly metamorphosed bullfrogs, prolactin, like aldosterone, stimulates Na^+ transport by increasing the density of functional ENaCs in the apical membrane and Na^+–K^+ pump activity in the basolateral membrane (Takada and Hokari 2007). The ability of prolactin to stimulate Na^+ transport via increased apical ENaC density has also been shown for adult treefrogs (*Hyla japonica;* Takada and Kasai 2003). This may be caused by changes in receptor expression in the skin as has been proposed for the kidney.

Insulin

Insulin stimulates Na^+ transport across the skin and urinary bladder (Herrera 1965; Cox and Singer 1977; Schoen and Erlij 1987). Studies with amphibian kidney cell cultures show that these effects of insulin and aldosterone both result from an increase the density of apical ENaCs (Blazer-Yost *et al.* 1998). Because glucose is not transported across the skin or bladder in appreciable amounts, why should insulin stimulate Na^+ transport? After a meal, plasma insulin levels are presumed to increase to stimulate nutrient assimilation and storage. Jorgensen (1992) demonstrated that following a meal toads will increase water absorption. A concomitant increase in Na^+ transport may allow the animals to maintain hydromineral balance at this time. Alternatively, insulin may increase glucose uptake in transporting epithelia and provide an increase in energy metabolism that can be used for active transport.

Sympathetic Nervous System

Isoproterenol stimulated water absorption by hydrated toads, *Scaphiopus couchi*, above that produced by hyperosmotic NaCl injection (Hillyard 1979), demonstrating the contribution of β-adrenergic stimulation to the hydro-osmotic response *in vivo*. Yokota and Hillman (1985) obtained similar results, showing both stimulation of water absorption by hydrated toads, *Bufo cognatus*, by isoproterenol and inhibition of water absorption in dehydrated toads by propranolol. The effect of isoproterenol on water uptake by hydrated *B. bufo* was greater than that of exogenous AVT, which might be due to simultaneous stimulation of water permeability and capillary perfusion (Viborg and Rosenkilde 2005; Fig. 3.3).

Studies with isolated skin and urinary bladder have produced a variety of responses to epinephrine, norepinephrine, sympathomimetics and inhibitors of adrenergic receptors (Foster 1974). Because norpinephrine, in low concentrations, stimulated water permeability across the isolated epidermis and the effect was blocked by propranolol, Rajerison *et al.* (1972) concluded the sympathetic nervous system in addition to AVT contributed to the hydro-osmotic response via β-adrenergic stimulation. Like AVT, this response is mediated by increased synthesis of cyclic AMP but appears to be independent of AVT (DeSousa and Grosso 1982).

At higher concentrations, the response to norepinephrine is more variable. In addition, when the water permeability of the isolated skin was stimulated with exogenous cyclic AMP and the β-adrenergic receptors were blocked with propranolol, norepinephrine treatment reduced water permeability. The α-adrenergic blocker, phentolamine prevented this inhibition and it was proposed that both α- and β-adrenergic receptors are present in the skin with α (presumably α 2) receptors inhibiting cyclic AMP synthesis (Rajerison *et al.* 1972).

3.2 Cutaneous Water Exchange: Skin Uptake and Evaporation

The exchange of water across the skin of amphibians is an important process with respect to their overall physiology because their skin is generally moist and quite permeable to water, as well as and ions and gasses (see Chapter 2). Amphibians do not drink to obtain water (although the waterproof frog *Phyllomedusa sauvagei* will drink rain droplets; McClanahan *et al.* 1994). Water is gained as preformed water in the food, but this is generally insignificant in magnitude compared to the rate of cutaneous water uptake (again, waterproof frogs such as *Phyllomedusa*, *Chiromantis* and *Hyperolius* may be exceptions).

The process of water exchange between the external environment and the body fluids depends on whether the external medium is water, soil or air. When the external medium is water, the diffusive exchange is determined by osmotic forces. The osmotic concentration of the medium (C; Osm) is generally low (most amphibians inhabit fresh water environments) compared to the body fluids (generally 200–300 mOsm/l) so there is a net osmotically-driven influx of water. The osmotic concentration gradient (ΔC_{osm}) is sufficient to explain the dynamics of water exchange for an aquatic amphibian (flux $\alpha\ \Delta C_{osm}$), although consideration of the water potential gradient ($\Delta\Psi$) provides an equivalent descriptor to ΔC_{osm}. When the external medium is soil, the combination of the osmotic concentration of interstitial soil water and the capillary force of the soil particles (the matric potential) determines the exchange of water. In theory there is a balance of water flux when the water potential of body fluid (Ψ_{plasma}) is equal to the water potential of the soil (Ψ_{soil}). When the external medium is air, water exchange depends on the relative humidity (RH) and temperature of the air, since these determine the concentration of water vapor in air; the rate of diffusion of water vapor (based on Fick's law) is determined by the difference in water vapor concentration (c_{wv}) rather than water potential.

3.2.1 Cutaneous Exchange in Water

Amphibians are able to absorb liquid water from their environment if it is available. Many caecilians, salamanders and a number of anurans are completely aquatic; they always have an osmotically-driven uptake of water across their skin and excretion of excess water is more of a physiological issue for them than water uptake (see Chapter 4). Semiterrestrial and terrestrial amphibians rely, to a great extent, on water uptake by cutaneous absorption from free water when it is available in their environment (e.g. ponds, puddles, moisture on the soil surface, vegetation or rocks). For example, the grassfrog *Rana pipiens* absorbs morning dew from moist grass (Dole 1967), and Archey's frog *Leiopelma archeyi* absorbs water from wet leaves (Cree 1989).

Most aquatic amphibians have smooth ventral and dorsal skin; they are completely submerged in water and need to minimize cutaneous water uptake rather than facilitate it. More terrestrial species, generally, have granular ventral skin (and sometimes lateral body and ventral thigh skin). This epidermal sculpturing facilitates water absorption by increasing the surface area for water absorption (Cohen 1952; Lillywhite and Licht 1974; see Chapter 1.3). Adpressing the granular skin against a moist substrate exposes the thin skin between the granules for water reabsorption. Cutaneous channels on the flanks and dorsum can move water from the ventral surface to these surfaces, increasing the overall wetted area for water absorption. Capillary forces due to the sculpturing channels might increase the force with which water can be absorbed from the environment, particularly soil. The aquatic eft form of the salamander *Triturus torosus* has smooth skin that is nearly devoid of tubercles whereas the terrestrial newt form has rough skin with many minute tubercles (Cohen 1952). The grey treefrog *Chiromantis xerampelina* and the reed frog *Hyperolius viridiflavus* have polygonal verrucae hydrophilicae ("warts") on the flank and thigh skin; these verrucae are the primary sites of water absorption (Drewes *et al.* 1977; Kobelt and Linsenmair 1986).

When sitting in water, amphibians experience an osmotic uptake of water based on the osmotic concentration difference between their body fluids and the external water. The osmotic concentration gradient (ΔC_{osm}) is sufficient to explain the dynamics of water exchange for an aquatic amphibian (flux $\alpha\ \Delta C_{osm}$), although conversion of osmotic potential to water potential provides a more equivalent approach to when we have soil or air as the external medium. Water potential (Ψ) equals $C_{osm}RT$, where R is the gas constant and T is the absolute temperature; Ψ has units of pressure, for example kPa (see Chapter 1). There is a balance of water flux when the body fluid osmotic concentration (C_{plasma}) is equal to the osmotic concentration of the medium (C_{medium}), and the body fluid water potential (Ψ_{plasma}) is equal to the environmental water potential (Ψ_{medium}).

The rate of water influx is equal to $P_{osm}A(C_{plasma}-C_{external})$ where P_{osm} is the osmotic permeability (ml H_2O/cm^2 mOsm) and A is the surface area for osmotic exchange (cm^2). Values for P_{osm} vary depending on the degree of terrestriality of the amphibian, and its physiological state, for example hydrated or dehydrated (Table 3.3). Permanently aquatic frogs tend to have a low osmotic permeability, limiting transcutaneous influx of water (e.g. *Xenopus laevis*). More terrestrial species (e.g. *Salamandra*, *Taricha*, *Bufo*) have a higher P_{osm}, particularly when dehydrated because of the action of the circulating hormone arginine vasotocin (AVT; see Chapter 2). The terrestrial caecilian, *Dermophis mexicanus*, has a low osmotic influx of 0.019 ml/g h, which is unaltered by AVT (Stiffler and Talbot 2000).

Interspecific differences in the rate of water uptake by various amphibians have been related to ecological differences; in general, terrestrial species absorb water faster than aquatic species (Rey 1937; Jorgensen 1950; Bentley *et al.* 1958;

Table 3.3. Cutaneous permeability of skin from various amphibians, illustrating lower values for aquatic species/caecilians and salamanders and higher values (in bold) for more terrestrial/anuran species, particularly under the influence of arginine vasotocin (AVT). Data taken or calculated from Bentley (1969), Bentley and Heller (1964, 1965), Brown and Brown (1977, 1980), Dicker and Elliot (1967), Goldenberg and Warburg (1977), Hillman (1974), Mullen and Alvarado (1976), Stiffler *et al.* (1990), Warburg (1971), Warburg and Goldenberg (1978)

Species	**Cutaneous Permeability ($\mu l\ H_2O\ cm^{-2}\ hr^{-1}$)**	
	Control	**AVT**
Caecilians		
Typhlonectes compressicauda	1	
Ichthyophis kohtaoensis	2	
Salamanders		
Necturus maculosus	3	4
Salamandra salamandra	3	2
***Notopthalmus viridescens* eft**	3	11
Amphiuma means	4	4
Siren lacertina	4	5
Ambystoma tigrinum larva	4	7
***Notopthalmus viridescens* newt**	4	21
Triturus alpestris	7	8
Triturus cristatus	7	7
Aneides lugubris	7	16
Triturus vittatus	8	14
Ambystoma tigrinum	12	11
Salamandra maculosa	13	14
Taricha torosa	16	27
Triturus vittatus	17	29
Anurans		
Ascaphus truei	4	
Rana pipiens	5	
Hyla regilla	6	
Xenopus laevis	8	9
Rana catesbeiana	13	61
Bufo boreas	16	
Bufo melanostictus	17	37
Rana ridibunda	22	39
Bufo marinus	28	85

Main and Bentley 1964). The aquatic caecilian, *Typhlonectes compressicauda,* has a lower rate of water influx (0.009 ml/g h) compared to the terrestrial *Ichthyophis kohtaoensis* (0.024 ml/g h; Stiffler *et al.* 1990). The higher rate of water uptake from a moist substrate by *Bufo bufo* compared to *Rana* spp. reflects the capillarity of tube-like grooves on its skin surface (Christensen 1974). Other morphological differences (e.g. extent of the pelvic patch) and physiological differences (rate of cutaneous blood flow, increased permeability of the skin by AVT) also contribute to differences in cutaneous uptake rate (see Chapter 2). Interspecific variation in water uptake rate does not seem to result from differences in cutaneous sodium uptake rate (Schmidt and Underhill 1964). Of a variety of Western Australian frogs, two burrowing species (*Arenophryne rotunda* and *Myobatrachus gouldii*) that are probably independent of free water (e.g. have direct development) have lower rates of cutaneous water uptake from free water, presumably reflecting the general availability of water in their fossorial environment (Withers and Roberts 1993; Cartledge *et al.* 2005).

The ventral pelvic skin ("pelvic patch") is the area of skin that is most responsible for water absorption by anuran amphibians (Dole 1967; Baldwin 1974; McClanahan and Baldwin 1969). It is highly vascularized (under neural and hormonal control) compared to the rest of the skin. More terrestrial anurans (e.g. *Bufo*) have a more vascular pelvic patch than semiterrestrial species (e.g. *Rana*) and aquatic species (e.g. *Xenopus*), and their pelvic patch skin is more responsive to water balance hormones such as AVT (Roth 1973; Christensen 1974; Bentley and Main 1972; see also Section 3.1 above). The nature and distribution of aquaporins is most likely a major factor in water uptake *in vivo*, along with epidermal sculpturing, capillary forces and circulation.

3.2.2 Cutaneous Exchange in Soil

Many anuran amphibians are adept at burrowing into the soil, because it generally provides a much more favorable microclimate than being above ground in air. Burrowing (fossorial) species generally have a globose body shape and short legs (Davies and Withers 1993). They typically have a large ventricle that would be associated with a high metabolic scope for digging, and a large liver presumably as an energy store (Withers and Hillman 2001). The few forward-burrowing frogs (e.g. *Hemisus, Myobatrachus, Arenophryne*) have reduced fingers so that their hand is broad to dig easily and/or thickened skin over the snout to protect the head as it is pushed into the substrate; their nasals or premaxillae may be modified to provide bony protection for the snout (Emerson 1976; Davies 1984; Burton 1986). The more typical, backward-burrowing frogs generally have pairs of digging structures, metatarsal tubercles, on their hind feet. These are raised and compressed tuberosities, which are often keratinized with a hard edge. There are two forms of backwards burrowing. Backwards-sliding burrowers remain facing the same direction as they push backwards into the ground at an angle, whereas circular burrowers corkscrew backwards into the soil (Sanders and Davies 1984).

The dynamics of cutaneous water exchange for frogs in soil is more complex than in water because of the effect of physical properties, particularly particle size, on the "water holding" capacity of soil (see Chapter 1). The effect of soil capillarity forces (τ) is summed with the osmotic potential of soil water (Π_S) to determine the overall soil water potential; $\Psi_{soil} = \tau - \Pi_S$. For soil that is saturated with water (at field water capacity), Π_S determines the soil water potential. Otherwise, τ is generally much more substantial than the osmotic potential in contributing to Ψ_{soil}. Soils with a small particle size (e.g. clays) generally have a more negative matric water potential than soils with a large particle size (e.g. sands; Fig. 3.5), and so they have a greater "suction" force, to draw water out of a buried amphibian. For example, a clay soil has to have a higher water content to maintain water balance for a buried amphibian (about 12% water by weight for 250 mOsm/l) than a sandy soil (about 2% water by weight).

The exchange of water between frog and soil is determined by the water potential difference; water is always transferred from a high water potential (0 or slightly negative Ψ) to a lower potential (lower, more negative Ψ). The water potential of a frog (Ψ_{frog}) is determined by the osmotic concentration of its body fluids; $\Psi_{frog} = -RTC_{osm}$ where Ψ is the water potential (MPa), R is the gas constant (8.31 L/kPa M), T is the absolute temperature (°K) and C_{osm} is the osmotic pressure (Osm; Nobel 1999). A frog is, in theory, at water equilibrium if its water

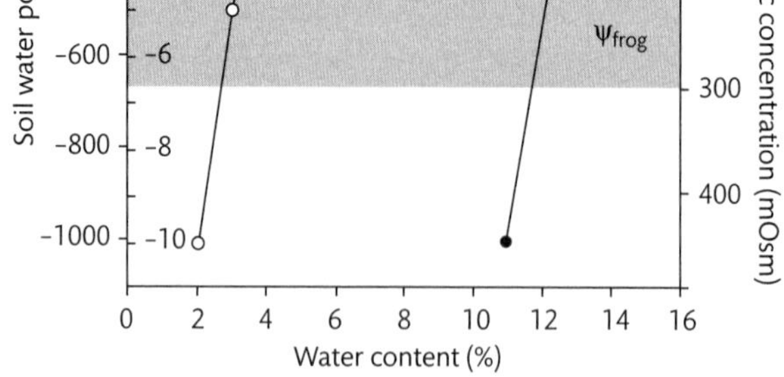

Fig. 3.5 Relationship between soil water potential (ψ_{soil}) and water content for sand, intermediate and clay soils; also shown is the equivalent osmotic concentration. Modified from Shoemaker *et al.* (1992), adapted from Ruibal *et al.* (1969).

potential equals the osmotic potential of the soil ($\Psi_{frog} = \Psi_{soil}$), and the frog can remove water from the soil if $\Psi_{frog} > \Psi_{soil}$, but loses water if $\Psi_{frog} < \Psi_{soil}$.

There are two potential mechanisms for water exchange between a frog and soil, water vapor exchange and liquid water exchange (Booth 2006). Water in the soil interstitium is a mixture of liquid water and water vapor, except in very dry soil (only vapor) and saturated soil (only liquid water). The rate of liquid water exchange is determined by the hydraulic conductivity of the soil and the water potential gradient. However, as soils dry their hydraulic conductivity decreases. For sandy soils, hydraulic conductivity is negligible if the Ψ_{soil} is more negative than about −0.1 MPa and so hydraulic loss of water is unimportant for frogs because their Ψ is more negative than −0.6 MPa (Fig. 3.5). Water loss from frogs in sandy soil would be determined by vapor phase exchange. For frogs in clay, hydraulic conductivity occurs until the water potential is very negative (e.g. −1.5 MPa). Therefore, hydraulic conductivity is theoretically important until frogs are considerably osmo-concentrated (about 600 mOsm/l). The rate of water vapor exchange depends on the water potential at the surface of the frog's skin (virtually 100% relative humidity, at skin temperature) and the water potential in the soil interstitial air spaces.

Water exchange by hydraulic conduction is much faster than by water vapor diffusion. The hydraulic conductivity (L_{soil}) is less than 10^{-10} m^2/s Pa for saturated clay and 10^{-7} m^2/s Pa for saturated sand (Nobel 1999), corresponding to water uptake fluxes ($= -L_{soil}\, \Delta\Psi_{soil}/\Delta x$, where Δx is the distance for the Ψ gradient, $\Delta\Psi_{soil}$) of about 36 and 36000 ml/min cm^2 for a water potential difference of −0.6 MPa (hydrated frog in saturated soil). The water vapor conductance is 100 mmol/m^2 s^1 (over 1 cm diffusion distance in air at 20°C; Nobel 1999), corresponding to a water vapor uptake ($= g_{water}\, \Delta N$, where ΔN is the mole fraction concentration difference) of about 1.4×10^{-4} ml/min cm^2 for the same water potential difference.

Species differences in water flux across the skin are apparent *in vitro* as well as *in vivo*. Fossorial anurans such as *Bufo cognatus* and *Scaphiopus couchii* have water uptake across the skin *in vitro* from water potentials as low as −1.4 MPa whereas the more aquatic *Rana pipiens* required a larger gradient (Hillyard 1976). Soil particles adhere more extensively to the skin of terrestrial toads compared to the aquatic frog, so there is a more effective pathway for capillary movement of water from the soil matrix to the skin. This probably contributes to terrestrial species being able to take up water from drier soils (e.g. Table 3.4).

Because many amphibians burrow into soil, there have been numerous studies of the dynamics of water exchange between soil and amphibians, particularly salamanders and anurans. In general, the water absorption threshold of amphibians (the soil water potential when the amphibian is in balance with the soil) is related to their extent of terrestriality. More terrestrial amphibians are able to absorb water from soils with a more negative water potential (Table 3.4). For many

Table 3.4 Water absorption threshold (WAT) of various amphibians, illustrating low values for aquatic compared to terrestrial species

Species	WAT (atm)
Plethodon cinereus	−1
Plethodon glutinosus	−1.0
Hyla cinerea	−1.2
Rana pipiens	−1.3
Desmognathus fuscus	−1.9
Pseudotriton ruber	−2.4
Ambystoma opacum	−2.4
Scaphiopus hammondi	−2.5
Bufo americanus	−2.5
Bufo cognatus	−2.5
Scaphiopus couchii	−2.7
Plethodon jordani	−2.8
Cyclorana platycephala	−3
Plethodon jordani	−3.1
Cyclorana alboguttata	−4
Siren lacertina	−5
Scaphiopus couchii (field)	≈−10
Notaden nichollsi (field)	≈−10
Arenophryne rotunda (field)	≈−10

Data calculated or taken from Cartledge 2006; Etheridge 1990; Ruibal *et al.* 1969; van Beurden 1984; Walker and Whitford 1970; Booth 2006; Spight 1976; Spotila 1972.

amphibians, their water absorption threshold is not similar to their body fluid osmotic potential (about −5 to −6 atm; Fig. 3.5. Rather, body fluid water potential is usually more negative than the soil water at equilibrium potential ($\Psi_{plasma} < \Psi_{external}$). This discrepancy might reflect a measurement artifact (particularly for soil water potential) or it might reflect incomplete contact of skin and soil during experiments.

The ecological importance of cutaneous water absorption from soil has been illustrated by some simple laboratory and field studies. Green-striped burrowing frogs (*Cyclorana alboguttata*) choose to burrow in moist sand rather than moist clay but moist clay rather than dry sand (Booth 2006). Woodhouse's toad (*Bufo woodhousei*) absorb water from moist sand surfaces after they emerge from their

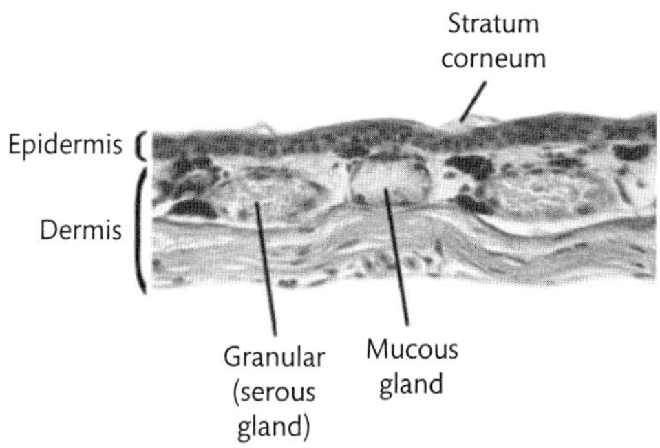

Fig. 3.6 Cross-section of skin of a typical frog *Kassina fusca* (Photo: R. C. Drewes).

burrows, using a water-absorbing posture that adpresses the ventral surface to the substrate (Stille 1951). Lee (1968) showed that the moaning frog (*Heleioporus eyrei*) loses up to 22% of its body mass while foraging at night, then burrows into the soil to regain the water deficit during the day. The sandhill frog (*Arenophryne rotunda*), which inhabits coastal sandhills on the arid west coast of Australia, dehydrates during nocturnal foraging then rehydrates by burrowing into the moist dune sand during the day (Roberts 1990).

3.2.3 Cutaneous Exchange in Air

For most amphibians, evaporative water loss (EWL) from their skin when in subsaturated air is equivalent to the rate of evaporation from a free water surface. EWL depends strongly on the relative humidity of the air and the temperature. In moving and/or dry air, EWL can be so high that it rapidly compromises water status, leading to immobilization and death within hours, or at most days. For example, a 1-g plethodontid salamander is estimated to lose 10% of its body mass by evaporation in less than 20 minutes, in moving air (1 m/s) at moderate temperature (20°C) and relative humidity (40%), whereas a 3-g plethodontid might take over 10 h to lose 10% of its body mass in still air at 90% RH (Feder 1983). A large amphibian (much larger than any plethodontid), such as the marine toad (*Bufo marinus*), might take 2–3 days to lose 10% of its body mass by evaporation in dry air at room temperature (calculated from Young *et al.* 2005).

Amphibians have highly permeable skin, typically with numerous mucous and serous glands and variable vascularization of the dermis. The skin epidermis (Fig. 3.6) consists of a stratified epithelium, as in other vertebrates, but does not have an outer layer of dead, cornified cells (unlike reptiles, mammals and birds). Rather, the outer layer of epidermis (stratum corneum) is a thin, single-cell (or a few cells) pavement of living keratinized cells (Drewes *et al.* 1977; Kobelt and Linsenmair 1986). The inner layer of the epidermis, the s. germinativum, consists of four to eight layers of cells. The epidermis is separated from the dermis by a basement membrane. Frequent mucous gland discharge, especially under desiccating conditions, maintains the skin moist and pliable, but at the expense of a high EWL. Some amphibians, particularly the more terrestrial anurans, modulate the rate of mucous gland discharge to keep the skin moist (Lillywhite 1971; Lillywhite and Licht 1975). The surface of the s. corneum can be covered by a thin, dried film of lipid or proteinaceous glandular secretions, and this film may reduce EWL from the skin surface (see below). The retention of thin and pliant skin by amphibians, regardless of how terrestrial they have become, may be related to constraints of genetic expression due to selection in aquatic larval stages, that is developmental canalization (Lillywhite and Mittal 1999).

Rates of EWL (e.g. mg/gh) have been measured for hundreds of species of amphibians, but direct comparison of these values is of little use because of the great variability in measurement techniques and ambient conditions (Shoemaker *et al.* 1992). It is more useful to compare the rates of cutaneous water loss for various amphibians as cutaneous resistance (R; s/cm) rather than absolute EWL (mg/h; which does not account for differences in cutaneous surface area or ambient conditions), mass-specific EWL (e.g. mg/gh; which partially accounts for differences in surface area), or surface-area-specific-EWL (e.g. $mg/cm^2 h$; which accounts for surface area differences but not difference in ambient conditions; see Chapter 1.3). Resistance is calculated as $R = \Delta C/EWL$, where ΔC is the water vapor concentration difference (mg/cm^3) and EWL is the evaporative water loss ($mg/cm^2 s$). A free water surface has a resistance of about 1 s/cm, and most anurans have a cutaneous R of about 1–3 s/cm (see Lillywhite 2006 for review), as do salamanders (Cohen 1952; MacMahon 1965; Spight 1968; Spotilla 1972; Feder 1983). Resistance accounts for ambient relative humidity and temperature conditions, as well as surface area, but it does not account for the thickness of the evaporative barrier. For example, the R of frog cocoon increases with cocoon thickness over time, although the resistance is the same per unit thickness (see Chapter 4).

Water vapor concentration is the appropriate potential to use in calculating water vapor flux (Monteith and Campbell 1980) rather than water potential (Toolson 1978, 1980). There is a balance of water flux across the skin when the body fluid water potential (Ψ_{plasma}) is equal to the air water potential (Ψ_{air}). The water potential of water vapor in air depends on the relative humidity (RH) and temperature of the air; $\Psi_{air} = [RT/V_w] \ln [RH/100]$, where V_w is the partial molal volume of water (see Chapter 1). However, the rate of diffusion, based on Fick's law, is determined by the difference in water vapor concentration (c_{wv}) or water vapor pressure (vp_{wv}) rather than the difference in water potential, so EWL α $(c_{frog} - c_{wair})$ and $R = (c_{frog} - c_{wair})/EWL$, where c_{frog} is the water vapor content of 100% saturated air at the skin temperature of the frog, and c_{wair} is the water vapor content of the ambient air.

Biophysical Effects of T_a, Relative Humidity and Wind

The biophysical effects of temperature (°C) and relative humidity (RH; %) on EWL are well understood. The driving force for evaporation is the water vapor pressure, or relative humidity, or water potential gradient between the skin and the ambient air. The skin surface of an amphibian is assumed to be an air film at 100% relative humidity (actually slightly less, about 99.5%, depending on the osmotic concentration of the body fluids), and the RH of the ambient air can vary from 0 to 100%.

Temperature has a marked and exponential effect on EWL because the water vapor pressure, content and potential of saturated air increase exponentially with T_a, so the water gradient driving evaporation increases and EWL increases at higher Ta and lower RH (Fig. 3.7; Damstra 1986).

A major difficulty remains in comparing the cutaneous resistance of amphibians even after accounting for RH and temperature. Wind speed (v, m/s) influences the rate of evaporative water loss from a free water surface because it determines the thickness of the boundary layer and hence the resistance of the boundary layer to evaporation (R_{bl}). This effect is very significant because the boundary layer resistance can equal or considerably exceed the morphological resistance of the skin. Boundary layer resistance depends on both the size and shape of the amphibian, as well as the ambient wind speed and if the air flow is laminar or turbulent (see

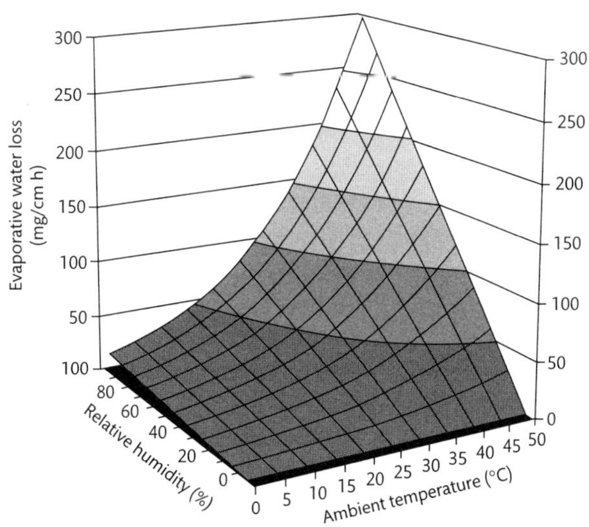

Fig. 3.7 Effect of ambient temperature and relative humidity on rate of evaporative water loss. Calculated for a resistance of 1 s/cm using the psychrometric equations of Parrish and Putnam (1977) for absolute humidity.

Chapter 1). The mass transfer coefficient (h_D), which is the inverse of boundary layer resistance, can be estimated for frogs (Mitchell 1976; Tracy 1976; Tracy *et al.* 1980) as

$$h_D = 0.3\ (D/SVL)\ (V\ SVL/v)^{0.6}$$

where D is the diffusivity of water vapor in air (15.4 cm^2/min) and v is the kinematic velocity of air (0.153 cm^2/s; Tracy and Porter 1973), SVL is the snout–vent length of the anuran (cm) and V is wind velocity (cm/s). The general relationship between snout–vent length (cm) and body mass (g) is SVL = a $mass^b$, where SVL (cm) = 2.38 $g^{0.31}$ for anurans, 3.93 $g^{0.34}$ for urodeles, and 11.3 $g^{0.33}$ for caecilians (calculated from Pough 1980), so

$$R_{bl} = 1/h_D = 3.3\ v^{0.6}\ (aM^b)^{0.4}\ V^{-0.6}\ D^{-1}$$

For anurans, $R_{bl} = 5.94\,g^{0.124}V^{-0.6}$ so for a 500-g frog the boundary layer resistance can vary from about 10 s/cm at a wind speed of 2 cm/s, to 1 s/cm at about 100 cm/s; for a 5-g frog, the corresponding wind speeds are about 1 and 40 cm/s (Shoemaker *et al.* 1992) (Table 3.5).

A high EWL is rapidly lethal for an amphibian in a terrestrial environment, but amphibians, and in particular anurans, are abundant even in some of the most hot and arid environments on Earth (Chapter 4.2). In fact, the world-wide distribution of amphibians would seem to be more limited by low temperature than water availability (Duellman and Trueb 1986). Most deserts of the world have at least a few anuran species, and the key to their success is more often than not a behavioral adaptation rather than a morphological or physiological one (see also Chapter 4).

Table 3.5 Cutaneous and boundary layer resistances for non-waterproof/medium waterproof frogs, in dry air at 0.25 cm/s; data from Young *et al.* (2005, 2006)

Species	R_{skin} (s/cm)	R_{bl} (s/cm)
Uperoleia sp nov	0	4.1
Crinia bilingua	0	2.9
Austrochaperina adelphe	0.1	3.0
Platymantis vitianus	0.2	2.8
Platymantis vitiensis	0.4	3.2
Bufo marinus	1.6	1.8
Notaden melanoscaphus	2.1	2.8
Limnodynastes ornatus	3.2	4.0

Water-Conserving Behaviors

One of the most common water-conserving behavioral adaptations of anurans to limit EWL is to restrict activity to periods that minimize EWL, that is they are active at low wind speed, high RH, and at night rather than during the day. A subtle example of behavioral avoidance of unfavorable ambient conditions is provided by the rainforest frog, the coqui (*Eleutherodactylus coqui*). It only emerges from its forest floor refuge on nights when the RH is greater than 87% (Pough *et al.* 1983). Many forest or stream-dwelling frogs are active by day, and avoid dehydration by inhabiting microclimates with very high ambient RH, or they rehydrate by returning to water or stream spray. The marine toad (*Bufo marinus*) uses retreat site selection and modulation of activity to avoid desiccation during dry periods (Dobkin *et al.* 1987). An extreme illustration of behavioral and physiological adaptation to desiccating conditions is the long-term inactivity of desert frogs that aestivate to avoid hot, dry periods (see Chapter 4).

Another characteristic behavior of anuran amphibians that minimizes EWL is the use of a water-conserving posture. Many species, particularly when they are exposed to potentially desiccating conditions, assume a characteristic posture with the body and chin adpressed to the substrate and the limbs tucked under the body (Fig. 3.8). The coqui adopts a water-conserving posture if there is no surface water (Pough *et al.* 1983). Many arboreal frogs perch in relatively exposed sites, and they characteristically adopt the water-conserving posture to minimize

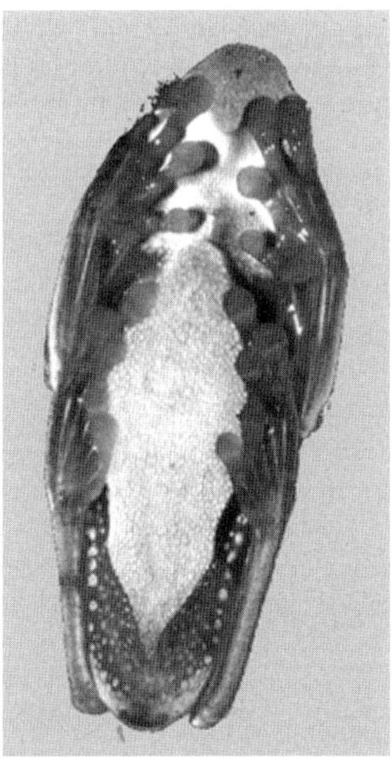

Fig. 3.8 Ventral view of a treefrog *Agalychnis callidryas* in its water-conserving posture. (Photo: R. C. Robert Drewes).

their EWL. Several elongate salamanders (plethodontids, *Amphuima*) tightly coil their body and tail to reduce their evaporative surface area (Ray 1958). The African treefrog, *Chiromantis xerampelina,* and the reed frog, *Hyperolius viridiflavus,* use the water-conserving posture to cover their flank and thigh skin, which are their main sites for water absorption (Drewes *et al.* 1977; Kobelt and Linsenmair 1986).

Some amphibians aggregate into groups to reduce their EWL. Aggregated individuals reduce their exposed surface area, hence EWL, by being in close contact with other individuals. Tiger salamanders (*Ambystoma tigrinum*) have a lower EWL when aggregated (Gehlback *et al.* 1969), as do *A. macrodactylum* (Alvarado 1967) and *Salamandra salamandra* (Warburg and Degani 1979). Aggregated frogs also experience a lower EWL than individual frogs (Johnson 1969).

Morphological Adaptations

Some amphibians have skin modified to reduce cutaneous evaporative water loss in air. Terrestrial frogs that bask or shelter in open areas and arboreal treefrogs that bask and are exposed to dry ambient air are particularly prone to evaporative desiccation. Many of these anurans have independently evolved a reduced cutaneous EWL, at least for the dorsal skin that is exposed when the frogs are in their water-conserving posture. No salamanders have been reported to have reduced cutaneous evaporative water loss, presumably as none of these species are as terrestrial or scansorial. The only EWL measured for a caecilian is 2.44 mg/cm^2 h for the terrestrial *Scolecomorphus kirki* (Steele and Louw 1988), which corresponds to an R of about 15 s/cm. More species need to be measured before any broad conclusions can be made about their EWL and R. Many caecilians have dermal scales, which might provide rigidity for locomotion (Duellman and Trueb 1994) but they might possibly reduce cutaneous EWL as well.

Waterproof Scansorial Frogs

Frogs that bask or shelter in open rocky areas or in trees are particularly prone to evaporative desiccation because they are exposed to relatively dry ambient air, often at a considerable wind speed (that reduces cutaneous boundary layer thickness). It is not therefore not surprising that many of these anurans have a reduced cutaneous EWL, and an increased resistance to evaporation, at least for the dorsal skin that is exposed when the frogs are in their water-conserving posture (Fig. 3.9). No salamanders have been reported to have a reduced cutaneous evaporative water loss, presumably as none of these species are so terrestrial or scansorial as these anurans.

Various terrestrial or scansorial anurans, in a variety of genera, have a moderately elevated cutaneous resistance values, from about 2 to 20 s/cm. The canyon treefrog, *Hyla arenicolor*, which basks on canyon rock walls, has a resistance

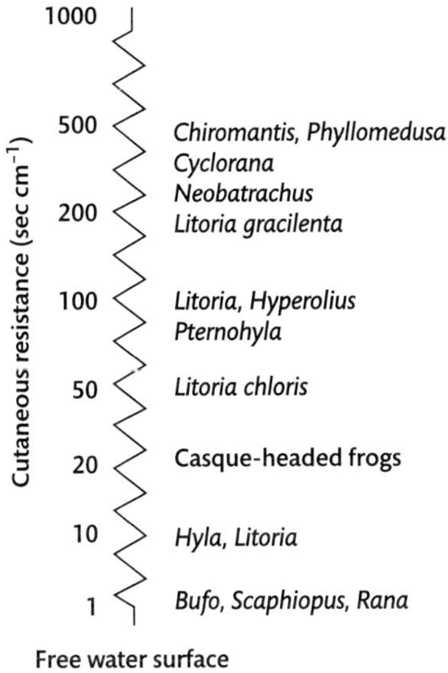

Fig. 3.9 Schematic summary of cutaneous resistance for anuran amphibians with waterproof skin or a cocoon. Modified from Buttemer (1990); see Lillywhite (2006) for summary of data.

of about 15 s/cm (Snyder and Hammerson 1993). The hylid, *Bokkermannohyla alvarengai,* basks daily on rocks; it has an EWL of 62 mg/h (Tattersall *et al.* 2006) which is about 14 s/cm. Various treefrogs (*Afrixalus, Agalychnis, Hyla, Litoria, Pachymedusa, Polypedates, Rhacophorus*) range in R from about 2 (i.e. little reduction in EWL) to 10 s/cm (Bentley and Yorio 1979; Wygoda 1984, 1988a; Withers *et al.* 1984; Buttemer 1990; Withers 1995; Lillywhite *et al.* 1997a, b; Amey and Grigg 1995; Young *et al.* 2005; Buttemer and Thomas 2003). Various casque-headed hylid frogs shelter in crevices, tree holes or bromeliads (phragmotic behavior) and use the top of their head to plug the opening of their retreat (Duellman and Trueb 1994). The exposed skin of their skull is co-ossified with the underlying dermal bones, and this reduces the rate of EWL compared to other parts of their body (Siebert *et al.* 1974; Andrade and Abe 1997; Navas *et al.* 2002), and hence their EWL while sheltered. The EWL of the co-ossified skin is about 39 compared to 73 mg/cm^{2}h for the normal skin in *Aparasphenodon brunoi* and 30 and 48 mg/cm^{2}h for *Corythomantis greeningi* (Andrade and Abe 1997). Calcification of the dermal layer may also contribute to reduced EWL in *Corythomantis greeningi* (Katchburian *et al.* 2001). Dermal osteoderms occur in some anurans, but more likely are protective dermal armor than a water-proofing barrier (Ruibal and Shoemaker 1984).

Some arboreal treefrogs have a considerably higher cutaneous resistance to evaporative water loss (20 to 400 s/cm) equivalent to that of some reptiles, birds and mammals, whose skin epidermis has a thick stratum corneum (see below). This has evolved independently in a number of genera from various families. The African treefrog (*Chiromantis xerampelina*; Rhacophoridae) was the first reported "waterproof" frog (Loveridge 1970). Shoemaker *et al.* (1972) subsequently described a South American treefrog (*Phyllomedusa*; Hylidae) as "waterproof". Various other species of *Chiromantis* are "waterproof", for example *C. petersi* (R = 347 s/cm; Drewes *et al.* 1977) and *C. rufescens* (R = 404 s/cm; Withers *et al.* 1984), as are various *Phyllomedusa* species 200–360 s/cm; Withers *et al.* 1984; Wygoda 1984) but not other genera of these treefrogs (Shoemaker and McClanahan 1975). African reed frogs (*Hyperolius*; Hyperoliidae) have reduced cutaneous EWL and high R (Loveridge 1976; Withers *et al.* 1982; Withers *et al.* 1984; Geise and Linsenmair 1986). Many Australian *Litoria* treefrogs are "waterproof", for example *L. gracilenta* at 118 s/cm (Withers *et al.* 1984) and *L. chloris* at 30 s/cm (Buttemer 1990).

The cutaneous "water-proofing" mechanism is only well understood for *Phyllomedusa* spp., which have a unique type of small but numerous skin glands that contains a lipid material. The frog's skin exudes a waxy secretion (e.g. after activity or handling) and the frog spreads it over the entire body surface using all four feet in a series of stereotyped wiping behaviors (Blaylock *et al.* 1976; Gomez *et al.* 2006). The waxy material is predominantly wax esters, with some free fatty acids and hydrocarbons (McClanahan *et al.* 1978). The skin of *Chiromantis* and *Hyperolius* does not contain lipid glands and they do not have a sterotypic wiping behavior; their mechanism of "waterproofing" is unclear (Drewes *et al.* 1977; Kobelt and Linsenmair 1986). The dorsal skin of *Chiromantis* has a thin outer band of unknown composition (Shoemaker *et al.* 1992); its role is not known but it has the appropriate location to function as a water-proofing barrier. Wygoda (1988) suggested that a dried mucous film on the skin of arboreal hylids reduces EWL, and Kobelt and Linsenmair (1986) suggested that dried mucous sealed the small spaces between the body and the limbs of reed frogs. In both instances, the nature of the "mucous" and a water-proofing role are unclear (Lillywhite and Mittal 1999). Christian *et al.* (1988) and Christian and Parry (1997) have suggested that *Litoria* wipes epidermal secretions of lipid and protein materials over the skin surface, and on drying these form a barrier to EWL. Lillywhite *et al.* (1997a, b) suggested a similar (though minor) role of wiped secretions of mucous and lipids from cutaneous mucous glands for the slightly waterproof Indian treefrog *Polypedates*. A dried layer of protein forms a vitreous barrier to EWL for eggs of an annual killifish (Podrabsky *et al.* 2001), suggesting a similar possible role of dried protein/ lipid secretions for frog skin. Other suggested mechanisms for waterproofing of frog skin include dermal layers of ground substance (Elkan 1968), lipids (Amey and Grigg 1995) and iridophores (Drewes *et al.* 1977; Withers *et al.* 1982; Kobelt and Linsenmair 1986), but the location of these layers underneath the living epidermal cell layers makes a role in EWL reduction unlikely. Any water-proofing layer presumably must be external to the outermost living epidermal cells, that is a surface film external to the stratum corneum.

High cutaneous resistance to EWL has thermoregulatory significance to these "waterproof" frogs (see section on thermoregulation). A low cutaneous EWL enhances the rate of heating by basking and reduces the rate of cooling, by lowering the rate of evaporative heat loss (Wygoda 1988b, 1989). It also enables a "waterproof" frog to remain for extended periods at a higher body temperature than other frogs without suffering a large concomitant water loss and potentially becoming dehydrated. Many waterproof frogs have the unusual habit (for amphibians) of basking in direct sunlight, and their skin color often blanches to a bright white as they heat up, including *Chiromantis* (Poynton 1964; Kaul and Shoemaker 1989; Fig. 3.10), *Hyperolius* (Withers *et al.* 1982; Kobelt and Linsenmair 1992), *Bokermannohyla alvarengai* (Tattersall *et al.* 2006), *Hyla* (Snyder and Hammerson 1993; King *et al.* 1994) and *Litoria* (Withers 1995), but *Phyllomedusa* do not seem to (McClanahan *et al.* 1994). White skin would reflect more incident radiation and prevent overheating. *Chiromantis* and *Phyllomedusa* are able to tightly

Fig. 3.10 Basking *Chiromantis petersii* blanched white. (Photo: R. C. Drewes).

modulate secretion of serous skin glands when hot, to control body temperature in a manner analogous to sweating in mammals.

Cocooned Frogs and Salamanders

Many frogs and a few salamanders aestivate for a long period, underground or in other refugia, to avoid harsh hot, dry seasonal conditions (see Chapter 4). Some of these species accumulate layers of shed skin while burrowed underground to form a cocoon, which becomes an effective barrier against EWL. No caecilians have been reported to form a cocoon to reduce cutaneous evaporative water loss, presumably as none of these species are particularly terrestrial or xeric-adapted.

Amphibians frequently shed the outer, single-cell squamous layer of their epidermis. Generally, this layer of shed skin is scraped from the body towards the mouth by the hands, and eaten to recycle its nutrients (dermatophagy; Weldon *et al.* 1993). The frequency of skin shedding varies with species and ambient conditions (e.g. increases with temperature and in constant darkness), but it is generally about every 7 days or so (Scharrer 1958; Taylor and Ewer 1956; Budtz 1979; Stefano and Donosa 1964; Withers 1995; McMaster 2006).

Burrowed amphibians continue to regularly shed their skin at about the same frequency. Some species remain sufficiently alert to ingest the shed skin, whereas others do not eat their shed skin but accumulate the layers to form a cocoon, which becomes an increasingly effective barrier against EWL over time as additional layers are added. The rate of skin shedding is higher for cocoon-forming species, about every 1 to 2 days (McClanahan *et al.* 1976; Withers 1995; McMaster 2006). Cocoon formation has independently evolved in a number of genera from various families of anuran amphibians, and a few sirenid salamanders.

The term cocoon was first used to describe thin peeling fragments of dorsal skin observed for spade-foot toads (*Scaphiopus*) when emerging from aestivation underground (Mayhew 1965), but spade-foot toads do not form the thick accumulations of shed epidermis apparent in true cocoon-forming species (see below) and their thin pieces of peeling skin probably just reflect incomplete shedding of a few recent layers (as is observed for other terrestrial anurans such as *Bufo marinus*). Lee and Mercer (1967) subsequently described cocoon formation for four genera of Australian frogs, *Neobatrachus* (*pictus*), *Cyclorana* (*platycephala* and *australis*), *Litoria* (*alboguttata*) and *Limnodynastes* (*spenceri*). Van Beurden (1984) further described the cocoon of *Cyclorana platycephala* but diagrammed only a few epidermal layers to the cocoon. Cocoon formation has been subsequently described for all *Neobatrachus* and *Cyclorana* (which now includes *alboguttata*) species examined (e.g. Withers 1995, 1998; Withers and Richards 1996) but not for any *Limnodynastes* spp. (including *spenceri*; Withers 1998). Formation of a cocoon of a similar structure has been reported for a variety of other anurans. McClanahan *et al.* (1976) described the cocoon of South American *Lepidobatrachus* and *Ceratophrys* (Leptodactylidae, Ceratophryinae) as having many layers of shed epidermis, corresponding to the continued accumulation of layers at nearly one every day. The north and central American burrowing hylid *Pternohyla fodiens* forms a multilayered cocoon (Ruibal and Hillman 1981), as does the hylid treefrog *Smilisca baudinii* (McDiarmid and Foster 1987) and the southern African bullfrog (*Pyxicephalus adspersus*; Ranidae) and treefrogs (*Leptopelis viridis* and *bocagei*; Arthroleptidae) (Schiotz cited in Loveridge 1976; Loveridge and Crayé 1979; Parry and Cavill 1979; Loveridge and Withers 1981; Grafe 2000). The salamanders *Siren intermedia*, *S. lacertina* and *Pseudobranchus striatus* also form a multilayered cocoon of shed epidermal layers, during summer aestivations (Gehlbach *et al.* 1973; Reno *et al.* 1972; Etheridge 1990).

The structure of the amphibian cocoon was originally described as a single layer of flattened cells, identical to the stratum corneum of the epidermis, but subsequent studies have shown the cocoon to be a series of accreted layers of keratinized squamous epidermal cells (van Beurden 1982, 1984) that eventually forms a multilayered cocoon of tens to hundreds of layers (Parry and Cavill 1979; McClanahan *et al.* 1976; Ruibal and Hillman 1981; Damstra 1983; Davies and Withers 1993; McDiarmid and Foster 1987; Reno *et al.* 1972; Withers 1995a; Withers and Richards 1995; Withers and Thompson 2000; McMaster 2006; Figs 3.11 and 3.12). The cocoon is initially a thin, plastic-like, transparent film but over time becomes a thick, opaque and paper-like covering. It completely envelops the frog (being fused across the eyes, mouth and cloaca) except for the external nares (Damstra 1983; McMaster 2006). The salamander, *Siren intermedia*, has been described to form a mucous cocoon (Reno *et al.* 1972), like the African lungfish (DeLaney *et al.* 1974), but its cocoon appears to have the same epidermal cell/ mucous structure as the anuran cocoon, as does the cocoon of *Siren lacertina* (Etheridge 1990). The capacity to form a cocoon has clearly evolved independently a number of times in various frog and salamander groups (see also Chapter 1.4). This is perhaps not surprising, as cocoon formation simply requires immobility and suppression of the normal response to eat the shed skin when it forms.

The number of layers in a cocoon depends on the length of aestivation, as the rate of shedding of skin layers seems to remain fairly constant throughout the aestivation period (McClanahan *et al.* 1983; Withers 1995; McMaster 2006). There are some differences in the rate of shedding between species, from <0.7 days per layer to 5 days per layer, but the thickness of layers is relatively constant across species, from about 0.4 to 0.7 μ (Table 3.6).

The physiological role of the cocoon is to reduce EWL from the epidermis (Lee and Mercer 1967; Withers 1998). Each layer of shed epidermis, once desiccated and cemented to adjacent layers by dried mucous, contributes an equivalent resistance to the cocoon (the total resistance of the cocoon is the sum of the resistance of each layer, since they function as resistors in series). McClanahan *et al.* (1983) calculated that each cocoon layer, added about daily, contributes about 2 s/cm of

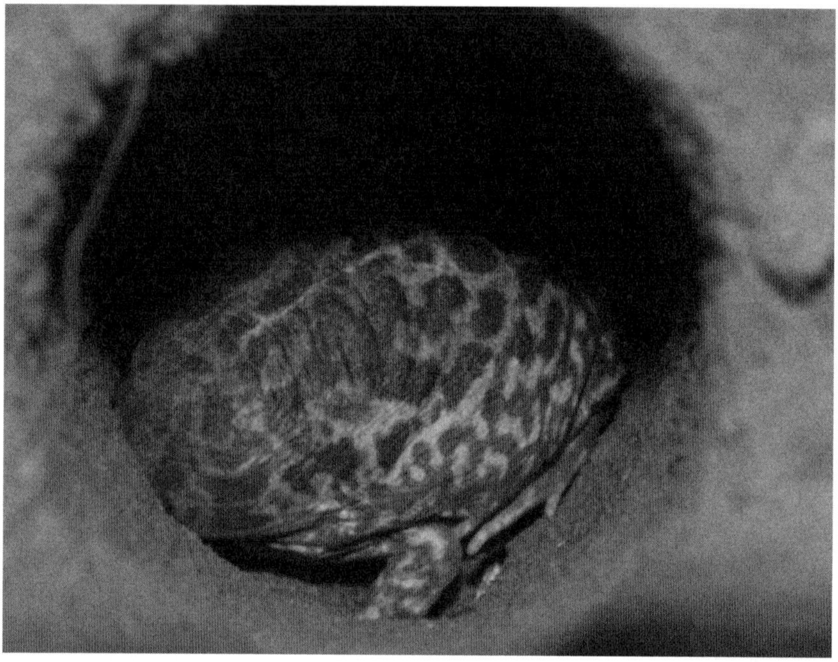

Fig. 3.11 Cocooned trilling frog (*Neobatrachus aquilonius*) in its burrow (loose sand removed from the burrow for clarity). (Photo: G. Thompson, P. Withers, V. Cartledge and K. McMaster).

resistance for *Lepidobatrachus*. The thickness and resistance of each cocoon layer is quite similar for various groups of cocooning frogs (Table 3.6). Resistance per layer varies from about 3 to 10 s/cm, and the resistance per thickness of cocoon varies from about 5.1 to 22.5 $\times$ 10^4 s/cm^2. Resistance to EWL varies widely for cocoons of different species, from <50 to over 400 s/cm (Table 3.5). This variability is primarily a function of the thickness of the cocoon, which reflects the number of layers present, that in turn is determined by the rate of shedding and the duration of cocoon formation. The very low EWL of cocooned aestivating frogs would ensure an extended survival period before EWL would deplete body water stores, for example many months, or even many years.

The amphibian cocoon is considerably hygroscopic, particularly at high relative humidity (Withers 1998; Fig. 3.13), so it is not immediately apparent why *in vivo* the cocoon does not absorb water from the epidermis, become highly hydrated and lose its structural integrity. The inner surface of the cocoon, which often intimately contacts the outer epidermis but is sometimes separated by a distinct air space, should be fully hydrated, even if the outside air was dry. However, if the RH gradient decreases linearly through the cocoon from 100% inside to 0%

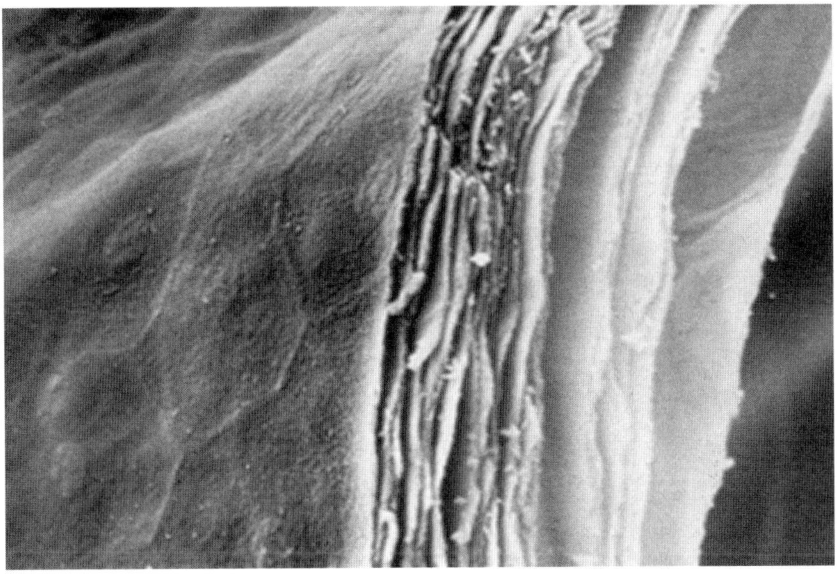

Fig. 3.12 Cocoon of the Australian water-holding frog *Cyclorana platycephala*, showing the single-cell squamous layers of shed epidermal cells. (Micrograph: P. Withers and T. Stewart).

outside (if the frog is in dry air), then the water content of cocoon material would decrease dramatically from the inside of the cocoon; only a thin part of the inner cocoon would be hydrated and structural integrity would be maintained by the remainder of the cocoon (Withers 1998).

3.3 Dehydration Tolerance

One of the unique capacities of amphibians is their ability to survive levels of water loss that would be lethal to other vertebrates. The limit of dehydration tolerance is usually referred to as the "critical activity point" (CAP), the amount of water lost at which an individual is unable to right itself when flipped on its dorsum (Ray 1958). It is usually expressed as the percentage of standard body mass (bladder empty; Ruibal 1962), lost through evaporation, at which the CAP is reached. Some researchers use the percent of body water lost to define CAP. For example if a frog was 80% water and lost 40% of its body mass to reach CAP, then CAP would be 40% as a percentage of body mass, but 50% of the body water (40%/80%). The level of dehydration tolerance tends to positively correlate with a species' primary environment; more-terrestrial species have a greater tolerance

Table 3.6 Comparison of physical properties of the cocoon with human skin and air (modified from Withers 1998)

	Thickness (μ)	Number of layers	Thickness per layer (μ)	Resistance s/cm	Resistance per layer s/cm	Resistance per cm s/cm
Human skin[1]	10.4	-	-	377	-	36.3×10^4
Cyclorana maini[2]	9.6	23	0.39	198	9.6	22.5×10^4
Cyclorana platycephala[2]	3.4	21	0.16	67	3.2	19.7×10^4
Pternohyla fodiens[3]	30	43	0.7	414	9.6	13.8×10^4
Pyxicephalus adspersus[4]	15	30	0.5	175	5.8	11.7×10^4
Neobatrachus kunapalari[2]	12.2	24	0.57	92	3.8	7.5×10^4
Neobatrachus sutor[2]	16.6	27	0.62	107	4.1	6.6×10^4
Lepidobatrachus llanensis[5]	19	38	0.5	116	3.1	6.3×10^4
Cyclorana alboguttata[6]	16	24	0.65	89	3.7	5.6×10^4
Neobatrachus pelobatoides[2]	8.0	16	0.57	41	2.6	5.1×10^4
Air[7]	-	-	-	-	-	4.2

Thickness, number of layers and resistance depend on the period of aestivation. Data from: [1]Scheuplein (1978a, b, c); [2]Withers (1998); [3]Ruibal and Hillman (1981); [4]Loveridge and Crayé (1979), Parry and Cavill (1979), Loveridge and Withers (1981); [5]McClanahan *et al.* (1976, 1983); [6]Withers and Richards (1996); [7]Withers (1992).

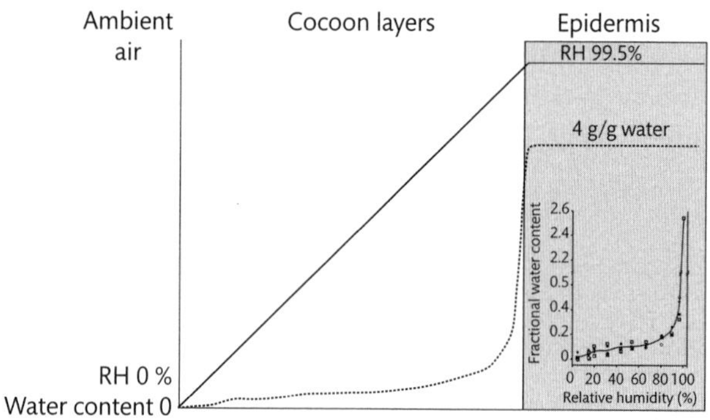

Fig. 3.13 Theoretical relationship between the relative humidity gradient across the cocoon and the water content of the cocoon layers, based on the hygroscopic properties of cocoon (inset). Modified from Withers (1998).

of dehydration than aquatic species (Thorson and Svihla 1943; Thorson 1955; Main and Bentley 1964; Schmid 1965; Farrell and MacMahon 1969; Ralin and Rodgers 1972; Hillman 1980). Anurans being more terrestrial tend to be more dehydration tolerant than urodeles.

These two observations suggest that natural selection has operated to favor enhanced dehydration tolerance as amphibians have come to occupy more terrestrial environments. Given the high rates of evaporation and high water turnovers of terrestrial species, it is not unexpected that periods can occur when bladder water reserves are exhausted and that natural selection would favor tolerance to water loss. Dehydration tolerance is not the result of inheritance of a single gene, but is polygenic (Layne *et al.* 1989). Interestingly, some researchers argue that dehydration tolerance may have served as a preadaptation for freeze tolerance in amphibians (see Chapter 4).

The distribution of dehydration tolerance for different families of amphibians indicates that that dehydration tolerance can be quite variable, even in the same family (summarized in Fig. 3.14 below). To put a human perspective on this figure, you would tolerate a dehydrational loss of body mass of 10–12% before death. Dehydration tolerance has only been determined for less than 1% of the 6100 plus amphibian species, so there is a limited phylogenetic perspective for this database. Values are also from a variety of different studies, and there is significant variation in experimental parameters between the studies that are known to influence CAP (see below). The general conclusions to be drawn from these data are: (1) tolerance

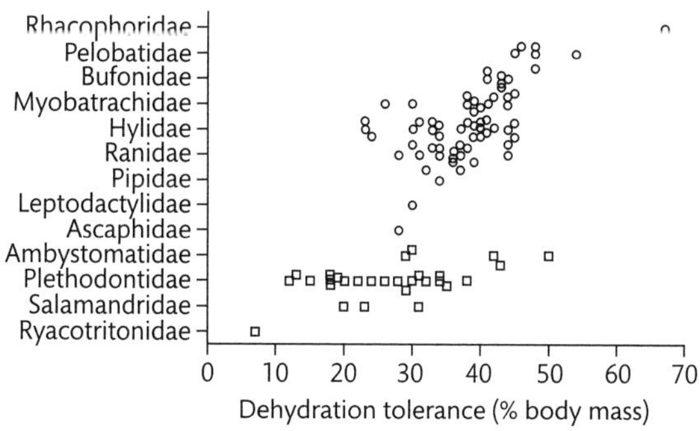

Fig. 3.14 Summary of dehydration tolerance for urodeles (squares) and anurans (circles). Urodele data taken from: Littleford *et al.* 1947; Ray 1958; Hutchison 1958; Whitford 1968; Spight 1968; Alvarado 1972, Larson 1971, Houck and Bellis 1972; Walters and Greenwald 1977; Marangio and Anderson 1977; and Gillis and Breuer 1984. Anuran data taken from: Thorson and Svihla 1943; Thorson 1955; Bentley *et al.* 1958; Main and Bentley 1964; Schmid 1965; McClanahan 1967; Krakauer 1970; Loveridge 1970; Johnson 1971; Heatwole *et al.* 1971; Ralin and Rogers 1972; Claussen 1973; Gillis 1979; Hillman 1980; Pough *et al.* 1983; Layne *et al.* 1989; and Newman and Dunham 1994.

of dehydration is greater for amphibians compared to other vertebrates; (2) there is a large range of dehydration tolerance in amphibians; (3) some families are more variable than others; and (4) species that are the most tolerant are characteristically the most terrestrial. Obviously, to look at adaptations involved in dehydration tolerance from a phylogenetic perspective, a fruitful approach would be to look within families that have the greatest range of tolerance. The urodele family Plethodontidae and the anuran Hylidae have the greatest ranges of dehydration tolerance.

Intraspecific variables known to influence dehydration tolerance are: (1) the rate of dehydration (Gillis 1979); (2) body size (Thorson 1955; Newman and Dunham 1994); (3) water content (Ralin and Rogers 1972); (4) population locality (Ralin and Rogers 1972; Larson 1971); (5) hybrid status (Layne *et al.* 1989); and (6) ontogenetic stage (Thorson 1955; Gillis and Breuer 1984). Consequently, specific attribution of a dehydration tolerance actually has a broad error band.

What are the physiological mechanisms that explain this interspecific differential tolerance of dehydration? Answering this entails first knowing what the proximate physiological issues are that dehydration presents to maintenance of a regulated homeostatic condition. As water is lost once bladder reserves are exhausted, coupled with the renal inability of amphibians to excrete solutes at a higher concentration than their own body fluids, some stresses can be predicted: an increase in body fluid concentration, a loss of plasma volume, an increase in blood viscosity, a loss of cell volume, and changes in the intracellular environment including an increase in oxygen free radical concentration. A summary of how these stresses would ramify to create cardiovascular and metabolic problems for an amphibian are delineated in Fig. 3.15.

First, there will be an increase in the body fluid osmotic concentration with dehydration. The concentration increases of the principle extracellular ions, Na^+ and Cl^- can be predicted from $C_iV_i = C_fV_f$ (Shoemaker 1964; Alvarado 1972; Hillman 1978). Since the extracellular osmotic concentration will equilibrate with the intracellular osmotic concentration, the osmotic concentration of both body fluid compartments will increase. This stress would be described as a hyperosmotic stress on all the cells of the body. Total osmotic concentration during dehydration is not as easily predictable, as some solutes are synthesized in association with metabolic processes and some extracellular ions are transported to the intracellular space. A good example of a synthesized solute is urea, the major nitrogenous waste product of most terrestrial amphibians (see Section 3.6). Any urea produced via protein catabolism during dehydration would not fit the $C_iV_i = C_fV_f$ paradigm (since it was not initially present) and fall above the predicted increase. Examples of the patterns of increase for a solute that is constantly added as dehydration continues and one that is removed during dehydration compared to the predicted concentration increase from concentration and volume are presented in Fig. 3.16.

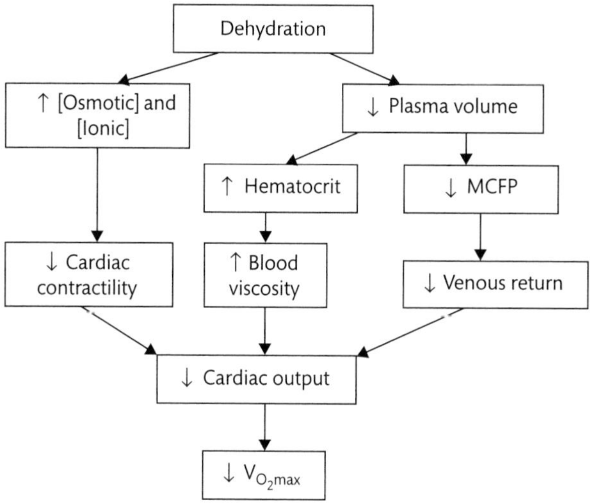

Fig. 3.15 Summary of the physiological stresses during dehydration in amphibians. MCFP, mean circulatory filling pressure.

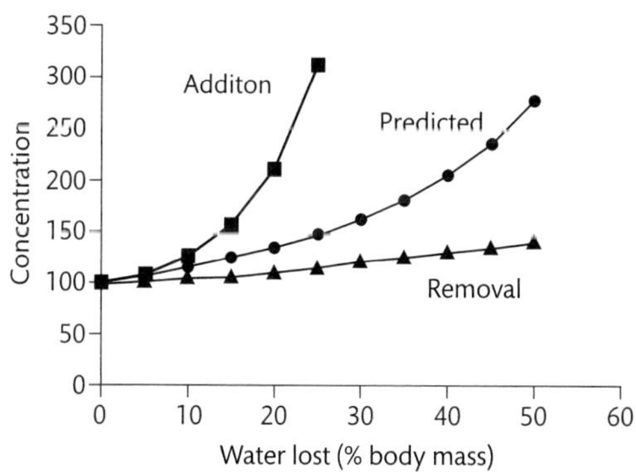

Fig. 3.16 The pattern of increase of a solute that follows the predicted $C_iV_i = C_fV_f$ pattern (circle), a solute which is removed during dehydration (triangle) and a solute which is added (1 unit/5% lost, square) during dehydration. C_i, initial concentration; C_f, final concentration; V_i, initial volume; V_f final volume.

Second, some of the water will be lost from the extracellular fluid space. This means that the partitioning of the lost extracellular fluids between the plasma and interstitial compartments takes on importance. As we saw in Chapter 2, blood volume (plasma volume and blood cell volume) determines, in part, venous return to the heart and its consequential effect on cardiac output via a Frank–Starling mechanism. If blood volume decreases, then the stress is known as hypovolemia and it would decrease venous return to the heart and decrease cardiac output. Obviously, the ability of the lymphatic system to mobilize interstitial fluids and the volume of interstitial fluid available to be mobilized will ameliorate the hypovolemic stress an amphibian experiences during dehydration.

Third, any loss of plasma volume by evaporation will increase the cellular fraction of the blood (hematocrit), as blood cells are not lost. As we saw in Chapter 2, blood viscosity is related exponentially to hematocrit. Blood viscosity is directly related to the vascular resistance to blood flow. This hyperviscous stress of dehydration is layered on top of the hypovolemic stress. Both stresses would independently compromise cardiac output. This will eventually lead to a decline in tissue P_{O_2} at rest as the dehydration tolerance point is reached. Consequently, the terminal stages of this compromised cardiac output will eventually lead to hypoxic cellular conditions then death (see Chapter 4).

Fourth, some of the water lost will come from the intracellular fluid compartment. Which cells lose water (osmometric strategy) versus which cells maintain volume (volume regulatory strategy) determines the proportion of this intracellular water loss. Data on the effect of dehydration on organ water contents (Fig. 3.17) show that some organs (brain, cardiac muscle, liver) lose less water

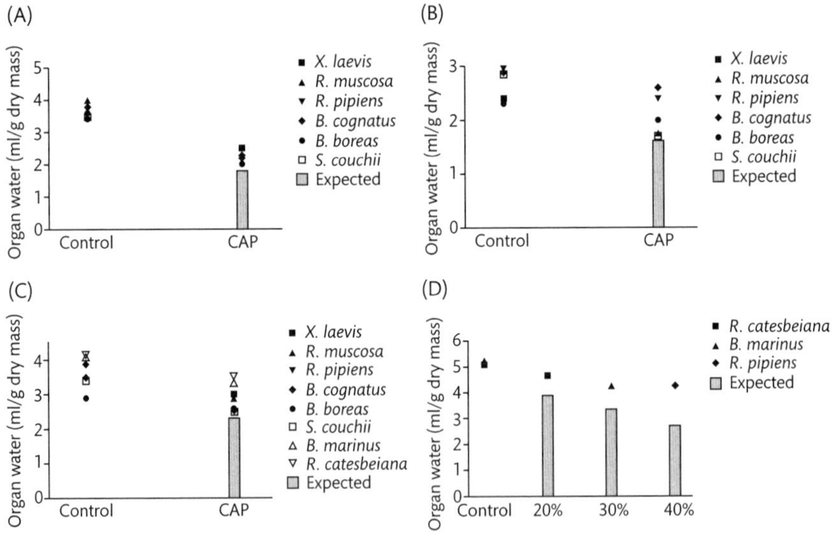

Fig. 3.17 Organ water content and the effect of dehydration at critical activity point or degree of dehydration on water content in (A) muscle, (B) liver, (C) heart and (D) brain from anuran amphibians. Expected is the water content if the organ loses in proportion to the whole organism. Data summarized from Hillman (1978, 1980, 1982, 1988); Churchill and Storey (1995); and unpublished data (liver).

than predicted if they lost water in proportion with the whole animal (Smith and Jackson 1931; Hillman 1980; Churchill and Storey 1996). These would be organs whose cells are volume regulating to a certain degree, and the presumption is that organ function would more likely be preserved. Skeletal muscle loses a greater fraction of its initial water (Smith and Jackson 1931; Hillman 1980, 1982; Churchill and Storey 1993); despite this, skeletal muscle functions effectively over a broad range of tissue water content (Hillman 1982). Jump distance is only impacted if anurans have lost significant amounts of water (>20% body mass; Beuchat *et al.* 1984; Preest and Pough 1989; Rogowitz *et al.* 1999). Therefore skeletal muscle, which has lost the largest fraction of its water, is still able to function normally to a substantial level with dehydration. Water loss from cells means that not all the water lost by evaporation has to come from the extracellular space. This is significant as 66% of the body water is intracellular. The overall dehydrational loss of water also means the intracellular osmotic concentration will track increases in the extracellular osmotic concentration. The specific solutes involved in this intracellular increase of solutes will vary dependent upon whether the cells principally volume regulate or volume conform (Lang *et al.* 1998). Of particular interest are the effects on organelle function, enzymatic function and the accumulation of oxygen free radicals.

Oxygen free radical (O_2^-) formation of H_2O_2 and its degradation in the presence of ferric ions (Fe^{3+}) leads to the formation of highly reactive hydroxyl radicals ($\cdot OH^-$). The formation of oxygen free radicals occurs when organisms are exposed to wide variations in O_2 tensions and metabolic activity. Dehydration, followed by rehydration, represents such a stress since it compromises systemic O_2 transport and upon rehydration there is an increase in P_{O_2} (Hermes-Lima and Storey 1998). There was an increase in the concentration of the H_2O_2 detoxifying enzymes, catalase in skeletal muscle and glutathione peroxidase in liver of dehydrated *Rana pipiens* (Hermes-Lima and Storey 1998). Hypoxia also leads to higher catalase activity in skeletal muscle of *R. pipiens* (Hermes-Lima and Storey 1996). These data suggest that enzymatic detoxification of H_2O_2 is involved in handling the consequences of both dehydrational and hypoxic stress at a cellular level.

Based on a blood pressure decline during dehydration for toads, the cardiorespiratory systems were the first systems hypothesized to fail during dehydration (Shoemaker 1964; McClanahan 1967). Further evidence for a cardiorespiratory insufficiency was predicted by measurements of V_{O_2max} during dehydration in *Xenopus laevis* (Hillman 1978; Fig. 3.18) where there is no effect of dehydration on standard V_{O_2}, but V_{O_2max} declines during dehydration, intercepting standard V_{O_2} at approximately the CAP. Organ energetics (ATP levels) at rest decline as the organismal metabolic capacity declines. Liver ATP begins to decline at 20% body mass loss for *Rana pipiens*, while total adenylates and energy charge are significantly decreased by 40% body mass loss (Churchill and Storey 1996). At CAP, resting whole animal lactate concentrations are elevated, indicating that

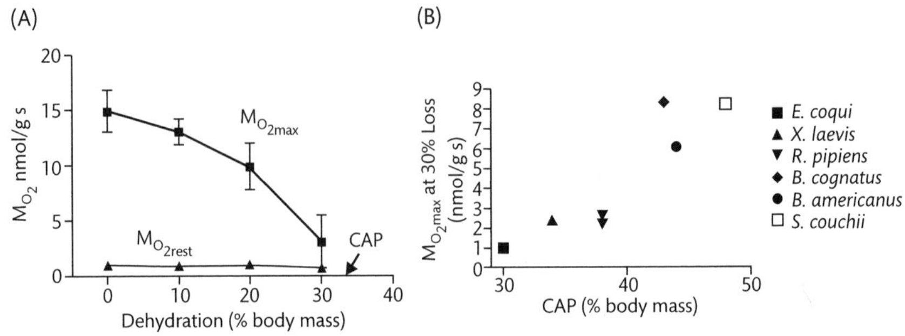

Fig. 3.18 (A) The effect of dehydration on resting and maximal rates of O_2 consumption for *Xenopus laevis* (modified from Hillman 1978) and (B) the maximal rates of M_{O_2} at 30% mass lost in a variety of species with different dehydration tolerances, the CAP (critical activity point). (Modified from Hillman 1991).

resting metabolic needs are being met by anaerobic metabolism (Hillman 1978). Anaerobic metabolism is a truly short-term solution given limited substrate (glucose) availability and the consequential deleterious effects of a metabolic acidosis associated with lactate production. This pattern of compromised $V_{O_{2}max}$ has been demonstrated for other anuran species during dehydration (Pough *et al.* 1983; Gatten 1987; Hillman 1991). The only difference is that resting V_{O_2} increases with dehydration for *Eleutherodactylus coqui* (Pough *et al.* 1983). Activity characteristically increases as dehydration occurs (Heatwole and Newby 1972; Putnam and Hillman 1977; Pough *et al.* 1983), presumably reflecting an animal's searching for a water source. A decline of $V_{O_{2}max}$ with dehydration is less clear for two species of salamander (Stefanski *et al.* 1989). The ability to maintain activity V_{O_2} during dehydration varies interspecifically, with more-dehydration-tolerant species maintaining a greater V_{O_2} as dehydration progresses (Fig. 3.18; Hillman 1991). This indicates that aerobic capacity is correlated with dehydration tolerance.

The greatest resistance to O_2 flux during $V_{O_{2}max}$ is the systemic O_2 transport step of the O_2 cascade (see Chapter 3). Consequently, evaluating the relative effects of the various stresses during dehydration are most important for the function of the cardiovascular system. At rest there is a progressive tachycardia with dehydration, which limits the heart rate scope possible with activity (Fig. 3.19; Hillman 1978). There is also a widening of the difference in arteriovenous O_2 content during dehydration at rest, limiting the scope in arteriovenous content during activity (Hillman 1978). Maximal blood flow rates decline in proportion to the decline in $V_{O_{2}max}$ and vary interspecifically (Fig. 3.19; Hillman 1987, 1991). The decline in blood flow rates are the result of a decline in stroke volume and an increase in vascular resistance, mediated via both an increase blood viscosity and vascular resistance (Hillman 1987). Dehydration does not elicit an increase in plasma cathepsin activity (Kimmel 1988). Cathepsins have been implicated as part of

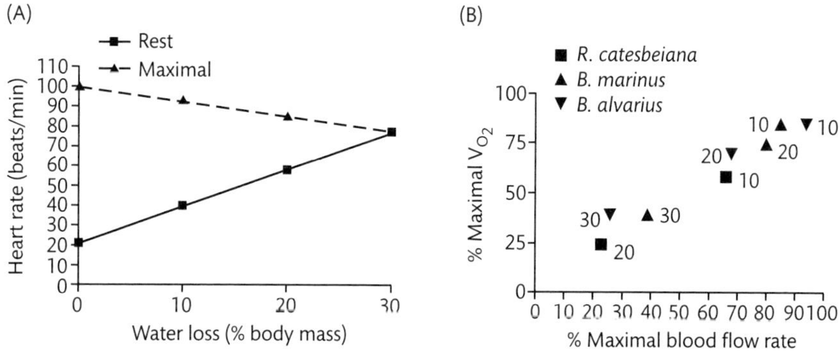

Fig. 3.19 (A) The effect of dehydration on resting and maximal heart rates in *X. laevis* (modified from Hillman 1978) and (B) the relationship between maximal blood flow rates (Q_{max}) and V_{O_2max} at different degrees of dehydration in three species (modified from Hillman 1991).

the myocardial depressant factors associated with circulatory shock in mammals (Kimmel 1988). Dehydration compromises systolic pressure and the rate of pressure development for both *B. marinus* and *R. catesbeiana* in a manner consistent with a hyperosmotic stress. (Hillman 1984). The maximum systolic pressure that can be developed is greater for *B. marinus* than *R. catesbeiana* (Hillman 1984). Given the interspecific consistency of a decline in cardiovascular capacity contributing to the decline in aerobic capacity, what stress on the cardiovascular system is the most important in accounting for the interspecific differences in cardiovascular capacity—hyperosmolality, hypovolemia or hyperviscosity?

The importance of these stresses to the differential maintenance of blood flow during dehydration can be gauged from the effect of each of the three stresses independently on maximal blood flow rates in species that differ in dehydration tolerance. If there were a significant interspecific difference in the effect of a stress on cardiac function, then it would prove useful to investigate further the physiological basis for this difference. Hyperosmolality of the extracellular fluids can lead to a redistribution of fluid from the intracellular fluid and/or or an increase in solute concentration, which can compromise organ function via a change in the intracellular environment. The effect of sequential salt loads on maximal blood flow rates (Q_{max}) in *B. marinus* and *R. catesbeiana* did not differ interspecifically, and did not fit the pattern of decline in V_{O_2max} in either species (Hillman and Withers 1988). This indicates that hyperosmolality is not the basis for the decline in Q_{max} during dehydration.

In both *B. marinus* and *R. catesbeiana*, maximal blood flow rate decreased in proportion to an increase in blood viscosity in both species (Hillman *et al.* 1985; Withers *et al.* 1991). This indicates that interspecific abilities to deal with hyperviscous stress do not explain differential dehydration tolerance.

The effect of hypovolemia induced by hemorrhage on Q_{max} demonstrated that the more terrestrial species, *B. marinus*, was able to maintain Q_{max} over a broader range of hemorrhagic blood loss, while Q_{max} of the semiaquatic *R. catesbeiana* was almost immediately compromised (Fig. 3.20; Hillman and Withers 1988). The toad was able to mobilize interstitial fluids to replace the volume removed by hemorrhage, hence maintain blood volume and cardiac output. This is also consistent with an interspecific differential ability to maintain hematocrit during dehydration in these two species (Fig. 3.20; Hillman *et al.* 1987). These data suggest that one component of differential dehydration tolerance is correlated with an ability to maintain blood flow over some level of dehydration, and plasma volume maintenance is one component of the ability to maintain blood flow. Plasma volume maintenance has two advantages from a cardiovascular perspective; venous return and cardiac output will be maintained and the stress of hyperviscosity will be delayed. What participates in a differential ability to maintain cardiac output under a hypovolemic challenge?

A variety of mechanisms have potential significance to the maintenance of Q_{max} during hypovolemia. One involves replacing lost plasma volume with interstitial fluid. The other involves sympathetic maintenance of enhanced cardiac output through direct effects on the heart and maintaining the central venous pressure during hypovolemia, which would assure uncompromised right atrial filling, and the maintenance of cardiac output via a Frank–Starling mechanism.

Are there interspecific differences in the ability to maintain central venous pressures? The effect of hypovolemia on central venous pressure can be minimized by having a more compliant vascular system and by having a developed sympathetic nervous system responses that can vary the vascular compliance when volume stressed. As we saw in Chapter 2, vascular compliance is defined as the change in

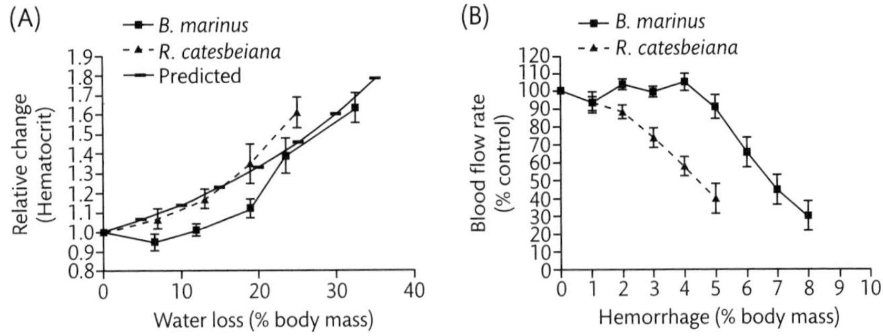

Fig. 3.20 (A) Proportional increase in hematocrit, a measure of blood volume, during dehydration; values are mean and SE. The predicted relationship is if blood volume followed the CV = CV (concentration times volume initial = concentration times volume final) relationship with dehydration. (B) The relationship of maximal blood flow rates with graded hemorrhage. Modified from Hillman *et al.* (1987) and Hillman and Withers (1988).

volume/ change in pressure. A loss of blood volume of 10 ml/kg in two circulatory systems of compliance 15 and 30 ml/kg kPa will lead to a net decrease in mean circulatory filling pressure of 0.67 and 0.33 kPa, respectively. The more compliant circulatory system has less compromise of mean circulatory filling pressure and eventually venous filling pressure. As we saw in Chapter 2, the two variables that determine vascular compliance are vascular distensibility and vascular volume. Vascular distensibility is the percent change in volume/change in pressure. If vascular compliance were an important component of dehydration tolerance, then we would expect more dehydration tolerant species to have increased vascular compliance that results from either increased vascular distensibility and/or vascular volume. Unfortunately, there are few data to evaluate this hypothesis. In a comparison of two species, *B. marinus* had a greater vascular compliance than *R. catesbeiana* (Hoagland 1997). This may result from a greater plasma volume of *B. marinus* compared to *R. catesbeiana* (DeGrauw 1998). Consequently, the hypothesis that enhanced dehydration tolerance is in part the result of increased vascular compliance needs to be evaluated on a broader comparative basis to be effectively tested, though the preliminary data suggest that vascular compliance may be a component of differential dehydration tolerance (Fig. 3.21).

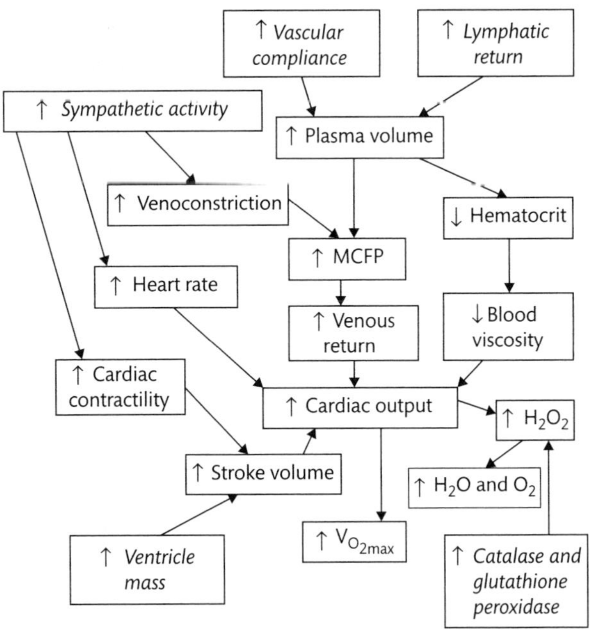

Fig. 3.21 Summary of various anatomic and physiological solutions (italicized) that can vary interspecifically and correlate with stabilizing cardiovascular and metabolic capacity during dehydration in amphibians. MCFP, mean circulatory filling pressure.

The ability of the vascular system to compensate for a decrease in vascular volume by stimulating heart rate, force of ventricular contraction, and venoconstriction to mobilize blood from venous storage are well-known mammalian responses during hypovolemic stress (Fig. 3.21). Amphibians have sympathetic innervation of both their heart and venous system, as well as adrenal chromaffin release of catecholamines (Morris and Nilsson 1994). There is an order of magnitude higher plasma concentration of epinephrine following hemorrhage for *B. marinus* than *R. catesbeiana* (Withers *et al.* 1988; Hillman *et al.* 1998). Adrenal release appears to be the predominant source for the plasma epinephrine concentration increase (Hillman *et al.* 1998). The much greater sympathetic response of the terrestrial *B. marinus* suggests that heart rate, strength of cardiac contraction and mobilization of venous blood by venoconstriction may be components of hypovolemic compensation. It also suggests that adrenal gland development may be a component of differential dehydration tolerance. Resting heart rates increase during dehydration, consistent with sympathetic stimulation (Hillman 1978). Dehydration compromises cardiac contractility, suggesting that sympathetic enhancement of contractility may be necessary (Hillman 1984).

The ability to mobilize interstitial fluid to stabilize blood volume is clearly a component of differential interspecific maintenance of Q_{max} (Hillman *et al.* 1987; Hillman and Withers 1988). The principal avenue of fluid return in an amphibian blood volume budget is lymph (Chapter 2). Increased lymphatic return of interstitial fluid would maintain plasma volume and prevent the increase in hematocrit associated with dehydration. How lymphatic return is increased is not completely understood. There are interspecific differences in the strength of lymph heart contractions, with *B. marinus* able to generate higher systolic pressures than *R. catesbeiana*, as well as higher lymph heart rates (DeGrauw and Hillman 2004). This suggests that interspecific differences in lymph heart output may be involved in differential lymph mobilization. A second possible mechanism is a differential ability to bring lymph to the lymph heart to be pumped into the venous circulation (Drewes *et al.* 2007; Hedrick *et al.* 2007).

The ability to increase cardiac output by increasing stroke volume is another potential anatomic solution to the maintenance of greater cardiac outputs during dehydration. One estimate of stroke volume is ventricular mass. Dehydration tolerance is correlated with ventricular mass in neobatrachian amphibians (Hillman 1980; Hillman *et al.* 2000; Withers and Hillman 2001). Increased ventricular mass may also be correlated with increased vascular compliance and greater peak systolic pressures.

Another piece of evidence germane to the understanding of comparative dehydration tolerance in amphibians is the fact that if species are salt loaded then there are interspecific differences in plasma sodium concentrations that are tolerated, and more dehydration tolerant species tolerate higher concentrations (Hillman 1980). This suggests that ionic and osmotic concentrations can play a role in the demise of regulated physiological function, independent of the compromise of cardiovascular function. It is important to recognize that an increase of plasma Na of 20 mmol/l in humans can induce severe nervous system dysfunction leading quickly to 50% mortality in adults (Engle and Jaeger 1954; Arieff *et al.* 1977). The exact cause of nervous system dysfunction is not clear but seems to be a combination of brain shrinkage inducing hemorrhage and electrolyte effects on excitable cell function (Arieff *et al.* 1977). Contrastingly, amphibians will tolerate sodium concentration increases of over 100 mmol/l, and still maintain normal behavior. The physiological mechanisms involved in maintaining nervous system function over such a wide rage of osmolalities and electrolyte concentrations remain to be elucidated.

3.4 Temperature

Temperature is one of the most important physical aspects of the environment to affect amphibians. Temperature effects range from short-term to long-term and determines in part global species richness patterns (Buckley and Jetz 2007) and influence virtually all physiological systems including metabolism, O_2 transport, enzymatic digestion, muscle contraction, solute transport, vision, hearing, development and metamorphosis, growth, reproduction, regeneration and even sex reversal. Thermal effects on physiological systems are reflected in thermal dependence of integrated physiological performances, such as locomotion, digestion and iono/osmoregulation.

All amphibians are ectotherms; their internal heat production from metabolism is so low compared to rates of heat exchange with the environment that metabolic heat has no significant effect on body temperature (T_b) (see Chapter 1). For example, the T_b of the large aquatic salamander *Amphiuma means* is within 0.1°C of water temperature (Bohlin and Beitinger 1979). Consequently, body temperature of most amphibians, particularly the aquatic or nocturnally-active species, is determined only by their environment, and generally is similar to ambient temperature (T_a).

Amphibians tend to be active at whatever suitable temperatures are available in their microhabitats. They are seldom able to regulate a relatively constant T_b, unlike many ectothermic reptiles (Huey and Slatkin 1976; Carey 1978). In cold environments, amphibians tend to be sluggish and can be in danger of extreme cooling or even freezing. In hot environments, amphibians can be in danger of exceeding their upper temperature limit, as well as dehydration from losing excessive water by evaporation.

The annual timing of activity by amphibians is influenced in part by temperature. For hylid frogs, breeding phenology is partially constrained by their thermal sensitivity (John-Alder *et al.* 1988). For temperate species, the higher-latitude species have the best low-temperature thermal performance, whereas southern species have the best high-temperature performance. Subtropical and tropical species have an even higher maximum thermal tolerance. It is likely that these thermal sensitivities of amphibian populations will be affected by climate change. For example, reproductive cycles can respond to climate change; spawning time of *Rana esculenta* and arrival time at ponds by *Triturus vulgaris* have become earlier by about 9–10 days per 1°C increase in T_a in Britain (Beebee 1995).

3.4.1 Thermal Environments

Amphibians are found in both aquatic and terrestrial habitats, in a wide range of thermal environments (e.g. Brattstrom 1970). Body temperature can range from subzero for a few high-latitude salamanders and frogs, to over 40°C for basking water-proof frogs (Fig. 3.22). There are various patterns for the thermal range of amphibians, including phylogeny, latitude and altitude effects.

There are phylogenetic patterns for the thermal range of amphibians, for example salamanders, as a group, are generally found at lower temperatures than frogs. In North and Central America, the thermal range for salamanders is about −2.0 to 30.0°C, compared to 3.0 to 35.7°C for anurans (Brattstrom 1963; Feder *et al.* 1982). Neotropical salamanders have higher minimum temperature tolerances than temperate (ambystomatid and plethodontid) salamanders (Feder and Lynch 1982); tropical anurans have a higher thermal regime than temperate-zone anurans, not only interspecifically but also intraspecifically (Brattstrom 1968, 1970b). Similar differences in thermal environment due to altitude are reflected by marked variation in thermal tolerance, for example tropical salamanders have

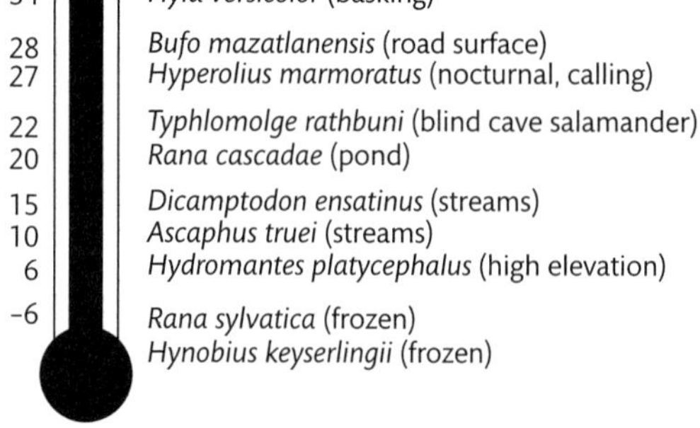

Fig. 3.22 Examples of body temperature range observed for various amphibians. Data from Brattstom (1963), White (1979), Passmore and Malherbe (1985), Kobelt and Linsenmair (1995), Shcherbak and Kovalynkh (1974), McClanahan *et al.* (1994), Schmidt (1982).

a lower mean T_b at higher elevation (Feder and Lynch 1982). There are altitudinal clines in thermal tolerance of populations of chorus frogs on the eastern face of the Rocky Mountains (Miller and Packard 1977).

Aquatic amphibians are generally buffered from the extremes of low and high temperatures that terrestrial amphibians can experience. The high heat capacity and thermal diffusivity of water mean that T_b cannot differ much from T_a, and an aquatic amphibian exchanges heat very rapidly with its environment (Chapter 1). Aquatic environments tend to be thermally stable over the daily cycle, for example the thermal springs frog *Somuncuria somuncurensis* has a near constant T_b of about 21°C (Cei 1969). However, shallow, ephemeral desert ponds can fluctuate markedly in temperature on a daily cycle, and developing eggs and larvae must cope with the same thermal regime. Adults and tadpoles of *Cyclorana cultripes*, *C. platycephala* and adult *Litoria rubella* have been found in ephemeral desert ponds at water temperatures up to 39.2°C (Main 1968). Many aquatic amphibians survive low temperatures, but few survive 0°C, which is the lowest T_b possible for an aquatic amphibian in freshwater, even though their body fluids would not freeze until about −0.35°C (depending on osmotic concentration); fresh water will freeze before the amphibian. A few amphibians survive freezing to moderate temperature, for example −5°C. Aquatic amphibians are unable to ameliorate high water temperatures because they cannot evaporatively cool. Heating and cooling in water are relatively fast compared to air. The time constant (time for half of the temperature change; τ) is higher for larger aquatic amphibians, that is cooling/ heating is slower (Table 3.7). The magnitude of the thermal time constant is essentially independent of the direction of temperature change, but it is dependent on body size/ surface area of the amphibian; larger individuals

Table 3.7 Comparison of cooling and heating constants for amphibians in water and air

	τ Cooling (min)	τ Heating (min)
Water		
Rana pipiens (54 g)	≈1.7	≈1.7
Necturus maculatus (133 g)	2.57	-
Amphiuma means (500 g)	3.88	3.61
Air		
Rana sphenocephala (5 g)	4.26	4.51
*Hyla cinerea** (5 g)	7.34	7.02
Hyperolius viridiflavus* (0.5 g)	≈0.4	-

*The treefrog *Hyla cinerea* and reedfrog *Hyperolius viridiflavus* have a reduced EWL. Data from Bohlin and Beitinger (1979), Anderson and Beitinger (1979), Wygoda (1988, 1989), Tracy *et al.* (1992), Kobelt and Linsenmair (1995).

have greater thermal inertia. Live mudpuppy salamanders (*Necturus maculosus*) cool faster when alive with their gills ligated than when dead, suggesting that blood circulation through their external gills is responsible for about half of their heat exchange (Beitinger and Anderson 1979).

The much lower heat capacity and thermal diffusivity of air mean that terrestrial amphibians exchange heat more slowly with their environment, and their T_b can differ much more from T_a than for aquatic amphibians. Heating and cooling are substantially slower, and the time constant is higher, for terrestrial compared to aquatic amphibians (Table 3.7). Amphibians that bask in sunlight (heliothermy) or in contact with warm objects (thigmothermy) can have a T_b substantially higher than T_{air}. A general thermal model suggests that the predicted difference between T_b and T_{air} is higher for larger basking amphibians, for example 40°C for 0.1 g and >50°C for 1000 g individuals (Stevenson 1985). However, their capacity for behavioral and physiological thermoregulation means that such large differences are not actually achieved, and indeed would be lethal for most amphibians.

A few salamanders are active at T_a near 0°C, for example *Hydromantes platycephalus* as low as −2°C (Brattstrom 1963), *Ambystoma* spp. at <1°C (Feder *et al.* 1982), *Eurycea multiplicata* at 0°C (Ireland 1976) and *Ensatina escholtzi* at 1°C (Stebbins 1954). A few anurans are active at slightly higher temperatures, for example *Bufo boreas* at 3.0°C (Brattstrom 1963) and *Bufo bocourti* at 3.5°C (Stuart 1951). Most amphibians cannot survive freezing temperatures, but there are some notable exceptions. Some can supercool to below their normal freezing point, for example juvenile spadefoot toads (*Scaphiopus bombifrons*) survive supercooling to −4°C, but not freezing (Swanson and Graves 1995). The anurans *Hyla crucifer, H. versicolor* and *Rana sylvatica* survive cooling to about −6°C by cooling slowly and freezing extracellularly but not intracellularly (see Chapter 4.3; Schmid 1982). The Siberian salamander *Hynobius keyserlingi* has been reported to survive frozen inside ice for about 90 years (Shcherbak and Novalyukh 1973; Berman *et al.* 1984).

Thermal microclimates are important for amphibians. Some terrestrial environments are relatively thermally stable over the day/night cycle; bromeliad frogs have a fairly constant T_b, similar to T_a, due to the thermal constancy of their floral environment (Feder 1982a; Feder and Lynch 1982). The strictly subterranean caecilian *Boulengerula boulengeri* migrates vertically in the soil, occurring at a higher frequency near the surface at temperatures greater than about 20°C (Measey and Barot 2006). The caecilian *Schistometopum thomense* varies in mass along a thermal (about 10°C) and altitudinal (about 1000 m) gradient (Measey and Dongen 2006), suggesting that the thermal environment they experience influences their biology. Foam nests provide a thermally stable and moist environment for tadpole and larval development of various anurans, for example *Physalaemus pustulosus* (Dobkin and Gettinger 1985) and *Rhacophorus arboreus* (Kusano *et al.* 2006).

Physical and Chemical Effects

Temperature is essentially a measure of the average velocity of atoms and molecules; particles in a solid, a liquid and a gas at the same air temperature (T_a) have the same average molecular velocity (v). Heat content is a more complex variable, depending on average particle velocity and kinetic energy ($= \frac{1}{2} m v^2$), molecular density (highest for solids, lowest for gases), and physicochemical properties of the particular atoms or molecules. Heat capacity *per se* does not affect the steady-state physiology of amphibians, although it influence the dynamics of non-steady-state heat exchange. Temperature on the other hand has a direct and substantial effect on physiological functions.

The effect of temperature on physical or chemical reactions is to exponentially increase the average molecular velocity, hence their energy level and the rate at which molecules come into contact. Two common measures of the effect of temperature on reaction rate are Q_{10} and activation energy (E_a; kJ/mole).

Q_{10} is the ratio of reaction rates at two different temperatures (R_1 at T_1 and R_2 at T_2) and 10°C is the standard temperature difference used to calculate Q_{10}. If the observed temperature difference is not exactly 10°C, then the Q_{10} can be calculated as $(R_2/R_1)^{10/(T_2-T_1)}$. Q_{10} is the most common temperature coefficient used to describe the effect of temperature on a physiological reaction rate. For example, the heart rate of the frog *Rana temporaria* has an exponential relationship with temperature *in vivo* and *in vitro* (Harri and Talo 1975a, b), equivalent to a Q_{10} of about 2.5 to 3 *in vivo* and *in vitro* (Fig. 3.23). In biological systems, Q_{10} is commonly in the range of 2 to 3 (see Fig. 3.31).

Activation energy (E_a) is an alternative temperature coefficient, based on thermodynamic theory of chemical and biochemical reactions. Since the rate of

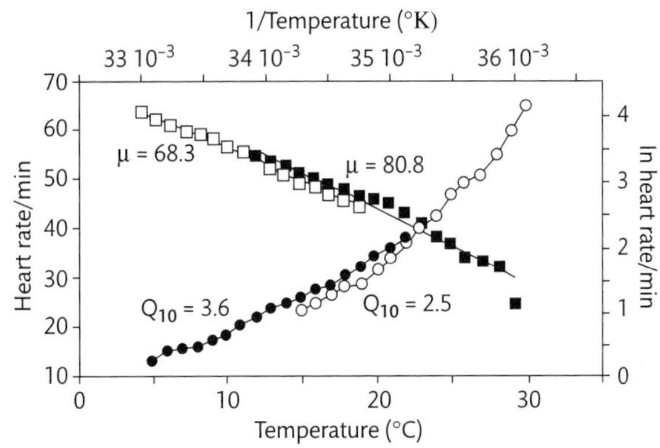

Fig. 3.23 Exponential relationship between heart rate and temperature for *Rana temporaria in vivo* (o) and *in vitro* (•), with Q_{10} (circles; temperature and heart rate scales) and Arrhenius plot μ coefficients (squares; 1/Temperature and ln heart rate scales). Data from Harri and Talo (1975a, b).

a chemical reaction (Rate) can be described as k' $e^{-Ea/RT}$ (Arrhenius' equation), then ln (Rate) = − (E_a/R)(1/T) + ln (k'). Consequently, a regression of ln(Rate) against (1/T) has a slope of $-E_a/R$, which is a temperature coefficient over the range of temperatures, and an intercept of ln(k'). Physiological processes are not simple one-reaction biochemical pathways but are complex sequences of many individual reactions, each with its own kinetic properties and thermal dependence, so the term critical thermal increment (μ) is often substituted for E_a when the concept of activation energy is applied to physiological systems, to reflect their greater complexity. μ is commonly in the range of 50 to 100 kJ/mole. For example, heart rate of the frog *Rana temporaria* has a μ of 68.3 *in vivo* and 80.8 kJ/mole *in vitro* (Fig. 3.23).

Physiological Effects

For biological functions, temperature has the expected physicochemical effect on reaction rate; Q_{10} is commonly 2 to 3 and μ is 50 to 100 kJ/mole. As temperature increases, reaction rates and physiological function rates increase exponentially but only to a critical upper thermal temperature at which enzymatic catalytic function becomes adversely affected; the high random thermal motion of molecules destabilizes the active site and ultimately denatures enzymes. For amphibians, these critical upper thermal temperatures for biochemical and physiological functions tend to be low, at about 35–40°C, compared to other tetrapods (typically >40°C for reptiles, birds and mammals).

Although all physiological functions in amphibians are temperature dependent, the thermal effects on some particular systems have profound effects on the survival and fitness of individuals. For example, the effects of temperature on muscle and locomotor systems have an impact on all activities, including feeding, predator avoidance and reproduction; the effects of temperature on digestion have ecological significance by controlling rates of energy assimilation; effects on membrane transport affect on iono- and osmoregulation and intracellular/ extracellular ion concentrations, hence nervous system function. Thermal tolerance varies for different life stages, and for various tissues and organs. For example, tadpoles are more sensitive to temperature than adults, and various adult tissues (Fig. 3.24).

Thermal effects on physiological systems are reflected in thermal dependence of integrated physiological performances, such as locomotion, digestion and iono/osmoregulation (Huey and Stevenson 1979). Each physiological function has an optimal temperature, and the range of temperatures over which an amphibian performs "well" is the thermal performance breadth. For example, the sprint speed of the toad *Bufo americanus* has an optimum at about 28°C and a thermal performance breadth (defined as at least 80% of the maximal sprint speed) from about 21 to 34°C (Tracy *et al.* 1993; Fig. 3.25). In general, we would expect different physiological systems to have similar, or overlapping, thermal performance

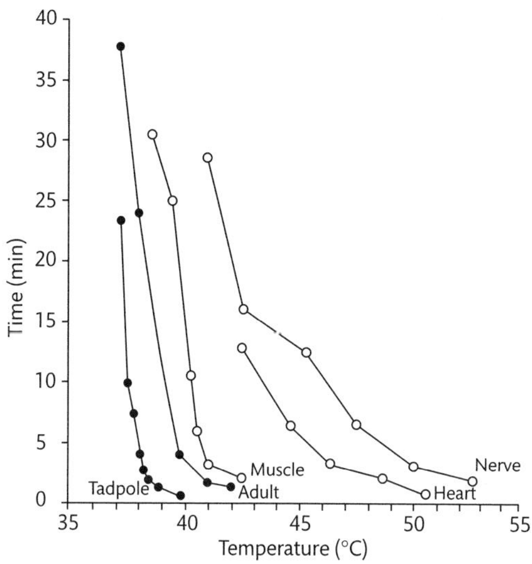

Fig. 3.24 Time to thermal heat death for tadpoles and adults, and adult muscle, heart and nerve, of the grass frog *Rana pipiens*. Data from Orr (1955).

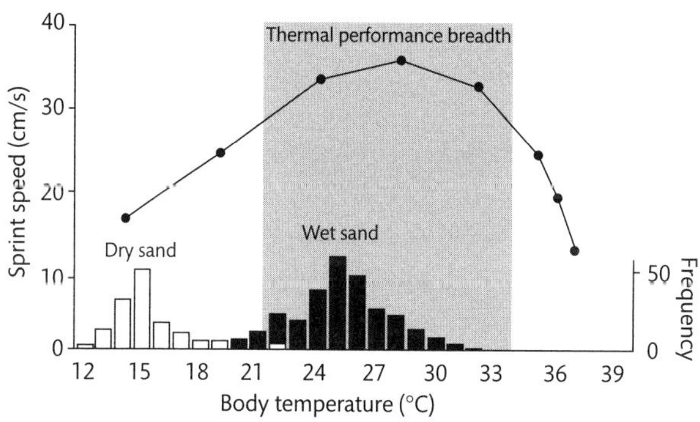

Fig. 3.25 Thermal performance breadth of sprint speed, and selected temperatures on wet and dry soil, for the toad *Bufo americanus* (modified from Tracy *et al.* 1993).

breadths. We would also expect the range of temperatures selected by an amphibian to match its thermal performance breadth. This is the case for *Bufo americanus* when normally-hydrated on wet soil, but not in dehydrating conditions of dry soil (Tracy *et al.* 1993; Fig. 3.25). Thermoregulatory behavior would seem to be coadapted with thermal physiology (e.g. Angilletta *et al.* 2002).

Frog muscle has been a classic model for muscle contraction studies, because it functions normally (albeit slowly) at low temperature, facilitating measurement of various *in vitro* parameters such as force and power (e.g. Hill 1938, 1970; Rome *et al.* 1992). Actions of muscles, such as rate of twitch and tetanic force generation, speed of shortening and power production, generally have a high Q_{10}, about 3–4 at low T_a and 2 at higher T_a. However, tension-related properties of muscle are nearly independent of temperature with a Q_{10} of 1.0 to 1.2 (Bennett 1984). For example, an increase in temperature from 10 to 20°C results in a relatively minor increase in force of contraction but large increases in velocity of contraction and mechanical power. Although the metabolic rate of muscle for a single contraction has a typical Q_{10} (2–3) the efficiency of contraction is nearly independent of temperature (Hill 1938).

In general, amphibian locomotion has a high thermal dependence, although the reason for the thermal dependence can be quite complex and variable (Rome *et al.* 1992). The effects of temperature have been described for swimming salamanders (Else and Bennett 1988) and frogs (Miller 1982), walking salamanders (Else and Bennett 1988), hopping toads (Londos and Brooks 1988) and jumping frogs (Hirano and Rome 1984; Huey and Stevenson 1979; Tracy 1979). Locomotor performance is generally poor at low temperatures, peaks at some intermediate temperature that is often between 25 and 30°C, then declines at higher temperatures. For example, jump distance of the frog *Rana pipiens* increases with temperature to a maximum of about 0.4 m at 25°C, then declines at higher temperatures; peak power generation shows a similar thermal dependence (Hirano and Rome 1984). Jump distance of the frog *Rana clamitans* has an optimal at about 20°C, but a wide thermal performance breadth from about 10 to 25°C (Huey 1975; Huey and Stevenson 1979; Fig. 3.25). Swimming speed is highest at 15°C for the mudpuppy *Necturus maculosus* and at 27°C for the aquatic frog *Xenopus laevis*, and these reflect thermal preferences of these species in the laboratory (Miller 1982). Thermal performance breadth is about 21 to 34°C for the toad *Bufo americanus* with a peak at 29°C (Tracy *et al.* 1993; Fig. 3.25). Jump distance of frogs (*R. pipiens*) increases linearly with peak power generation and temperature (Hirano and Rome 1984) up to about 25°C, above which distance and power decline. This might reflect failure of neuromuscular function to fully recruit all muscle fibers for locomotion. The relationship between jumping distance and T_a of the striped marsh frog (*Limnodynastes peronii*) varies over a geographic range, with populations from cooler temperatures performing better at lower temperatures (Wilson 2001).

The contractile properties of muscles *in vitro* can explain thermal dependence at some temperatures, but alone are generally not sufficient (Rome *et al.* 1992). Recruitment of anaerobic muscle fibers can limit the sustainability of locomotion, especially at low temperatures. The maximum rate of power generation can limit muscle performance. For swimming, the locomotor muscles must overcome drag so the power generating capacity of muscle might be limiting. During swimming at low temperatures (<8°C), the muscle twitch contraction/ relaxation times of the frog *Hyla crucifer* appear to limit limb stroke frequency but at higher temperatures the limiting factor may be power generation (John-Alder *et al.* 1989). For walking on land, drag is minimal compared to the power required for accelerating and decelerating the limbs, and the requirements for cyclical muscle contraction and relaxation may be limiting. The stride frequency of *Ambystoma* salamanders walking at 20°C is set by the muscle twitch contraction/ relaxation time, although at 10°C the contractile time (324 ms) is greater than the time for one half of a stride (203 ms; Else and Bennett 1987); this discrepancy suggests that *in vitro* muscle function measures do not necessarily apply *in vivo*.

Temperature influences a variety of feeding and digestive functions. Feeding is a complex interaction of sensory detection, behavioral motivation and muscle performance. For toads (*Bufo americanus*) the optimal temperature for feeding on mealworms is 20°C; feeding rate decreases at lower and higher temperatures (Stevens 1988). The thermal dependence of this feeding response is greater than that for locomotor performance and metabolic rate (Renaud and Stevens 1983). The green treefrog *Hyla cinerea* has a higher feeding rate at higher T_a (Freed 1980).

3.4.2 Thermoregulation

The body temperature (T_b) of many animals cannot often be defined by a single measurement because of internal thermal gradients, such that the exact value of "body" temperature depends on where it is measured. The relatively small size of amphibians generally limits their capacity to establish internal thermal gradients, so deep core body temperature ($T_{b,core}$) can generally be measured by a thermometer or thermocouple inserted into the cloaca ($T_{b,cloacal}$). However, the high cutaneous EWL of most amphibians can result in skin temperature ($T_{b,skin}$) being substantially lower than deep core temperature. There are sometimes small thermal differences that are difficult to explain. For example, the painted reed frog (*Hyperolius marmoratus*) has T_b about 3.1°C higher than T_{air} while calling from exposed sites on reeds overhanging the water's edge (Passmore and Malherbe 1985); T_b is however 1.0°C lower than the T_{water}. Their slightly higher T_b than T_{air} is not due to the metabolic heat production of calling; females have a similar T_b elevation.

Some amphibians, particularly small nocturnally-active species, thermoconform to their thermal environment, that is their T_b is essentially the same as the mean thermal environmental temperature. Many amphibians, particularly day-active species, can use both behavioral and physiological mechanisms to maintain T_b near a preferred value ($T_{b,pref}$). The primary source of heat for warming above ambient temperature is solar radiation, and the primary means of cooling below ambient temperature is evaporation (e.g. Brattstrom 1970).

Field and Laboratory T_b

There are many measurements of T_b for amphibians when active or inactive in the field (Fitch 1956; Brattstrom 1963, 1979; Hutchison and Dupré 1992). However, it is important to recognize that correct interpretation of the significance of field T_b measurements relies on the context in which the measurements were made and the methods that were used (e.g. Avery 1982; Hutchison and Dupré 1992). Behavior, physiology and recent thermal history (e.g. time since emergence, exposure to solar radiation or wind, skin chromatophore dispersion, etc.) all affect T_b. For example, Heath's (1964) "beer-can thermoregulation" experiment illustrated the danger in interpreting thermal data as indicating behavioral thermoregulation, and low variability of T_b does not necessarily indicate tight thermoregulatory control as animals in a thermally-stable environment can passively have a low T_b variability (e.g. Huey and Slatkin 1976).

The T_b of an amphibian in the field is the result of a complex set of thermal inputs and outputs; inputs include conduction, convection and radiation (but not metabolism); outputs can include conduction, convection, radiation and evaporation (see Chapter 2). Aquatic environments are generally more thermally-constraining than terrestrial environments. Amphibians are often active in nature at a characteristic T_b that is often presumed to reflect their thermal preference but can also reflect thermal constraints of the environment. For example, a terrestrial amphibian might not be able to behaviorally warm to its preferred T_b if the environment does not provide suitable conditions, for example a frog that basks to elevate T_b cannot warm as much in shady or overcast conditions.

Field T_bs of generic or family groups of amphibians show some general correlations with climatic conditions (Table 3.8). No thermal data are available for caecilians. Temperate, aquatic plethodontid salamanders have the lowest mean $T_{b,field}$ whereas aquatic cryptobranchids, amphiumids and sirenids have the highest, reflecting their high environmental temperatures. The lowest $T_{b.field}$ for a salamander is −2.0°C for *Hydromantes platycephalus* walking on snow (Brattstrom 1963, 1970a), and the highest is 30.0°C for the tropical *Bolitoglossa occidentalis* (Feder *et al.* 1982). For frogs and toads, cool-environment aquatic frogs have a low $T_{b,field}$ (e.g. *Ascaphus*) whereas warm-environment aquatic frogs have a higher $T_{b,field}$ (e.g. *Gastrophryne*). T_b of *Bufo canorus* in the field is 2.2°C (Karlstrom 1962) and T_b is 35.7°C for a "non-waterproof" frog *Gastrophryne olivacea* (Fitch 1956) and up to 43°C for a "waterproof" frog *Hyperolius viridiflavus* (Kobelt and Linsenmair (1992). The bullfrog *Rana catesbeiana* has a field T_b of 26–33°C while diurnally active, achieved by changes in posture and location; radiation is used as a heat source and pond water as a thermal sink (or sometimes source), and evaporation is used for cooling (Lillywhite 1970).

Preferred T_b measured in the laboratory is sometimes similar to field T_b, suggesting that the thermal environment is sufficient for these amphibians to attain their preferred T_b. For example, juvenile *Rana catesbeiana* have essentially the same T_b in a laboratory thermal gradient as in the field (Lillywhite 1971a). However, the mean or even range of T_bs observed for a species in the field often differs from that measured in the laboratory, and so laboratory "preferred" T_b

Table 3.8 Summary of field body temperatures for amphibians of various generic and family groups, from a range of thermal environments; modified from Duellman and Trueb (1994)

	Habitat	Field Temperature (°C)	
		Mean	Range
Caecilians	No data available		
Salamanders			
Plethodontids	temperate aquatic	11.3	2.0 to 22.0
	temperate terrestrial	13.5	-2.0 to 26.3
	tropical terrestrial	14.2	1.8 to 30.0
Ambystomatids	temperate terrestrial	14.5	1.0 to 26.7
Salamandrids		16.0	4.5 to 28.4
	tropical terrestrial	19.0	10.5 to 30.0
Cryptobranchids, amphiumids, sirenids	aquatic	20.1	8.0 to 28.0
Anurans			
Ascaphus	aquatic	10.0	4.4 to 14.0
Rana	terrestrial	21.3	4.0 to 34.7
Scaphiopus	terrestrial	21.4	12.2 to 25.0
Hylids	terrestrial	23.7	3.8 to 33.7
Bufo	terrestrial	24.0	3.0 to 33.7
Leptodactylids	terrestrial	24.7	22.0 to 28.0

does not necessarily equate to field T_b. For example, the laboratory $T_{b,pref}$ of six plethodontid salamander species was 1.1 to 5.7°C lower than $T_{b,field}$ (Feder *et al.* 1982). Laboratory preferred T_b alters with respect to many conditions, such as feeding state, developmental stage, reproductive state, oxygen and carbon dioxide levels, time of day, season and acclimation temperature. Nevertheless, $T_{b,pref}$ is usually nearer to the upper thermal tolerance limit than the lower limit. The $T_{b,pref}$ of Australian tropical *Litoria* is correlated with their skin R and ecological habitat (arboreal > terrestrial > aquatic; Tracey and Christian 2005).

Behavioral Thermoregulation

Many species of amphibians are capable of some behavioral control of their body temperature. In both natural and artificial situations, many diurnal amphibians also behaviorally or physiologically thermoregulate by using sunlight as a heat source, by heliothermy or thigmothermy. Even nocturnal amphibians have some behavioral control of T_b while in their retreat during the day, although when active at night their T_b usually closely approaches the ambient air temperature or is less if they are cooled by evaporative water loss.

Aquatic amphibians have very limited scope for physiological thermoregulation, dependent on the range of water temperatures available in their environment. Tadpoles of the tailed frog *Ascaphus truei* avoid T_a >22°C; first-year tadpoles tend to congregate in water at <10°C and second-year tadpoles at >10°C (de Vlaming and Bury 1970). Larval *Ambystoma tigrinum* select 16–31°C, *Rana catesbeiana* 12–33°C, and *Rana pipiens* 13–35°C (Lucas and Reynolds 1967). For juvenile bullfrogs, thermal history affects $T_{b,pref}$ (e.g. 22.3°C at T_a = 10) but not feeding or photoperiod; they maintain T_b about 27.6°C in an aquatic thermal gradient if acclimated to 25°C (Lillywhite 1971).

Basking behavior has been widely observed for various aquatic amphibians, particularly tadpoles. Aggregations of tadpoles are often attributed to thermoregulatory advantages, although there can be other reasons such as taxic responses to water currents, light, oxygen levels, chemicals, food or predators. Thermoregulatory aggregations often form early in the day in shallow parts of water bodies. Thermoregulatory basking behavior has been reported for some temperate, montane and arid-zone frogs (Hutchison and Dupré 1992; Tattersall *et al.* 2006). The frogs *Acris crepitans* and *Rana pipiens* are heliothermic (Brattstrom 1963). The bullfrog *Rana catesbeiana* both basks and adjusts its posture to maximize solar heat gain (Lillywhite 1970). However, evaporative cooling enhanced by high temperatures and exposure counteracts warming and causes T_b to stabilize at a lower level than could be achieved otherwise. Selecting optimal basking sites, to minimize convective and evaporative heat losses, and maximize exposure to sunlight, enables *Bufo debilis* to have a T_b higher than T_a (Seymour 1972). Diurnal *Atelopus* bufonids, particularly montane species, bask in sheltered clumps of grass and attain T_bs 0.5 to 4.6°C higher than adjacent temperatures (Duellman and Treub 1994). Waterproof frogs, such as the canyon treefrog *Hyla arenicolor* (Snyder and Hammerson 1993), and *Bokkermannohyla arhengei* (Tattersall *et al.* 2006) bask during the day in exposed sites. Very waterproof frogs such as *Phyllomedusa*, *Chiromantis* and *Hyperolius* bask in direct sunlight and attain much higher T_bs, of 35–40°C or more.

After feeding, there is generally a behaviorally derived increase in body temperature. The elevated T_b may facilitate digestive efficiency through better mechanical breakdown of food, faster enzymatic secretion and digestive action, and faster absorption of nutrients. A postprandial behavioral hyperthermia has been reported for *Hyla cinerea* (Freed 1980), *Bufo boreas* (Lillywhite *et al.* 1973; Carey 1978), *Pseudoeurycea smithii* (Feder 1982c), but not for *Bolitoglossa occidentalis* (Feder 1982) or juvenile *Rana catesbeiana* (Lillywhite 1971a). The stimulus for postprandial hyperthermia is probably not mechanical distension of the stomach, but a chemical stimulus from the food, gut secretions or nutrients (Lillywhite *et al.* 1973).

Hydration influences thermoregulation, particularly for terrestrial species (Farrell 1971; Tracy 1976). The marine toad *Bufo marinus* selects a lower preferred T_b in a thermal gradient with dry air (Malvin *et al.* 1989); the toad *Bufo valliceps* selects a T_b 6°C lower when dehydrated (Williams and Wygoda 1993). Increased temperature increases mucous gland discharge in *Rana catesbeiana* (Lillywhite 1971b). CT_{max} is lower for dehydrated ambystomatid salamanders (Claussen 1977).

Behavioral thermal selection and physiological regulation by "waterproof" frogs is also influenced by their hydration state. *Chiromantis xerampelina* has a preferred T_b of about 35–38°C in a thermal gradient when normally hydrated, but $T_{b,pref}$ is reduced to <30°C when they are water restricted; *Phyllomedusa sauvagei* select lower T_bs but show a similar reduction in $T_{b,pref}$ when dehydrated (Shoemaker *et al.* 1989). Both *C. xerampelina* and *P. sauvagei* allow their T_b to increase above that normally defended by increased EWL when dehydrated, presumably to conserve water (Fig. 3.26).

Physiological Thermoregulation

Amphibians in general have limited physiological capacity for thermoregulation. Enhanced cutaneous evaporative water loss enables many amphibians to keep their skin moist, and cool at high T_a or when basking, but the maintenance of skin hydration to prevent epidermal cell desiccation and death is probably the primary role of increased glandular secretion at high T_a rather than evaporative cooling *per se*. For example, the bullfrog *Rana catesbeiana* has a periodic synchronous discharge of cutaneous mucous glands, with the frequency of discharge increasing with temperature (Lillywhite 1971a). Basking hylid frogs discharge copious secretions compared to nocturnal frogs, as do basking ranids compared to aquatic ranids (Lillywhite and Licht 1975). Control of sympathetic innervation of

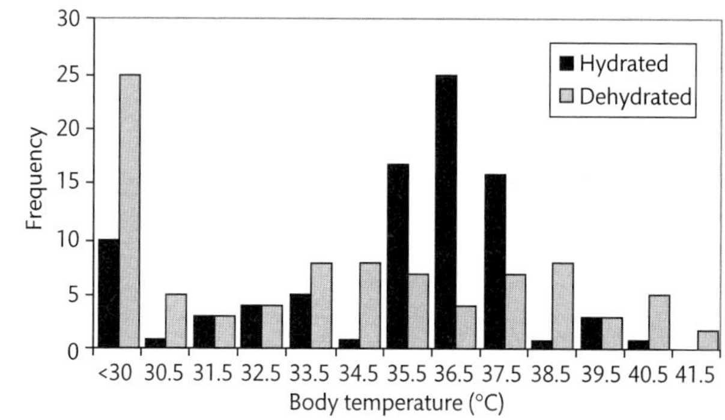

Fig. 3.26 Effect of hydration status on behavioral thermoregulation by the "waterproof" frog *Chiromantis xerampelina*. Data from Shoemaker *et al.* (1989).

mucous glands seems to be located in the anterior hypothalamus, which also controls physiological thermoregulation in other terrestrial vertebrates. Brattstrom (1970b) suggested that the Australian treefrog *Litoria caerulea* may control cutaneous EWL to prevent overheating. "Waterproof" frogs such as *Phyllomedusa*, *Chiromantis* and *Hyperolius* use mucous gland discharge to prevent overheating at high T_b (Fig. 3.27).

Skin color is also involved in the interplay between ambient temperature, body temperature and water balance for basking frogs. Many amphibians alter their skin color for a variety of reasons, including illumination, temperature and background crypsis (Tonosaki *et al.* 2004). For some basking frogs, skin reflectance in the near-infrared provides crypsis against the green foliage background, and may also have some thermoregulatory significance (Schwalm *et al.* 1977). A unique melanosomal pigment pterorhodin has been reported for some anurans (Bagnara 2003). Color change involves the dispersion/ aggregation of dermal melanosomes, for example when the skin is dark on a dark background the iridophore pigments are aggregated, reducing reflectance, and melansomes are dispersed (e.g. Nielsen 1978). Color change is controlled principally by circulating MSH (melanosome stimulating hormone), which disperses melanosomes, and melatonin, which aggregates melanosomes (Oshima 2001; Tonosaki *et al.* 2004). Chromatophores

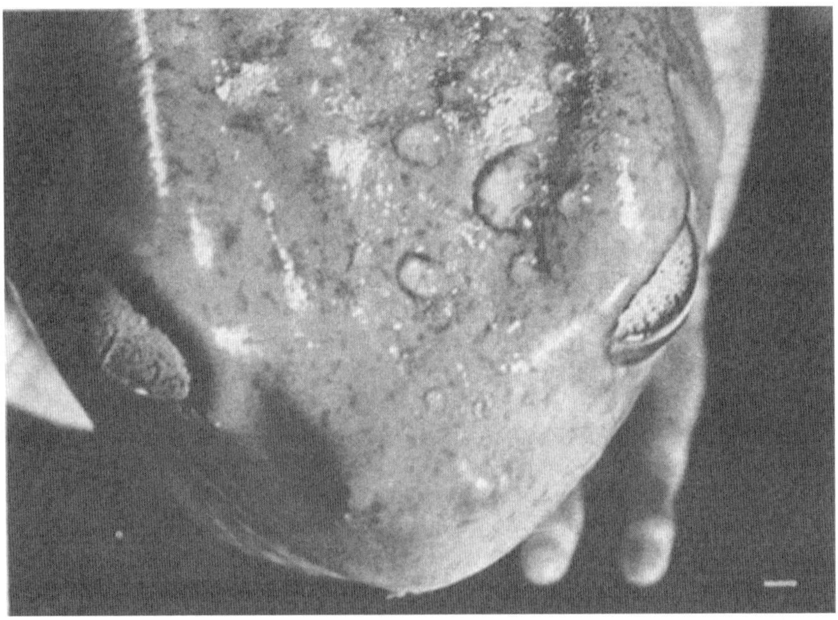

Fig. 3.27 Droplets of glandular discharge on the head of *Phyllomedusa sauvagei* for evaporative cooling at a T_a of about 39°C. From Shoemaker *et al.* (1987).

may also be responsive to adrenaline, acetylcholine, nervous control and direct effects of light (Bagnara and Hadley 1973; Iga and Bagnara 1967).

Many basking frogs blanch to a white color in sunlight and/or at high ambient temperatures. This is often associated with a concomitant reduction in EWL, but is not a mechanism for EWL reduction (see Chapter 3.2). Changes in skin reflectance accompanying color change have been measured for some terrestrial anurans (Carey 1978; King *et al.* 1994). The slightly waterproof South American hylid *Bokermannohyla alvarengai* has a light skin color when basking, in both the field and laboratory (Tattersall *et al.* 2006; Fig. 3.28). Exposure to sunlight rather than high T_a *per se* seems to be the trigger for their increased skin reflectivity. Similar blanching to an intense white color is common for the basking "waterproof" frogs, *Chiromantis* (Passmore and Carruthers 1979; Kaul and Shoemaker 1989; Hutchison and Dupré 1992; McClanahan *et al.* 1994) and *Hyperolius* (Kobelt and Linsenmair 1986; Schmuck and Linsenmair 1988; Withers *et al.* 1982). Presumably some thermal benefits of being hot (e.g. rapid digestion) are responsible for selection of high T_bs by these basking frogs.

Thermal Tolerance

Amphibians, like other animals, have a range of temperatures within which they can survive, grow and reproduce; this is the range of thermal tolerance. The activity temperature range is the normal range of T_bs that an amphibian experiences during routine "normal" activities (Cowles and Bogert 1944; Pough and Gans 1982). This range is often summarized as mean activity temperature, to facilitate

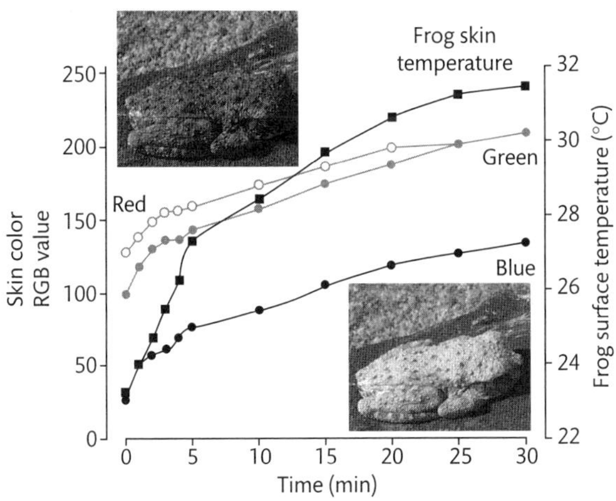

Fig. 3.28 Effect of basking on skin color (red value, open circle; green value, closed circle; blue value, triangle) and skin temperature for *Bokermannohyla alvarengai* (modified from Tattersall *et al.* 2006).

comparison between populations and species. It varies for different species, depending on many factors such as latitude, altitude and taxonomic affinity. The mean activity temperature also varies seasonally and with experimental changes in temperature in the laboratory. For aquatic species the mean activity temperature tends to be narrower and/or lower than for terrestrial species, reflecting the differences in their thermal environments.

Temperatures both below and above the range of tolerance result in impaired function; the further from the range of tolerance the greater is the impairment. These ranges of reduced function are called the zones of resistance. There is a lower voluntary minimum and an upper voluntary maximum temperature that define the range of temperatures that an amphibian will voluntarily tolerate, for example in a laboratory thermal gradient. At the lower limit of the zone of resistance is the lower lethal temperature, which results in rapid death; this is termed the critical thermal minimum (CT_{min}). At the upper limit of the zone of resistance is the upper lethal temperature, which results in rapid death; this is termed the critical thermal maximum (CT_{max}). An alternative definition and non-lethal methodology for CT_{min} and CT_{max} is the temperature when the individual is ecologically dead, for example unable to right itself. A variety of thermal "endpoints" can be used to define the CT_{min} and CT_{max}, including the lack of a righting response, onset of spasms, heat rigor and other endpoints (Fig. 3.29), but onset of muscle spasms is recommended as a standard endpoint in conjunction with loss of righting response (Lutterschmidt and Hutchison 1997).

Measurements of CT_{min} and CT_{max} (lethal and ecological) have been made for many amphibian species. The lowest CT_{min} for salamanders is −2.0°C (*Hydromantes platycephalus*) and the highest CT_{max} is 38°C (*Ambystoma mabeei*; Brattstrom

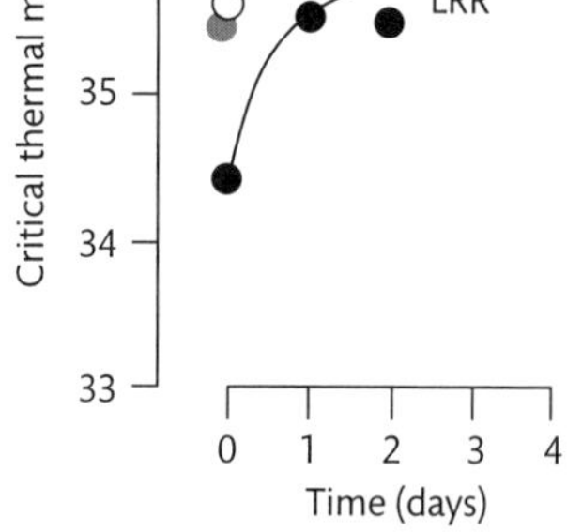

Fig. 3.29 Critical thermal maximum measured by three different endpoints; loss of righting response (LRR), onset of spasms (OS) and heat rigor (HR), and changes in CT_{max} with acclimation time, for the salamander *Ambystoma jeffersonianum*. Modified from Claussen (1977).

1963, 1968). For anurans, CT_{min} is as low as 3.0°C (*Bufo boreas*) and CT_{max} is 42.5°C for *Hyla smithii* (Brattstrom 1968) and about 43°C for *Hyperolius viridiflavus* (Kobelt and Linsenmair 1992).

CT_{min} and CT_{max} values of amphibians tend to be well outside the range of field T_bs. There is clearly a considerable "safety factor" to avoid lethal temperatures. For example, the hylid treefrog *Smilisca baudinii* has a CT_{min} of 5.0°C and a CT_{max} of 40.4°C, but the T_b when active in the field ranged from 21.2 to 28.8°C; comparable values for the marine toad *Bufo marinus* were CT_{min} 11.0°C, CT_{max} 41.8°C and $T_{b,field}$ 22.0 to 27.0°C (Brattstrom 1963). The tropical rainforest *Taudactylus diurnus* has a CT_{max} of 31.1°C where T_a ranges from 13.5 to 22.0°C in nature (Johnson 1971). Lower-latitude *Bufo cognatus* has a CT_{max} 2.34°C higher than the higher-latitude *Bufo hemiophrys* (Schmidt 1965). The bullfrog has a field T_b when active of 26–33°C and CT_{max} of 38.2°C (Lillywhite 1970).

Thermal Acclimation

Amphibians, being ectothermic, often nocturnal, with a high EWL, and having a limited capacity for behavioral thermoregulation, are substantially influenced by ambient temperature. Not surprisingly, they generally have a strong adaptive response to a sustained change in ambient temperature. When the change in ambient temperature is due to natural seasonal changes, adaptive responses are called acclimatization, whereas adaptive responses are called adaptation when the change in ambient temperature is artificial and laboratory-based. For example, the lower critical temperatures (CT_{min}) of North and Central American anurans are strongly correlated with the latitude of their northernmost occurrence (Brattstrom 1968), indicating a natural environmental acclimatization of CT_{min}. Similar adaptive acclimation of CT_{min} is apparent for many species in the laboratory.

Thermal acclimation has been extensively studied in urodele and anuran amphibians; nothing is known of thermal acclimation by caecialians. The observed physiological changes that accompany thermal acclimation are generally adaptive, for example CT_{max} increases at high acclimation temperatures and CT_{min} decreases at low temperatures. Both the magnitude and rate of acclimation are important aspects of the phenomenon. For example, salamanders generally acclimate in 24–72 h. There are also examples of amphibians with limited or essentially no capacity for thermal acclimation; these species often have very restricted distributions and are exposed to limited changes in their thermal environment. For example, anurans that are found on mountaintops in eastern Australia experience little environmental temperature change and have essentially no capacity for thermal acclimation (Brattstrom 1968, 1970). Tropical anurans generally do not show acclimation, whereas temperate species do (Feder 1982).

Acclimation changes in CT_{max} generally follow a hyperbolic time-course, whether the acclimation is to a higher or a lower temperature, but the rate of change can differ with the direction (e.g. Fig. 3.30). The time for 50% change in CT_{max} (½ AT) is often used to quantify the rate of change for such hyperbolic relationships (Claussen 1977). Acclimation is generally quite rapid for increasing acclimation temperature (½ AT about 2 days) despite substantial variation in the magnitude of the $\Delta CTM/\Delta T$ (Table 3.9). However, reverse acclimation is often considerably slower, especially for some salamanders. For example, thermal acclimation of newts (*Notophthalmus viridescens*) has a ½ AT of about 1 day for acclimation to a higher T_a but about 2 days for acclimation to a lower T_a (Figure 3.30).

The amount by which CT_{max} changes (ΔCT_m) varies between species and populations, depends on the magnitude of the change in acclimation temperature, and often varies with the direction of temperature change (Table 3.9). For salamanders, the ΔCT_{max} is greater (in absolute terms) for decreasing acclimation temperature (ΔT), as is the half-acclimation time (½ AT) and acclimation response ratio ($ARR = \Delta CT_{max}/\Delta T$); the difference between increasing and decreasing acclimation temperatures on these three variables is significant (multivariate t-test, $P = 0.003$). For anurans, the ΔCT_{max} and ½ AT are the same as for salamanders, but the $\Delta CT_{max}/\Delta T$ is significantly higher ($P = 0.004$). ARR tends to be higher for species with a wide geographical or altitudinal range. Acclimation of CT_{min} is often quite different in magnitude and time-course from that for CT_{max}, and this may have adaptive value in allowing differential thermal responses to warmer and cooler thermal environments (Layne and Claussen 1982).

There are differences in CT_{max} or CT_{min}, and in the magnitude of thermal acclimation, with direction of thermal change and season. Body size does not seem to influence thermal acclimation. Developmental stage can influence thermal tolerance and acclimation in some species (e.g. larger individuals of *Rana pipiens* have a higher CT_{max}; Siebel 1970) but not others (e.g. larval *Ambystoma tigrinum* have

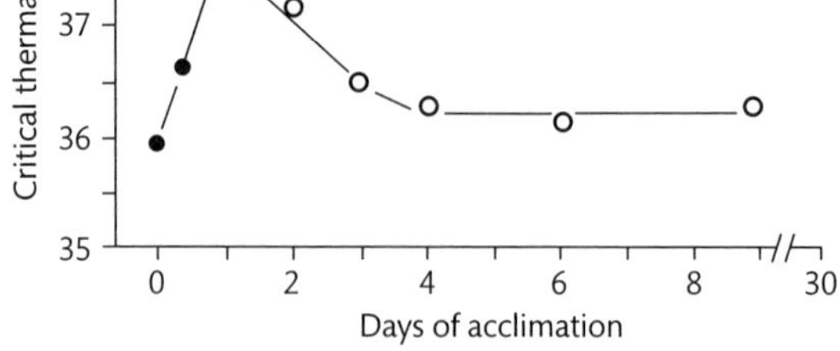

Fig. 3.30 Thermal acclimation of the newt *Notophthalmus viridescens* from 4 to 20°C, and 20 to 4°C. Redrawn from Hutchison (1961).

similar magnitude and rate of thermal acclimation as adults; Nietfelt *et al.* 1980). Water balance can also influence thermal acclimation, as does photoperiod. There can be both daily variation in CT_{max} (with CT_{max} highest during the times of highest T_a) and seasonal variation (with CT_{max} highest during warmer parts of the year). CT_{max} of the scansorial hylids *Litoria gracilenta* and *L. caerulea* are highest during the day (Johnson 1971, 1972). All of these influences on thermal acclimation suggest that the mechanisms for thermal acclimation are complex and variable.

Fever

Like all other vertebrate groups, amphibians develop a fever in response to infectious or inflammatory pyrogens (Kluger 1979; Bicego *et al.* 2007). Because amphibians are ectothermic, their fever response involves behavioral selection of higher body temperature in a thermal gradient (rather than endogenously-elevated T_b as in endothermic vertebrates). The green treefrog elevates preferred T_b by more than 2°C, for over 12 h, in response to injection of pathogenic bacteria (Kluger 1977), *Rana pipiens* by 2.6°C and *R. catesbeiana* by 2.7°C (Casterlin and Reynolds 1977), and *R. esculenta* by about 5°C (Myhre *et al.* 1977). Injection of plasma from previously infected frogs causes *R. esculenta* to increase T_b by 2–5°C, and injection of prostaglandin (PGE_1) into the diencephalon (forebrain) elicits a profound hyperthermia of 6–10°C within minutes (Myhre *et al.* 1977). PGE1 injected into the midbrain also causes behavioral fever in the mudpuppy *Necturus maculosus* (Hutchison and Erskine 1981).

Table 3.9 Summary of thermal acclimation parameters, change in CT_{max} (ΔCT_{max}), half acclimation time (½ AT) and acclimation response ratio (ΔCTM/ΔT), for caecilians, salamanders and anurans with increase and decrease in acclimation temperature

	Increased acclimation temperature	Decreased acclimation temperature
Caecilians	No data available	
Salamanders		
ΔCT_{max} (°C)	2.67 ± 0.44	–3.35 ± 0.40
½ AT (days)	0.88 ± 0.37	1.64 ± 0.32
ARR ($\Delta CT_{max}/\Delta T$)	0.14 ± 0.02	0.19 ± 0.02
Anurans		
ΔCT_{max} (°C)	4.20 ± 0.68	-
½ AT (days)	1.05 ± 0.20	-
ARR ($\Delta CT_{max}/\Delta T$)	0.26 ± 0.03	-

Values are mean ± se. Data from Claussen (1977) and Layne and Claussen (1982).

Fever in amphibians results from exogenous pyrogens (bacteria, endotoxins) stimulating the release of endogenous pyrogens and prostaglandins as the mediators of the behavioral pyretic response (Myhre *et al.* 1977; Bernheim and Kluger 1976; Hutchison and Erskine 1981; Bicego *et al.* 2002a, b, 2007). Prostaglandins mediate a central nervous system response, possibly via the preoptic area, that results in behavioral fever. The development and extent of fever represents a balance of the endogenous pyrogenic response and endogenous antipyretic agents. Arginine vasotocin seems to have an antipyretic action in amphibians (Bicego-Nahas *et al.* 2000).

Hypoxia and Hypercapnia

Hypoxia is a potent inducer of hypothermia in a wide range of organisms, from protozoans to mammals, including amphibians (Wood 1995; Bicego *et al.* 2007). Low oxygen concentrations can impact on thermoregulation by amphibians, whether it is determined environmentally (hypoxic hypoxia) or internally (anemic hypoxia). For ectotherms such as amphibians, the thermoregulatory depression is reflected by behavioral selection of lower ambient temperatures. Tiger salamanders (*Ambystoma tigrinum*) select T_b about 2–3°C lower when hypoxic (pO_2 = 37 torr) compared to normoxic (127 torr; Dupré and Wood 1988). Marine toads (*Bufo marinus*) select a T_b about 18°C when hypoxic (pO_2 = 60 torr) compared to 29°C when normoxic (Riegel and Wood 1988; Malvin and Wood 1991), and 18°C when anemic hypoxic (Wood *et al.* 1989). Similarly, cold-acclimatized *Rana temporaria* choose a lower temperature when hypoxic (1.9°C) than normoxic (6.8°C; Tattersall and Boutilier 1997). An endogenous hypoxia, caused by anemia, also elicits a hypothermic response (T_b = 10°C) compared to non-anemia (T_b = 24.5°C) in *Bufo marinus* (Wood *et al.* 1989).

The physiological significance of hypoxic hypothermia is to lower metabolic rate through a Q_{10} effect, to decrease the critical pO_2 where V_{O_2} becomes dependent on ambient pO_2, to decrease the pO_2 threshold for hypoxia-induced hyperventilation, and to left-shift the haemoglobin-O_2 dissociation curve and thereby facilitate O_2 loading onto hemoglobin (Hicks and Wood 1985; Wood and Hicks 1985). A reduced oxygen demand is probably an important consequence of hypoxic lowering of T_b that maintains a favorable matching of reduced O_2 demand with reduced O_2 availability.

Increased carbon dioxide levels can have a similar thermoregulatory depression as hypoxia, although hypercapnia is not an associated consequence of hypoxic hypercapnia with elevation or aquatic hypoxia, or anemic hypoxia. A hypercapnia of 60 torr pCO_2 decreases preferred T_b from about 29°C to 21°C (Riedel and Wood 1988). The physiological role of this hypercapnic hypothermia might by related to acid–base balance and the maintenance of relative alkalinity (Hutchison and Dupré 1992). Lowering T_b during hypercapnia would increase pH by about 0.10 pH units/°C (Howell *et al.* 1970) and thereby offset the decrease in pH by hypercapnia.

3.4.3 Advantages of Ectothermy

For ectothermic amphibians, body temperature is determined by passive thermal exchange with their environment, despite various behavioral and physiological specializations for thermoregulation (e.g. Brattstrom 1970). The relatively small size of living amphibians precludes any thermal effects of metabolism by gigantothermy (cf large seaturtles, dinosaurs; Paladino *et al.* 1990). None appear to use thermogenic tissues and vascular countercurrent heat exchange to regulate the temperature of any internal organs (unlike spatially endothermic fishes, for example; Dickson and Graham 2004).

Ectothermy is one of a suite of behavioral and physiological traits that are related to the low-energy turnover life style of amphibians. The metabolic rate of amphibians is low, typical of other ectothermic animals (see Chapter 3.6). Amphibians have relatively large energy stores (for example, many anurans have extensive abdominal, axillary and inguinal fat pads). Consequently, amphibians survive extended periods of starvation, particularly when aestivating or hibernating (see Chapter 4.5). This low-temperature, low-energy lifestyle of amphibians is in marked contrast to the high-temperature, high-energy lifestyle of endothermic mammals and birds. Nevertheless, many of the behavioral and physiological characteristics of amphibians that support their low-energy lifestyle should not necessarily be considered "primitive" but may in fact be adaptations (Pough 1980). The small size of many amphibians (i.e. <5 g body mass) allows ectotherms to exploit a size niche that is precluded for endothermic birds and mammals. The elongate body shape of salamanders and caecilians has no thermal disadvantage (compared to an endotherm) and allows them to exploit this shape niche without endothermic competition. The low metabolic rate of amphibians makes them better suited to survive in ephemeral environments; they can survive extended periods of inactivity in response to shortage of food, water, oxygen or extreme ambient temperatures.

Plethodontid salamanders (lungless salamanders) are an excellent example of a large group of amphibians that excels at this low-energy lifestyle (Feder 1983). Plethodontids are generally small salamanders, primarily found in North and South America (and two species of *Hydromantes* in southern Europe). Lunglessness and reliance on cutaneous gas exchange eliminates the metabolic costs of developing, maintaining and ventilating lungs, and any buoyancy disadvantages of lungs for aquatic species. Salamanders have a low metabolic rate compared to other ectothermic animals, including anurans (Feder 1976a, b; Pough 1980; see Section 1.6). Lungless salamanders are able to develop the remarkable feeding strategy of a projectile tongue since their hyoid apparatus is freed from a respiratory ventilation role (Duellman and Trueb 1994). Lunglessness has some severe restrictions, however. Reliance on cutaneous gas exchange is a severe limitation to rates of gas exchange for large individuals. Development of a waterproof skin is precluded by the reliance on the skin for gas exchange. Small size and low resistance to EWL precludes these salamanders from heliothermic thermoregulation, and restricts them to cool, moist environments and relatively short foraging times when ambient conditions are hygrically favorable. Nevertheless, plethodontids salamanders are a relatively widespread and successful group. This is due to their extraordinarily low energy requirements, from an individual and population point of view, and their capacity to survive extended periods of unfavorable conditions. The annual energy requirement of plethodontids is small compared with that of other amphibians, other ectotherms, and particularly with endotherms (Pough 1980; Feder 1983). Low metabolic rate means that a high proportion of ingested energy can be allocated to growth and reproduction. Fat energy stores enable these salamanders to survive starvation for extended periods, and to allocate these energy stores to reproduction when conditions are favorable.

3.5 Metabolism

The catabolic processes associated with the production of ATP and NADH and the anabolic processes associated with work, growth and biosynthesis utilizing the energy of ATP and NADH are together categorized as metabolism. The catabolic processes are characteristically divided into aerobic and anaerobic metabolism.

3.5.1 Aerobic Metabolism

As was discussed in Chapter 2, the rate of aerobic metabolism is usually measured indirectly by measuring the rate of O_2 consumption. The rate of O_2 consumption can vary from a minimum rate known as standard rate ($V_{O_2standard}$) to a maximum rate (V_{O_2max}). All the activities of an active amphibian will be supported by rates of aerobic metabolism that fall somewhere within this metabolic continuum from standard to maximal. Most of the literature reports metabolic rates as a volume of O_2 consumed per unit time, which is the basis for the designation V_{O_2}. We have described gas fluxes in Chapter 2 as molar fluxes, hence the rate of metabolism would be M_{O_2}. The flux conversion is straightforward since for any gas at STPD, 1 mole = 22.4 l.

Standard

The standard rate of metabolism is specified at a given temperature in animals that are non-reproductive, postabsorptive (not processing food) and at rest. An objective definition of rest can be problematic as we saw with "resting" heart rate in Chapter 2. Standard conditions actually exist for only a small proportion of time during an amphibian's life. Consequently, the ecological applicability of standard V_{O_2} is limited. It does not even represent a true minimum rate of metabolism, since states of even lower metabolic rates occur during aestivation and hibernation

(Chapter 4). These states are commonly described as metabolic depression or hypometabolism. Rates of O_2 consumption that do not rigidly control either the non-reproductive or postabsorptive criteria for a standard metabolic designation, but are determined in resting animals are usually referred to simply as resting rates of metabolism (V_{O_2rest}). The majority of data available are reported under the less strict designation as V_{O_2rest}. An even less restrictive definition for the rate of metabolism is known as routine V_{O_2}. This measure allows for the average daily processes associated with life and represents the true cost of existence. Routine V_{O_2} consequently has more relevance from an ecological energetic perspective, since it reflects the daily cost of living and acknowledges that life is more than rest. There is an extensive literature on various rates of metabolism for amphibians that has been recently and thoroughly reviewed by Gatten *et al.* (1992). There are three major variables that influence standard or resting V_{O_2}: temperature, body mass and processing food. Other variables can also contribute as 15 to 25% of intraspecific variation in *Rana pipiens* standard V_{O_2} has been correlated to variation in kidney size, thyroxine concentration and liver mitochondrial content (Steyermark *et al.* 2005).

Temperature Effects

Amphibians are ectothermic vertebrates, and as such their rate of metabolism will increase with body temperature. This increase in rate of metabolism, or any other physiological rate influenced by temperature, is usually quantified by the parameter known as the Q_{10}. The Q_{10} for resting metabolism would be defined as the V_{O_2rest} at a body temperature of X + 10°C/V_{O_2rest} at a body temperature of X°C, essentially the ratio of metabolic rates at 10°C increments. This ratio is generally about 2 for V_{O_2rest} in ectothermic vertebrates, so metabolism doubles for every 10°C increase in body temperature. The Q_{10} for V_{O_2rest} can be quite variable in amphibians as can be seen in Fig. 3.31, but averages about 2. Much of this variation may derive from the difficulty in repeatability of V_{O_2rest} determinations influencing ratios (Packard and Boardman 1988, 1999). Interestingly, the Q_{10} for V_{O_2rest} does not appear to vary systematically with temperature in salamanders or anurans, either inter- or intraspecifically (Fig. 3.31).

Mass Effects

The rate of standard and/or resting metabolism at a given body temperature is different when expressed per unit mass for amphibians that differ in body mass. Larger amphibians have lower mass-specific rates of aerobic metabolism. This is characteristically determined by plotting the log of absolute rate of metabolism (volume or moles per time) against log of body mass. If the mass-specific rate of metabolism (volume or moles per time and mass) was not influenced by body mass, the relationship between log of absolute metabolic rate and log body mass would increase with a slope of 1, and the relationship would be described as

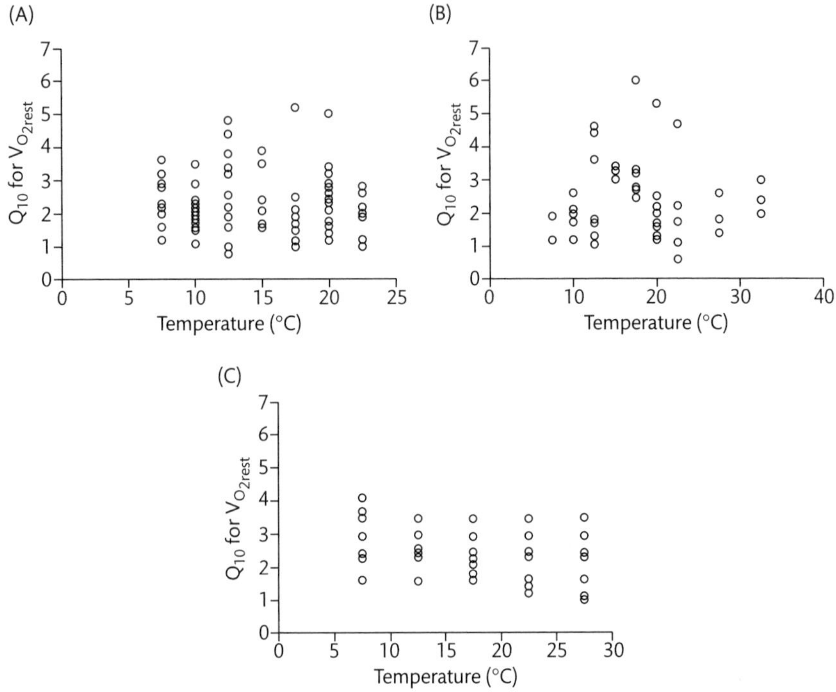

Fig. 3.31 Values reported for Q_{10} over various temperature intervals in urodeles and anurans: (A) interspecies urodeles; (B) interspecies anurans; (C) *Bufo boreas*. The temperature is the mean temperature of the Q_{10} interval. The data for *Bufo boreas* represent two subspecies acclimated to different initial conditions (Carey 1979). Data are from Whitford and Hutchison 1967; Packard 1971; Dunlap 1971; Fitzpatrick 1973; Guimond and Hutchison 1968; Fitzpatrick and Atebara 1974; Feder 1976; Pitkin 1977; Carey 1979; and Brown and Fitzpatrick 1981a, b.

isometric (body mass[1]). If the relationship has a slope different from 1 then the relationship of these two variables is allometric in nature, and would indicate that body mass influences the rate of metabolism besides the effect of simply more tissue. The elevation of the relationship at the same temperature does not appear to differ between caecilians, urodeles and anurans (Gatten *et al.* 1992). The slope of the relationship between log of absolute metabolic rate and log body mass for both urodeles and anurans at a variety of different temperatures run from 0.79–0.84 with substantial overlap of their confidence intervals (Gatten *et al.* 1992). A subsequent analysis for all amphibians at 20°C found an exponent of 0.88 (White *et al.* 2006). This holds in both intra- and interspecific comparisons. Therefore as amphibians grow or are inherently larger than other species, their mass-specific rates of resting V_{O_2} will be lower and this effect will be proportional to a value between body mass$^{0.8}$ and body mass$^{0.88}$. From a practical perspective this means that one has to use regressional analyses (analysis of covariance) in order to make comparisons of standard V_{O_2} for species that differ in body mass, or correct mass-specific rates to the appropriate exponent if standard statistical comparisons are made (see Chapter 1.7). The use of mass-specific ratios or percentages must be interpreted with extreme caution, as ratios can be very misleading (Packard and Boardman 1988, 1999; Hayes 2001). What actually accounts for the change in mass-specific metabolism of organisms with varying body mass is not understood, but has been the subject of extensive study and conjecture: there is even a question of whether a universal explanation exists (Bokma 2004; White *et al.* 2006). It is beyond the scope of this treatise to delineate potential explanations but good summaries of the controversy and variables involved can be found in Hochachka *et al.* (2003), West *et al.* (2002), and Brown *et al.* (2005).

Feeding Effects

The effect of feeding on resting metabolic rates can be substantial and is known as the specific dynamic action (SDA) of food. It includes aerobic energetic expenditures entailed in ingestion, digestion, absorption and assimilation of foodstuffs. Feeding is characteristically intermittent in amphibians and meal size varies interspecifically (see Chapter 2). The pattern of an SDA response is an increase in metabolic rate post food ingestion that reaches a peak 7 to 48 h later and then declines back to standard within 3 to 7 days (Wang *et al.* 1995; Powell *et al.* 1999; Sievert and Bailey 2000; Secor 2001; Secor and Faulkner 2002; Grayson *et al.* 2005; Secor and Boehm 2006). The integral of this response above standard V_{O_2} is the SDA response. The duration of the SDA response runs from 3 to 7 days, and is shorter with small meal sizes and/or increased body temperatures (Secor and Faulkner 2002; Secor and Boehm 2006). The peak V_{O_2} during an SDA response increases with meal size but averages less than 25% of V_{O_2max} (Secor and Faulkner 2002). Consequently a relatively small proportion of maximal rates of aerobic capacity are utilized even at the peak aerobic SDA response. To make interspecific comparisons of SDA across species of different body mass it is frequently expressed per unit mass. It is important to recognize that this may introduce error since a mass-specific presentation does not totally account for mass effects because of allometric differences in the variables involved (Beaupre 2005). The purpose here is to demonstrate that meal size (over a variety of body mass, food types and body temperatures) has a major effect on the magnitude of the SDA in Fig. 3.32. These values do not to account for interspecific differences. The strong effect of meal size on SDA is clear, and it does not appear to differ between urodeles and anurans. About 10–20% of the ingested energy is normally used in the physiological processes involved in the SDA response (Fig. 3.32). The SDA cost for some species is 60% of ingested energy (Secor *et al.* 2007). The SDA for frogs processing an amino acid/ peptide solution was only 25% of the SDA for frogs

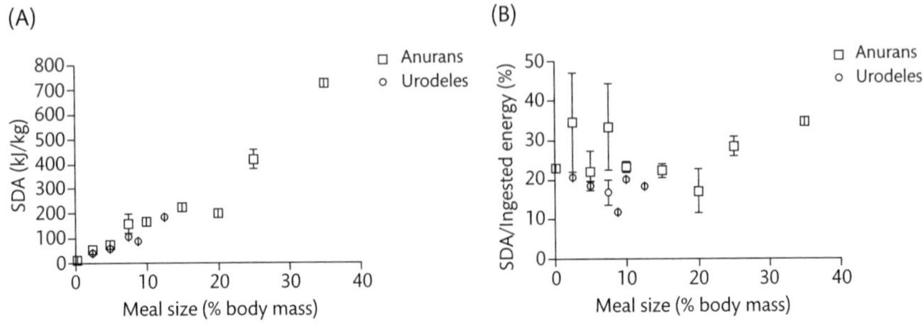

Fig. 3.32 The effect of meal size on (A) the magnitude of the integrated specific dynamic action and (B) its proportion of the total energy ingested. Data compiled from Wang *et al.* 1995; Powell *et al.* 1999; Secor 2001; Secor and Faulkner 2002; Grayson *et al.* 2005; Secor 2005; Secor and Boehm 2006; and Secor *et al.* 2007. The data encompass various species, meal types and body temperatures. SDA, specific dynamic action.

consuming mice of equivalent protein content (Powell *et al.* 1999). Interestingly, the amount of protein contained in a meal is a significant source of variation in the magnitude of the SDA in other vertebrates (Borsook 1936; Kleiber 1961). It appears that SDA in amphibians is more influenced by the total energy of a meal, rather than the meal's specific chemical constituents (Powell *et al.* 1999; Secor and Faulkner 2002; Secor and Boehm 2006), though the degree of metabolic depression preceding feeding and the meal type contribute to variation in SDA (Secor 2005; Secor *et al.* 2007).

Maximal Metabolism

The maximal rate of M_{O_2} sets the upper limit for aerobic metabolism, and as a rate-limiting process is fundamentally different from any of the aerobic processes we have discussed so far, where variables can add to the rate and variance of metabolism. Maximal metabolism has been determined in several different ways. One method measures the rate of metabolism with increasing workload on a treadmill until a maximum is reached where increased work is not matched by increased V_{O_2}. The other is to exploit behavioral reflexes that will keep the animal working at levels that achieve maximal M_{O_2}. The most common behavior exploited to keep an animal maximally active is the righting reflex, where an individual animal is rotated on its back in the metabolic chamber, and it immediately flips back over much like a wrestler avoiding a pin (Seymour 1973). Obviously, by remaining on its back normal escape behaviors are impossible and poison skin glands on the dorsal surface are not in play. The flipping technique gives higher values when done manually, than when using motorized rotation at a constant speed (Walsberg 1986). Comparisons of the manual flipping and treadmill techniques show excellent correspondence in V_{O_2max} determinations (Walton and Anderson

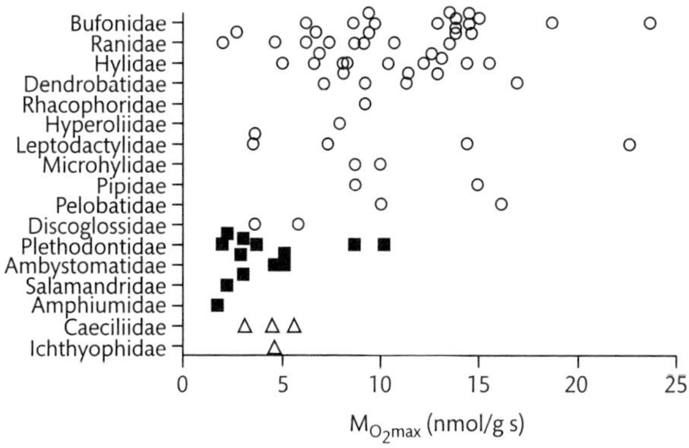

Fig. 3.33 The rates of maximal O_2 exchange at 20–25°C in a variety of amphibian species, measured using non-manometric techniques. Data are principally abstracted from the summary of Gatten *et al.* (1992) and Smits and Flanigan (1994).

1988). The treadmill technique has the advantage of being able to determine a metabolic cost of physical transport (see below, moles of O_2/mass × distance). The manual flipping technique has the advantage of relative quick determination of M_{O_2max} that is useful in analyses of variables that limit maximal rates of V_{O_2}. There have been attempts to use electrical stimulation to induce exercise and determine V_{O_2max} (Bennett and Licht 1973; Feder 1977; Turney and Hutchison 1974). Normally this approach was coupled to the measurement of V_{O_2} using manometric techniques. Since heat is generated during activity, temperature in the metabolic chamber increases, violating a fundamental condition (constant temperature) for manometric determination of V_{O_2} (Hillman *et al.* 1979). A substantial fraction of the amphibian data on V_{O_2max} is saddled with artifacts created by an activity-induced thermal transient during manometric V_{O_2} determinations. Unusual aspects of this thermal transient are that V_{O_2} during activity falls to zero and maximal V_{O_2} occurs after activity ("delayed development of V_{O_2max}"). We purposefully have not considered manometric data in our summary and analyses of V_{O_2max}. Examples of the range in V_{O_2max} from a variety of different families are given in Fig. 3.33. It is interesting to note V_{O_2max} (M_{O_2max}) can be variable up to a 20-fold difference, with the variation resulting from both interindividual and interspecific differences. Anurans also tend to have higher M_{O_2max} than urodeles and caecilians.

Just as with resting and standard M_{O_2}, rates of maximal metabolism are influenced by temperature (Fig. 3.34). Unlike resting metabolism there is a clear decrease in the temperature sensitivity of V_{O_2max} as the temperature increases when viewed either inter- or intraspecifically. There is decreasing variation in Q_{10} as the temperature increases, indicating a maximum rate is being approached.

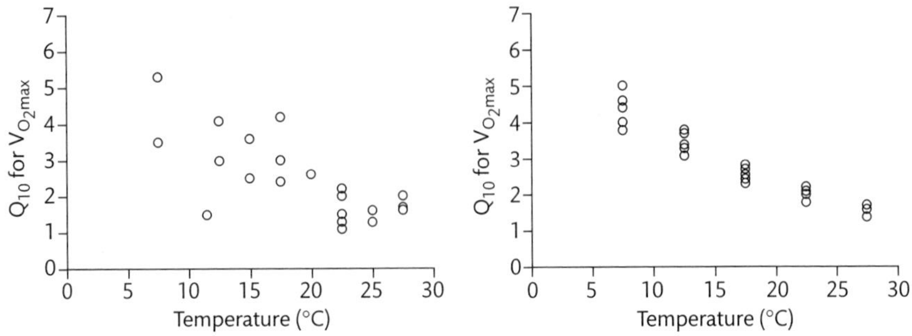

Fig. 3.34 The Q_{10} for maximal rates of oxygen exchange at different body temperatures (A) for different species and (B) for two subspecies of *Bufo boreas* acclimated to different initial temperatures (Carey 1979).

3.5.2 Limits to Maximal Rates of Oxygen Exchange

The concept of a rate-limiting step in the series exchange of O_2 from the environment to the mitochondria contradicts what evolutionary biologists would predict. From an evolutionary perspective, if O_2 exchange is the only function of the convective and diffusive steps of the respiratory and cardiovascular processes, then each step in the series process should be equally limiting. Otherwise non-limiting steps are functionally over-designed; this is a hard case to argue from a selective viewpoint. The conundrum is, what is the selective advantage of having an O_2 exchange capacity at an exchange step that exceeds maximal rates of O_2 exchange? This implies a wasted or unnecessary structural capacity. This led to a concept known as symmorphosis, which explicitly states that structural elements of the respiratory system are designed to meet demand at V_{O_2max}, and there should be no excess capacity in any step (Weibel *et al.* 1991). Symmorphosis promised to be a unifying principle for understanding anatomic structure (Weibel *et al.* 1991). There is an extensive literature dealing with symmorphosis (Jones and Lindstedt 1993), and the idea still has some credibility as a point of analysis for understanding variation in anatomic structure. Much of the testing of symmorphosis using mammalian examples has concluded that the pulmonary system is over-designed (excess O_2 exchange capacity) in comparison to the other series steps involved in O_2 exchange (Gehr *et al.* 1981). The fundamental problem with this approach is that the series functional chain for O_2 influx is also responsible for CO_2 efflux (see Chapter 2). If CO_2 efflux does not match O_2 influx, there will be a respiratory acidosis to contend with ($H_2O + CO_2 \leftrightarrow H_2CO_3$). The driving force and conductances are different for O_2 and CO_2 at each exchange step in the series chain. Therefore selection for V_{CO_2max} also needs also to be considered in the analyses of form and function of the four series steps in CO_2 exchange.

The simplistic focus on only one flux (O_2) in the functional analysis of each gas exchange step precludes a meaningful analysis of the functional requirements necessary to eliminate CO_2. Most structural elements of the body are involved in multiple functions, so in retrospect the analysis of a single functional purpose for any anatomic feature, a criterion of testing symmorphosis, is destined to invalidate the symmorphosis hypothesis.

Another approach to evaluating rate-limiting processes in the exchange of gasses is to experimentally manipulate driving forces, conductances (resistances) or evaluate the allometry of these structures in relation to allometry and modification of V_{O_2max}. If each of the four exchange steps are equally resistive, the null hypothesis from both evolutionary theory and the theory of symmorphosis, the effect of doubling the resistance of any one step would lead to a 20% decrease in overall flux ((1 + 1 + 1 + 1)/(1 + 1 + 1 + 2) = 0.8). If the flux changed by less than 20% or greater than 20% then the conclusion would be the four series resistances to overall O_2 flux are not equal. This approach has been utilized in describing the various series O_2 exchange elements and their effect on modifying V_{O_2max} in amphibians in a series of work described below.

Respiratory exchange, the combination of lung ventilation and diffusive exchange, has been assessed in amphibians in a couple of different ways. The first was to look at the allometry of V_{O_2max} interspecificically and intragenerically and compare that to the allometry of respiratory capillary length in a variety of anurans (Hillman and Withers 1979). The prediction was that V_{O_2max} and respiratory capillary length should scale with body mass with the same exponent and interspecific differences in V_{O_2max} (elevation) should be matched by differences in respiratory capillary length. Neither the slopes nor elevational differences matched, indicating that respiratory capillary length does not match maximal metabolic rates (Hillman and Withers 1979). Similarly, using ablational techniques, removing one lung from *X. laevis* should decrease V_{O_2max} in proportion to the change in O_2 conductance (V_{O_2max} = cutaneous flux + pulmonary flux to V_{O_2max} = cutaneous flux + 0.5 pulmonary flux). The control V_{O_2max} was 1.13 ml O_2 and maximum cutaneous flux was 0.03 ml, hence we would expect the respiratory resistance to increase by 1.13/0.58 or 1.9 times. This would lead to a decrease in V_{O_2max} of 28% if all the resistances were equal. The actual decrease was 11% of the control value following surgery, indicating the respiratory system represents less than 25% of the overall resistance to maximal O_2 flux in anurans (Hillman and Withers 1979). The effect of temperature on a diffusionally-limited process is predicted from effects of temperature on the diffusion coefficient for O_2. This is normally related to the ratio of absolute temperature differences for 10°K increments. For instance, from 20 to 30°C it would increase the rate of diffusion by 303°/293°K or 1.03 times. The data above on the Q_{10} for V_{O_2max} over the same temperature interval was about 1.6, therefore V_{O_2max} cannot be limited by a diffusional process. Another approach was to determine V_A and V_{O_2max} under progressive hypoxia (Withers and Hillman

1983). The predictions are that hypoxia would decrease the diffusional driving force for gas exchange and V_A should not increase if it was operating at maximal rates during normoxic conditions. In fact V_{O_2max} did not decline until the ambient and intrapulmonary P_{O_2} had declined to 10.9 kPa (equivalent to an altitude of 4500 m) and V_A increased by about twofold during hypoxic exercise (Withers and Hillman 1983). These data taken together indicate that the respiratory ventilation and perfusion steps represent less than half the total resistance to maximal O_2 flux, and from a symmorphosis perspective are over-designed for maximal rates of O_2 flux and hence do not normally limit V_{O_2max}.

The next step in series that might limit V_{O_2max} is the cardiovascular transport step. Cardiac transport as a physiological process (heart rate) should have a Q_{10} of 2 or greater, which is higher than the Q_{10} for V_{O_2max} at elevated body temperatures (Carey 1979), but when coupled to increased mixing in the ventricle at higher body temperatures (see Chapter 2), a Q_{10} value of less than 2 would be predicted. V_{O_2max} generally scales isometrically with body mass in anurans (Hillman and Withers 1979; Gatten *et al.* 1992). Ventricle mass also scales isometrically with body mass in anurans (Withers and Hillman 2001) and interspecific differences in V_{O_2max} correlate with interspecific differences in ventricle mass (Hillman 1976). Intraspecific variation in hemoglobin concentration in *Rana pipiens* correlated with variation in V_{O_2max} (Hillman 1980). The product of hemoglobin concentration and ventricle mass explained 64% of the intraspecific variation in V_{O_2max} in juvenile *Rana catesbeiana* (Walsberg *et al.* 1986). Limiting maximal heart rate with various doses of the β-blocker propranolol demonstrated a direct correlation of V_{O_2max} and maximal heart rate in *Scaphiopus intermontanus* (Hillman 1982). The product of blood flow rate and variation in hemoglobin concentration induced by erythrocythemia were linearly correlated with V_{O_2max} in *Bufo marinus* (Hillman *et al.* 1985). Variation in V_{O_2max} was linearly related to blood flow rates during dehydration in multiple species (Hillman 1991). These data taken individually and together all support the concept that systemic oxygen transport provides the principal limitation to V_{O_2max} in anurans. Blood flow rate to muscle has been argued to be the best predictor of muscle oxygen consumption (Marsh and Ellerby 2006). The only investigation that failed to find such a link is Longphre and Gatten (1994).

The next step in O_2 transport is the tissue diffusive step from the capillaries to the mitochondria. The Q_{10} for V_{O_2max} is too high for a diffusion limitation of maximal rates of O_2 flux. There are no other data to evaluate the contribution of tissue diffusion to V_{O_2max}.

The final step that could limit V_{O_2max} is the number and activity of the mitochondria. The activity of the mitochondrial enzyme citrate synthase is used as an estimate of mitochondrial capacity. Interspecific differences in V_{O_2max} are correlated with interspecific differences in skeletal muscle citrate synthase activity (Putnam and Bennett 1983). Intraspecific variation of V_{O_2max} was not correlated with intraspecific citrate synthase activity (Walsberg *et al.* 1986). Osmotic effects on mitochondrial function, the only mitochondrial effect possible during dehydration, do not correlate with the pattern of decline in V_{O_2max} during dehydration (Hillman *et al.* 1991). Consequently, the data are still not conclusive as to the role played by the mitochondria in limiting V_{O_2max} at an organismal level. It should be clear from the data in Fig. 2.45 that amphibians cannot perfuse all of their skeletal muscle at once. This is consistent with your mother's admonishment "do not swim until an hour after eating, lest you cramp up while swimming and drown". Not all of an organism's mitochondria can be simulataneously supplied with O_2, therefore the total number of mitochondria in an organism would not normally limit V_{O_2max}. At an individual muscle level, an organism's ability to supply O_2 may be greater than the mitochondrial capacity to consume the O_2, hence the number of mitochondria could limit the maximal metabolic capacity of an individual muscle.

The various parameters developed from the experimental approaches described above have been used to develop a computer model of both CO_2 and O_2 fluxes in amphibians (Withers and Hillman 1988). The model clearly predicted much of the empirical data described above. The model indicates that cardiovascular transport is the rate-limiting step to V_{O_2max} in anurans, so selection for enhanced aerobic capacity would normally entail an increase in arterial O_2 content via a decrease in mixing, or an increase in stroke volume (Fig. 3.35). On the other hand CO_2 transport and tissue P_{CO_2} levels are primarily determined by alveolar ventilation, systemic blood flow, and the separation efficiency in the ventricle (Fig. 2.35). This analysis indicates different rate limiting steps for the maximal fluxes of O_2 and CO_2. Therefore each of the four series transport steps have different resistances for O_2 and CO_2, therefore not the equality necessary for a rigorous symmorphosis argument. This explains why the respiratory system appears overdesigned from the perspective of O_2. The structure–function relationships for the respiratory system are evolutionarily shaped by selection around CO_2 fluxes. Consequently, the focus in most of the literature on only O_2 fluxes in series for the respiratory and cardiovascular systems has led to some very confusing conclusions for evolutionary biologists and proponents of symmorphosis.

More recently, a model for O_2 exchange has been developed for amphibians and reptiles that have the potential for shunting and mixing in the ventricle (Wang and Hicks 2002).

3.5.3 Anaerobic Metabolism

The principal pathway for the generation of ATP anaerobically in vertebrates is the breakdown of glucose to form two molecules of lactic acid, two molecules of H_2O, and two ATP molecules. The source of the glucose can be from either intracellular glucose or the breakdown of glycogen. In comparison, the aerobic breakdown of

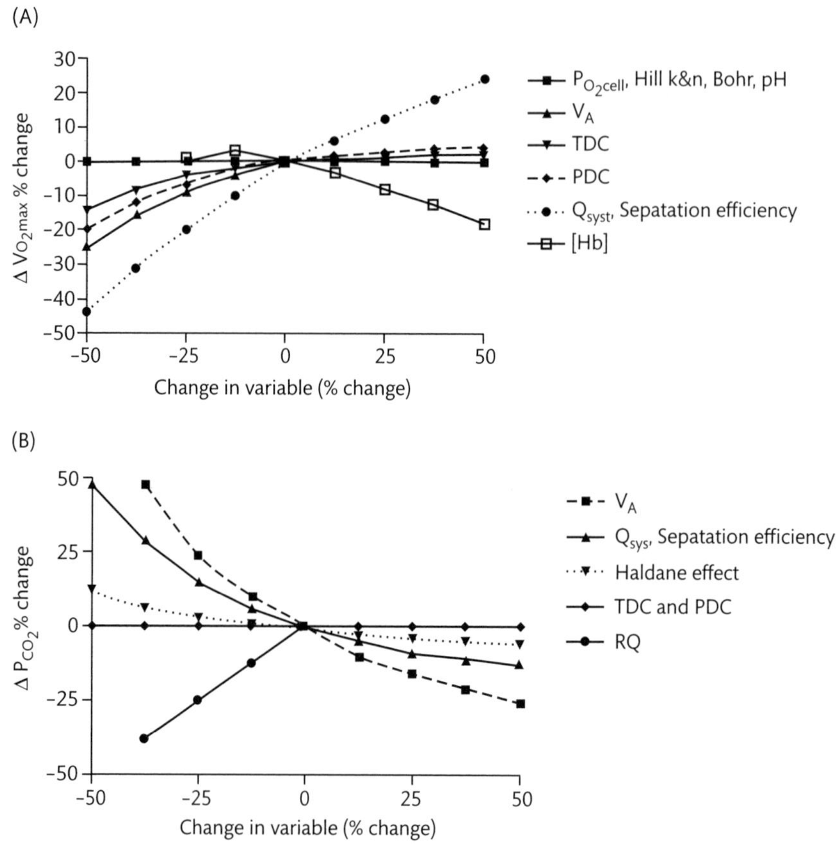

Fig. 3.35 The predicted change in (A) V_{O_2max} or (B) the P_{CO_2} difference at various steps in the series exchange process as a consequence of varying the parameters associated with each step in O_2 and CO_2 exchange; from Withers and Hillman (1988). Recognize these predictions utilize physiological and anatomical parameters specified in their model, and the simulations might vary if different control parameters are specified. TDC, tissue diffusion capacity; PDC, pulmonary diffusion capacity; RQ, respiratory quotient.

a glucose molecule can maximally synthesize 38 ATPs. Given the inefficiencies of proton leaks, etc. (see Chapter 4) the real number of ATPs synthesized during aerobic activity is probably closer to 20–30 ATPs. Whatever the number of ATPs generated with aerobic metabolism that one uses, there is a decreased efficiency (5–10%) of converting the chemical potential energy of the glucose molecule into the physiologically usable form of energy (ATP) with anaerobic metabolism. Functionally, this means that to generate the same number of ATP molecules, 10 to 20 times more glucose would have to be consumed. Another disadvantage of glycolytic anaerobic metabolism is the generation of two H^+ associated with the two lactic acids. This acidosis is known as a metabolic acidosis, and is more complicated to compensate for than the respiratory acidosis associated with the formation of carbonic acid with an increase in P_{CO_2}. This decline in pH can be disruptive of intracellular processes, and may contribute to the muscle fatigue that is frequently associated with intense activity in amphibians (see below).

Alternative anaerobic pathways for metabolism other than fermentation of glucose also appear to exist in amphibians. Liver and muscle (skeletal and cardiac) appear to be able to catabolize certain amino acids anaerobically during freezing. Aspartate fermentation to succinate with the accumulation of alanine and glutamate conversion to glutamine are fermentative catabolic reactions that appear to occur during anaerobic freezing periods (Churchill and Storey 1996). The contribution of fermentative metabolic pathways of amino acids to ATP generation during activity has never been assessed.

The obvious advantage of anaerobic metabolism is that no O_2 is required for ATP synthesis, so it can be a source of ATP when the flux of O_2 to cells is either interrupted or insufficient to supply the energetic needs of cells. This is most likely to occur in muscle cells during exercise since the work output, hence demand for O_2 is large, and because as muscles contract the intramuscular pressure increases. Blood flow to muscle is dependent upon the pressure difference between the intramuscular pressure and the mean arterial pressure generated by the heart. As more skeletal muscle motor units are recruited, the work and strength of a muscle contraction increases. Intramuscular pressure increases with the strength of a muscle contraction. As intramuscular pressure increases, blood flow to a muscle will therefore decrease as the pressure difference decreases. This creates an imbalance, as the demand for ATP increases with an increased strength of a muscle contraction, the O_2 delivery via blood flow decreases. This deficit in ability to supply O_2 means that aerobic ATP generation will decrease because of insufficient O_2 being available, and CP and anaerobic generation of ATP will have to fill this ATP gap. In amphibians the threshold for lactate accumulation begins at 40–60% of V_{O_2max} (Taigen and Beuchat 1984). This transition to the use of anaerobic metabolism means that the rate of energy utilization is not sustainable long term and fatigue will occur.

3.5.4 Muscle Fatigue and Fiber Types

The process of fatigue is most studied in skeletal muscle cells (fibers) as they are the cells that are responsible for the increase in metabolism with activity. Fatigue is a decline in both the amount of tension a muscle and/or a muscle fiber can maintain and a slowing of both the velocity of shortening and relaxation. It is important to recognize that not all muscle fibers are the same. Skeletal muscle fibers are generally classified based on the rate at which the can generate tension. If they generate tension quickly they are described as fast twitch, if the rate of tension generation is slow they are slow twitch fibers, and if they can generate tension but not a twitch they are described as tonic fibers. There are gradations of fast in skeletal muscle fibers. The cellular characteristic that determines this rate of tension generation is the rate of cycling of the myosin ATPase heads (myosin heavy chain) in their binding to the actin filaments, movement when bound to the actin, release of the actin binding in the presence and breakdown of ATP with the recocking of the head. This cycle is repeated to achieve macroscopic shortening of the muscle fiber. This rate of cycling is frequently described as myofibrillar ATPase activity (Edman *et al.* 1988). Obviously, the faster this cycle occurs the faster the shortening and rate of tension generation occurs. The frog, *R. pipiens*, expresses four types of myosin heavy chains, three associated with fast twitch forms and a tonic form (Lutz *et al.* 1998a, b, 2002). There can be coexpression of these isoforms in a single fiber. There are minimally four muscle fiber types in *Rana pipiens* contributing most of the variation in the velocity of fast twitch fibers (Lutz *et al.* 1998a, b). The large hind limb extensor muscles of *R. pipiens*, which power jumping, have a preponderance (86%) of the Type 1 fastest twitch form of the myosin heavy chain isoform to generate power (Lutz *et al.* 1998). In the clawed frog, *Xenopus laevis,* there are five fiber types, the additional form being a slow twitch rather than just a tonic fiber at the slow end of the fiber force generation spectrum (Lannergren and Hoh 1984; Rowlerson and Spurway 1988). Functionally, this classification, based upon the velocity of shortening, implies a coupling of the rate of ATP utilization with the velocity of shortening, since the velocity is determined by the myofibrillar ATPase activity.

The metabolic characteristics of muscle fibers can vary from principally glycolytic to quite aerobic (Putnam and Bennett 1983). The fastest fibers tend to be glycolytic in enzymatic profile while the slower fast twitch fibers have greater aerobic capacity. Aerobic fast twitch fibers also have about half the cross-sectional diameter of anaerobic fibers (Putnam and Bennett 1983). This will obviously decrease the diffusional distances, and hence the resistance, to diffusional gas exchange. How, then, does the rate of ATP generation relate to this rate of ATP utilization? The time for fatigue to occur in individual muscle fibers is directly correlated with the aerobic character of a fiber measured either via mitochondrial density (Stary *et al.* 2004) or the succinate dehydrogenase activity, a mitochondrial enzyme (van der Laarse *et al.* 1991). The ratio of myofibrillar ATPase activity to succinate dehydrogenase activity, a metric of the rate of ATP utilization/ ATP generation is directly related to the length of time it takes to fatigue in different fast twitch muscle fibers (van der Laarse *et al.* 1991). Therefore the ability to aerobically supply ATP will determine fatigue resistance of individual muscle fibers. Do muscle fibers normally reach their aerobic potential?

The range of M_{O_2max} determined for individual fast twitch muscle fibers isolated from *X. laevis* leg muscle range from 25 to 150 nmol O_2/g s and was directly correlated with succinate dehydrogenase activity (van der Laarse *et al.* 1989).

These rates greatly exceed the organismic M_{O_2max} for *X. laevis* of between 10 and 15 nmol O_2/gs. Part of the disparity can be explained by the proportion of the organismal body mass that is muscle, about 40% (Putnam 1979), and in part by the proportion of that muscle that is fast oxidative versus fast glycolytic fiber types, 20% oxidative versus 80% glycolytic (Putnam and Bennett 1983). Weighting for these muscle variables, and assuming no other organs are oxidatively metabolizing the organismal M_{O_2max}, would be 0.4 (25 (0.8) + 150 (0.2)) = 20 nmol O_2/gs. Clearly, even under this best case scenario, with 60% of the body mass not consuming O_2 during activity, the metabolic capacity of the muscle alone is about two times greater than the measured organismal M_{O_2max}. This again suggests that the organismal mitochondrial capacity does not limit organismal M_{O_2max}. This conclusion is reinforced in a subsequent study on isolated *X. laevis* skeletal muscle fibers, where the critical P_{O_2} that limits fiber M_{O_2max} is about 10 kPa, about equivalent to arterial P_{O_2} and minimally two times greater than muscle P_{O_2} at rest (van der Laarse *et al.* 2005). It seems clear from these data that muscle fibers cannot reach their M_{O_2max} based on limitations of the circulatory system. This limitation will be exacerbated during activity as the increase in intramuscular pressure impedes muscle blood flow (see above).

Clearly, then muscle fatigue will start to occur as long as anaerobic metabolism is being utilized to supply the ATP to support the activity, nominally when levels of organismic locomotor activity reach 40–60% of V_{O_2max} (Taigen and Beuchat 1984). What changes do we see in the intracellular concentrations of organic molecules during anoxic exposure and during activity? A good synopsis for the concentrations of the principal cytosolic constituents in resting frog skeletal muscle cells can be found in Godt and Maughan (1988). From an energetic and fatigue viewpoint, the molecules that attract the most interest are glycogen, lactate, creatine phosphate, phosphate, H^+, ATP, ADP and AMP. Creatine phosphate is synthesized at the mitochondrion in the following fashion:

Creatine + inorganic phosphate (P_i) + H^+ → phosphocreatine

Phosphocreatine is important because it can phosphorylate ADP:

Phosphocreatine + ADP → ATP + creatine + H^+

Phosphocreatine represents a storage molecule for phosphorylated energy that can rephosphorylate the ADP without any immediate aerobic or anaerobic metabolism occurring. The quantity of phosphocreatine found in the cytoplasm of skeletal muscle cells is enough to power about 15 muscle fiber contractions in *Xenopus laevis* (Stary and Hogan 2000) or to swim about 250 body lengths at maximal velocity (Miller and Sabol 1989). Since the ADP availability for this reaction to occur results from the breakdown of ATP to ADP and P_i, as the phosphocreatine reaction is occurring there is a net increase in the intracellular concentration of both H^+ and P_i. Given the use of anaerobic metabolism and its by-products of lactate$^-$ and H^+, there should also be an increase in the intracellular concentrations of H^+ and lactate$^-$. The use of NMR or wet chemical methods with muscles during activity or during anoxia have documented an increase in cytoplasmic creatine, P_i, lactate$^-$, ADP and H^+ concentrations and a decrease in glycogen and ATP concentration (Cerreteli *et al.* 1972; Ambrosoli 1973a, b; Boutilier *et al.* 1986; Binzoni *et al.* 1990; Krause and Wegener 1996; Vezzoli *et al.* 1997; Vezzoli *et al.* 2003; Hsu and Dawson 2003; Stary and Hogan 2005). Most of the glucose utilized for anaerobic metabolism in amphibian skeletal muscles arises from intramuscular glycogen stores, and not the liver (Fournier and Guderly 1993). This increase in concentration of a variety of solutes leads to an increase in the intracellular osmotic concentration and uptake of water to balance the intra- and extracellular osmotic concentrations (Boutilier *et al.* 1986). The accumulation and increase in intracellular concentrations of the by-products H^+ and P_i have both been historically implicated as agents involved in skeletal muscle fatigue.

The increase in intracellular H^+ (decrease in pH) is handled by the muscle cell in several different ways: buffering, lactate transport and H^+ extrusion. Skeletal muscle cells have a temperature-dependent buffering capacity of between 30 and 70 μmol/g pH unit (Curtin 1986; Vezzoli *et al.* 2003). This buffering can be augmented by the exchange of intracellular Cl^- for extracellular HCO_3^- and Na^+ (Putnam and Roos 1986). There are membrane exchangers that will move intracellular H^+ in exchange with extracellular Na^+ (Abercrombie *et al.* 1983; Putnam *et al.* 1986). This will raise the intracellular Na^+ and increase the activity of the Na–K ATPase, in its attempt to maintain the concentration difference for Na^+ across the membrane. This Na–K ATPase activation has been implicated as the cause of the increase in V_{O_2} over standard that follows an activity period (Pinz and Portner 2003). Finally, there are monocarboxylate transporters that move lactate$^-$ and H^+ across the cell membrane in response to concentration differences (Mason and Thomas 1988). This facilitated diffusion dominates simple diffusion of the lactic acid across the membrane (Mason and Thomas 1988) This transport of lactate$^-$ and H^+ to the cell exterior when intracellular concentrations are elevated has been implicated in mammals as a mechanism to move lactate to other cells, which can then use lactate as a substrate for aerobic metabolism as well as removing H^+ (Brooks 2000). In mammals over 80% of the generated lactate is quickly aerobically oxidized. In amphibians, less than 20% of the lactate is oxidized (Connett 1979; Withers *et al.* 1988; Gleeson 1991), the remaining lactate is involved in gluconeogenesis and amino acid synthesis in the muscle itself (Fournier and Guderly 1992). The amphibian lactate shuttle does not fit the mammalian paradigm since so little of the lactate is oxidized, but it may instead function from a mass action perspective, preventing the accumulation of product (lactate) that could slow the utilization of substrate (glycogen) during anaerobic ATP generation and act to prevent intracellular acidification. Despite these regulatory mechanisms, there is a decrease in the intracellular pH with activity in amphibian skeletal muscle fibers (Renaud 1989; Stary and Hogan 2005). This decrease in pH may contribute to aspects of fatigue like decreasing the velocity of shortening (Curtin and Edman 1994) or the binding of calcium to troponin (Blanchard and Solaro 1984), but most of the recent literature dismisses low pH as the cause of muscle fatigue in amphibian skeletal muscle (Putnam 1979; Renaud 1989; Stary and Hogan 2005).

An increase in phosphate concentration is associated with a reduction in the calcium sensitivity of the myofilaments (Westerblad *et al.* 2002). Functionally, fatigue is also eventually associated with a decline in the intracellular concentration of calcium during stimulation and a lower affinity of the troponin for calcium from control stimulated states (Kabbara *et al.* 2000; Stary and Hogan 2000; Usher-Smith 2006; Westerblad *et al.* 1997, 2000). Since calcium is the trigger for contraction and its interaction with troponin determines the interaction of the myosin cross bridge with the actin, tension is bound to decrease in this situation. The mechanism for the decline in intracellular calcium concentration is a combination of less release and uptake by the sarcoplasmic reticulum.

Understanding fatigue is important because the accumulation of lactic acid has been demonstrated in various levels of locomotion (Miller 1983), chorusing frogs (Pough and Gatten 1984), during hibernation (see Chapter 4) and during predatory–prey interactions (Feder and Arnold 1982). Consequently, anaerobic metabolism and the fatigue process is obviously a component of an animal's field biology. There is wide interspecific variation in aerobic capacity (see above) and anaerobic metabolism (Bennet and Licht 1973; Putnam 1979a, b; Taigen *et al.* 1982; Taigen and Pough 1985). Unfortunately, the mechanisms underlying interspecific differences in anaerobic metabolism are poorly understood. Enzyme differences are important (Bennet 1974; Mendiola and De Costa 1990; Marker and Gatten 1993), though there has not been a systematic study that relates the variables involved in anaerobic capacity: phosphocreatine concentration, glycolytic capacity and intracellular buffering to variation in anaerobic metabolism.

Understanding what physiological, behavioral and environmental correlates are associated with this variation has drawn considerable interest in the literature. The two extremes are species that do not fatigue and species that quickly fatigue. These extremes are respectively characterized as high aerobic, low anaerobic capacity and low aerobic, high anaerobic capacity (Bennet and Licht 1974; Putnam and Bennet 1981). In order to understand the potential correlates we need to first put various activities (feeding, locomotion and calling) in perspective on the aerobic capacity continuum.

3.5.5 Locomotion

Feeding was covered above in our discussion of specific dynamic action and represents an activity that operates at less than 25% of aerobic capacity. The measurement of the costs of locomotion is straightforward in its design, having an animal move on a treadmill while M_{O_2} is determined. Though straightforward in its design the logistics are not easy, in part because some species crawl, some walk, others hop and many jump. Consequently there is a paucity of data in comparison to other vertebrate groups. The data are summarized in Fig. 3.36. It is clear M_{O_2} increases with speed in all the species investigated, except *Desmognathus quadramaculatus* (Full *et al.* 1988). Lateral undulation by salamanders is significantly less expensive than anuran locomotion (Gatten *et al.* 1992). The cost of transport is approximately 10–300 µmol O_2/g km (Gatten *et al.* 1992). Toads, *Bufo woodhousei* for instance, move between 0 and 0.06 km/h when naturally foraging, consequently the locomotory M_{O_2} is less than 40% of M_{O_2max} (Walton 1988). At these speeds it is unlikely that the anaerobic threshold has been exceeded and fatigue is not part of the picture.

3.5.6 Vocalization, Amplexus and Nest Building

Reproductive activity has also drawn interest from an oxidative capacity perspective. Amplexus, calling and even foam nest building are activities that would normally exceed the anaerobic threshold of 40–60% of M_{O_2max}. The most expensive reproductive activity appears to be calling on the part of the male. Females preferentially pair with males with higher calling rates (Whitney and Krebs 1975;

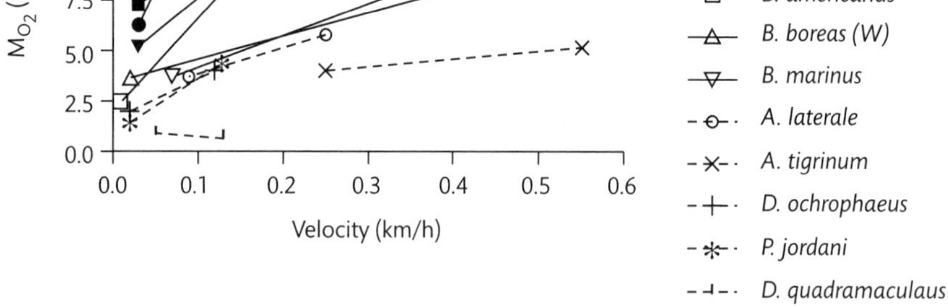

Fig. 3.36 The measured relationship of mass specific rate of O_2 consumption at different treadmill speeds in a variety of amphibian species. Data for the various species are from: Full *et al.* (1988); Walton and Anderson (1988); Walton (1993); Walton *et al.* (1994); Baudinette *et al.* (2000).

Sullivan 1982; Schwartz *et al.* 1995; Forsman and Hagman 2006) and males that spend more time in a breeding chorus (Arak 1983; Jacobsen 1985; Gerhardt *et al.* 1987; Sullivan and Hinshaw 1992). Calling is an honest indicator of male genetic quality in poison dart frogs, as females who mate with high calling performance males obtained more offspring per egg than with poor callers (Forsman and Hagman 2006). Despite the selective advantage of calling, males generally only spend a third of the breeding season in choruses (Murphy 1994). Part of the reason for this paradox is that predation can be high in choruses (20%, Murphy 1994), but the metabolic costs are so high that substrate depletion can limit chorus participation (Murphy 1994; Bevier 1997). The rate of oxidative metabolism is related to call frequency (Fig. 3.37). As is clear from these measurements, the rate of aerobic metabolism can be very high at high calling rates, exceeding not only the anaerobic threshold but also M_{O_2max} determined during locomotion. Interestingly, there was no accumulation of lactate during these calling bouts (Taigen and Wells 1985), a conundrum resulting from work on locomotor energetics (Taigen and Beuchat 1984). The trunk muscles (internal and external obliques) are principally responsible for generating the calls by forcing air from the lungs through the larynx to the vocal sacs (Girgenrath and Marsh 2003). These muscles are an order of magnitude larger in males than females, and represent about 12% of a male's body mass (Marsh and Taigen 1987). The muscles are entirely oxidative and the mitochondrial volume represents up to 20% of the fiber volume as opposed to 3–7% of the volume of locomotory skeletal muscles, the capillary density is also about three times greater (Marsh and Taigen 1987). This development appears to be under the

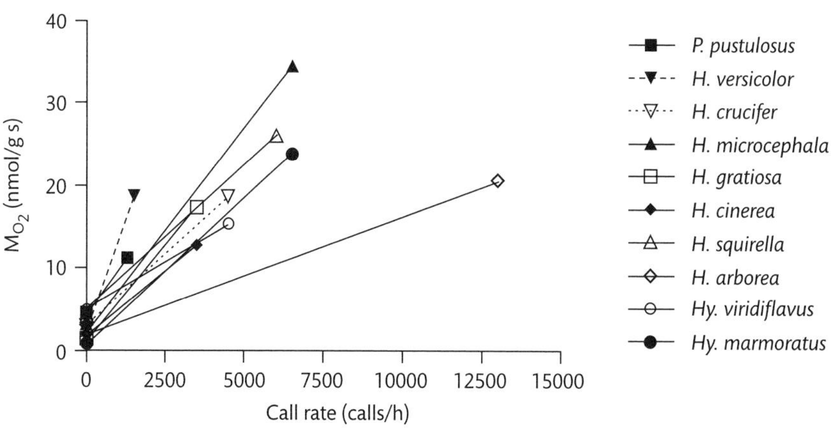

Fig. 3.37 The relationship between the frequency of mating calls and the rate of O_2 consumption in anurans. Data are from: Bucher *et al.* 1982; Taigen and Wells 1985; Taigen *et al.* 1985; Wells and Taigen 1989; Prestwich *et al* 1989; Grafe *et al.* 1992; Grafe 1996; Grafe and Thein 2001.

control of testosterone (Girgenrath and Marsh 2003). Interfamilial trunk muscle mitochondrial volume is proportionately higher in species with higher calling rates (Ressel 2001). Obviously, calling represents a very different situation than locomotion, especially in the lower proportion of body mass the active muscles represent as well as their extraordinary metabolic capacity. If a large fraction of the cardiac output can be shunted to such a small fraction of the animal's mass, the normal limitations to maximal M_{O_2} may be different and revolve around the aerobic capacity of that muscle. Not only are the trunk muscles more developed in calling males, but also the forelimb muscles involved in amplexus (Dorlochter *et al.* 1994; Oka *et al.* 1984; Melichna *et al.* 1972; Yetka and Blackburn 1992). Not only is the mass of muscle enhanced, but the muscle contraction characteristics are modified (Peters and Aulner 2000). The metabolic rate of amplectant males of *Hyla versicolor* has been measured at 2 nmol/g s or about 30% of the locomotory M_{O_2max} (McLister 2003). Given the high cost of calling coupled to the relatively low cost of amplexus, it appears that reproductively successful amplectant males actually spend less energy per night than calling males and extend the number of nights they can participate in chorusing activity (McLister 2003). This serves as a positive feedback loop on reproductive success, as originally proposed by Dyson *et al.* (1998). Finally, the metabolic cost of whipping up a foam nest in *Physalaemus pustulosus* has also been measured and is equivalent to rates of M_{O_2} associated with high calling rates (Bucher *et al.* 1982).

Given the intra- and interspecific differences in M_{O_2max} and the differences in muscle characteristics that relate to aerobic and anaerobic metabolism, it is a logical and attractive extension to relate this variation to differences in organismal performance. The linkage between metabolic performance and muscle characteristics with organismal performance has generally eluded description. Attempts to organize aerobic and anaerobic metabolic capacity around aspects of natural history have only been successful in very general terms. For instance, predatory mode (active searcher) and non-jumpers were the only behavioral aspects that correlated with aerobic metabolism (Taigen *et al.* 1982; Taigen and Pough 1983, 1985). Dehydration tolerance is another feature that correlates with aerobic metabolism (Hillman 1991). High-rate callers of *Pseudacris crucifer* had higher muscle citrate synthase and β-hydroxyacl-CoA dehydrogenase enzyme activities and larger ventricles and hemoglobin concentrations than low-rate callers (Zimmitti 1999). Interindividual differences in the sprint performance of *R. pipiens* were not related to differences in lactate accumulation or lactate dehydrogenase activity of the gastrocnemius muscles (Marker and Gatten 1993). Attempts to relate interindividual differences in locomotor aerobic capacity in *B. woodhousei* and their call rate, a linked variable to reproductive success (see above), has been equally unsuccessful (Sullivan and Walsberg 1985). Interindividual aerobic capacity differences and locomotor performance or field foraging behavior have also been unsuccessful (Rogowitz and Sanchez-Rivoleda 1999; Walton 1988). The inability to correlate

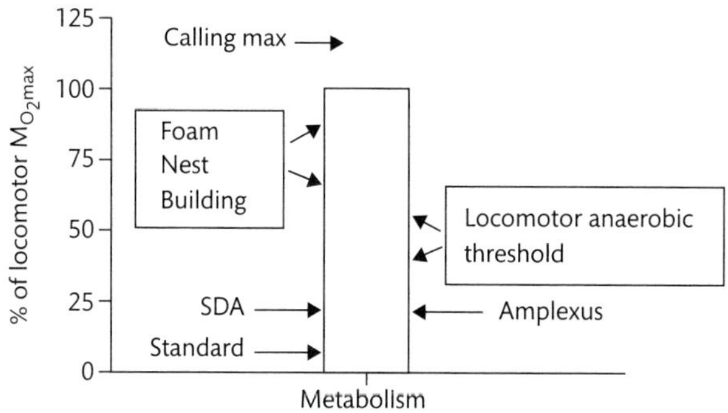

Fig. 3.38 A summary of the costs of various activities as proportions of locomotor M_{O_2max}.

interindividual animal behavioral differences with their aerobic or anaerobic capacities may indicate either that these variables do not set limits to normal behavior or that they interact in complex ways with other physiological variables. Single factor explanations for the significance of variation in metabolic patterns may be inappropriate (Taigen *et al.* 1982). Understanding the source of interindividual and interspecific variation in aerobic and anaerobic capacities may help sharpen the ability to unravel these associations.

A summary of the relative magnitude of the various parameters that influence metabolism expressed relative to locomotor M_{O_2max} is presented in Fig. 3.38.

3.6 Nitrogen Excretion

The use of proteins (amino acids) as a substrate for aerobic metabolism entails a deamination of the amino acids being catabolized, generating ammonia. Some of this ammonia is of further value from an anabolic synthesis of amino acids perspective (i.e. transamination reactions) but most is not and is excreted as a waste product. The process of nitrogen excretion occurs principally via the kidneys in anurans (Balinsky and Baldwin 1961; Shoemaker and McClanahan 1975, 1980), but it can also take place by diffusion across the skin and gills. These diffusional routes actually dominate renal losses in aquatic situations, for example the urodele *Necturus maculosus* (Fanelli and Goldstein 1964) and caecilian *Typhlonectes natans* (Stiffler and Manokham 1994).

3.6.1 Rates of Nitrogen Excretion

The rate of nitrogen excretion averages about 0.7 µmol N/g h (Fig. 3.39), and does not vary between the orders of amphibians. As resting metabolism is about

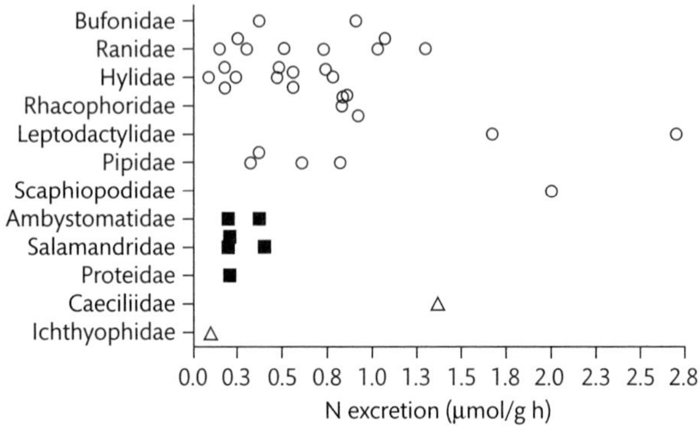

Fig. 3.39 The reported rates of N excretion for amphibians from a variety of different families. Data are from: Cragg *et al.* 1961; Fanelli and Goldstein 1964; Shoemaker and McClanahan 1973, 1975, 1980; Jones 1980; Cree 1985; Stiffler and Manokham 1994; Stiffler and Talbot 2000; Loong *et al.* 2002; Weng *et al.* 2004; Grafe *et al.* 2005.

2.2 μmole O_2/g h, there is about a 3:1 molar flux of O_2:N at rest in amphibians. This N efflux is sensitive to water availability (Shoemaker and McClanahan 1975, 1980). The actual excretion rate is dependent upon urine formation rate in anurans. The rate of N excretion increases in acidotic conditions (Stiffler 1991). The rate of N excretion is also sensitive to feeding, and increases post feeding (Shoemaker and McClanahan 1975; Cree 1985). Temperature increases the rate of N excretion in bullfrog tadpoles (Ashley *et al.* 1968). There have been no systematic studies of the effects of body mass on N excretion rates.

Probably the most interesting aspect of N excretion for amphibians from an environmental perspective is the chemical form in which it can be excreted. There are three principal chemical forms of N for excretion: ammonia (NH_3; or ammonium NH_4^+), urea ($CO(NH_2)_2$) and uric acid ($C_5H_4N_4O_3$ or its base urate $C_5H_3N_4O_3^-$). The tradeoffs of the water required for excretion, the energy of synthesis, the toxicity, excess base generated as HCO_3^- during protein catabolism and the contribution to osmotic concentration interact to determine the proportion of one chemical form versus another, and this is one of the central historic stories of environmental physiology. We shall first examine ammonia (and ammonium) excretion since this is surely the ancestral state for amphibians as aquatic organisms. The removal of the α-amino group from amino acids forms NH_3. NH_3 combines with water to form the weak base NH_4OH. The pK of ammonium hydroxide is 9.5, therefore the ammonia exists predominantly as NH_4^+ at physiological pHs. Both forms (NH_3 and NH_4^+) exist, and "ammonia" is routinely used as a generic term to include both ammonia and ammonium. The catabolism of protein also leads to an excess of base in the form of HCO_3^- being formed relative to protons (Atkinson and Camien 1982; Atkinson 1992), so there are two issues involved with protein catabolism from an excretory perspective; N excretion and neutralizing the excess bicarbonate. In a water environment the ions (NH_4^+ and HCO_3^-) can be excreted in the water as ions. This is not as easy in air.

3.6.2 Ammonia as a Nitrogen Waste

The most common amino acid catabolism pathway is an aminotransferase reaction forming glutamate:

α-amino acid + α-ketogluturate → glutamate + α-keto acid

The glutamate is then taken up in the mitochondrion where it is oxidatively deaminated by glutamate dehydrogenase to form NH_4^+ and α-ketoglutarate, which is used in the TCA cycle.

glutamate → NH_3 + α-ketoglutarate

The ammonia is disruptive of normal cellular function at low concentrations and described as "toxic" as a consequence. Glutamate can be converted to glutamine (an additional amino group) in the glutamine synthetase reaction, essentially binding up the NH_4^+ that was generated:

glutamate + MgATP + NH_4^+ → glutamine + MgADP + P_i

Glutamine is an effective, non-toxic form to transport ammonia.

Ammonia is highly toxic. The toxicity of ammonia derives from a variety of potential mechanisms. Most entail the glutamate dehydrogenase reaction above. There is accumulation of glutamine and depletion of α ketogluturate and ATP at high ammonia levels. This loss of the tricarboxylic acid cycle intermediate (α-ketoglutarate) and ATP with the accumulation of glutamate (also involved as a neurotransmitter) are disruptive, particularly for the nervous system. Ammonia also influences intracellular and intraorganelle pH. Ammonia as NH_3 alkalanizes and as NH_4^+ acidifies. Another potential problem is diffusion of NH_3 across the inner mitochondrial membrane and binding with H^+ to form NH_4^+, which effectively reduces the pH gradient across the inner mitochondrial membrane and eliminates the electromotive driving force for ATP synthesis (Campbell 1997). Regardless of ammonia's mechanism of toxicity, the biological imperative is to prevent the ammonia concentration from reaching a toxic intracellular concentration. This requires either ammonia excretion at very low concentrations, combining ammonia with glutamate to form glutamine or using ammonia as a substrate for the synthesis of urea or uric acid, both of which are less toxic anabolic products of ammonia. The advantage of both these synthetic processes is that they liberate protons that will balance the excess bicarbonate production associated with protein catabolism.

3.6.3 Urea as a Nitrogen Waste Product

Urea ($CO(NH_2)_2$) is formed in the ornithine–urea cycle, generally in the liver but also in the kidney. The general form of the ornithine–urea cycle is as follows:

$$\textbf{glutamic acid} + NAD^+ + H_2O \rightarrow \alpha\textbf{-ketoglutaric acid} + NH_3 + NADH + H^+$$

The free ammonia is then converted by carbamyl phosphate synthetase to form:

$$NH_3 + CO_2 + 2\,ATP + H_2O \rightarrow \textbf{carbamyl phosphate} + 2\,ADP + P_i$$
$$\textbf{carbamyl phosphate} + \textbf{ornithine} \rightarrow \textbf{citrulline} + H_3PO_4$$

The carbamyl phosphate is then combined with ornithine to form citrulline and H_3PO_4; this reaction is catalyzed by ornithine transcarbamylase. The carbamyl phosphate synthetase and ornithine transcarbamylase are enzymes present in the mitochondria, preventing ammonia from disrupting mitochondrial function. The citrulline is then transported from the mitochondrion to the cytosol in exchange with ornithine.

An additional amino group is added to the carbonyl group of citrulline in the cytosol, catalyzed by arginosuccinate synthetase:

$$\textbf{citrulline} + \textbf{aspartic acid} + ATP \rightarrow \textbf{arginosuccinic acid} + AMP + PP_i$$

The arginosuccinic acid is cleaved by arginosuccinase to form arginine and fumaric acid.

The fumarate is combined with H_2O to form malate, which is converted to oxaloacetate. The oxaloacetate combines with glutamate to regenerate the aspartic acid. The arginine is broken down by arginase to form urea and ornithine, which is exchanged for citrulline across the mitochondrial outer membrane.

$$\textbf{arginine} \rightarrow \textbf{ornithine} + \textbf{urea}$$

Therefore both the aspartic acid and ornithine are regenerated, which is the basis for its description as a cycle. The two amino groups associated with urea enter the cycle at the level of carbamyl phosphate and aspartic acid. Urea is a neutral molecule, and a relatively non-perturbing solute of enzyme function in some anurans (Grundy and Storey 1994). Other vertebrates that accumulate urea as a solute to balance osmotic relations generally accumulate additional solutes to counterbalance the perturbing effect of urea on proteins (Yancey and Somero 1979; Somero 1986). The best-described counterbalancing solute to urea is trimethylamine oxide (Yancey and Somero 1979, 1980; Somero 1986). Amphibians that accumulate high concentrations of urea to balance their osmotic relationships do not appear to accumulate non-perturbing molecules to counterbalance the perturbing effects of urea (Withers and Guppy 1996). Amphibian species that do not estivate appear to be more sensitive to the perturbing effects of urea (Cowan and Storey 2002). It thus appears that the accumulation of urea in saline and dry soil conditions (Chapter 4) may have led to the evolution of urea-insensitive proteins rather than the co-accumulation of counteracting solutes. It appears that urea may actually be required to culture embryos of *Gastrotheca riobambae*, a species that is ureotelic as it develops in the pouch on the back of the female (del Pino *et al.* 1994).

3.6.4 Uric Acid as a Nitrogen Waste Product

The third form of nitrogen excretion is the synthesis of the purine, uric acid. It starts in most vertebrates with the intramitochondrial formation of glutamine via the glutamine synthetase catalyzed reaction of glutamic acid and ammonia, described above. The glutamine is then transported to the cytosol where the rest of the uric acid biosynthesis occurs. A good summary of uric acid biosynthesis can be found in Campbell (1991, 1997). Amphibians start the biosynthetic process of uric acid differently from reptiles and birds, as the glutamine synthetase is found principally in the cytosol and not the mitochondria (Campbell *et al.* 1984). This may reflect the fact that carbamyl phosphate synthetase, the ancestral mechanism for detoxifying ammonia in ureotelic species, still dominates as the principal detoxification of intramitochondrially generated ammonia from amino acid catabolism.

3.6.5 Optimizing the Form of Nitrogen Waste

The advantage of ammonia as a terminal N excretory product is that there is no further expenditure of metabolic energy associated with the anabolic reactions to modify its chemical structure. The obvious disadvantage is that it requires a lot of water to excrete it in order to prevent accumulation of concentrations toxic to metabolism.

Urea is not appreciably disruptive of biological processes at relatively moderate concentrations, therefore less water is needed to excrete it. Because of its low toxicity and electroneutrality, urea also can be accumulated as an additional solute to raise the animal's body fluid osmotic concentration and create an appropriate osmotic concentration to passively absorb water from saline solutions or the soil (see Chapter 4). Since urea is quite permeable to cell membranes it will accumulate as an osmolyte in both the intracellular and extracellular spaces, avoiding cell volume disruption (Wood *et al.* 1989). Urea synthesis also generates two protons to titrate the excess bicarbonate associated with protein catabolism. The major disadvantage is that four moles of ATP are required for the synthesis of each mole of urea.

Uric acid has a very low solubility in physiological solution. This means that uric acid does not act osmotically, since it is effectively not in solution. Uric acid is then the most economical N excretory form, from the perspective of water required for excretion. Synthesis of a molecule of uric acid releases four protons for titrating the bicarbonate excess associated with protein metabolism. However disadvantages include the requirement of the equivalent of 24 moles of ATP energy per mole of uric acid synthesized *de novo* (Atkinson 1977), the difficulty of moving a paste rather than a solution through the ureters, and the fact that uric acid as an anion carries cations (Na^+ and K^+) with it as urate salts. This entails a fixed cation loss for every mole of uric acid excreted, which could disrupt ion balance, especially for aquatic species. The urinary bladder of uricotelic reptiles contains cilia in the neck of the urinary bladders that are absent in ureotelic reptiles and amphibians (Bolton and Beuchat 1991). These cilia have been conjectured to function as a mucociliary transport system that moves particulates, like precipitated urate salts, from the bladder to the cloaca for excretion (Bolton and Beuchat 1991). There are no data suggesting that a mucociliary system exists in the urinary bladder of uricotelic anurans.

The predominant chemical form of N excretory product that is excreted by amphibians reflects the interplay of these advantages and disadvantages. If the principle N excretory form is ammonia, the species is described as ammonotelic. If the principal N excretory form is urea it is known as ureotelic. Finally, if most of the N excretory product is in the form of urate salts it is described as uricotelic. These N excretory patterns have strong environmental correlates developmentally, seasonally and interspecifically. A schematic distribution of the principal N excretory products for amphibians with access to water (Fig. 3.40) shows that the

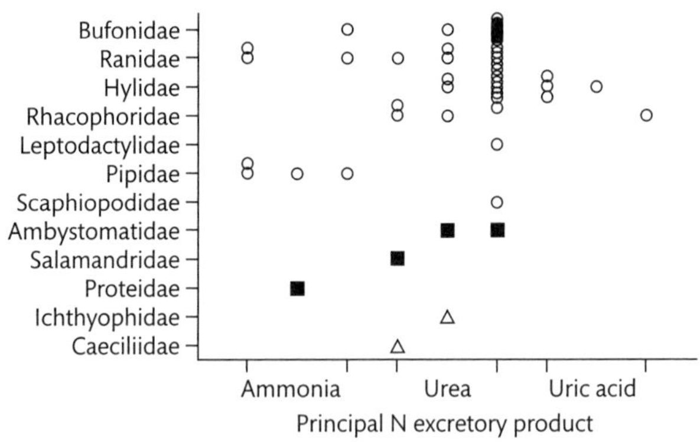

Fig. 3.40 The principal form of N excretion product for amphibians with access to water reported from a variety of different families. The percentages of each pattern were broken into three categories: 50–67%, 68–84%, and greater than 84%. Data are from: Cragg *et al.* 1961; Fanelli and Goldstein 1964; Schmid 1968; Shoemaker and McClanahan 1973, 1975, 1980; Jones 1980; Cree 1985; Rovedati *et al.* 1988; Wood *et al.* 1989; Stiffler and Manokham 1994; Stiffler and Talbot 2000; Loong *et al.* 2002; Weng *et al.* 2004.

predominant excretory mode is ureotely. Uricotely has only been described in two families, the Hylidae and Rhacophoridae.

The evolution of ureotelism from ammonotelism has implications for how N is transported and distributed throughout the body. As we saw above, urea distributes evenly between the intra- and extracellular fluids based on concentration. The distribution of ammonia is more complicated as it can exist as NH_3 or NH_4^+. The electrochemical driving force for ammonia distribution is determined by both the membrane potential (NH_4^+), and the pH difference across the cell membrane (both NH_3 and NH_4^+; Wood *et al.* 1989). The distribution of ammonia at equilibrium inside versus outside the cell decreases as an amphibian moves from being principally ammonotelic (ratio 17) to ureotelic (ratio 9; Wood *et al.* 1989). Amphibians lie between fish (ratio 30; extreme ammonoteles) and mammals (ratio 3; extreme ureoteles) in their equilibrium ratio (ammonia inside/ammonia outside; Wood *et al.* 1989). As amphibian species excrete a greater fraction of their N as urea there appears to be a reduction in the permeability of the cell membrane to NH_4^+, increasing the ratio of the permeability of NH_3/permeability of NH_4^+ (Wood *et al.* 1989).

From a developmental perspective, amphibians generally make the transition from an aquatic world as a larva to a terrestrial world as a metamorph. For larvae, water enters diffusionally so the kidneys are continuously excreting water and the animal is in a large enough volume of water to effectively dilute the ammonia excreted and allow for the elimination of bicarbonate across its skin and gills. Many anuran larvae are herbivorous so protein intake is low and the demands of nitrogen excretion and bicarbonate titration are low. The characteristic N excretory pattern for most amphibian larvae is ammonotely (Brown *et al.* 1959). Most amphibian larvae do not even express urea cycle enzymes until late in development (Paik and Cohen 1960). With metamorphic induction by the hormone thyroxine, amphibians begin expressing urea cycle enzymes (Dolphin and Frieden 1955; Paik and Cohen 1960) and are ureotelic as metamorphs and adults. The interpretation of this developmental transition is that so long as water is present in large quantities in the aquatic environment, there is no selective pressure for a less toxic form of N or the need for freeing up protons to titrate the bicarbonate at the cost of increased expenditure of metabolic energy in forming urea. There may be an additional advantage in being able to exchange NH_4^+ for cations and HCO_3^- for anions. The exceptions to the ammonotelic larval pattern actually reinforce our understanding of the process. Ureotelic larvae have been described for two foam-nesting anurans, *Leptodactylus bufonius* and *Hemisus marmoratus* (Shoemaker and McClanahan 1973; Grafe *et al.* 2005), for offspring developing in brood pouches as in *Gastrotheca riobambae* (Alcocer *et al.* 1992), as well as for larvae of *Salamandra salamandra* which are contained *in uteri* (Schindelmeiser and Greven 1981). Ureotelism develops early in tadpoles where drying of ponds is a common occurrence in their biology (Martin and Cooper

1972; Jones 1980). The unifying environmental variable for these larvae is a more limited and unpredictable volume of surrounding water. The significance of their ureotelic larval pattern is that ammonia would be much more likely to increase to toxic levels in their immediate environments, therefore ureotely would be selectively advantageous. An unexplained pattern has been observed for N excretion by tadpoles of *Phyllomedusa sauvagei*, a species that is uricotelic as an adult. The larvae are principally ureotelic, even though they are entirely aquatic, and begin uric acid synthesis during metamorphosis (Shoemaker and McClanahan 1982).

Adult amphibians can shift from ammonoteley to ureotely if environmental water availability is decreased (see Chapter 4). The clawed frog, *Xenopus laevis*, becomes ureotelic if the osmotic concentration of its water is increased (Scheer and Markel 1962; Funkhouser and Goldstein 1973). Dehydration will also induce this shift (Balinsky *et al.*1961).

Seasonal differences have also been correlated with a change in the proportion of N excreted as ammonia. Salamanders that move from terrestrial habitats, where they are ureotelic, to breeding sites for prolonged breeding activity shift to increased ammonia excretion, although the pattern remains ureotely during the breeding season (Nash and Funkhauser 1959; Weng *et al.* 2004).

Amphibians normally excrete only a very small fraction of their N as uric acid. Historically, it was simply assumed that amphibians could not be uricotelic, and that what small amounts of uric acid that were excreted were simply a product of purine metabolism. An African rhacophorid treefrog, *Chiromantis xerampelina,* was discovered to be uricotelic (Loveridge 1970). This discovery was published in a relatively obscure journal, not because of the insignificance of the discovery, but as a consequence of the fact that reviewers in more respected journals refused to believe the results. Subsequently, another species of *Chiromantis*, *C. petersii,* and a variety of South American phylomedusine treefrogs have been shown to be principally uricotelic (Shoemaker *et al.* 1972; Shoemaker and McClanahan 1975; Drewes *et al.* 1977). All these uricotelic frog species are arboreal, and have very high cutaneous resistance to evaporative water loss and low rates of water turnover. It seems clear that the benefits of being uricotelic as an amphibian only balance the costs when water turnovers are low. Shoemaker and McClanahan (1975) have calculated that the amount of water required to excrete as urate salts the N ingested from a mealworm diet for *Phyllomedusa sauvagei* (their most uricotelic species) to be 3.8 ml H_2O/kg day. A related ureotelic species requires about 60 ml H_2O/kg day (Shoemaker and McClanahan 1975). Assuming an ammonia tolerance of 0.006 mol/l, an extreme 50% lethal concentration for *Xenopus laevis* tadpoles (Schuytema and Nebeker 1999), a frog would require about 4000 ml H_2O/kg day to avoid reaching lethal ammonia levels if ammonotelic with the same N input. So, the water savings for the transition from ammonotely to ureotely is 3940 ml H_2O/kg day, while the savings from ureotely to uricotely only amounts to 56 ml H_2O/kg day at an additional cost of about 3 ATP/ N excreted. This indicates about a 70-fold greater water savings associated with a transition from ammonotely to ureotely compared with the ureotely to uricotely transition.

3.7 Kidneys and Nephrostomes

Historically, the amphibian kidney has been an important model for studying vertebrate kidney function. Consequently, amphibian renal function has been the object of study for general physiology rather than comparative or environmental physiology. Part of this relates to the large size of the functional kidney unit, the amphibian nephron. Micropunture samples and measurements from inside various nephron segments that were impossible or difficult in mammals, which have smaller nephrons, were more easily obtained from amphibians. Other reasons for the use of amphibians as models relates to their ready availability and their less charismatic status among terrestrial vertebrates; their study is less objectionable from a societal perspective. The evolutionary and ecological variability of amphibian renal physiology has only secondarily been a topic of basic research resulting in far less literature.

3.7.1 Anatomy of the Kidney

The kidneys are paired organs that lie retroperitoneally in the subvertebral lymph sac. The ventral surface faces the pleuroperitoneal space and has connections via ciliated funnels with the lymph contained in this space. If these funnels connect to renal tubules the funnel is described as a nephrostome. If the ciliated funnel empties into a renal portal vein it is known as a ciliated peritoneal funnel. The cilia of peritoneal funnels beat at a frequency of 7 Hz, generate a pressure of 0.3 kPa and move 0.5 ml/h in *B. marinus* (Morris 1981). The cilia of true nephrostomes can generate pressures between 0.8 and 1.1 kPa (White 1929a).

The kidneys are supplied with blood via both renal arteries and hepatic portal veins. The renal arteries are branches from the dorsal aorta, while the renal portal veins carry blood from the posterior part of the animal. The renal arteries enter the kidneys medially, then branch into renal arterioles which branch further into capillary tufts known as glomeruli. Hence there is an afferent arteriole going to the glomerulus and an efferent arteriole carrying blood from the glomerulus, which then connects to peritubular capillaries. The peritubular capillaries also contain branches of the renal portal veins which enter the kidneys from a dorsolateral direction. These anastomoses between venous and arterial supplies means the peritubular capillaries carry both arterial and venous blood to the interstitium of the kidney. This peritubular capillary blood drains via efferent renal veins to the postcava where it enters the sinus venosus.

The functional unit of the kidney is known as the nephron. A nephron is comprised of the following elements in series: a glomerulus, Bowman's capsule surrounding the glomerulus, a slender ciliated neck region of the proximal convoluted tubule, a large-diameter proximal convoluted tubule, a thin ciliated intermediate segment, a slender distal convoluted tubule and finally a collecting tubule which drains into the collecting duct. The neck region may or may not also have a nephrostome connected to it. If it does it is known as a primary nephron (Fig. 3.41B), if not a secondary nephron (Fig. 3.41C). True nephrostomes are ciliated funnels opening to the pleuroperitoneal space (Fig. 3.41A) and are present in adult caecilians (Brauer 1902; Sakai *et al.* 1986; Mobjerg *et al.* 2004), urodeles (Sakai and Kawahara 1983) and some anurans (Withers unpublished). Interestingly, in the most comprehensive survey of 50 species of caecilian kidneys, Wake (1970) reported that nephrostomes were absent, though previous and subsequent workers have documented their presence (Brauer 1902; Sakai *et al.* 1988; Carvalho and Junqueira 1999; Mobjerg *et al.* 2004). The nephrostome has been estimated to contribute between 2 and 11% of the tubular fluid (White 1929a; Stiffler *et al.* 1980).

The combination of a glomerulus and its surrounding Bowman's capsule is known as the renal corpuscle, or in some literature as the Malpighian corpuscle.

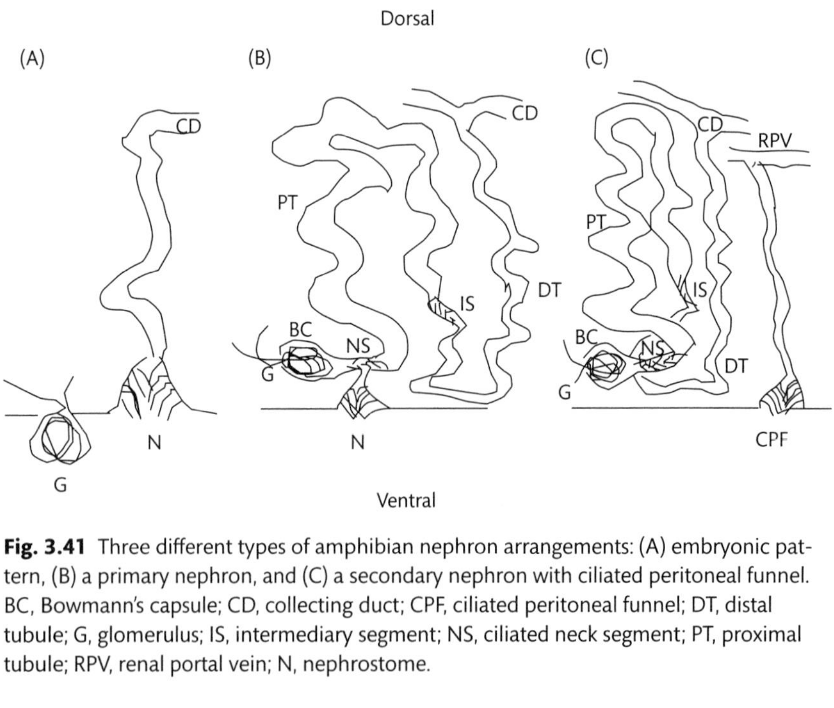

Fig. 3.41 Three different types of amphibian nephron arrangements: (A) embryonic pattern, (B) a primary nephron, and (C) a secondary nephron with ciliated peritoneal funnel. BC, Bowmann's capsule; CD, collecting duct; CPF, ciliated peritoneal funnel; DT, distal tubule; G, glomerulus; IS, intermediary segment; NS, ciliated neck segment; PT, proximal tubule; RPV, renal portal vein; N, nephrostome.

Bowman's capsule is drained via the ciliated neck segment. The cilia of the neck region contribute to maintaining the hydrostatic pressure difference, which moves fluid (urine) from Bowman's capsule to the proximal convoluted tubule. The distal end of the proximal convoluted tubule is connected to a second ciliated tubular segment known as the intermediate segment, which is also partially responsible for contributing to the pressure difference that moves the fluid from the proximal to the distal convoluted tubule. The collecting ducts from all the nephrons drain into the ureter. The kidney is comprised of thousands of nephrons.

Among the three amphibian orders the nephrons are arranged in the kidney in characteristic ways. The corpuscles and distal tubules are generally ventromedially located in caecilian and urodele kidneys and in the ventral half of anuran kidneys. The primary nephrons with attached nephrostomes, are always associated with the ventral kidney surface in the three orders. The proximal convoluted tubules and collecting tubules are characteristically in the lateral and dorsal kidney regions in caecilians and urodeles and the dorsal region of anuran kidneys (Huber 1917; Dawson 1924; Barch *et al.* 1966; Stoner 1977; Clothier *et al.* 1978; Sackin and Boulpaep 1981; Taugner *et al.* 1982; Guggino *et al.* 1988; Farrias *et al.* 1998; Gealekman and Warburg 2000; Mojberg *et al.* 2004).

The glomerulus is composed of branched capillary loops. The glomerular capillaries from lumen to exterior are composed of a fenestrated endothelial cell layer, a basement membrane, which provides the selectivity for molecule movement, and an epithelial cell layer on the outer surface characterized by podocytes (Pak Poy 1957). The podocytes have long radial extensions that branch into numerous secondary foot processes, known as pedicels. These pedicels interdigitate and form filtration slits on the basement membrane. The basement membrane is thicker than in higher vertebrates (Schaffner and Rodewald 1978; Taugner *et al.* 1982). Primary urine is the fluid formed by filtration across these barriers and the rate of fluid formation is the glomerular filtration rate. The filtrate is similar to plasma except that plasma proteins do not cross the glomerular membranes.

Bowman's capsule is composed a parietal layer of squamous cells overlying the glomerulus and is continuous with the epithelial layer of the rest of the tubule. The space between the glomerulus and capsule is known as the nephrocoel. The capsule is drained by the neck segment, which is a narrow structure lined by cuboidal ciliated cells. This neck segment then joins the proximal tubule. The proximal tubule is lined by cells having brush borders on their apical surfaces. There are some data to support a correlation of increased microvilli number with the degree of terrestriality in salamanders (Maunsbach and Boulpaep 1984). The cells have tight junctions on their apical-lateral surface and numerous mitochondria on their basal lateral surface. The iso-osmotic uptake of organic and inorganic molecules occurs in this segment. Approximately 30–50% of the filtrate volume is absorbed in this segment of the nephron (Walker and Hudson 1937; Bott 1962). This is also where the process of secretion occurs, where some molecules are actively transported from the renal interstitium into the tubule. The circulatory supply for the proximal convoluted tubule is mainly by peritubular capillaries from the renal portal vein. At its distal end the proximal convoluted tubule connects to the ciliated intermediate segment, which is very similar anatomically to the neck segment.

The distal convoluted tubule follows the intermediate segment. It can be histologically differentiated into an early diluting segment and a late distal region. The early segment comes in close contact with the renal corpuscle, and has interdigitating intercellular junctions with an invaginated basal membrane with numerous mitochondria. The distal cells are larger and have well developed basal and lateral membranes, like in the early distal cells. The function of the distal tubules is to absorb sodium and chloride, with limited passage of water, hence the term diluting segment. The collecting tubule and collecting duct are generally not implicated in solute uptake but may be involved in potassium secretion in toads (Mobjerg *et al.* 2002, 2004).

3.7.2 Developmental Aspects of the Kidney

The kidney develops embryologically from the mesomere in an anterior to posterior direction. The first section to develop is known as the pronephros. Excellent anatomical descriptions and literature summaries of the pronephros can be found in Fox (1963), Christensen (1974) and Mobjerg *et al.* (2004). The pronephros is composed of nephrostomes draining into an aglomerular tubule. There is a glomus that projects into the coelomic cavity in proximity to the nephrostomes, but the glomus is not enclosed by a capsule connected to the tubule as in the case of a primary nepron. The pronephros functions through larval development, then it degenerates. Removal of one pronepros leads to an increase in size of the remaining pronepros; removal of both sides leads to an edematous death of the embryo (Howland 1916). Obviously, the functioning of the pronephros as an osmoregulatory organ is crucial to larval survival. The pronephros developmentally forms the ureter that will also be used by subsequent developmental derivates of the kidney.

The next kidney section to form developmentally is the mesonephros. It is characterized by multiple nephrons per body segment. The final section to develop is the metanephros. Normally the metanepric section is characterized by multiple, secondary nephrons per body segment. The adult amphibian kidney is composed of both the meso- and metanephric kidney, hence is frequently referred to as an opisthonephric kidney (mesonephros + metanephros). There is some controversy as to whether the amphibian kidney is simply mesonephric or opisthonephric. Given the presence of both primary neprons (with nephrostomes) and secondary (without nephrostomes) being present in the adult kidneys, the distinction does not bear on functional interpretations.

3.7.3 Nephron Function

Understanding environmental correlates of nephron function entails first understanding the anatomical and physiological processes involved in urine formation. Crucial to these analyses are understanding the variables that can modify the fluxes of urine formation, urine movement and fluid reabsorption in neprons, as well as physiological control systems which can modify these variables.

Rate of Urine Formation in the Corpuscle

The flux of fluid across the glomerular capillary beds is a derivation of the Starling Fluid flux equation that was introduced in Chapter 2.

$$\mathbf{Flux} = F_c\,(P_g - P_{Bc} - \pi_g)$$

where F_c = filtration coefficient across the glomerular capillary, P_g = glomerular capillary pressure, P_{Bc} = pressure in Bowman's capsule and π_g = the colloid osmotic concentration of the plasma in the glomerular capillary. The reflection coefficient for protein and the colloid osmotic concentration of Bowman's capsule are missing from the conventional Starling equation since the basement membrane is assumed to be impermeable to protein, therefore the reflection coefficient is 1 and there is no protein escaping into Bowman's capsule.

It is important to recognize that blood flow to an individual nephron can be intermittent (Richards and Schmidt 1924; Forster 1942). Also there appears to be intraglomerular capacity to shunt blood between the afferent and efferent arterioles, bypassing the glomerular capillary beds (Bieter 1930). During perfusion, values for P_g are obviously dependent upon afferent and efferent arteriole pressures and have been measured between 0.7 and 2.7 kPa (Hayman 1927, 1928; White 1929c; Grandchamp and Boulpaep 1974). Measured values of P_{Bc} range from 0.02 to 0.43 kPa (White 1929b, c). Values of π_g for amphibians fall within a range of 0.7–1.4 kPa (Landis and Papenheimer 1963). These data taken together indicate that in perfused glomeruli there will be a net efflux of fluid from the glomerulus into Bowman's capsule.

Fluid Reabsorption from the Nephron into the Interstitial Space

The volume flux of urine from the proximal convoluted tubule to the renal interstitial space is the result of iso-osmotic solute transport from this segment. The two most prevalent solutes in the filtrate, sodium and chloride, diffuse from the lumen across the apical membranes of the cells lining the lumen. This entry raises the osmotic concentration of the intracellular compartment and water passively follows. The sodium that entered the cells is then actively transported by the Na–K ATPase of the basolateral membrane into the interstitial space, raising the osmotic concentration of the interstitial space. Water then osmotically follows this solute flux from the intracellular to interstitial space. In conjunction, organic molecules like glucose and amino acids are recovered across the apical membrane via active transport processes and water passively following these fluxes. An excellent review of the myriad of channels and transporters that have been found in the renal tubules can be found in Uchiyama and Konno (2006). Diffusion across the basolateral membrane allows recovery into the interstitial space along with water. The result of this net volume flux into the interstitial space is an increase in interstitial pressure. The magnitude of the pressure increase is dependent upon the absolute volume flux and renal interstitial compliance. The osmotic concentration of the fluid at the end of the proximal tubule will be the same as in the beginning because water flux follows solute flux, but the urine volume has decreased.

Fluid reabsorption from the distal convoluted tubule is different in the sense that organic molecules destined for reabsorption should have been completely reabsorbed in the proximal tubule (Walker and Hudson 1937), and organics destined for excretion have been added by secretion or not reabsorbed from the primary filtrate. There is again an active transport of sodium and chloride in the distal tubule though there is limited water permeability in this segment hence water does not follow solute movement (Walker and Hudson 1937; Bott 1962; Guggino *et al.* 1988). The net result is a decrease in the tubular osmotic concentration with sodium and chloride uptake and limited water movement, hence the term "diluting segment".

Fluid Reabsorption from the Renal Interstitial Space

Obviously, the long-term renal recovery of solutes and water from the tubules requires movement from the renal interstitial space into the peritubular capillaries for return to the body. Again this can be described as a modified Starling fluid flux:

$$\mathbf{Flux} = \mathbf{F_c}\,(\mathbf{P_{pt}} - \mathbf{P_{ist}} - \pi_{pt})$$

Where F_c is the filtration coefficient for the peritubular capillaries, P_{pt} is the peritubular capillary pressure, P_{ist} is the renal interstitial fluid pressure and π_{pt} is the peritubular oncotic pressure. The F_c for the peritubular capillaries will be much greater than the F_c of the glomerulus, as there is a much greater surface area for exchange in the peritubular capillaries. The pressure in the peritubular capillaries will also be much lower than the glomerular pressure since 70% of the blood flow is venous with lower pressure than arterial when the kidneys are perfused at normal arterial and venous pressures (Deeds *et al.* 1977). The π_{pt} should be similar to the glomerular condition, since it is venous drainage from the posterior animal mixed with glomerular drainage. The P_{ist} should be greater that the P_{Bc} since the renal interstitium is less compliant than Bowman's capsule. Consequently, the overall driving force should be a peritubular influx.

The P_{ist} will also influence the amount of fluid that is eventually reabsorbed from the proximal tubule. The translocation of fluid to the interstium from the proximal tubule increases the interstitial pressure and decreases the pressure in the tubule. This P_{ist} increase opens up the lateral spaces between the cells lining the tubule. This pressure difference can open the tight junctions and move fluid from the renal interstitium back into the proximal tubule. This process creates a futile cycle where the cells spend ATP to move solute and water, which then reenters the tubule through the compromised tight junction on an interstitial pressure basis. This is a principal mechanism for increasing the renal loss of solute and water during periods when the animal is volume loaded (Bentzel *et al.* 1968, 1969; Grandchamp and Boulpaep 1974; Bentzel and Reczek 1978).

Another consequence of fluid reabsorption is that the urine can never be more concentrated than the total osmotic concentration of the renal interstitial space. Since peritubular blood flow is high and interstitial fluid equilibrates diffusionally with peritubular concentrations, the result is that that the most concentrated urine formed is iso-osmotic with plasma. When this condition arises during dehydration, filtration essentially stops. There is no advantage in removing plasma from a hypovolemic amphibian.

Urine Movement through the Nephron to the Environment

The flux of urine from one nephron series compartment to the next and eventually to the environment is assumed to be principally the result of hydrostatic pressure differences in each step of the process. As fluid can be removed from the proximal and distal tubules as well as the urinary bladder, the intratubular pressure will vary dependent upon the compliance of each segment and the mean volume of fluid within the segment. Additionally cilia, smooth muscle contraction and external organ forces can all serve to modify these pressures. As we saw above, the pressure in Bowman's capsule is about 0.25 kPa. Is this P_{Bc} sufficient to move fluid through the remaining nephron segments without any additional pressure source? White (1929b), using the Poiseuille–Hagen flow formula, calculated the resistance of each nephron segment using length and diameter measurements. The pressure required to move the filtrate volume through each series resistance of the various segments of the nephron were 0.12 kPa across the neck segment, 0.03 kPa through the proximal convoluted tubule and 0.15 kPa through the distal convoluted tubule. Obviously the total pressure difference is close to the average P_{Bc}, which does not include the two pressure generating ciliated neck and intermediate segments nor the nephrostomes of primary nephrons. In a second set of measurements White (1929b) determined the ciliated neck segment was capable of generating a pressure of 0.4–0.57 kPa and the nephrostomes a pressure of 0.8–1.1 kPa. These additional pressures, in concert with P_{Bc}, account for the driving force necessary to move fluid through the nephron. Once the urine enters the ureters smooth muscle contraction, gravity and lung inflations compressing the ureters generate sufficient pressure to move urine to the cloaca. Cloacal smooth muscle contractions generate cloacal pressures that exceed bladder pressures (Martin and Hillman 2009). Urination from the bladder to the exterior entails hypaxial muscle contractions that generate pressures to forcefully expel the urine (Martin and Hillman 2009).

3.7.4 Organ Level Metrics of Renal Function

The three metrics normally used to quantify and describe kidney function are the glomerular filtration rate (GFR), renal plasma flow (RPF) and urine flow rate (UFR). The UFR is directly measured as the volume of urine collected per unit time from direct catheterization of the ureters or periodic draining of the urinary bladder. The GFR is determined by the clearance of a substance that is filtered at the glomerulus and neither secreted or reabsorbed in the tubules. Inulin and creatinine are the most common substances used in this regard (Forster 1938). The GFR = $[\text{inulin}]_{\text{urine}}$ UFR/$[\text{inulin}]_{\text{plasma}}$. The RPF is determined by measuring the clearance of a chemical that is filtered at the glomerulus and then completely secreted by the tubular cells from the peritubular capillaries in one pass of the circulation. The most common solute used for this purpose is para-aminohippuric acid (PAH). The RPF = $[\text{PAH}]_{\text{urine}}$ UFR/$[\text{PAH}]_{\text{plasma}}$. The ratio of GFR/RPF is known as the filtration fraction, the amount of plasma filtered per volume of plasma delivered to the kidney. The clearance of any solvent or solute can be calculated as [urine] UFR/[plasma].

The normal variables that influence metabolism also influence GFR. The presumption is that the kidneys participate in eliminating wastes and hence the rate of metabolism should reflect the rate of waste generation. The GFR is temperature sensitive with a Q_{10} of 2–3 (Adolph 1927). The GFR also increases with body mass and scales allometrically with body mass with an exponent of 0.73–0.89 (Yokota *et al.* 1985; Pruett *et al.* 1991) for all amphibians. These values falls within the range of the mass exponents measured for the relationship of metabolism with body mass.

3.7.5 Renal Responses to Environmental Conditions

Generally, the UFR matches osmotic water exchange in aquatic situations (see Chapter 4; Shpun and Katz 1995). Deviations from this relationship are most striking at low temperatures where UFR is less than osmotic water entry (Jorgensen 1991). This seems to be an indifference to a volume-expanded state at low temperature rather than a failure of the kidneys (Jorgensen 1950, 1991). The UFR generally decreases with aerial exposure and a lack of osmotic water entry. A fresh water environment, where water influx exceeds outflux, creates a hypo-osmotic, hypervolemic physiological condition. When water efflux exceeds influx in air a hyperosmotic, hypovolemic physiological condition is created. Hemorrhage creates an iso-osmotic, hypovolemic condition.

3.7.6 Control of Renal Function

The control systems that match renal output to state of hydration can be divided into negative feedback hormonal and neural regulation, and physical force changes associated with the volume changes. We will first start with the effect of physical changes induced by differences in hydration status.

Physical Forces

When water uptake exceeds water loss it will increase the extracellular fluid volume (hypervolemia) and dilute the extracellular solutes (hypo-osmotic). From a purely physical perspective this imbalance in water flux will increase the venous pressure and decrease the colloid osmotic concentration of the plasma. This affects the kidney at both the glomerulus and peritubular capillaries. The GFR will increase because increased venous return will increase cardiac output, and consequently the pressure in the afferent arteriole. The decrease in π_{plasma} will also increase GFR. The increase in venous pressure would increase peritubular capillary pressure. Couple this to the decline in π_{plasma} and less fluid will enter the peritubular capillaries raising the renal interstitial pressure. This increase in interstitial pressure will lead to fluid return into the proximal tubule through opening of the tight junctions in the paracellular pathway (Bentzel *et al.* 1968, 1969; Bentzel *et al.* 1970; Grandchamp and Boulpaep 1974; Bentzel and Reczek 1978). The net result is that from a purely physical perspective urinary excretion will increase as a result of the increase in GFR and decrease in tubular reabsorption. Obviously this will tend to compensate for the originally expanded volume perturbation.

There also appears to be a shunting of venous return from the posterior part of the animal from the hepatic portal venous pathway to the renal portal pathway with hypervolemia (Killorn and Toews 2001). Flow via the renal portal pathway increases 2.4-fold, representing almost 60% of the posterior blood flow in comparison to only 40% in a normovolemic situation (Killorn and Toews 2001). This will augment the decreased reabsorption of tubular fluid by increasing peritubular capillary pressure.

When water loss exceeds water gain the extracellular volume will decrease (hypovolemia) and the extracellular osmotic concentration will increase (hyperosmotic). If there is a hemorrhage of the circulation hypovolemia can also be induced but it will be iso-osmotic. Physically, hypovolemia will tend to decrease both venous and arterial pressures, which will increase tubular reabsorption and decrease GFR respectively. Consequently increases and decreases of the extracellular fluid volume will tend to increase or decrease UFR respectively, in part adjusting gain and loss of fluid from the circulatory system without any true negative feedback control system involved.

There are also a variety of hormones involved in osmo- and ionic regulation by the kidneys, in particular atriopeptins, arginine vasotocin, angiotensin, aldosterone, mesotocin, glomerulopressin and sympathetic innervation.

Atriopeptins

The release of atriopeptins, in response to a hypervolemic induced stretch of the cardiac atria, was discussed in Chapter 2. There is evidence that intrarenal atriopeptins may also act in an autocrine fashion in adaptation to seasonal changes (Fenoglio *et al.* 2004). Atriopeptins decrease renal resistance to blood flow (Meier *et al.* 1999; Donald *et al.* 2002; Toop and Donald 2004; Donald and Trajanovska 2006). This raises afferent arterial pressure enhancing GFR. A lower renal resistance will also increase peritubular capillary pressure, decreasing tubular reabsorption. The net effect is that more urine is formed, translocating vascular volume to the urinary bladder. Atriopeptins may also inhibit angiotensin II stimulation of aldosterone (Lihrmann *et al.* 1988; Kloas and Hanke 1993). Atriopeptins are thought to prevent overstretching the striated cardiac muscle and maintaining an appropriate overlap of the actin and myosin during rapid rates of rehydration (Donald and Trajanovska 2006).

Arginine Vasotocin

The release of AVT occurs in response to hypovolemic and hyperosmotic stimuli (Chapter 2; Nouwen and Kuhn 1985; Konno *et al.* 2005). AVT decreases renal blood flow, hence GFR, and peritubular capillary pressure, hence increases tubular reabsorption (Bakker and Bradshaw 1977; Pang *et al.* 1980; Toews *et al.* 1989; Boyd and Moore 1990). There are data from terrestrial species to suggest they distal tubule water permeability can be augmented, which would increase the tubular reabsorption of urine and osmotic concentration of the urine formed (Sawyer 1957; Henderson *et al.* 1972). This appears to be mediated by aquaporins (AQP-t2) channels in the connecting tubule and collecting ducts (Uchiyama and Konno 2006). It also increases tubular sodium uptake (Jard and Morel 1963). The end result is a decrease in UFR, which is also more concentrated.

Renin, Angiotensin II and Aldosterone

The release of renin by the juxtaglomerular apparatus at the junction of the renal corpuscle and distal tubule is controlled by the amount of sodium chloride reaching the distal tubule. If insufficient sodium chloride reaches the distal tubule, the interstitial concentrations following salt absorption by the tubule is decreased and renin is released by the juxtaglomerular apparatus (Persson 1981; Persson and Persson 1981; Persson *et al.* 1988). The renin is eventually converted to angiotensin II, whose renal effect is to constrict the efferent arteriole of the glomerulus (Sokabe *et al.* 1972). This raises the glomerular capillary pressure and decreases the peritubular capillary pressure and provides an autoregulatory loop to maintain GFR and increase tubular reabsorption (Henderson *et al.* 1972). The UFR may not change (Henderson *et al.* 1972) but more primary filtrate is formed and waste excretion is increased. An increase in angiotensin II also has the effect of stimulating the release of aldosterone (Konno *et al.* 2005). Aldosterone leads to an increase in the activity and synthesis of transporters for sodium and chloride in the distal tubule, conserving sodium and chloride (Guggino *et al.* 1988; Gnionsahe *et al.* 1989).

Mesotocin

Mesotocin is a neurohypophysial hormone that acts as a diuretic in a dose-dependent fashion (Pang and Sawyer 1978; Stiffler 1981). The diuretic effect seems to result from afferent arteriole dilation, increasing GFR (Pang and Sawyer 1978; Stiffler 1981). There are essentially no data on stimuli that lead to its release.

Glomerulopressin

Glomerulopressin may be another diuretic substance that is released by the liver. There is some controversy as to the existence of a renal–hepatic control loop mediated via a diuretic principle (Laug *et al.* 1992). The control seems to be that an infusion of amino acids into the hepatic portal vein, reflective of a postprandial state, leads to an increase in GFR in mammals. In amphibians, the release of glomerulopressin leads to an increase in UFR (Uranga 1969, 1971) and an increase in lymphatic return to the circulation mediated by prostaglandins (Bonetto *et al.* 1981).

Sympathetic Innervation

The last control system is innervation of the vasculature via the sympathetic nervous system. Hypovolemic states induce sympathetic activation via a baroreceptor-mediated reflex (see Chapter 2). Sympathetic activation in the case of the kidneys leads to an increase in renal vascular resistance decreasing blood flow, GFR and UFR, helping to stabilize blood pressure (Richards and Schmidt 1924; Adolph 1936; Gallardo *et al.* 1980) as well as stimulation of transporters in the proximal tubule (Morgunov 1987; Abdulnour-Nakhoul 1994).

The variety of hormonal and neural control systems involved in the control of renal function illustrates how glomerular intermittency can be regulated. It is always important to recognize that on the surface such elaborate renal control may seem unnecessary, as the urine will end up in the bladder where both water and solutes can then be reabsorbed. It may more appropriately reflect the importance of fast acting, redundant control systems of renal function given the high plasma turnovers in amphibians.

3.7.7 Environmental Correlates of Renal Function

There are potential anatomical and physiological variables that could account for differential kidney function in amphibians. As we saw above, the processes for urine formation and reabsorption are multivariate, although the literature frequently approaches an explanation for interspecific variation in kidney structure and function from a univariate perspective.

Anatomically, a major correlate invoked is the size and number of glomeruli with the environment. These variables seem to be excellent gross morphological measures when classes of vertebrates are compared in terms of the amount of water environmentally available (Marshall and Smith 1930). The hypothesis is that glomerular size and number relate to glomerular filtration rate (GFR). Amphibians are descendants of fresh water teleost ancestors. A fresh water environment leads to both passive water intake and solute loss, hence the amphibian kidney should be able to produce a high volume, dilute urine. Amphibians have large glomeruli and long tubules, in comparison to fish and reptiles (Marshall and Smith 1930; Marshall 1932). Amphibians tend to have greater GFRs than reptiles and GFRs equivalent to fresh water fish (Yokota *et al.* 1985). Amphibians have approximately the same number of glomeruli as reptiles and fresh water fish, though their single nephron filtration rate is greater than those two groups (Yokota *et al.* 1985). These data suggest a greater glomerular surface area in amphibians if the driving forces are equivalent between the classes.

The correlation of glomerular size and GFR fails when intraclass variation has been assessed. Glomerular size can vary with kidney region (Dawson 1951). Anurans tend to have glomeruli of smaller diameter than salamanders, implying less surface area for filtration in anurans. Despite this, the GFRs of terrestrial anurans tend to be greater than that of urodeles (Pruett *et al.* 1991), exactly the opposite of the expected relationship. Terrestrial anurans tend to have greater UFRs and GFRs than more aquatic species (Vondesaar and Stiffler 1989; Pruett *et al.* 1991) and a greater capacity to increase UFR (Hillman and Schimpf 1994), yet their glomerular diameters are similar. A good summary of glomerular number among different species can be found in Geakleman and Warburg (2000). The conclusion at this stage must be that glomerular size does not correlate with either GFR or UFR, though aquatic species tend to have lower GFRs, UFRs and renal capacity than terrestrial species. There are insufficient data to draw any environmental correlates with glomerular number in amphibians.

Anatomical differences may exist at a microscopic level for glomeruli. A correlation has been drawn between the slit pore area and complexity of glomeruli with habitat in anurans. More terrestrial species have greater slit pore area and podocyte complexity than aquatic anuran species (Richter and Splechtna 1996; Chan *et al.* 2002). The presumed functional advantage is to increase surface area for glomerular filtration even with the same glomerular capillary diameter and glomerular diameter. This explanation is consistent with the analyses that terrestrial species generally have greater glomerular filtration rates than aquatic species (Vondesaar and Stiffler 1989; Pruett *et al.*1991). This greater glomerular filtration rate has been interpreted as a mechanism in terrestrial species to renally compensate for faster rates of water uptake when water is available. The function of the increased complexity of podocyte architecture may also relate to increased mechanical support of the capillary wall (Takahashi-Iwanaga 2002).

On the tubular side of the nephron, the transport capacity of the proximal tubule appears more developed in a terrestrial species (*Ambystoma*) compared to an aquatic species (*Necturus*) (Maunsbach and Boulpaep 1984). This is consistent with greater GFRs and UFRs in the terrestrial species Acclimation to more dilute environments led to an increase in the transport capacity (Na–K ATPase) of the distal tubule in *Ambystoma*. The increase is a consequence of an increase in number of basolateral transporters and has been interpreted as reflective of an ability to produce a more dilute urine in an aquatic environment, where ion loss is a potential environmental stress. (Morgunov *et al.*1989). The channel density of the distal tubule is dynamic also responding to pH changes and mechanical distension (Hurst and Hunter 1990; Sattlin *et al.* 2006).

There is still much to be learned from a systematic survey of anatomical and physiological renal function from amphibian species that span a wide array of environmental conditions.

3.8 Developmental Plasticity

As noted in Chapter 1, the classic pattern for the amphibious lifestyle is transition from an aquatic larva to an adult that can live, to varying degrees, a terrestrial existence. Denver (1997) noted that there is "considerable variation both between and within species in the duration of the larval period" and that "the timing of metamorphosis is the central characteristic in the life history of an amphibian". A variety of biotic and abiotic factors are known to affect the timing of metamorphosis and the response to environmental factors that results in different phenotypes (e.g. the ability to accelerate metamorphosis) has been termed "phenotypic plasticity" (Newman 1992). For example, species such as *Scaphiopus couchii* that breed in ephemeral ponds accelerate metamorphosis as the pond dries up (Newman 1992) while laboratory experiments have shown *Bufo americanus* tadpoles metamorphose more rapidly in the presence predatory dragonfly larvae (Skelly and Werner 1990). The proximate mechanism for phenotypic plasticity in amphibian developmental rates has been proposed to be a neurohormonal stress response that is "a phylogenetically ancient developmental trigger that allows the developing organism to assess habitat quality and respond accordingly" (Denver 1997). These factors are summarized in Fig. 3.42 adapted from Denver (1997). Sensory input activates or inhibits pathways that either sustain growth in the larval phenotype (prolactin and growth hormone) or accelerate metamorphosis to the adult phenotype (thyroid hormones and corticosteroids). The physiological roles of these hormones have been studied most extensively in larval bullfrogs, *Rana catesbeiana*, because of their large size and availability.

One of the major changes that occurs during metamorphosis is the loss of gills and the development of the skin as the primary surface for ion and water exchange

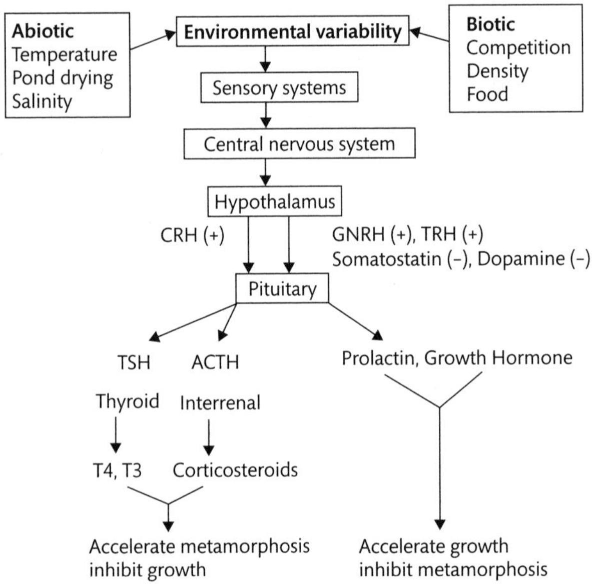

Fig. 3.42 Larval amphibians are able to detect changes in their environment related to survival. Prolactin and growth hormone accelerate larval growth to give a larger adult that is more fit while Thyroid hormones and corticosteroids accelerate metamorphosis to allow larvae to escape from unfavorable aquatic habitats. After Denver (1997).

in the environment (Alvarado and Moody 1970; Cox and Alvarado 1979). The following is a discussion of the structure and function of the larval skin, the changes that occur during metamorphosis and interaction of hormones, identified above as mediating plasticity, in the development of the adult phenotype for Na^+ transport.

3.8.1 Structure and Transport Properties of the Larval Skin

Skin of larval frogs consists of two cell layers (Weed 1934; Figure 3.43): an outer or apical cell layer that forms the high resistance barrier and a basal cell layer that is not electrically coupled to the apical cells. The skin is interspersed with skein cells that contain bundles of intermediate filaments. During metamorphosis, the adult skin layers are derived from cell division of the basal cells (Robinson and Heintzelman 1987).

The larval frog skin contributes relatively little to total water and ion exchange *in vivo*, most being across the gills (Alvarado and Moody 1970). With regard to the low water permeability, Willumsen *et al.* (2003) were unable to obtain RNA that indicated expression of aquaporins in the ventral skin of larval *Ranas catesbeiana*.

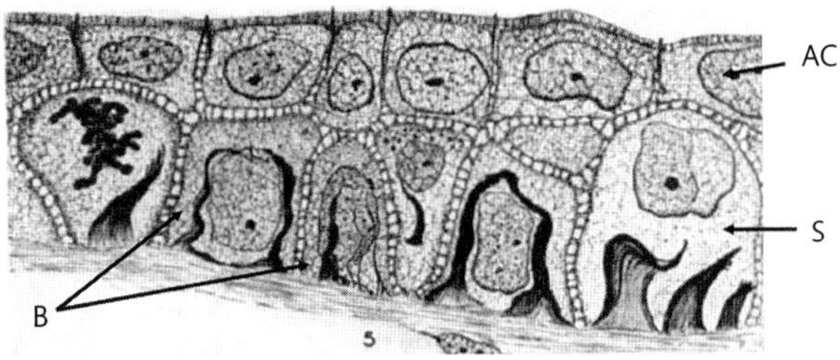

Fig. 3.43 Drawing of the skin of a fully developed *Rana clamitans* tadpole, prior to metamorphosis, showing the general ranid pattern with an outer apical cell layer (AC) forming the barrier for salt and water change and underlying basal cells (B) and skein cells (S) that are characterized by thick bundles of intermediate filaments (from Weed 1934).

Early studies on ion transport properties of isolated larval skin in Ussing chamber preparations showed a low electrical resistance and no potential difference prior to the climax stages of metamorphosis (Taylor and Barker 1965). Thus, little functional significance was given to the skin. This view changed when Cox and Alvarado (1979) found that larval skin carefully mounted in a chamber to prevent damage to the integrity of the apical cell layer has an extremely high resistance and a small, but measurable, short circuit current. The antibiotic nystatin permeabilized the apical membrane of the apical cells and greatly stimulated the short circuit current, indicating that the larval epithelium had adult-like transport capability in the basolateral membrane (Cox and Alvarado 1983). Hillyard *et al.* (1982a) subsequently showed the transepithelial current was limited by a channel in the apical membrane of the apical cells that is non-selective for the alkali metals and is stimulated by amiloride that inhibits Na^+-selective channels in the apical membrane of adult frog skin. Amiloride stimulation is transient, as is apical cation transport stimulated by ATP and acetylcholine (Cox 1993, 1997) to levels that approach or exceed that of the adult skin. It remains to be seen whether these responses result from separate channels or from a common channel that responds to the three ligands. In favor of a common channel, all three responses show little selectivity for the alkalai metal cations and become refractory to higher doses after stimulation with smaller doses (Hillyard and Van Driessche 1989). In favor of separate channels, Jensik *et al.* (2001) cloned and sequenced RNA from larval bullfrog skin that corresponds with a purinergic (P2X) receptor that had a similar pattern of transient ATP activation when expressed in *Xenopus* oocytes. Antibodies against the acetylcholine (nicotinic) receptor labeled the apical membrane of the larval skin and transient stimulation by acetylcholine was inhibited by acetylcholine receptor antagonists (Takada *et al.* 1996). The function of the larval cation channel or channels remains to be determined. Cox (1997) has suggested the channels might serve chemosensory, nociceptive or mechanosensory functions.

Takada *et al.* (2006) have recently shown that mRNA for all three subunits of the Na^+ selective epithelial sodium channel (ENaC) is expressed in the larval skin, prior to the onset of amiloride blockable Na^+ current, and the subunits have the same amino acid sequence as the adult skin. However, they remain inactive. The onset of amiloride-blockable Na^+ transport versus retention of the larval pattern thus provides a model to study physiological regulation by hormones accelerating metamorphosis (e.g. thyroid and steroid hormones) or promoting larval growth (e.g. prolactin and growth hormone).

3.8.2 Changes at Metamorphosis

The apical cell layer is retained during larval stages but during metamorphosis it is shed and replaced by principal cells that are derived from the basal cell layer (Robinson and Heinzelman 1987). Hillyard *et al.* (1982b) showed that the increase of active Na^+ transport during the climax stages of metamorphosis corresponded with an increase in functional ENaC expression as the adult pattern of epithelial cells developed. There is also a shift in the distribution of Na^+–K^+ ATPase during metamorphosis that can be demonstrated by the binding of radiolabelled ouabain, an inhibitor of the enzyme (Robinson and Mills 1987a, b). Prior to metamorphosis the enzyme is located primarily in the outer third of the skin, corresponding to the basolateral membrane of the apical cells. During metamorphosis, the enzyme becomes located in the inner two-thirds of the skin, corresponding to the basolateral membranes of the spinous and germanitive cell layers that form a functional syncytium with the granular cells.

3.8.3 Hormonal Control of the Development of the Adult Skin

Studies on hormonal control of ion transport commonly use the Taylor–Kollros (1946) staging system to delineate when amiloride-sensitive transport begins. In this system, the stage I begins with the first appearance of the hind limb bud and proceeds to climax stages XX–XXI when the front legs emerge and finally to stage XXV when the tail is completely reabsorbed. A number of studies have demonstrated that levels of thyroid hormone (T_4), triiodothyronine (T_3), mineralocorticoid (aldosterone) and glucocorticoid (cortisone) increase markedly during the climax stages (Regard *et al.* 1978; Kikuyama *et al.* 1986; Jaffe 1981). Taylor and Barker (1965) observed that the adult pattern of Na^+ transport, indicated by an increase in transepithelial potential, occurred at stage XXI in spontaneously metamorphosing *R. catesbeiana* but appeared earlier when metamorphosis was accelerated by adding thyroxine to the aquaria where the larvae were kept. It was concluded that thyroxine is the proximate stimulus for the development of the adult pattern of transport. The expression of functional ENaCs was stimulated by injecting larvae with T_3 over a 3-week period or if mineralocorticoid and glucocorticoid antagonists were given to the larvae (Takada 1990). On the other hand, after the appearance of functional ENaCs, aldosterone was able to stimulate sodium transport (Takada 1990; Hillyard and Van Driessche 1991).

The role of the mineralocorticoid, aldosterone and glucocorticoids in the development of the adult pattern of Na^+ transport was further studied with an isolated skin preparation that had been treated with the calcium ion chelator, EDTA, which causes the apical cells to dissociate from the skin (Takada *et al.* 1995a, b). The skin is then maintained in culture medium and treated with different hormone combinations for several days and the pattern of transport that arises from cell proliferation of the basal cell layer is observed as to whether it is of the larval (amiloride, acetylcholine or ATP stimulated) or adult (amiloride inhibited) phenotype. Of interest, a mixture of aldosterone, hydrocortisone and corticosterone stimulated the development of the adult pattern of Na^+ transport while thyroid hormone was without effect (Takada *et al.* 1995a, b). Furthermore, if the tissue was cultured with the corticoid mixture in the presence of prolactin, the larval pattern of transport developed (Takada *et al.* 1996). This is consistent with the proposed role of prolactin in prolonging the larval condition as a "juvenile hormone" in ranid frogs (Etkin and Gona 1967; Brown and Frye 1969), although it should be noted that transgenic *Xenopus laevis* expressing additional prolactin do not show delayed metamorphosis and some question is raised as to whether prolactin is a juvenile hormone an all anurans (Huang and Brown 2000). It is also important to note that prolactin stimulates amiloride-sensitive Na^+ transport across the skin of adult frogs by increasing the number of active ENaCs in the apical membrane as opposed to its effect on the larval skin (Takada and Kasai 2003).

Takada and Kasai (2003) and Takada *et al.* (1999) suggested that the role of thyroid hormones in stimulating the development of amiloride-sensitive Na^+ transport *in vivo* is to inactivate an inhibitor of the initial expression of ENaCs by aldosterone and have proposed that growth hormone might be that factor since the expression of growth hormone mRNA is greatest in the larval state and is inhibited by thyroxine during the climax stages of metamorphosis (Buckbinder and Brown 1993). In summary, the proposed roles for thyroid hormones and corticostroids in promoting adult characteristics and prolactin and growth hormone in maintaining larval characteristics can be applied to the changes in the transport properties of the skin during metamorphosis.

4

Physiological Adaptations to Extreme Environments

4.1 Salt and Osmotic Sensitivity

4.1.1 Low Salt Environments

Many aquatic species occur in streams or ponds with extremely low Na^+ and Cl^- concentrations. Krogh (1937) showed that *Rana esculenta* are able to independently absorb Na^+ and Cl^- from solutions as dilute as 10^{-5} M. This study is often cited to support the assumption that cutaneous salt absorption is essential for maintaining the plasma Na^+ and Cl^- concentrations. However, in this same study Krogh observed that frogs sprayed with distilled water survived for up to 12 weeks. More recently, McAfee (1972) kept *Rana pipiens* without food in flowing deionized water for 30–45 days without appreciable loss of Na^+. Like Krogh (1937), he suggested the role for cutaneous absorption is minimal for aquatic species with sufficient food available. In a similar study, Bentley (1973) maintained three anuran species for 15 days in dilute tap water (0.25 mmol/l NaCl) with 10 μM amiloride, an inhibitor of epithelial Na^+ transport. Plasma Na^+ concentration of the aquatic and semiaquatic species (*Xenopus laevis* and *Rana pipiens*) was not significantly reduced but declined by approximately 10% in the more terrestrial toad, *Bufo marinus*. As noted in Chapter 3, NaCl absorption from very dilute media requires the energetic contribution of both the Na^+–K^+ pump in the basolateral membrane of principal and MR cells and active H^+ transport by the V-ATPase in the apical membrane of MR cells. The density of MR cells increases when frogs (*R. temporaria*) or toads (*Bufo bufo* and *B. marinus*) are exposed to dilute NaCl solutions (Voute and Meier 1978; Budtz *et al.* 1995; Devuyst *et al.*1991). This is may explain the capacity of *R. temporaria* to maintain plasma Na^+ and Cl^- concentrations but appears to be insufficient for bufonids species that appear to lose these ions in dilute media. Isotopic measurements with intact animals and isolated skin have shown terrestrial species to have a lower affinity for NaCl absorption than aquatic species in conjunction with a greater level of water permeability (Greenwald 1972; Mullen and Alvarado 1976). Alternatively, the reduction of plasma Na^+ concentration in terrestrial species could result from excessive water gain since they display greater levels of cutaneous water permeability under control conditions (Table 3.3).

4.1.2 Hypersaline Environments

Amphibians are usually found in freshwater habitats where salt retention is of primary concern. However, many species have been reported in brackish water that for a few may approach the salinity of seawater (Neil 1958). Field observations and laboratory. experiments that determine salinity tolerance use equivalent NaCl concentrations, often in parts per thousand (g/L = ppt). Table 4.1 gives the millimolar concentrations and equivalent osmotic concentrations over the range of salinities where amphibians are known to occur, using an osmotic coefficient of 0.93 for NaCl. The term brackish water covers a range of NaCl concentrations defined by the European Environment Agency as between 5 and 18 ppt.

Balinsky (1981) identified 61 species in 13 families of anurans and 12 species in four families of urodeles that inhabit or tolerate brackish water. Three anuran species, *Rana cancrivora, Bufo viridis* and *Xenopus laevis,* have been shown to tolerate salinities in excess of what is commonly considered brackish. The most saline-tolerant frog, *R. cancrivora,* inhabits mangrove swamps in southeast Asia and can be acclimated to salinities of 29 ppt NaCl (Gordon *et al.* 1961). In the laboratory, frogs showed a progressive rise in plasma Na^+, Cl^- and urea as the external salinity was increased from freshwater to 250, 500 and 800 mOsm/l (Fig. 4.1). The plasma osmolality was regulated to be greater than the external solution while plasma K^+ concentration remained relatively stable.

A similar pattern of elevated plasma Na^+, Cl^- and urea concentrations was observed with *Bufo viridis* and *Xenopus laevis. B. viridis* occur in wide range of habitats from the middle eastern deserts to Scandinavia were it has been found in

Table 4.1 Summary of NaCl concentration values in units of salinity, molarity and osmotic concentration

Salinity (ppt NaCl)	NaCl concentration (mmol/l)	Osmotic concentration mOsm/l	
35	598	1113	Sea water
30	513	954	
25	427	795	
20	342	636	
15	256	477	↑
10	171	318	| Brackish water
5	85	159	↓
1	17	32	
0.5	9	16	
0.1	2	3	Fresh water

salinities of 20 ppt (Gislen and Khoury 1959). Katz (1973) was able to acclimate *B. viridis* to NaCl concentrations as high as 800 mOsm/l (c. 25 ppt) by gradual exposure to increasing salinity. In arid regions, *B. viridis* also accumulates urea to offset water loss while burrowed in dry soil (see below). *X. laevis* is an entirely aquatic frog that may encounter drying ponds and, like *B. viridis*, is able to accumulate urea and endure elevated Na^+ and Cl^- concentrations to offset water loss to substrates with low water potential. *X. laevis* is also saline tolerant and can be acclimated to NaCl concentrations as high as 600 mOsm/l.

The capacity of *R. cancrivora*, *X. laevis* and *B. viridis* to tolerate elevated salinity is mediated by a greater tolerance of elevated Na^+ and Cl^- concentrations in the extracellular fluid and the ability to accumulate urea. Katz (1973) observed that the ability of *B. viridis* to accumulate urea was highest during the summer that is consistent with observations, noted in Chapter 3, that corticosteroid levels are greatest during this time (Jolivet-Jaudet *et al.* 1983). The accumulation of urea is a function of synthesis and retention. The accumulation of urea, and also free amino acids, during exposure to elevated salinity has been linked to upregulation of ornithine urea cycle (OUC) enzymes as an osmoregulatory response (reviewed by Balinsky 1981). In contrast, Jorgensen (1997) suggests the upregulation of OUC enzymes is a response to stress and that non-stimulated rates of urea synthesis are adequate to explain the accumulation seen in saline-acclimated animals. Rather, it is the balance between urea synthesis and excretion that allows accumulation. Wright *et al.* (2004) acclimated juvenile *R. cancrivora* from fresh water to 20 and 25 ppt NaCl and found the OUC enzyme carbamoyl phosphate synthase to be upregulated and also the excretion of urea to be unchanged despite the greater

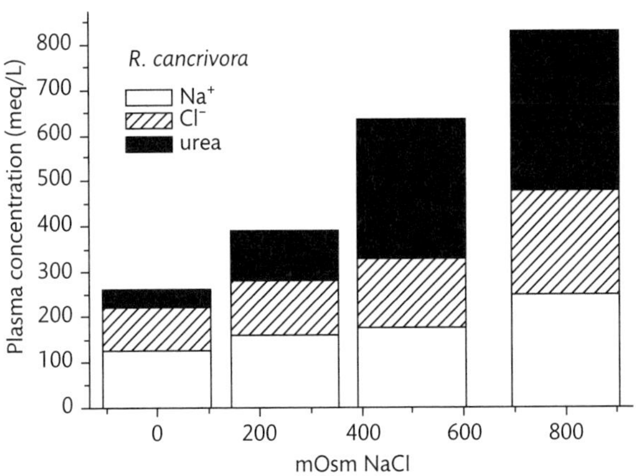

Fig. 4.1 Increasing plasma Na^+, Cl^- and urea concentrations of *R. cancrivora* acclimated to NaCl solutions with salinities of 0, 250, 500 and 800 mOsm/l. A similar pattern is seen with *B. viridis* and *X. laevis* (plotted from tabular data in Balinsky 1981).

gradient for loss to the outside, suggesting an increase in cutaneous absorption and hepatic synthesis of urea. Studies with isolated skin have demonstrated active urea influx in the aquatic frog *Rana esculenta* and the toads *B. bufo* and *B. viridis* (Garcia Romeu *et al.* 1981; Katz *et al.* 1981). Of interest, active urea absorption across the skin was only observed in saline-acclimated *R. esculenta* and increased as the NaCl concentration was increased from 240 to 400 mOsm/l. Active urea uptake by toad skin was observed in freshwater and saline-adapted animals.

In addition to cutaneous uptake, urea retention is mediated by a reduction in urea excretion by the kidneys and absorption across the urinary bladder. Urea excretion can be evaluated as the ratio of urea concentration in the urine relative to the plasma (U/P ratio) where a higher ratio represents greater excretion (Schmid 1968) while a value at or below unity indicates retention of urea in the body if water is absorbed across the bladder (Jorgensen 1997). U/P ratios of 0.5, 0.3–1.0 and 0.94 have been reported, respectively, for *R. cancrivora*, *B. viridis* and *X. laevis* (Schmidt-Nielsen and Lee 1962; Katz *et al.* 1984; McBean and Goldstein 1970). The equilibration of urine with plasma appears to be mediated by a facultative urea transport protein (UT) that has been localized in the apical membranes of cells in the early distal tubule and urinary bladder of *B. marinus* (Konno *et al.* 2006). These results are consistent with earlier observations that cystectomized *R. cancrivora* are unable to accumulate urea (Chew *et al.* 1972).

Although considered less tolerant, a number of anuran species have been anecdotally observed to occur in habitats where brackish water might be encountered (Niel 1958). *Bufo marinus*, as its name implies, is found in estuarine areas (Zug and Zug 1979) and can tolerate salinities of 15 ppt (Liggins and Grigg 1985). *Bufo calamita* inhabit estuarine habitats in costal regions of the Baltic sea and Great Britain and are able to tolerate salinities of 16–17 ppt for up to 4 days (Mathias 1971). Sinsch *et al.* (1989) found that inland populations of *B. calamita* were less tolerant of elevated salinity than coastal populations. However, Sinsch *et al.* (1992) attributed this to seasonal variation and noted a maximal level of NaCl tolerance to be 454 mOsm/l (c. 14 ppt) in active, feeding toads. *Bufo bufo* is considered more of a mesic species but is able to tolerate salinities of 150 mmol/l NaCl (Ferriera and Jesus 1973).

The capacity of urodeles to tolerate elevated salinity by accumulating Na^+, Cl^- and urea has been investigated in *Ambystoma tigrinum* and several species of *Batrachoseps* (Plethodontidae). *A. tigrinum* is found in arid regions and is able to endure drying of ponds by burrowing. As with other burrowing species, urea accumulates and contributes to the lowering of water potential so the animals retain water in dry soil (Delson and Whitford 1973). Romspert and McClanahan (1981) were able to acclimate *A. tigrinum* to NaCl solutions with osmolality has high as 450 mOsm/l during which the plasma urea concentration increased from 15 to 182 mmol/l along with increases in Na^+ and Cl^- to maintain the osmolality above that of the external solution. Licht *et al.* (1975) collected *Batrachoseps relictus* from a coastal area of California where exposure to seawater was a likely event. Its capacity to acclimate to salinity was compared with that of *B. attenuatus* from inland areas. Tolerance was judged by survival of animals placed in shallow solutions of FW or diluted SW. Both species tolerated direct exposure 30% SW (c. 300 mOsm/l); however, *B. relictus* survived longer when placed directly into 40 or 50% SW (c. 400 and 500 mOsm/l) and recovered better from brief exposure to full SW as might occur during a tidal event. After 24 hour's exposure to 50% SW the plasma urea concentration of *B. relictus* rose modestly from 38 to 57 mmol/l while that of *B. attenuatus* rose from 19 to 35 mmol/l. Under the experimental conditions the animals were able to drink, which appeared to be a cause of excessive salt accumulation and death. This was addressed by Jones and Hillman (1978) who gradually acclimated *B. attenuatus* and *B. major* to elevated salinity. The animals were placed on paper towels saturated with NaCl solutions that included FW and 250 mmol/l NaCl (c. 465 mOsm/l). After 25 days all animals had survived and showed an elevation of plasma urea from 48 mmol/l in FW compared with 218 mmol/l in NaCl. Clearly, acclimation to elevated salinity requires gradual exposure. Similar examples can be found in the anuran literature. *Bufo viridis* that were acutely transferred to 500 mOsm/l NaCl solutions survived for only 2–4 days (Gordon 1962; Katz 1973).

From the above examples, it is obvious that behavioral avoidance will benefit species that are found in habitats where available hydration sources include those with elevated salinity. As shown in Chapter 3, dehydrated toads will exhibit water absorption response behavior (WR) in which the seat patch is pressed to a wet surface. If the surface is moistened with NaCl solutions the frequency of initiating WR behavior is sustained until the osmotic concentration for water absorption decreases to near zero (Fig. 4.2). Brekke (1991) showed a similar avoidance of hyperosmotic urea solutions and to lightly moistened filter paper with a low water potential. Nagai *et al.* (1999) tested the hypothesis that the skin serves a chemosensory function that allows toads to detect and avoid unfavorable hydration sources in a manner analogous to taste cells in the tongue of vertebrates that drink orally. It was found that spinal nerves make connections with epithelial cells in the skin and that exposure of the skin to NaCl concentrations at or above 200 mmol/l produced a burst of activity in afferent nerve fibers that corresponds with the behavioral avoidance. Toads consistently avoided WR behavior on 200–250 mmol/l NaCl and the neural response increased dramatically between 200 and 300 mmol/l NaCl. The sensory ability of bufonids that have been studied (Brekke *et al.* 1991; Nagai *et al.* 1999; Maleek *et al.* 1999) correlates well with the ranges for acute physiological tolerance of salinity that are summarized by Balinsky (1981). The neural response was minimal at lower NaCl concentrations (less than 200 mmol/l) that most species can tolerate.

There are examples of amphibians capable of detecting dilute salt solutions. *Rana temporaria* embryos and larvae are more sensitive than adults to elevated

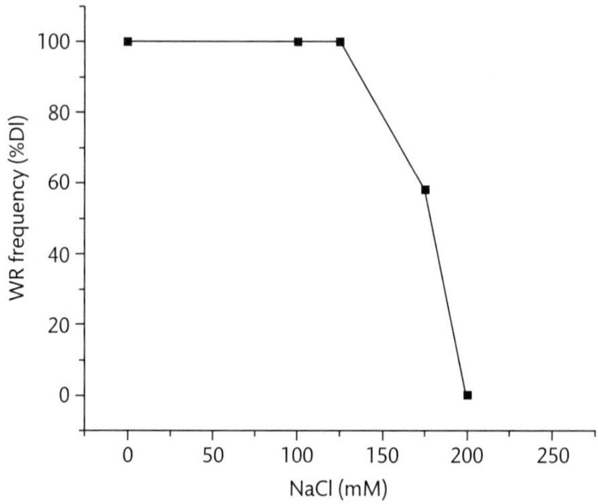

Fig. 4.2 Dehydrated *B. punctatus* will consistently initiate the water absorption response (WR) behavior on surfaces moistened with deionized water and with NaCl solutions up to approximately 125 mmol/l, which is approximately iso-osmotic with the plasma. WR behavior declines rapidly at higher concentrations.

salinity. Survival is reduced by 20% in salinities of 2.3 ppt (56 mOsm/l) and further declines to about 80% with 130 mOsm/l (4.5 ppt) NaCl solutions (Viertel 1999). Because the larvae have little or no ability to change habitat, the adults must select an appropriately dilute body of water for oviposition. The greatest salinity where oviposition was observed in natural habitats was 0.9 ppt indicating the frogs select sites well within the larval salinity tolerance. Reboreda and Segura (1984) were able to record neural responses from isolated legs of *Bufo arenarum* that were transferred between deionized water and hyperosmotic NaCl solutions. A response was recorded during transfer from either solution to the other, indicating the existence of water and salt-sensitive fibers. Recent experiments with intact toads have obtained a neural response to deionized water that is much slower in developing than the response to salt, which might explain why it was not observed in earlier studies (Hillyard *et al.* 2007). Salt-specific and water-specific fibers have been characterized in the glossopharyngeal nerve from frog tongue and are connected to salt and water-sensitive taste cells in the tongue (Sato and Beidler 1975; Okada *et al.* 1993). The skin of adult amphibians doesn't have identifiable taste cells (Fox 1984) so characterizing the sensory pathway remains an area for future study.

The chemosensory function of the skin has also been shown to integrate physiological responses to salinity. Petriella *et al.* (1989) showed that *B. arenarum* immersed in 115 or 400 mmol/l NaCl solutions for 5 min reduced urine production before any measurable change in plasma concentration was observable.

This response was inhibited by the guanethidine, indicating the response is mediated as a sympathetic reflex. Viborg and Rosenkilde (2004) found that the increase in cutaneous blood flow to the seat patch observed in dehydrated toads (Chapter 3) required skin contact with a dilute solution; dehydration alone was insufficient. The increase in blood flow was mimicked by the β-adrenergic agonist isoproterenol and inhibited by the antagonist propranolol, indicating a sympathetic reflex in response to water potential receptors in the skin. Thus, behavioral and physiological mechanisms rely on the ability of the skin to detect dilute as well as concentrated solutions.

4.2 Arid Environments

Most amphibians lose water freely from their skin by evaporation (Section 3.2), they are unable to produce urine that is hyperosmotic to their blood (Section 3.7), and they have limited tolerance to high ambient temperatures (Section 3.4). Consequently, we might not expect amphibians to cope well with the characteristically low water availability and high temperatures of arid environments. Nevertheless, many deserts of the world have an abundant, if not diverse, amphibian fauna, consisting predominantly of anurans.

Arid and semiarid environments are characterized by low water availability, often in conjunction with high ambient temperatures ("hot" deserts) but sometimes low ambient temperatures ("cold" deserts). For example, under Köppen's climate classification scheme, an arid climate (category B) occurs where $P_{ann} < 10\ P_{th}$ (P_{ann} is annual rainfall, mm, and P_{th} is a function of the mean annual temperature, T_{ann}, °C; Kottek *et al.* 2006). Steppe climate (BS) is $5\ P_{th} < P_{ann} < 10\ P_{th}$ and desert climate (BW) is $5\ P_{th} < P_{ann} < 10\ P_{th}$. Arid climates are also categorized as hot (h, $T_{ann} \geqslant 18$) and cold (k, $T_{ann} < 18$). Desert and steppe climates generally occur about 30° above and 30° below the equator, and include the major deserts of the world (Fig. 4.3).

The deserts of Australia have species of burrowing *Neobatrachus*, *Cyclorana*, *Notaden*, *Uperoleia*, *Pseudophryne*, *Limnodynastes* and arboreal frogs *Litoria* (Cogger 2000; Seymour and Lee 1974; Tyler 1999). North American deserts have burrowing anurans (*Scaphiopus*, *Spea*, *Bufo*) and treefrogs (*Hyla*, *Pternohyla*); the salamander *Ambystoma mexicanum* is also found throughout the southern deserts, at least where there is permanent water. South American arid and semiarid regions, particularly the Cerrado-Caahinga-Chaco have a diverse amphibian fauna, including 98 anurans and a single aquatic caecilian (*Chthonerpeton perisodus*); many anurans burrow and aestivate in termitaria (*Dermatonotus*, *Elachistocleis*) or rodent burrows (*Bufo*, *Leptodactylus*), or form a cocoon (*Lepidobatrachus*, *Ceratophrys*; Duellman 1999). The cold Patagonian deserts have *Bufo*, *Pleurodema* and *Zachaenus*, and the very arid Atacama has just *Bufo* and a poison-arrow frog

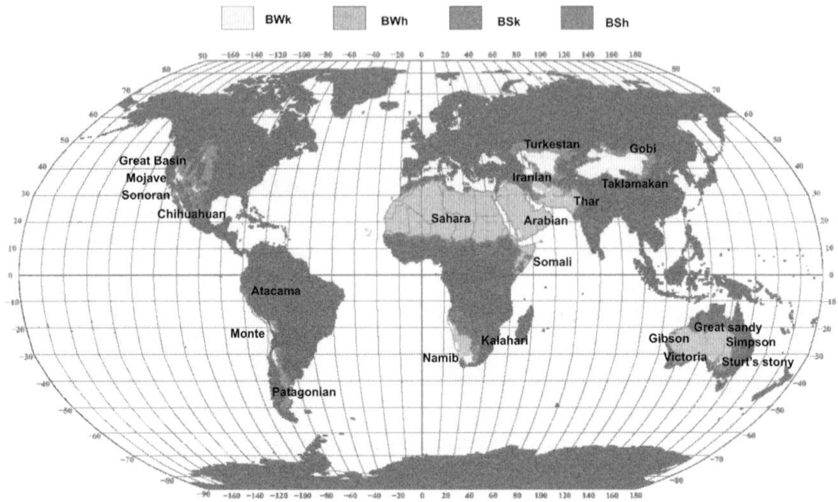

Fig. 4.3 World map of the Köppen–Geiger climate classification, showing cold semiarid (BSk), hot semiarid (BSh), cold arid (BWk) and hot arid (BWh) zones; major deserts of the world are indicated. Climate zones modified from Kottek *et al.* 2006.

Colostethus (Duellman 1999). Deserts of southern Africa have various burrowing species (*Pyxicephalus*, *Tomopterna*) and scansorial species (*Leptopelis* and *Chiromantis*), and sub-Saharan Africa has the aquatic *Xenopus* as well as *Bufo*, *Hoplobatrachus* and *Ptychadena* (Poynton 1999). North African deserts similarly have *Xenopus*, *Bufo*, *Hoplobatrachus* and *Ptychadena* as well as *Tomopterna* and *Rana* (Borkin 1999). The Sinai and Arabian deserts have a few anuran species, including *Bufo*, *Ptychadena*, *Rana* and *Hyla*. The "desert" salamanders *Salamandra salamandra* and *Triturus vittatus* are found in very xeric environments (Warburg 1988). The Eurasian and central Asian deserts similarly have a diverse anuran fauna, of *Bufo*, *Batrachuperus*, *Euphlyctis*, *Tomopterna*, *Hoplobatrachus*, *Paa*, *Rana*, *Hyla* and *Pelobates* (Borkin 1999).

The adaptations of these various amphibians from the semiarid and arid zones include behavioral avoidance of the harsh environmental conditions as well as specific physiological adaptations that increase desiccation tolerance, increase thermal tolerance, and decrease cutaneous EWL (e.g. Warburg 1988; Loveridge 1977; Seymour and Lee 1974; Shoemaker *et al.* 1992; Rome *et al.* 1992; Hutchison and Dupre 1992; Pinder *et al.* 1992; McClanahan *et al.* 1994). Bladder urine is an important store of water for amphibians to buffer changes in body fluid solute and osmotic concentrations during dehydration (see Section 3.1). The low concentration of urine in the bladder (very dilute compared to body fluids) and the large volume of urine stored in the bladder (up to 50% of body mass) provides a substantial water store to buffer changes in body fluid solute and osmotic concentrations during aestivation. Over time, Na^+, Cl^-, urea and osmotic concentrations increase in both plasma and urine, but the urine urea and osmotic concentrations increase much more than Na^+ and especially Cl^- in absolute mM difference. The urine-to-plasma (U/P) ratio is 0.5–0.9 for total osmotic concentration and 0.6–1.9 for urea. The amount of urine remaining in the bladder is low (e.g. McClanahan 1972), indicating that the bladder stores are becoming depleted as well as isosmotic. Most amphibians have a high body water content, about 77–83% of total body mass (bladder empty), and have a high tolerance to dehydration, particularly the more terrestrial species. Desert amphibians store a considerable amount of relatively dilute urine in their bladder (see Section 3.1); up to 50% or even more of their body mass can be bladder urine (Bentley 1966; Mayhew 1968; Main and Bentley 1964).

4.2.1 Diurnal Retreat

The vast majority of these desert amphibian species are not able to withstand the high temperature and desiccating environment of a desert during the day, or during the dry season. Behavioral selection of an appropriate sheltering site to avoid dehydration and hyperthermia is consequently important; most frogs would be unable to survive even 12 h exposure to typical desert daytime conditions. The typical response of a number of amphibians to dehydrating conditions is increased activity, to seek out a more favorable location, although dehydration eventually compromises locomotor capacity and the amphibian becomes physiologically incapacitated at its debilitating critical activity point (Heatwole *et al.* 1971; Heatwole and Newby 1972; Putnam and Hillman 1977; Moore and Gatten 1989).

Frogs seek refuge on a daily basis either underground by burrowing into the soil, by moving deep into cracks in rocks or by sheltering in trees under bark or in hollows (e.g. Loveridge 1977; McClanahan *et al.* 1994). Diurnal basking behavior is relatively uncommon in amphibians (Hutchison and Dupré 1992; see Chapter 3) due to their permeable skin and hence high rate of evaporative water loss, but some anurans do bask during the day for thermoregulation (Bradford 1984; Carey 1978; Freed 1980; Lillywhite *et al.* 1973, 1998; Muths and Corn 1997; Pearson and Bradford 1976; Sinsch 1989; Vences *et al.* 2002; Tattersall *et al.* 2006). Only a few amphibians, notably waterproof scansorial frogs, are able to bask during the day in exposed sites (Loveridge 1970; Shoemaker *et al.* 1972; see Section 3.4). Most desert frogs are nocturnal; they emerge from their refuges on suitable nights to forage, etc.

4.2.2 Seasonal Aestivation

Many amphibians in arid environments escape the hot dry season by remaining underground or in tree hollows or under bark for extended periods of time (aestivation; see Section 4.4). Many fossorial species burrow underground, where they retain water balance with the soil; other fossorial or arboreal species form a protective cocoon that reduces cutaneous evaporative water loss.

An extended period of summer dormancy is called aestivation. For example, the water-holding frog *Cyclorana platycephala* can remain buried underground for over 5 years if favorable surface conditions do not occur (van Beurden 1980), even though it is more typical for these frogs to be active each year after heavy cyclonic rains. The period of aestivation is a well-defined physiological state, especially for cocoon-forming species (see below). Aestivating frogs have considerably reduced activity. Non-cocooning species, such as North American spadefoot toads (*Scaphiopus*, *Spea*) and the Australian spadefoot (*Notaden* spp.), will remain sufficiently active to consume their skin whenever it is shed (see Section 3.2) and presumably are sufficiently mobile underground that they can burrow deeper as the soil dries, to follow the moisture and remain in water balance (McMaster 2006). Cocoon-forming species become immobile and do not consume their shed epidermis, which accumulates to form a water-proofing cocoon (see below). The metabolic rate of aestivating frogs is typically depressed to about 20–25% of their normal standard metabolic rate at that temperature (Guppy and Withers 1999; see Section 4.5). This allows endogenous energy stores to sustain metabolism for about five times longer, so that the amphibian can survive the lengthy aestivation period.

The African clawed toad (*Xenopus laevis*) is normally aquatic, like other pipids, but it has been reported to burrow in drying mud (Alexander and Bellerby 1938; Tasman 1956; Loveridge 1977). During this time, *Xenopus* accumulate urea but do not become dehydrated (Balinsky *et al.* 1967). However, in the laboratory *Xenopus* kept in mud failed to aestivate as the mud dried and hardened (Jokumsen and Weber 1980). This suggests that *Xenopus* "aestivating" in the field remain in sufficiently moist mud to maintain water balance, but as they do not form a cocoon they are unable to survive in dried mud. Interestingly, *Xenopus* are remarkably tolerant of exposure to hyperosmotic solutions, and accumulate urea as a balancing osmolyte (Section 4.1).

4.2.3 Fossoriality

Most "desert" amphibians escape from the hostile environment by burrowing underground. During favorable conditions, they burrow during the day and are active at night. During extended periods of unfavorable conditions, they remain underground and aestivate (Section 4.5). Many of these fossorial species continue to ingest their shed skin and hence do not form a cocoon. Some of these fossorial "desert" amphibians form a cocoon (see Section 3.2) during aestivation; a few salamanders (*Siren* spp.) also form a cocoon. Over time, the cocoon becomes a thick paper-like covering (Fig. 4.4). Cocoon-forming frogs tend to shed their skin

Fig. 4.4 B. Aestivating Australian trilling frog (*Neobatrachus aquilonius*) with a thick, parchment-like cocoon covering the entire outer body surface except for the external nares. (Photo: G. Thomson, P. Withers, V. Cartledge, K. McMaster).

at a higher frequency than non-cocooning species, about every 1 to 3 days, both normally and during aestivation.

The underground environment is typically more hospitable to an amphibian than the surface conditions. The thermal environment is relatively stable and benign for fossorial frogs. Temperature under the ground is lower than the daytime surface air temperatures (e.g. Fig. 4.5A), and the daily temperature cycle becomes more attenuated with greater depth underground. Annual soil temperature changes are also reduced with increasing depth (Fig. 4.5B). Soil temperature generally remains within tolerable limits for amphibians except near the surface. Even in extreme deserts, soil temperature at depth is 25–30°C (e.g. van Beurden 1980; Thompson *et al.* 2005) which is well within the thermal limits of desert amphibians (see Section 3.4). The temperature in *Scaphiopus* burrows varies from 17–22°C in autumn, 5–15°C in winter and 24–25°C prior to emergence in summer (Ruibal 1962; Shoemaker *et al.* 1969). The average temperature in aestivating *Cyclorana platycephala* burrows was 15°C with daily variations of

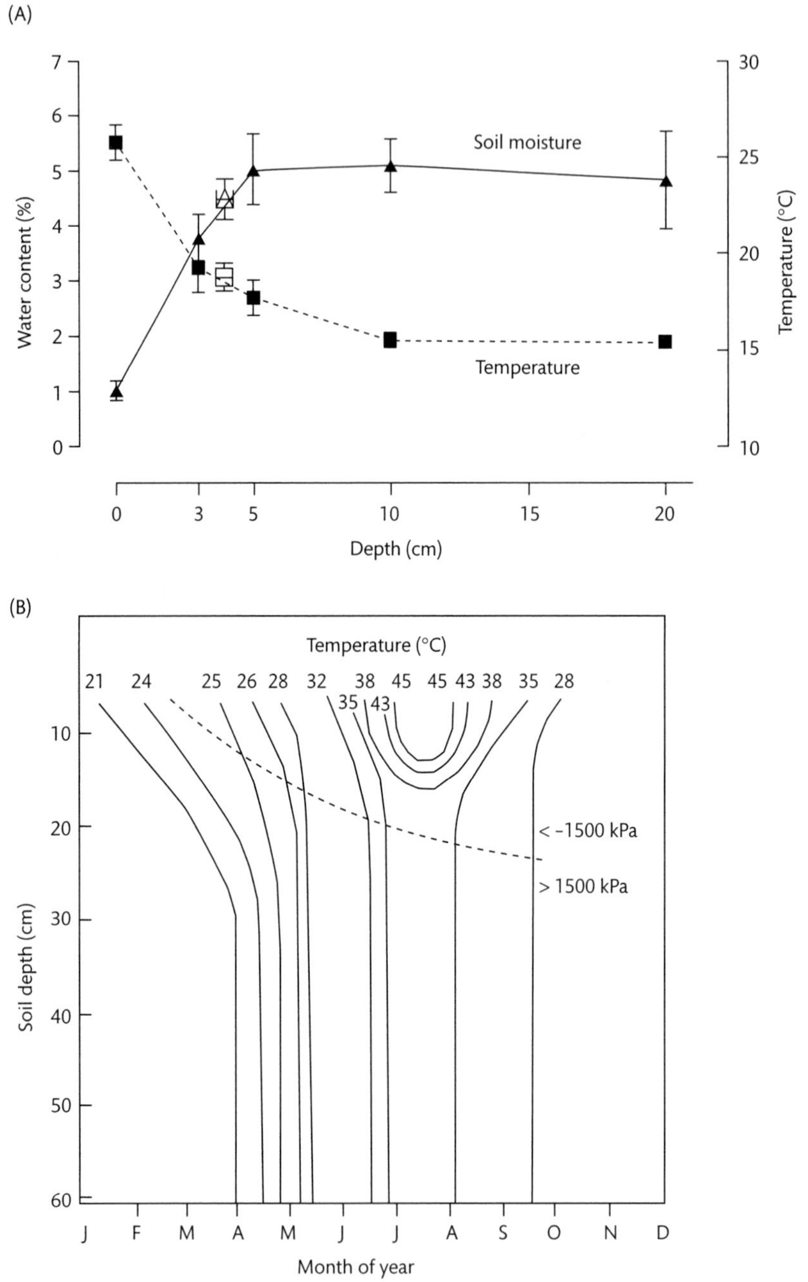

Fig. 4.5 (A) Soil water content (%) and temperature (°C) profiles for sand dunes where the sandhill frog (*Arenophryne rotunda*) occurs. The open triangle and square indicate the average depth and moisture/ temperature location of frogs. Values are mean ± SE. From Cartledge *et al.* (2006). (B) Annual temperature cycles at various depths underground in a desert wash at Palm Desert, California; dotted line indicates the isopotential for –1500 kPa (about 600 mOsm/l at 25°C). Modified from Muth (1980).

about 7°C (van Beurden 1980). Temperatures range from 21–27°C for the salamander *Siren intermedia* burrowed in dry mud (Gehlbach *et al.* 1973). Nevertheless, soil temperatures in the top 10–20 cm of exposed sites can exceed 30°C during the hottest part of the year; the highest recorded burrow temperature was 39°C for a shallow *Scaphiopus* burrow in summer, although lower soil temperatures are found in more sheltered locations. High soil temperatures would be disadvantageous to burrowed frogs since metabolic rate, even though depressed during aestivation (see Section 4.4) would be increased because of the Q_{10} effect, and this would reduce the period of time that frogs could remain aestivating underground.

Humidity is generally high underground, reducing the rate of evaporative water loss, although this depends on the soil moisture level and temperature. For example, coastal sand dunes in Western Australia provide a favorable environment for the sandhill frog *Arenophryne rotunda*. Sand has a high moisture content after rainfall, and is saturated by 5 cm depth (5%); the temperature is stable by 10 cm depth at 15°C (Fig. 4.5A). In summer, the surface sand is much drier and sandhill frogs burrow much deeper to reach similar moisture levels (Tyler *et al.* 1980; Roberts 1990). In more extreme arid environments such as the Gibson Desert in Western Australia, soil water content is lower (1–3%) and underground temperatures are higher (25–30°C) in sand dunes where the desert spadefoot *Notaden nichollsi* burrows to 1 m or more, and claypans where the desert trilling frog *Neobatrachus aquilonius* burrows to about 30 cm (Thompson *et al.* 2005; Cartledge *et al.* 2006). In an exposed site (desert wash), the hygric environment is favorable for burrowed amphibians (>−1500 MPa) at depths near the surface in the wet season and below about 20 cm in the dry season (Fig. 4.5B).

Frogs that burrow in sandy soils generally do not form a cocoon and dig deeper as the dry season progresses, following the moisture profile. As the frogs burrow deeper, their burrow is typically loosely filled, and the frogs remain surrounded by loosely-packed sand (rather than an air space, as in some cocoon-forming frogs; see below). For example, *Scaphiopus* are burrowed at about 20 cm depth at the start of the dry season, are at 50 cm or more deep in mid-winter, then return closer to the surface in early summer in anticipation of summer rains (McClanahan 1967; Ruibal *et al.* 1969; Seymour 1973b). The day-time burrows of *Scaphiopus* during the summer activity period are much shallower, about 6–8 cm (Seymour 1973b). Similarly, the sandhill frog *Arenophryne* burrows deeper as the dry season progresses, to remain in noticeably moist soil (Tyler *et al.* 1980; Roberts 1984, 1990) whereas active wet-season burrows are only about 4 cm deep in the moist surface sand (Cartledge *et al.* 2006). Southern African *Breviceps* and *Pyxicephalus* burrow 80–150 cm during aestivation (Mayhew 1968). Australian sand-burrowing frogs such as *Heleioporus* and *Notaden* burrow 80 to 150 cm, or even more (Mayhew 1968; Thompson *et al.* 2005; Fig. 4.6).

(A) (B)

Fig. 4.6 Excavated burrow of (A) desert spadefoots (*Notaden nichollsi*) from sand dunes (burrow at back, hole for excavation in front; burrow was loosely-filled with sand before excavation) and (B) a trilling frog (*Neobatrachus aquilonius*) from a clay pan (burrow depth about 30 cm; vertical burrow). (Photos: G. Thomson, P. Withers, V. Cartledge, K. McMaster)

In contrast, cocoon-forming frogs burrow in heavier clay soils, which become rock-hard when dry. They cannot move after they have formed their cocoon, without disrupting its physical integrity and are unable to follow the water profile deeper as the soil dries out. However, the cocoon provides protection against desiccation *in situ* (see below). The Australian cocoon-forming *Cyclorana* and *Neobatrachus* generally burrow into harder clay soils, to only about 30 cm (Main *et al.* 1959; van Beurden 1980; Thompson *et al.* 2005), but *Neobatrachus aquilonius* have been excavated from depths of as much as 120 cm in claypans (Thompson *et al.* 2005). Interestingly, the cocoon-forming *N. aquilonius* has also been excavated from moist sandy sites without a cocoon (Cartledge *et al.* 2006). Cocoon-forming frogs therefore appear to have the behavioral flexibility to delay cocoon-formation and follow the soil moisture profile as do non-cocooning frogs, prior to their eventually forming a cocoon. In the laboratory, cocoon-forming frogs will delay the onset of cocoon formation until the soil dries out and becomes a desiccating rather than a rehydrating environment (personal observations; Cartledge 2006). Booth (2006) found that *Cyclorana albiguttata* did not begin to form a cocoon until after 3 months, when the soil environment became desiccating relative to the water potential of the frog.

Cutaneous–Soil Water Exchange

Fossorial amphibians are able to readily rehydrate from soil interstitial water, providing the osmotic balance is favorable (and if they do not form a cocoon). Many laboratory studies have demonstrated the capacity to absorb water from moist soil for a wide variety of salamanders and anurans (see Chapter 3). Amphibians in the field have been shown to rehydrate while burrowed underground in moist soil during their resting period (e.g. *Heleioporus eyrei*, Lee 1968; *Scaphiopus* toads, McClanahan 1972).

The water balance between a buried frog and its surrounding soil environment is complex, depending on, firstly, the osmotic balance between soil water and the frog and, secondly, on the rate at and mechanism by which water moves through the interstitial spaces of the soil (see Section 3.2). The driving force for water movement between soil and a buried frog is the difference in water potential (ψ; units of pressure e.g. MPa) between the soil and the frog, but the relationship between soil water potential (ψ_{soil}) and water content is a complex function of soil particle size (see Fig 3.5). Sandy soils do not hold water as "tightly" as clay or silt soils, so the water potential of sandy soil is more favorable for reabsorption by a burrowed frog compared with the same percent water content as clay or silt soils, that is frogs in sandy soil can reabsorb water from relatively dry sand (water content of about 0.05%) whereas frogs in clay soil can only reabsorb water from wetter soil (about 0.25%; Fig. 4.7).

Liquid water exchange and water vapor exchange are the potential mechanisms for water exchange between a frog and soil because water in the soil interstitium is a mixture of liquid water and water vapor, except in very dry soil (only vapor) and saturated soil (only liquid water; Section 3.2; Booth 2006). The rate of liquid water exchange is determined by the soil hydraulic conductivity; this decreases as soils dry. Hydraulic conductivity is negligible for sandy soils if ψ_{soil} less than about -0.1 MPa, so hydraulic loss of water is unimportant for frogs in sand because ψ_{frog} is about -0.6 MPa and water loss is determined by vapor phase exchange (Fig. 4.7). For frogs in clay, hydraulic conductivity occurs until the water potential is very negative (e.g. <-1.5 MPa) so it is theoretically important even when frogs are considerably osmoconcentrated (about 600 mOsm/l). This is probably, at least in part, why frogs that burrow and aestivate in clay generally create an air space around their body, to break any liquid water connection between their skin and soil water (Section 3.2; Booth 2006); it would also prevent them from becoming solidly encased in dried clay.

The sandhill frog, *Arenophryne rotunda,* is able to absorb water from sand with a moisture content of 5% but not from 1 or 2% (Cartledge *et al.* 2006). Sandhill frogs could burrow into sand at 5% water content in the field and are in

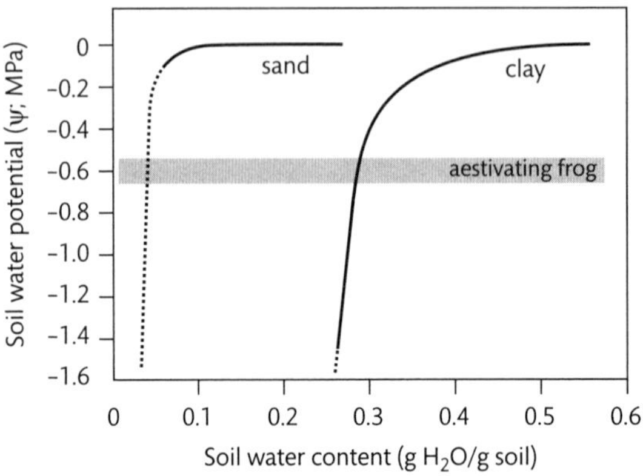

Fig. 4.7 Relationship between soil water potential and soil water content for sandy and clay soils; approximate Ψ range is shown for water exchange by hydraulic conduction (solid line) and water vapor diffusion (dotted line). Ψ is about -0.6 MPa for normally-hydrated frogs and -1.6 MPa for frogs aestivating in the field (see text). Modified from Booth (2006).

positive water balance (Table 4.2) and while burrowed during the day replenish any water losses experienced during foraging on the dune surface at night. For *Scaphiopus*, the soil water tension is equivalent to about 60–200 mOsm just after burrowing, and rises to 150–200 mOsm during the dry season; soil water can be absorbed by the frog since this is less than the osmotic potential of its body fluids (Shoemaker *et al.* 1969; Ruibal *et al.* 1969). In early summer, the soil is much drier (about 450–600 mOsm equivalent) and the spadefoots must have an increased body fluid osmotic concentration to remain in balance. Similarly, the Australian spadefoot *N. nichollsi* can maintain positive water balance in moist sand but not in drier sand, in the field (Table 4.2). This desert sand dune frog will burrow to over 2 m depth to remain in moist sand (Thompson *et al.* 2005). The trilling frog, *N. aquilonius,* burrows in clay pans, to about 30–120 cm depth (Thompson *et al.* 2005), where it can be near water balance in favorable years but not in drier years (Table 4.2). However, the trilling frog is a cocoon-former; its protective cocoon (see Chapter 3) increases cutaneous resistance to prevent desiccation under conditions of unfavorable water balance with the soil.

Accumulation of Solutes

Amphibians that aestivate in soil with a higher ψ than their body fluids can retain a positive water balance and absorb water from the soil to replace any lost by urination or evaporation when active. Urinary excretion would prevent urea (synthesized from ammonia formed by protein catabolism) from accumulating in

Table 4.2 Summary of balance of osmotic forces in the field for burrowed North American spadefoot toads (*Scaphiopus couchii*), Australian sandhill frogs (*Arenophryne rotunda*), Australian desert spadefoots (*Notaden nichollsi*) that do not form a cocoon, and Australian cocooned trilling frogs (*Neobatrachus aquilonius*), by comparison of body fluid osmotic concentration (OC_{frog}) and equivalent osmotic pressure (Ψ_{frog}), with the actual soil water potential (Ψ_{soil})

	OC_{frog} (mOsm/l)	Ψ_{frog} (kPa)	Ψ_{soil} (kPa)	Balance
A. rotunda†	293	-727	-200	positive
S. couchii†				
March	300	-743	-371	positive
June	450	-1114	-1485	negative
N. nichollsi†				
sand dune 2003	351	-869	<-1000	negative
sand dune 2004	271	-671	-200	positive
*N. aquilonius**				
clay pan 2003	324	-803	<-1000	negative
clay pan 2004	211	-522	-600	near balance

† non-cocooning; *cocooned. 1000 mOsm/l ≡ -24.4 atm ≡ -2.475 MPa at 25°C. Data from Shoemaker *et al.* (1969) and Cartledge *et al.* (2006a, b).

the body fluids, but would also result in ion depletion (especially Na^+ and Cl^-). Dilute urine stored in the bladder is also an important water store. Consequently, urination might be avoided or at least minimized even if water balance with the soil was positive. For example, spadefoot toads buried in relatively moist soil in the laboratory slowly accumulate urea and ions in their body fluids, and their total osmotic concentration increases (McClanahan 1972; Fig. 4.8). The minor mass loss after 17 months of aestivation presumably reflects dry tissue loss (e.g. fat body) rather than water loss.

As soil progressively dries out during aestivation, Ψ_{soil} decreases to less than Ψ_{frog} and water balance becomes negative. Non-cocooned frogs could move deeper underground, to retain contact with moist sand and avoid osmoconcentration. Alternatively, the total body fluid osmotic concentration can be increased such that Ψ_{soil} is similar to Ψ_{frog} and at least neutral if not positive water balance is maintained. Body fluid osmotic concentration could be increased by loss of water to the soil, but a better strategy would be to increase body fluid osmotic concentration by accumulation of additional solutes. This pattern of osmotic change over time is well illustrated by *Scaphiopus couchii* burrowed in the laboratory (McClanahan 1972; Fig. 4.8). As soil dries, the body fluid osmotic concentration increases further to retain water balance, by increases in urea concentration (Fig. 4.8). Body

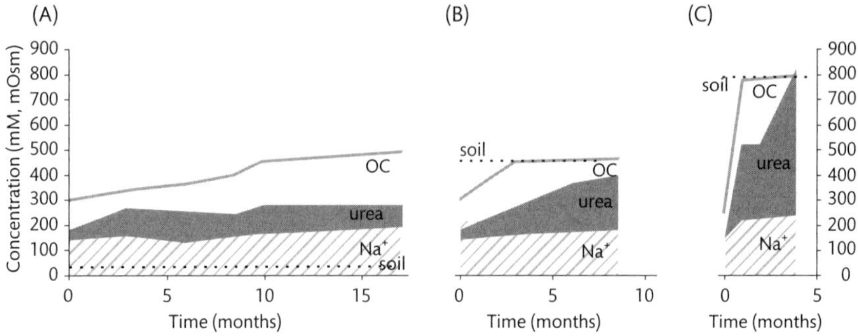

Fig. 4.8 Changes in plasma Na^+ (hatched), urea (shaded) and total osmotic concentrations (OC) of the spadefoot toad (*Scaphiopus couchii*) during aestivation in soil of varying water content; (A) 8.8, (B) 5.5 and (C) 5.1 ml/100 g. The equivalent water potential, expressed in equivalent osmotic concentration, is given by the dotted line in each panel. Data from McClanahan (1972).

mass declines slightly over time, presumably reflecting loss of metabolized tissue mass (e.g. fat body) rather than loss of water. A similar pattern of increasing ion but especially urea and thus osmotic concentration over time has been observed in the field in *Scapiopus couchii* (Shoemaker *et al.* 1969).

Many other burrowed (non-cocooned) amphibians show similar increases in body fluid solute concentration (particularly urea), and hence Ψ, over time. The Australian spadefoot (*Notaden nichollsi*) has a higher osmotic concentration in drier sand in the field, due to higher Na^+, Cl^- and urea concentrations (Cartledge *et al.* 2006). The toad *Bufo viridis* initially loses mass, then stabilizes after a few days when burrowed in the laboratory (Katz *et al.* 1986; Katz and Hoffman. 1990; Hoffman and Katz 1997). After 7 weeks, its osmotic concentration increases to over 1000 mOsm, reflecting changes in Na^+ (160 mmol/l) and Cl^- (205 mmol/l) but especially urea (495 mmol/l; Katz and Hoffman 1990). Urine voidance ceases, minimizing urinary urea, ion and water loss. Burrowed tiger salamanders (*Ambystoma tigrinum*) exceed 500 mOsm/l osmotic concentration after 9 months burrowed in soil, reflecting a large increase in urea concentration (>200 mM; Delson and Whitford 1969). This ureo-osmoconforming strategy (see Withers 1998) is not restricted to burrowed amphibians, occurring in chondrichthyeans, the coelacanth, the crab-eating frog *Rana cancrivora* and other amphibians acclimated to hypersaline media (Section 4.1). Some of these burrowing amphibians acclimated to hypersaline media (400–850 mOsm/l) have an increased osmotic concentration to retain water balance, and this involves large increases in urea, for example *Xenopus laevis*, *Scaphiopus couchii*, *Bufo viridis* and *Ambystoma tigrinum* (see Katz 1992). Other non-arid amphibians also show a ureo-osmoconforming response to hypersaline media, for example the crab-eating frog *Rana cancrivora*, treefrog *Hyla arborea*, and a plethodontid salamander *Batrachoseps* spp. (see Katz

1992). Many anurans show a lesser osmotic/ urea accumulation response to desiccating burrow conditions, with only moderate increases in osmotic concentration (e.g. to 250–400 mOsm/l) and small changes in urea (e.g. 10–50 mmol/l), for example *Bufo bufo*, *B. regularis*, *Pelobates syriacus*, *Rana ridibunda*, *R. pipiens*, *R. temporaria* and *R. esculenta* (Katz 1992).

An increase in urea concentration over time is important in many aestivating amphibians for maintenance of water balance with the soil, and it is clear that the urea concentration increases faster for frogs in drier soil. The retention of urea from background levels of protein catabolism during aestivation is apparently not sufficient to explain the magnitude of the increased urea concentration, nor is osmoconcentration by loss of body water, since Na^+ and Cl^- do not increase so dramatically, for example *Scaphiopus couchii*; (McClanahan 1972; Fig. 4.8), *Bufo viridis* (Katz and Hoffman 1990) and *Ambystoma tigrinum* (Delson and Whitford 1969). Similar, rapid accumulation of urea has been reported for amphibians acclimated to hypersaline media (e.g. *X. laevis*). Consequently, the increase in urea is often attributed to accelerated rates of urea synthesis (Balinsky *et al.* 1981; McClanahan 1972; Jones 1980b; Hoffman and Katz 1999; Delson and Whitford 1973). However, Barker Jorgensen (1997) suggested that the data for accelerated urea synthesis by *Scaphiopus*, *Bufo* and other ureo-osmoconforming amphibians are equivocal, with elevated urea concentrations reflecting changes in balance between synthesis and excretion.

Urea becomes an important balancing osmolyte for burrowed, non-cocooned amphibians, whether or not its synthesis is accelerated as a response to soil water balance. Urea levels in aestivating amphibians depend on the length of the aestivation period, and the soil conditions; concentrations from 100 to 500 mM are not unusual. These high urea concentrations approach or exceed those of marine chondrichthyean fishes, whose enzymes are adapted to resist the inimical effects of high urea on protein structure and function and beneficial effects of counter-acting solutes such as betaine and sarcosine (see Hochachka and Somero 2002). In aestivating amphibians, counter-acting solutes do not appear to coaccumulate with urea (Withers and Guppy 1996) and the inimical kinetic effects of urea are relatively minor, particularly compared with the effects of high ion concentrations (Grundy and Storey 1994; Cowan and Storey 2002; see Section 4.4).

The cocoon of amphibians seals their entire body surface, including the eyes, mouth and cloaca, but not the external nares which remain open for pulmonary gas exchange. A cocooned frog is therefore isolated from the external environment (apart from gas exchange) and must survive its aestivation period on internal energy and water reserves. Energy reserves are extended considerably by metabolic depression (see Section 4.4). By the time a cocooned is formed, water balance with the soil is most likely negative and the essential role of the cocoon is to minimize cutaneous EWL. Frugal use of bladder water reserves is equally important for survival.

For cocooned amphibians, body fluids and bladder water are the only water stores to balance loss by cutaneous and pulmonary evaporation; metabolic water production is a minor source of water for aestivating cocooned frogs. The inevitable loss of water by pulmonary and cutaneous avenues during aestivation means that the body fluids will become concentrated, particularly after the bladder reserves are depleted or become equivalently concentrated as body fluids. For example, body fluids of the aestivating, cocooned *Lepidobatrachus llanensis* become concentrated over time, particularly with respect to Na^+ and Cl^- (McClanahan *et al.* 1972). Urine solute concentrations increase even more dramatically than plasma over the course of extended aestivation, and eventually are about the same concentration as body fluids. Urine volume also declines markedly. Similar changes in plasma and urine solutes have been shown for other aestivating frogs in the laboratory, including *Cyclorana platycephala* and *Neobatrachus aquilonius* (Cartledge 2006; Cartledge *et al.* 2006; Table 4.4). Field measurements of *N. aquilonius* show no significant increases in plasma Na^+, Cl^-, K^+ or total osmotic concentration, with aestivation in relatively moist conditions (in sand, non-cocooned), but substantial increases were measured under drier conditions for cocooned frogs in clay pans (Cartledge *et al.* 2006).

Cocooning amphibians do not have the adaptive need to accumulate solutes to balance soil water potential during aestivation, unlike non-cocooning frogs. The cocoon seems to not be formed until the soil water potential becomes desiccating, and then the high diffusion resistance of the cocoon retards EWL. They will accumulate urea over time to detoxify ammonia formed by protein metabolism, and progressive dehydration by evaporative water loss will increase the urea concentration, but there would seem to be no selective pressure for cocoon-forming amphibians to accelerate urea synthesis during aestivation. Consequently, they generally do not accumulate urea to the high concentrations measured for urea-tolerant, non-cocooning amphibians (Withers and Guppy 1996; Table 4.3). Balancing/non-perturbing solutes do not co-accumulate during aestivation (Withers and Guppy 1996) and high concentrations of urea have lesser adverse effects on enzyme function for aestivating frogs (Fuery *et al.* 1997; Cowan and Storey 2002).

4.2.4 Scansoriality

Many amphibians are found in trees, on rocks and other above-ground locations. Particularly in semiarid and arid environments, a scansorial habit potentially exposes an amphibian to lethal desiccating conditions, but it also provides the potential to exploit heliothermic behavioral thermoregulation. Many scansorial amphibians remain nocturnal and shelter during the day from the desiccating and high temperature conditions, seeking shelter in tree-hollows, under bark and inside vegetation (e.g. bromeliads, etc). Some, however, have become more diurnally active. Not surprisingly, diurnal scansorial amphibians have a suite of

Table 4.3 Urea levels of amphibians that do not form a cocoon (NC) with low or high tolerance, cocooning species (C), and species acclimated to hypersaline media (HS)

Species	Condition	Urea (mmol/l)	Reference
Bufo bufo	NC low	10	Schoffeniels and Tercafs (1965)
Rana temporaria	NC low	13	Ackrill *et al.* (1969)
Thoropa miliaris	HS	13	Abe and Bicudo (1991)
Rana pipiens	NC low	20	Jungreis (1971)
Pelobates syriacus	NC low	20	Katz (1992)
Rana esculenta	NC low	22	Garcia-Romeu *et al.* (1981)
Siren intermedia	C	32	Asquith and Altig (1986)
Lepidobatrachus llanensis	C	35	McClanahan *et al.* (1976)
Bufo regularis	NC low	43	Katz (1992)
Siren lacertina	C	49	Etheridge (1990)
Rana ridibunda	NC low	53	Katz (1992)
Notaden nichollsi	NC high	56	Cartledge *et al.* (2006a)
Neobatrachus aquilonius	C	66	Withers and Guppy (1996)
Cyclorana platycephala	C	76	Withers and Guppy (1996)
Pyxicephalus adspersus	C	81	Loveridge and Withers (1971)
Bufo woodhousei	NC high	120	Jones (1982)
Neobatrachus kunapalari	C	130	Withers and Guppy (1996)
Hyla arborea	NC high	143	Katz (1992)
Xenopus laevis	HS	155	Schlisio *et al.* (1973) Romspert (1976)
Heleioporus albopunctatus	NC high	163	Withers and Guppy (1996)
Neobatrachus pelobatoides	C	163	Withers and Guppy (1996)
Ambystoma tigrinum	NC high	182	Delson and Whitford (1973)
Neobatrachus sutor	C	195	Withers and Guppy (1996)
Batrachoseps sp.	HS	218	Jones and Hillman (1978)
Cyclorana maini	C	224	Withers and Guppy (1996)
Rana cancrivora (adult)	HS	350	Gordon *et al.* (1961)
Bufo viridis	NC high	500	Hoffman *et al.* (1988)
Scaphiopus couchii	NC high	584	McClanahan (1972)

behavioral and physiological adaptations that in concert allow survival in hygric and thermal conditions that would be lethal for most amphibians. These adaptations for diurnal scansoriality have evolved independently a number of times in anuran amphibians (with notable absence in salamanders and caecilians), and to differing extents between typical non-waterproof frogs to extremely waterproof xeric frogs. The suite of inter-related behavioral and physiological adaptations

Table 4.4 Plasma and urine solute and osmotic concentrations for aestivating anurans while burrowed, with the change in urine concentration (Δ urine) and final urine-to-plasma (U/P) ratio

	Initial		Final		Δ_{urine}	U/P_{final}
	Plasma	Urine	Plasma	Urine		
Bufo viridis[1]						
Na^+	114	14	189	44	30	0.23
Cl^-	89	15	195	17	2	0.09
Urea	64	69	149	266	197	1.79
OC	340	145	636	600	455	0.94
N. nichollsi[2]						
Na^+	127	20	154	43	23	0.28
Cl^-	92	2	118	2	0	0.02
Urea	20	37	56	105	68	1.88
OC	266	76	351	165	89	0.47
N. aquilonius[3]						
Na^+	73	3	195	172	169	1.19
Cl^-	63	2	98	9	7	0.09
Urea	7	25	66	84	59	1.27
OC	220	48	324	286	238	0.88

[1]7 weeks, Katz 1990; [2]about 8 months in 2004, Cartledge *et al.* 2006a; [3] cocooned in clay-pan for about 18 months in 2003, Cartledge *et al.* 2006a.

include reduced cutaneous evaporative water loss, basking behavior, capacity for color change (particularly blanching to a white color), and the ability to increase EWL when heat-stressed ("sweating"). Many other physiological systems are influenced by these changes, for example cutaneous gas exchange, kidney function and color change. A few waterproof frogs have even switched their nitrogenous waste disposal strategy to uricotely. None, however, have completely freed their reproductive strategy from the requirement for free water (although some fossorial frogs have direct development).

The primary behavioral switch that would initiate the suite of physiological adaptations of waterproof frogs is basking in exposed sites on trees or rocks. Exposure, even in shade, predisposes an amphibian to desiccation and warming to air temperature, and basking directly in sunlight further increases the risk of desiccation and elevation of T_b above ambient air temperature. The two main physiological changes required to allow basking are reduced EWL and elevated thermal tolerance, and not surprisingly these are interconnected in the graded transition from a typical frog to a very waterproof frog (Table 4.5). Resistance

Table 4.5 Change in cutaneous water loss (resistance; s/cm) and thermal tolerance in sequence from typical frogs to waterproof frogs

Species	Resistance (s/cm)	CT_{max} (°C)
Limnodynastes fletcheri	1.5	29.1
Rana pipiens	1.6	36.9
Bufo marinus	1.7	41.8
Litoria meiriana	3.9	38–40
Scaphiopus spp.	5	40.3
Bufo cognatus	5	41.3
Litoria aurea	6.9	35.1
Litoria rubella	9.6	38.5–41.5
Litoria rothii	14.2	40.8
Litoria caerulea	15	40.0
Litoria chloris	30	40
Litoria bicolor	80	41.8
Litoria gracilenta	118	39.8
Hyperolius viridiflavus	178	44
Chiromantis spp.	400	>41 ??
Phyllomedusa spp.	364	>42 ??

Data from Lillywhite (2006), Tracy *et al.* (2008), Shoemaker *et al.* (1989) Giese and Linsenmeir (1986b) Brattstrom (1968, 1970) Buttemer (1990).

to evaporative water loss increases from that for a free water surface and typical amphibians (about 1 s/cm) to over 200 s/cm in very waterproof species. This is essential for a scansorial amphibian to avoid rapid dehydration to their lethal limit since ambient air in semiarid/ arid environments can have a very low humidity. There is a general correlation between cutaneous EWL and selected temperature; the very waterproof *Chiromantis*, *Phyllomedusa* and *Hyperolius* have a higher CT_{max} than less-waterproof species such as *Rana* and *Limnodynastes* (Table 4.5); within *Litoria*, the less-waterproof species have a slightly lower CT_{max} than the more waterproof species (Tracy and Christian 2005).

One mechanism for reduction of EWL is well understood—*Phyllomedusa* spp. use a stereotypic wiping behavior to cover their body surface with a waxy secretion from cutaneous glands, and *Litoria* spp. might do the same. Other waterproof frogs do not wipe and do not have a waxy surface covering—dried cutaneous protein/ mucous secretions might provide their waterproofing barrier. The thermal consequence of exposure to the atmosphere is warming to ambient air temperature, which in many semiarid and arid environments would easily exceed the lethal thermal tolerance of a typical amphibian (30–35°C; see Section 3.4). Basking in direct sunlight would elevate skin and core body temperature above ambient air temperature, and further elevate the rate of cutaneous water loss.

Scansorial frogs that bask in direct sunlight often have a remarkable capacity to change color to a grey or sometimes brilliant white. The African tree frog *Chiromantis xerampelina* often basks in very exposed sites in direct sunlight, and can blanch to a very bright white (Shoemaker *et al.* 1989; McClanahan *et al.* 1994; Fig. 3.10). Some *Hyperolius* species also blanch to a very bright white when exposed to desiccating conditions (Withers *et al.* 1982; Kobelt and Linsenmair 1986), as does the Australian desert tree frog, *Litoria rubella* (Withers 1995). This pattern of blanching to white sometimes correlates directly with a decrease in EWL, suggesting that there is a common mechanism for EWL reduction and skin blanching, for example dermal iriodiophores (Drewes *et al.* 1977; Bentley and Yorio 1977; Withers *et al.* 1982; Kobelt and Linsenmair 1986). However, it is difficult to envisage how the waterproofing barrier could involve living cells such as iridophores (or layers of waxy material in subepidermal glands), so the tight correlation of reduced EWL with color change appears to not be a mechanistic link but rather a coadaptive response to different physiological consequences of a specific environmental challenge.

Waterproof frogs will select a higher body temperature (T_b) in a thermal gradient than non-waterproof frogs. For example, *Chiromantis* will select T_bs about 36–37°C and sometimes > 40°C, and *Phyllomedusa* will select 33–40°C and sometimes > 41°C (Shoemaker *et al.* 1989). Interestingly, both genera will select a lower T_b when dehydrated. The adaptive increases in selected T_b and CT_{max} of waterproof frogs means that they must be able to thermoregulate T_b below their CT_{max} at high ambient air temperatures. Enhanced evaporative cooling is the major strategy for thermoregulation by terrestrial animals in the heat, and waterproof frogs use the same strategy. They control the discharge of cutaneous mucous glands, like normal frogs (cf Lillywhite 1971; Lillywhite and Licht 1975) to regulate EWL and hence T_b. Both *Chiromantis* and *Phyllomedusa* will regulate T_b at about 40–42°C at higher T_a by proportional increases in EWL (Shoemaker *et al.* 1987). For *Chiromantis xerampelina*, the control of EWL is achieved by modulation of sympathetic outflow influencing β-adrenergic receptors of cutaneous mucous glands (Kaul and Shoemaker 1989). *Litoria chloris* and *L. xanthomera* have a reduced resistance to EWL at T_a >40°C (Buttemer 1990; Buttemer and Thomas 2003). *Litoria* spp. at high T_a (43–45°C) have a sufficiently low R that they can keep T_b below their CT_{max}, but the more waterproof species that increase R at lower T_a would reach or exceed their CT_{max} if they didn't adaptively reduce R (Fig. 4.9; Tracy *et al.* 2008). However, once waterproof frogs reduce their R to evaporatively cool, they reduce the selective advantage of being waterproof, that is water conservation, and then rely on bladder water stores to maintain hydration, like non-waterproof frogs.

Only two genera of amphibians are known to excrete their nitrogenous waste as uric acid, rather than the typical form of nitrogenous waster for aquatic amphibians

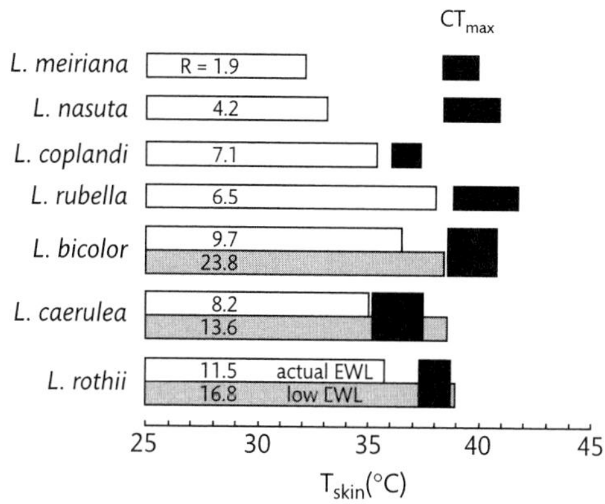

Fig. 4.9 Relationship between cutaneous resistance to EWL (R; s/cm), Skin surface temperature (T_{skin}, white bars; °C) is maintained lower than the critical thermal maximum (CT_{max}, black bars; °C) by various *Litoria* treefrogs at high ambient temperature. For *L. caerulea* and *L. rothii*, T_{skin} approaches or exceeds CT_{max} if their normally-low evaporative water loss (grey bars) is not increased by enhanced evaporation. The cutaneous resistance values (sec/cm) are indicated, to show the reduction in R that maintains T_{skin} below the CT_{max} for these species. Data from Tracy et al. (2008).

(ammonia) or terrestrial amphibians (urea). Not surprisingly, these were the first two genera of frogs reported to be waterproof, *Chiromantis* (Loveridge 1970) and *Phyllomedusa* (Shoemaker *et al.* 1972). The water-savings accruing from uricotely are presumably only of adaptive significance for a waterproof frog; a non-waterproof frog would lose so much more water by cutaneous evaporation than urinary excretion that there would be no selective pressure to conserve urinary water loss. Nevertheless, uricotely is not necessarily adaptive for all waterproof frogs. *Hyperolius nasutus* is a small, very waterproof frog that is conspicuously not uricotelic (Withers *et al.* 1984). Despite their high cutaneous resistance, they are unable to survive desiccating conditions for more than about 14 days because of their small size and high surface-to-volume ratio. However, urea (as their main nitrogenous waste) would not accumulate to lethal levels over this time, so there is no survival advantage to being uricotelic, and incurring the additional metabolic cost of synthesizing uric acid rather than urea (Withers *et al.* 1984). Similarly, Giese and Linsenmair (1986) have argued that there is no selective advantage for *Hyperolius viridiflavus* to be uricotelic despite apparently having the biochemical machinery for uricotely—these frogs accumulate uric acid in their chromatophores, connective tissue and liver during the dry season when metabolic rate is much reduced (Geise and Linsenmeir 1986). The small phyllomedusine *P. hypochondrialis* has about the same resistance to EWL as its larger congeners such as *P. sauvagei*, but excretes less of its nitrogen waste as uric acid (Shoemaker *et al.* 1975), perhaps also because of a lesser adaptive advantage for uricotely in small frogs.

Waterproof skin reduces EWL but presumably also reduces the diffusional exchange of other gasses, including O_2 and CO_2. In *Hyperolius viridiflavus*, there is a significant continuous loss of CO_2 across the skin in the wet season, when EWL is high, but virtually no cutaneous CO_2 loss in the dry season when there is a marked increase in CO_2 excretion via discontinuous respiration (Giese and Linsenmier1986b). When cutaneous EWL is low during the dry season, the proportion of respiratory water loss is high, at about 15% of total EWL. The cutaneous gas exchange of *Chiromantis xerampelina* and *Phyllomedusa sauvagei* is reduced concomitant to their low cutaneous EWL, being similar to the low cutaneous gas exchange of reptiles (Stinner and Shoemaker 1987). Interestingly, the blood gas status (blood pCO_2, HCO_3^- and pH) of these frogs is similar to that of "normal" non-waterproof frogs, suggesting that they hyperventilate to avoid hypercapnia in the absence of significant cutaneous CO_2 loss.

Being waterproof and relieved of excessive EWL has apparently not relieved the selective pressures on the role of bladder water as a water store that is so important for non-waterproof frogs. The bladder store of dilute urine is used to preserve normal hydration levels during dehydration. For example, the waterproof *Phyllomedusa* has a bladder volume of about 20–30% of body mass (Shoemaker and Bickler 1979) and a low urine concentration. Consequently, bladder urine serves as a water store that buffers EWL for *Phyllomedusa* (Shoemaker and Bickler 1979) just as it does for more typical anurans, (Shoemaker *et al.* 1992). The tolerance to water loss of waterproof frogs is high (e.g. up to 50% for *Hyperolius viridiflavus*; Giese and Linsenmeir 1986b).

4.3 Hibernation Adaptations

Hibernation is winter dormancy. Two different environmental stresses, a lack of food and low ambient temperatures, characterize the conditions amphibians must contend with in order to survive the winter dormancy. Food is generally not available during winter, so metabolism can only be supported by energy reserves. The advantage for ectotherms of experiencing low ambient temperatures during winter is that body temperature, and hence rate of metabolism, will be low. Consequently, the rate of depletion of energy reserves will be slow and the probability of death by starvation is diminished at these low temperatures. The obvious danger of hibernating at low temperature for an amphibian is the potential for the environmental temperature to get low enough to freeze and kill it. This danger is exacerbated when a species distribution encompasses higher latitudes and/or higher elevations where freezing temperatures are more likely.

The environmental conditions experienced by an amphibian in a temperate zone winter differ between aquatic and terrestrial environments. Ice-covered water mechanically prevents aerial breathing, hence gas exchange can only occur across the skin and/or gills. Aquatic environments have a much lower O_2 capacitance than terrestrial environments (Chapter 1). When water is covered by ice there is a much slower rate of O_2 exchange with the environment, so aerobic metabolic activity of organisms living in the water and substrate are much more likely to deplete the O_2 and create hypoxic (low P_{O_2}) conditions. Hypoxia is much more likely to develop in shallow bodies of water and when snow is deep, shading the water and preventing photosynthetic generation of O_2. Consequently, physiological and behavioral mechanisms involved in dealing with hypoxia are important for winter survival in aquatic environments. A variety of data suggest that winterkill resulting from hypoxia is an important source of mortality for aquatic amphibians (Bannikov 1948; Manion and Cory 1952; Bradford 1983; Licht 1991; Pasanen and Sorjonen 1994). Tadpoles are more tolerant of hypoxia than adults, and have a lower critical P_{O_2} (Bradford 1983). This differential hypoxia tolerance is probably the result of higher surface area/mass (gills and skin) and a lower hemoglobin P_{50} for tadpoles compared to submerged metamorphs and adults. Submergence in fresh water also drives diffusional water uptake and ion loss across the integument during the winter, potentially creating ionic/osmotic problems. An advantage of an aquatic environment is that the amphibian is very unlikely to be exposed to water temperatures low enough to cause death by freezing. An aquatic environment will also have less variation in temperature than terrestrial environments at the same mean temperature due to the high specific heat capacity of water (Chapter 1), leading to lower rates of metabolism, lower rates of depletion of energy reserves and an extension of the possible length of time hibernation is possible (see below).

An amphibian hibernating in a terrestrial environment has the advantage of remaining normoxic, hence aerobic metabolism will not be compromised. However the probability of both experiencing a greater variation in ambient temperature and being exposed to subzero temperatures is greater in a terrestrial winter environment. The increased temperature variation has the effect of raising the average rate of metabolism and decreasing the length of time that the amphibian can survive on its chemical energy reserves. The other potential difficulty with a terrestrial environment is that that ambient temperatures can drop significantly below 0°C. This increases the risk of freezing. Dehydration is yet another physiological problem for amphibians over-wintering in a terrestrial environment.

The higher M_{O_2} of a frog at a variable temperature derives from the Q_{10} relationship of metabolism and temperature. Take a hypothetical situation where one frog experiences a constant daily 5°C environmental temperature, whereas another spends half the day at 0°C and the other half at 10°C. The mean temperature of 5°C is the same for both frogs. If the Q_{10} is 2, and the metabolic rate at 0°C is 2 (arbitrary units) then the metabolic rate at 10°C will be 4 ($4 = 2(2^1)$). The average metabolic rate will be 3 ($(4 + 2)/2$) for the variable frog and lower at 2.82 for the constant 5°C frog ($2.82 = 2\,(2^{0.5})$). Because the relationship of metabolism to temperature is usually exponential, metabolism will be greater with increased variation despite having the same mean temperature. This effect increases as the Q_{10} for metabolism increases. The preceding example used a typical Q_{10} of 2, but it has been demonstrated to be 4.3 at these low temperatures (Tattersall and Boutilier 1997).

Temperate zone amphibians hibernate in three major environmental situations: aquatic, subterranean terrestrial, and shallow terrestrial under leaf duff and bark. The choice of hibernation sites varies interspecifically and has a strong behavioral component (Licht 1991). Metamorphs can also vary from adults in their choice of hibernation site (Laitinen and Pasanen 1998; Lamoureux and Madison 1999). The choice of over-wintering sites is not absolute for the adults of any species. Site preferences in the lab or field border on 70–90% for one site versus the other (Licht 1991; Pasanen and Sorjonen 1994; Laitinen and Pasanen 1998; Lamoureux and Madison 1999; Holenweg and Reyer 2000). This lack of absolute intraspecific fidelity for an aquatic or terrestrial over-wintering site means that during any given winter there will be individuals of the same species in both terrestrial and aquatic over-wintering sites. This may represent an evolutionary bet-hedging strategy. The intraspecific proportion of individuals choosing more terrestrial sites can also be related to age of frogs, with new metamorphs choosing aquatic sites and adults choosing terrestrial sites (Laitinen and Pasanen 1998; Lamoureux and Madison 1999). Small frogs have a higher surface area/ volume ratio, which would enhance cutaneous gas exchange in an aquatic environment but enhance the rate of evaporative water loss in a terrestrial environment. Both consequences would favor aquatic hibernation sites for juveniles. It is not clear whether this aquatic–terrestrial shift represents an ontogenetic shift of site preference or an age-related distribution of over-wintering sites as a consequence of higher rates of mortality for individuals selecting aquatic sites. Use of subterranean sites requires the ability to dig or the presence of fissures in the soil, so small size or well-developed metatarsal tubercles would be anatomical correlates of this pattern. The advantages of being able to penetrate deep in the soil are that temperature variation is lower, temperature is less likely to fall below freezing and there is a lower likelihood of dehydrational stress. This minimizes the possibility of freezing mortality, but since the soil temperature is higher, energy reserves will be expended faster. This pattern is characteristic of toads and many salamanders. Some frogs (tree frogs and some *Rana*) hibernate predominantly under bark or at the interface of soil and leaf litter. This entails a greater risk of both freezing and dehydration. Some species that occupy superficial terrestrial sites have some capacity to tolerate freezing of extracellular fluids (see below). The aquatic over-wintering environment is the preference of most frogs and obviously aquatic salamanders. This has the advantage of a low likelihood of freezing, a more predictable rate of expenditure of energy reserves via temperature selection, balanced by the risks of the development of hypoxia, hyperhydration and the loss of electrolytes.

This begs the question, is an aquatic or terrestrial environment the better situation for an amphibian to survive the winter? The answer, of course, is it depends. From a distributional perspective it appears that selection for terrestrial hibernation sites (both superficial and subterranean) characterizes the most extreme high latitudinal distributions for anurans. The most northerly distributed anurans tend to over-winter in superficial terrestrial locations, exactly the sites most likely to freeze. What might be the selective advantages of these over-wintering sites besides escaping the potential threat of hypoxia? Terrestrial sites allow greater adult mobility during more equitable weather in late fall and early winter. This gives the individuals more latitude in finding appropriate thermal and hydric microenvironments. Terrestrial sites will be the first hibernation sites to thaw, hence breeding activity can occur early in the spring. Given the short growing season at high latitudes, this gives a temporal advantage to the larvae of these species.

4.3.1 Aquatic Hibernation and Hypoxia

There is strong evidence that winter can serve as the "hammer of death" for aquatic hibernators. Winterkill due to hypoxia is well known for fish (Greenbank 1945; Barica and Mathias 1979) and populations of *R. temporaria* (Bannikov 1948; Pasanen and Sorjonen 1994), *R. muscosa* (Bradford 1983), and *R. pipiens* (Manion and Cory 1952; Licht 1991). These data clearly establish the strong selective pressure operating for hypoxia tolerance and the importance of understanding how amphibians cope with hypoxic, aquatic conditions over winter.

The physiological problems that develop during hypoxic exposure essentially ramify from a decreased diffusional flux of O_2 influx across the skin, the only available exchange surface. Therefore the overall O_2 flux is decreased during submergence. This can compromise aerobic metabolism under hypoxic conditions, and decrease the rate of ATP production, as can be seen in Fig. 4.10.

Amphibians that over-winter in aquatic environments have behavioral choices in both the temperature and P_{O_2} of the water they occupy at any time. The likelihood of hypoxia developing is less likely near the surface, and in moving water, whether from seeps or streams, since convective movement of non-hypoxic fresh water will replenish the O_2. Adult green frogs (*Rana clamitans*) leave ponds in late fall to occupy sites with flowing water (stream or seep sites) that will remain oxygenated through the winter (Lamoureux and Madison 1999). Recent metamorphs of *R. clamitans* remain in ponds to over-winter. Moving to flowing water during winter also seems to be a strategy of *R. muscosa* to prevent hypoxic stress (Matthews and Pope 1999). Bullfrogs (*R. catesbeiana*) select inlets of underground tributaries (Stinner *et al.* 1994). The European common frog (*R. temporaria*) when confronted with hypoxic conditions in a thermal gradient, shift preferred temperature from 6.8°C in normoxia, to 1.9°C in hypoxia (Tattersall and Boutilier

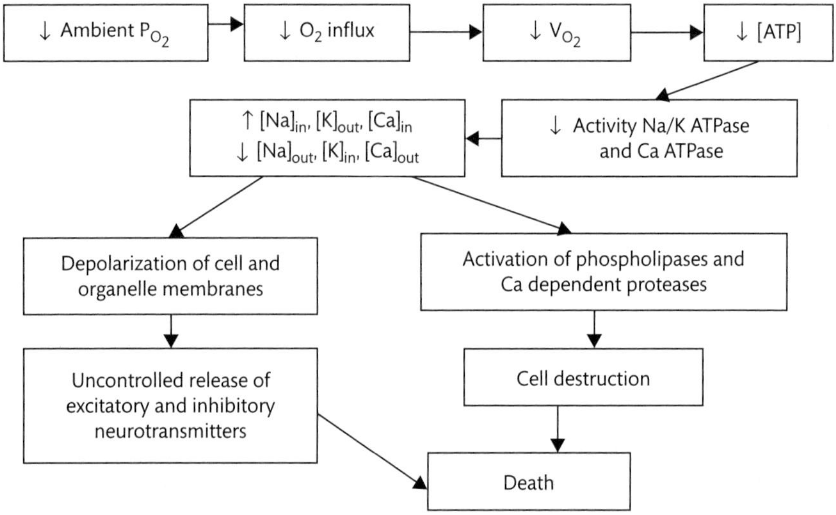

Fig. 4.10 A summary of the negative physiological consequences of hypoxic submergence in amphibians (see text).

1997). This reduces metabolism by 80% and postpones the shift from aerobic to anaerobic metabolism. The European common frog (*R. temporaria*) also selects a higher P_{O_2} in an oxygen gradient when submerged (Tattersall and Boutilier 1999). These results imply that submerged amphibians can behaviorally exploit heterogeneous oxygen environments, as well as lowering metabolism (demand for O_2), by behaviorally selecting lower temperatures in hypoxic environments. Behavior is the first key to surviving the winter for submerged amphibians, and from a physiological perspective this requires coordinated movement at low temperatures. Basically, the central nervous system and skeletal muscles have to work at low temperatures in a winter pond for the animal to exploit the thermal and oxygen heterogeneity of its environment.

The physiology of aquatic hibernation is equally fascinating. The cardiorespiratory implications of aquatic submergence need to be considered first in order to set the stage for understanding the tissue P_{O_2} conditions. First, skin is the only gas exchange surface, and the aquatic environment boundary layers can significantly reduce the cutaneous O_2 conductance in still water. Normally the skin only receives about 10–20% of the pulmocutaneous blood flow and its venous drainage (the blood with the highest P_{O_2}) returns and is mixed with systemic venous drainage. Since there is no pulmonary gas exchange, mixing within the heart essentially becomes irrelevant. Complete mixing of right and left side blood occurs in *Bufo marinus* at low temperature and resulting low heart rate (Hedrick *et al.* 1999). Systemic and cutaneous P_{O_2}s are basically equal. Cutaneous blood flow with low

P_{O_2} would maximize cutaneous O_2 diffusional uptake. This would be especially important in hypoxic aquatic situations where the blood P_{O_2} has to be below the water P_{O_2} for O_2 to diffuse into the animal. Consequently, the maximal P_{O_2} of the systemic blood will always be lower than the P_{O_2} of the aquatic environment. We might expect blood P_{O_2} to be low and cutaneous blood flow to be high during aquatic submergence to increase the rate of uptake from the water. The problem is that the blood supplied to the skin is the same P_{O_2} as systemic blood; a low blood P_{O_2} and high flow helps on the uptake side but compromises the systemic O_2 delivery side. Consequently, a left to right shunt is not necessarily favorable. We saw in Chapter 2 that systemic arterial P_{O_2}s are normally in a range of 8.7–12 kPa, leading to hemoglobin saturations of 85–95%. Measurements of blood P_{O_2} from normoxic submerged frogs range from 0.4 to 1.2 kPa (Ultsch *et al.* 2004; Stewart *et al.* 2004). Given normal blood O_2 loading curves this compromises O_2 carrying capacity, that is 5–10% hemoglobin saturations at these P_{O_2}s. Couple this to low blood flow rates resulting from the low heart rates at low temperature, and systemic O_2 transport is severely limited. Submergence consequently severely limits aerobic metabolic capacity, even in normoxic waters.

As we will see below, aerobic metabolism is very low during hibernation, but how much aerobic scope is there during submergence? We can estimate this from the data of Ultsch *et al.* (2004) who determined the critical P_{O_2} (P_c) for resting metabolism of submerged *R. pipiens*. The V_{O_2} at P_c was 0.05 nmol/g sec, the P_{aO_2} at this point was 1.7 kPa and using the limiting mitochondrial P_{O_2} of 0.1 kPa (St-Pierre *et al.* 2000), the tissue diffusing capacity equals $V_{O_2}/\Delta P_{O_2}$ or 0.05/(0.17–0.1) = 0.7 nmol/sec g kPa within the error of our determination in Chapter 2 of 0.9 nmol/sec g kPa. Consequently, at rest in normoxic water the factorial scope for further metabolic activity is dictated by the difference in P_{aO_2} in fully saturated water and at P_c divided by the P_{aO_2} at P_c ((0.4 − 0.17)/0.17 = 1.4, data taken from Ultsch *et al.* 2004). When this is compared to factorial increments of 6–10 for amphibians with all modes for respiratory exchange available, it becomes clear that hibernating *R. pipiens* are operating close to the limits of aerobic metabolic capacity at rest during hypoxic exposure. The only way to provide a measure of metabolic capacity from both an absolute and factorial perspective is to depress V_{O_2}, which is what we will delineate in the sections below.

The metabolic demand for O_2 decreases during hibernation in two ways, a Q_{10} driven decrease in metabolism and an hypoxic induction of a hypometabolic state. The Q_{10} driven decrease in metabolism is the result of the processes delineated in Fig. 4.11.

There are two things to note in this scheme: there will be a reduction in metabolism and this reduction is matched by compromised O_2 flux in terms of decreased blood flow rates, decreased O_2 loading of hemoglobin and decreased respiratory O_2 conductance. Consequently, there is concordance between the decreased rate of aerobic ATP synthesis and metabolic ATP utilization still maintaining

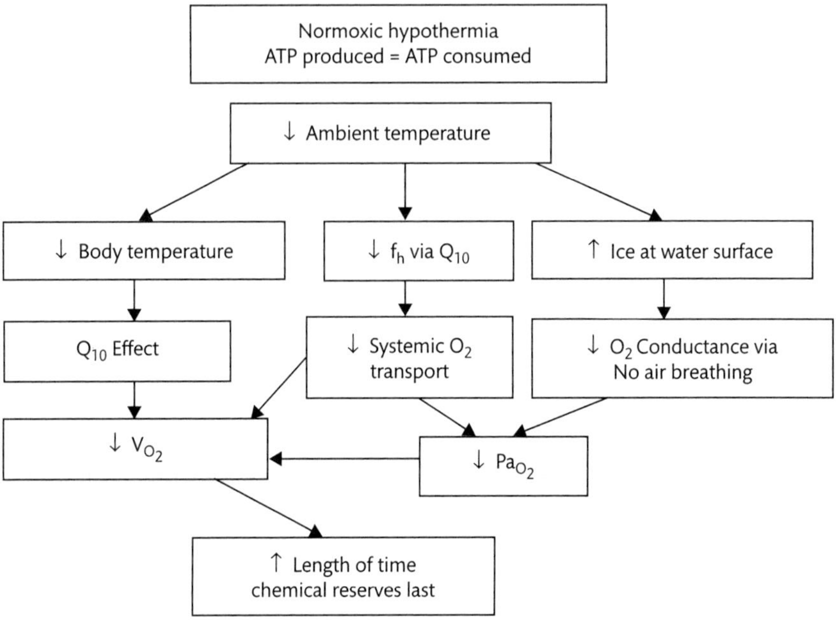

Fig. 4.11 A summary of the physiological consequences of hibernation at low temperature for amphibians (see text).

a homeostatic cellular energy balance. There is little capacity to increase aerobic ATP generation.

The second component of the physiology of aquatic hibernation is the induction of a hypometabolic state, below that normally associated with the Q_{10} effect on metabolism (Donohoe *et al.* 1998). Metabolic depression will increase the length of time energy reserves will last as well as provide some hypoxic protection. Standard V_{O_2} declined by 40% after 3 months of submergence in 3°C aerated water compared to standard V_{O_2} prior to submersion for *R. temporaria* at 3°C (Donohoe *et al.* 1998). This hypometabolism was intensified with a 75% reduction from standard V_{O_2} if the frogs were submerged in progressively hypoxic water from 20 to 4.5 kPa (Donohoe and Boutilier 1998). The 40 and 75% reductions in aerobic metabolism increases the length of time energy reserves will last by two and four fold. The substrates utilized to drive V_{O_2} during a period of submergence were a combination of fat and glucose at the start, to only glucose at the end, in both normoxic and hypoxic (4.5 kPa) submergence. The source of the glucose at the beginning of hibernation was liver glycogen stores, and both liver and skeletal muscle glycogen stores at the end of 90 days submergence (Donohoe and Boutilier 1998; Donohoe *et al.* 1998). Anaerobic metabolism was used in the first month of hypoxic submergence based on an increase in lactate concentration

(Donohoe and Boutilier 1998). The use of anaerobic metabolism during the first month of hypoxic submergence increased the rate of glycogen utilization in skeletal muscle, cardiac muscle and liver compared to normoxic submergence. The use of anaerobic metabolism with its accumulation of lactate and depletion of glycogen has been demonstrated in other amphibians exposed to anoxia (Rose *et al.* 1965; Christiansen and Penney 1973). Intracellular levels of ATP, phosphocreatine, AMP, and ADP did not change over time in either normoxic or hypoxic submergence, indicating that normal homeostatic intracellular energy status was maintained. The ability to reduce aerobic metabolic needs, and the utilization of anaerobic metabolism, at the same time systemic O_2 transport is compromised, allowed the energy status of the cells to be maintained.

How is this hypometabolism achieved? To understand hypometabolism one has to understand a chemical energy budget at a cellular level—essentially, how is the ATP partitioned at rest and what avenues of O_2 utilization can be minimized and still maintain cellular integrity? Unfortunately, there are insufficient data to discuss this solely for amphibians, though there is an extensive understanding for mammals (see Rolfe and Brown 1997; Section 4.5). Consequently we will assume the avenues of ATP expenditure are apportioned similarly between mammals and amphibians to start the discussion. The actual partitioning varies with organ; we will use skeletal muscle cells since they represent about 40% of the body mass in amphibians (Putnam 1979), and their V_{O_2} conforms to tissue P_{O_2} (Boutilier 2001). Consequently as the P_{O_2} declines during hypoxia their metabolic rate will decline, and this would make them likely candidates to contribute to the hypometabolic state. About 86% of the total O_2 consumed is by the mitochondria, the other 14% of O_2 being utilized by a variety of peroxisomal oxidases associated with fatty acid oxidation, endoplasmic reticulum oxidases, and some involved with the production of superoxide and hydrogen peroxide (the free radical problem). Of the mitochondrial O_2 consumption in skeletal muscle cells at rest, up to 50% of the resting O_2 consumption can be associated with a "proton leak" from the intermembrane space across the inner mitochondrial membrane back to the mitochondrial matrix (Rolfe and Brand 1996). The other 50% of mitochondrial O_2 consumption is coupled to the production of ATP. At rest in skeletal muscle, about 40% of the ATP generated is utilized by the Na-K-ATPases, Ca-ATPases and myosin ATPases, which maintain the concentration differences of these ions between the intracellular and extracellular spaces. The remaining 60% is used in the synthesis of protein, mRNA, glycogen, urea and everything else. Obviously, during skeletal muscle activity the fraction utilized by the ATPases increases dramatically to accommodate the action potentials, sarcoplasmic reticulum Ca fluxes which initiate the interaction of actin–myosin and the cycling of actin–myosin cross bridges during shortening.

The O_2 consumption of skeletal muscle cells associated with the "proton leak" is of particular interest during hibernation since it is a futile cycle from the perspective of generating useful energy for the cell in the form of ATP. This proton leak proceeds in the same direction as the flux through the mitochondrial F_1F_0-ATPase but is uncoupled from the synthesis of ATP (Rolfe and Brand 1996). This futile cycle of proton pumping into the intermembrane space and diffusion back to the matrix and not through the F_1F_0-ATPase creates a large inefficiency of the coupling between O_2 consumption and ATP synthesis. As Boutilier (2001) correctly cautioned, the production of 38 ATP per glucose is only true when mitochondria are operating maximally at P_{O_2} levels greater than 20 kPa, in the presence of saturating substrate and ADP concentrations (state 3 mitochondria). The reality is that these special conditions are very far from normal resting values of P_{O_2}, ADP and probably substrate concentrations in skeletal muscle (Chapter 2). In the frog, *Neobatrachus kunapalari*, skeletal muscle represents 35% of its body mass and accounts for 60% of its standard V_{O_2} and 85% of its minimal V_{O_2} (Flanigan and Guppy 1997). Obviously, if 50% of this V_{O_2} was involved with the ATP futile proton leak, it would be hard to sustain the rest of cell functions that require ATP, let alone maintain chemical energy reserves to extend survival time. Is the proton leak decreased during hypoxic exposure? The answer is yes, by reducing the rate of state 4 mitochondrial respiration. This is achieved by decreasing the rate of substrate oxidation by the electron transport system, which decreases the mitochondrial membrane potential but the inner mitochondrial membrane proton conductance does not change (St-Pierre *et al.* 2000). The F_1F_0-ATPase, normally the ATP synthase, will run backwards consuming ATP, acting as an ATPase to pump H^+ from the matrix to the intermembrane mitochondrial space if O_2 is in limited supply, exactly the mitochondrial situation during hypoxia. This backward flux would tend to reestablish the membrane potential or proton motive force in the mitochondria at the expense of consuming potentially limiting intracellular ATP concentrations (St-Pierre *et al.* 2000b; Boutilier and St-Pierre 2000). During hypoxia the activity of the mitochondrial F_1F_0-ATPase from amphibian skeletal muscle is inhibited. The inhibition of the F_1F_0-ATPase also prevents its action as an ATPase during anoxic conditions. This allows the use of ATP generated by the mitochondria and anaerobic metabolism for other cellular functions in lieu of maintaining the membrane potential across the inner mitochondrial membrane. Inhibition of the F_1F_0-ATPase also decreases the proton leak by decreasing the membrane potential which drives the proton leak (St-Pierre *et al.* 2000a, 2000b; Boutilier and St-Pierre 2002). Membrane proton conductance does not appear to change with hypometabolism.

Amphibian mitochondria change upon hypoxic exposure. The hypoxia-induced hypometabolism in amphibians is reflected as reductions in the rates of state 3 (maximal) and state 4 (resting) rates of V_{O_2} of isolated mitochondria (St-Pierre *et al.* 2000c). This is a consequence of a reduction in the activity of the electron transport chain (St-Pierre *et al.* 2000a; St-Pierre and Boutilier 2001). This reduction in mitochondrial resting and maximal rates of V_{O_2} would lead to a decreased rate of aerobic ATP production by the mitochondria. This change in V_{O_2} did not occur in mitochondria isolated from frogs submerged in normoxic water. Hypoxia also induced a greater O_2 affinity of the mitochondria (St-Pierre *et al.* 2000c). This allows mitochondrial function at lower cellular P_{O_2}s, exactly the conditions associated with hypoxia. The stresses and rates of hypoxic hypometabolism are reflected in the intrinsic aerobic characteristics of the mitochondria, a reduced rate of metabolism and ability to function at reduced P_{O_2}.

Hypoxic submergence also decreases the enzyme activity in both the tricarboxylic acid (TCA) cycle and glycolysis (St-Pierre and Boutilier 2001). The activity of citrate synthase, a key regulatory enzyme of the TCA cycle, is decreased following hypoxic submergence. So not only are the activities of mitochondrial enzymes of the electron transport change decreased (St-Pierre *et al.* 2000), but also a regulatory enzyme of TCA cycle which supplies substrate to the electron transport chain has decreased activity with hypoxia. This will slow the rate at which ATP is generated aerobically. The activity of lactate dehydrogenase was also significantly lower in frog skeletal muscle during hypoxic submergence (St-Pierre and Boutilier 2001), which will reduce the rate of ATP synthesis via anaerobic metabolism and will limit the synthesis of its by-product lactic acid. The decreased lactate dehydrogenase activity would favor oxidative utilization of the substrate pyruvate where the yield of ATP/glucose molecule is obviously greater. The increased efficiency would slow the rate of depletion of glycogen stores and limit the accumulation of a metabolic acid. Amphibians have limited buffering capacity to deal with metabolic acids in comparison to reptilian hypoxic hibernators (Stewart *et al.* 2004).

Another source of metabolic savings during aquatic hibernation is a decrease in the ATP spent maintaining the concentration differences of Na, K and Ca between the intracellular and extracellular compartments. The electrical potential across the cell membrane is a consequence of the diffusional efflux of K from cells and a smaller influx of Na. The diffusional flux, creating the membrane potential, is based on both the conductance for these ions and the concentration differences for these ions across the cell membrane. The maintenance of homeostatic sodium and potassium concentrations is dependent upon the Na-K-ATPase and the energy required is provided by oxidative metabolism (Sick and Kreisman 1981). Ions would normally not move through the non-ionic lipid bilayer that makes up the cell membrane because they are charged. The conductance, or permeability, for the ions is determined by protein channels specific for each ion. The concentration differences are established and maintained by the Na-K ATPase for Na and K, and by Ca-ATPase for Ca. If the diffusional flux of these ions is decreased, a decreased expenditure of ATP will be needed to move the ions back against their concentration differences and metabolism can be reduced. There are two potential avenues for decreasing the diffusional flux of these ions during aquatic hibernation and a developing hypoxia—decreasing the concentration differences and/or decreasing the conductance. There is strong evidence that both mechanisms are involved with the development of hypometabolic states as organs are exposed to hypoxic conditions. The first is decreasing the concentration difference of Na and K. It is clear from early studies on amphibians that simple exposure of amphibians to low temperature led to a decrease in extracellular Na concentration, principally the result of osmotic water uptake that was not excreted by the kidneys (Jorgensen 1950; Jorgensen *et al.* 1978; Miller *et al.* 1968). It was also established that lowered temperature also leads to the loss of K from skeletal muscle (Miller *et al.* 1968), but the metabolic significance of these effects was not discussed. The concentration difference across the cell membrane for both Na and K will be decreased, which will decrease the rate of diffusional flux of each of these ions, decreasing the need for ATP expenditure to pump the diffusing ions back against their concentration difference. More detailed, recent work found essentially the same result for Na and K in skeletal muscle during both normoxic and hypoxic (8 kPa) submersion (Donohoe *et al.* 2000). Interestingly, cardiac muscle maintained its intracellular K concentration, implying that maintenance of the K concentration difference might be important for muscle that still must maintain the same activity to convectively move the blood and its limited O_2 content. The other potential mechanism to decrease the normal diffusional flux of these ions is to decrease the number of ion channels present or open in the membrane for these ions. This is referred to as channel suppression or channel arrest. Donohoe and his colleagues present evidence that both K and Na conductance as well as Na-K-ATPase activity is decreased following submersion in hypoxic conditions (Donohoe *et al.* 2000). It seems clear from these data that a contributing element of the hypometabolism of skeletal muscle associated with normoxic and hypoxic submergence results from a decreased expenditure of ATP via the Na-K-ATPases responsible for maintaining the ion concentration differences between the extracellular and intracellular compartments of skeletal muscles.

A summary diagram of the array of cellular and biochemical changes associated with hypoxic submergence in amphibian skeletal muscle that are involved in both the hypometabolic response and allow extended life at low temperature is presented in Fig. 4.12.

The effects of hypoxia and anoxia on organ system function extend beyond relating them to environmental physiology but also their use as experimental models for understanding stroke and ischemia in organ function. Obviously, during a stroke the brain becomes hypoxic, just as regions of cardiac muscle becomes hypoxic/ anoxic following coronary artery blockage. Both processes are major sources of mortality for humans, and the organ stress and mechanism of injury are the same for frogs exposed to hypoxic submergence. The advantage that ectothermic organ system models present to biomedical researchers is that their tissues rate of metabolism is very much lower, hence the processes of dysfunction are spread over a longer period of time. This allows a more detailed analysis of the

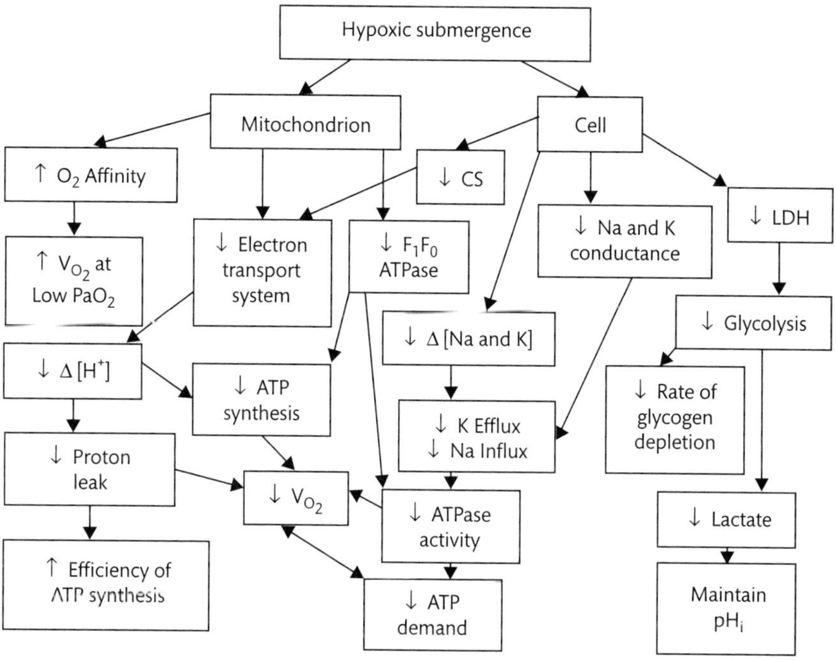

Fig. 4.12 A summary of the cellular and biochemical adaptations of amphibian skeletal muscle with hypoxic submergence at low temperature (see text).

mechanisms involved in compensation. It is referred to as "slow death". The two most popular models are amphibian brains and their hearts. These experiments are generally run at room temperature so their application to understanding the processes associated with hypoxic submergence at 3°C is problematic.

The amphibian heart has a remarkable capacity to maintain function and recover from hypoxic exposure. Isolated cardiomyoctes at 3°C from hypoxic acclimatized frogs have a much lower ATP turnover when exposed to hypoxia, indicative of a profound, yet reversible, metabolic depression compared to cardiomyoctes from control frogs (Currie and Boutilier 2001). Isolated toad hearts, perfused at 21°C with hypoxic buffer (<2 kPa) for 30 min at physiological filling and diastolic afterload pressures, ceased generating physiological outputs, but upon reoygenation 50–90% of their function returned (McKean *et al.* 1997). Interestingly, when exposure to hypoxia occurred under conditions of lowered preload and afterload the hearts continued to maintain a cardiac output and actually showed enhanced outputs upon reoxygenation. The data indicate that toad hearts will continue to maintain cardiac output under conditions of lowered demand and they recover from hypoxia quite well (McKean *et al.* 1997). The arterial pressure of bullfrogs declines with temperature (Rocha and Branco 1997,

1998). Consequently, it may be reasonable to infer the normal cardiovascular conditions for hypoxic submergence would entail a lowered diastolic afterload on the heart, exactly the conditions the *in vitro* heart maintained cardiac output. Just as we saw with skeletal muscle, hypometabolism is a key to balancing reduced O_2 availability to support some cardiac output during hypoxic exposure. Future work looking at the regulation of preload and afterload during hypoxic submergence at low temperature would certainly prove interesting as they relate directly to controlling the work, hence metabolic need, of the heart.

The amphibian brain also can function and recover following over 30 h of anoxia at 5°C (Hermes-Lima and Storey 1996). The amphibian brain is dependent upon the Na-K-ATPase to maintain the gradients of Na and K by oxidative metabolism (Sick and Kreisman 1981). The amphibian brain can withstand exposure to anoxia at 25°C for 3–5 h (Lutz and Reiners 1997; Knickerbocker and Lutz 2001) before significant quantities of K starts to escape and Ca starts to enter. It appears that even with significant Ca entry, over 70% of brain cells survived when measured 12–16 h post oxygenation (Hedrick *et al.* 2005). This suggests that the mammalian brain model, which ascribes the increase in intracellular Ca as activating irreversible proteolytic activity, may not apply to amphibians. The mechanism for these differences is unclear. Anoxia decreases cellular ATP levels and induces a loss of intracellular K (Knickerbocker and Lutz 2001). This is followed by a sharp increase in K loss with the release of the neurotransmitters dopamine, aspartate, taurine and GABA, and a decrease in glutamate (Lutz and Reiners 1997; Milton *et al.* 2003). The decrease in glutamate with anoxia is completely the opposite of the mammalian brain response to anoxia and has been implicated in the greater hypoxia tolerance of frog brains in comparison to mammalian brains (Lutz and Reiners 1997). The increase in the excitotoxic transmitter dopamine separates the frog brain from the more anoxia-tolerant turtle brain (Milton *et al.* 2003). Increased perfusion of the frog brain during anoxia, in lieu of other organs, does not seem to contribute to the maintenance of central nervous system function during anoxia (Soderstrom-Lauritzen *et al.* 2001). The amphibian brain demonstrates a unique pattern of responses in comparison to more anoxia tolerant reptiles and less anoxia tolerant mammals. The brain is quite anoxia tolerant and seems to recover from short-term anoxia quite well. There are no data on functional correlates of nervous system function during and following hypoxia.

A problem that has not been addressed in aquatic hibernators under hypoxic, low temperature conditions is how kidneys continue to excrete water. Obviously, osmotic water entry will continue to occur under aquatic hypoxic conditions. The maintenance of kidney function is critical to prevent ion loss while maintaining water excretion, or the osmotic concentration will continue to become hypo-osmotic. The maintenance of filtration pressures and ion reabsorption under these hypoxic conditions is critical to maintaining a homeostatic osmotic internal environment and is essentially unstudied.

4.3.2 Shallow Terrestrial Hibernation and Freezing

If an amphibian over-winters in a terrestrial environment, the threat of hypoxia is gone, but the threat of freezing and dehydration "loom" on the osmotically hostile and more thermally extreme terrestrial "horizon". Does winter also operate as a selective force for shallow terrestrial hibernators like hypoxic winterkill in aquatic ones? It is harder to visually discover evidence for death of amphibians burrowed under duff than in ponds. The only way to document the potential importance of temperature and dehydration as a source of mortality for terrestrial hibernators is via long-term studies of marked amphibian populations. By measuring winter survivorship over multiple years and then correlating rates of mortality to the severity of winter conditions, one can demonstrate the importance of winter weather to survivorship. A marvelous study of survivorship of both *Rana esculenta* and *R. lessonae* from Switzerland populations over 6 years provides unique insights into the evolutionary biology of terrestrial hibernation (Anholt *et al.* 2003). Both species are shallow terrestrial hibernators (Berger 1982). Over the 6-year study, winter survivorship varied from 10 to 80% for both species. The first principle component, accounting for about 60% of the variation in survivorship, included winter minimum temperature, average low temperature and its standard deviation. Other measures of average high temperature, average mean temperatures and their standard deviation loaded in a negative direction. A composite principal component measure of weather variables found that 90% of the variation in over-winter survival was accounted for in both species. Survival was lowest in winters with low minimum temperatures that were coupled to high and variable maximum temperatures (Anholt *et al.* 2003). The lowest survival was the winter with the lowest minimum low temperature (12.9°C), implicating freezing as a selective factor. Why should high and variable temperatures during winter also negatively impact survival? There are a variety of potential explanations. High and variable body temperatures will increase metabolic expenditure via a Q_{10} mechanism, which will increase the rate of expenditure of energetic reserves (see above). High and variable air temperatures will encourage movement as both species have been described as moving in the field even when the temperature is close to 0°C (Holenweg and Reyer 2000). Warm temperature may encourage frogs to change hibernation sites and become trapped by the subsequent colder weather in sites that are both less insulated and more dehydration prone. The data do indicate that minimum temperature and energy reserves may both be significant sources of variation in over winter survival in shallow terrestrial environments. It is clear that winter weather conditions can be a strong selective element in the biology of terrestrial hibernators.

Freezing

The physiological problems of shallow terrestrial hibernation principally derive from the stresses brought on by freezing. Data for an exposed site with little snow cover to insulate the soil from the extremes in low air temperature common in the continental environment of South Dakota are given in Fig. 4.13 (from Swanson and Graves 1995) and provides an example of the potential for freezing. Average temperature for many months is below freezing at shallow depths, but is ameliorated by soil depth. The minimum temperatures at most depths down to a meter could lead to a freezing event for an amphibian in any winter month; again the severity decreases with depth. These data emphasize that environmental conditions leading to freezing of amphibians are a real threat. The addition of significant snow accumulation can create a microenvironment much closer to 0°C by minimizing convective exchange with air temperatures lower than the snow and creating a microenvironment with more infrared radiative input than the winter sky.

The physical variables involved in freezing derive from the interaction of water and solute. As solute is added to a solution there is an increase in the osmotic concentration, a freezing point depression, a lowering of the vapor pressure and an increase in the boiling point (these are the colligative properties of a solution). Consequently, the temperature at which fluid will freeze is related to its osmotic concentration. A 1 osmolal solution will freeze at −1.86°C compared to 0°C for pure water. Given that amphibian body fluids are 0.25 osmolal, we would expect them to freeze at $0.25\ (-1.86) = -0.47$°C. If an amphibian were to sequester additional solutes it would freeze at a lower temperature than −0.47°C. Solute accumulation provides some resistance to freezing but this is not significant when put in the perspective of the lowered ambient air temperatures possible during winter.

The problems of freezing are multifaceted. Water in solution will preferentially freeze first. This will concentrate the unfrozen solutes as water is removed from the liquid state by freezing and create a hyperosmotic stress for cells adjacent to the ice as the solutes are left behind. These cells will lose water and shrink, leading to high ionic strength inside the cells. As water is frozen there is a hypovolemic stress, since there is less liquid fluid to pump through the circulatory system as well as in

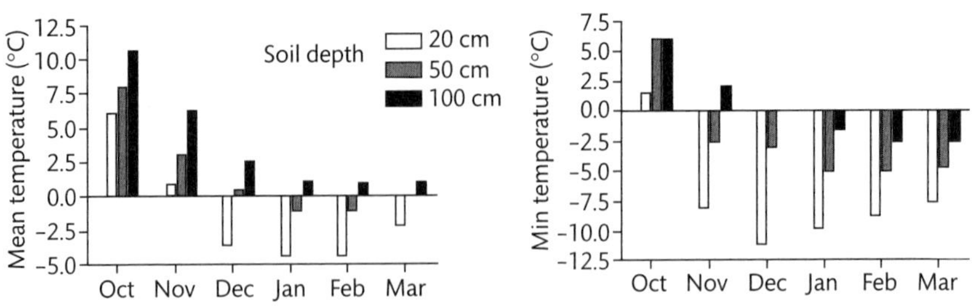

Fig. 4.13 A summary of (A) the mean and (B) minimum soil temperatures at various depths during the Fall and Winter in South Dakota (Swanson and Graves 1995).

increase in hematocrit and blood viscosity, as the lymphatic fluid and plasma form ice crystals. This is reminiscent of what we delineated for dehydrational stress, with the major difference being that only a very low metabolism to support life is needed at these low temperatures. A third problem is mechanical/ structural damage associated with formation of the ice crystals. If the ice forms inside cells, then cell membranes can be ruptured and the differences in concentration of solutes from the intracellular and extracellular spaces will be lost. If the ice crystals form in extracellular spaces then they can mechanically rupture the vasculature and organs. Upon thawing, the extracellular space will now be hypo-osmotic to the cells. This can also lead to cell rupture as a consequence of the swelling resulting from water entering from the diluted extracellular fluid resulting from ice melting. Couple these issues to evaporative water loss in the terrestrial environment and you also introduce the additional physiological problems of dehydration that we discussed in Chapter 3. Imagine a partially frozen amphibian that is beginning to freezer burn as a consequence of the evaporative water loss, and it can be easily seen why freezing has been long thought to be a terminal process in vertebrates.

Schmid (1982) delineated an amazing capacity of certain amphibians to tolerate freezing. Three species of terrestrial, hibernating northerly latitude frogs (*Rana sylvatica*, *Hyla versicolor* and *H. crucifer*) were capable of recovering from being frozen for 5–7 days at −6°C, environmental conditions that led to the death of two other northerly latitude species (*R. pipiens*, *R. septentrionalis*) that normally over-winter in water. He found that at least one species (*H. versicolor*) accumulated high concentrations of glycerol and inferred that this contributed to freeze tolerance (Schmid 1982). This observation set in motion a variety of subsequent studies concerning the physiology of and mechanisms involved in freeze tolerance in amphibians. Not only is the information of interest to environmental physiologists, but it may also delineate mechanisms that could be important in the cryopreservation of vertebrate organs for future use, a process of significant medical interest but currently not possible.

The pattern of an amphibian's body temperature during a freezing event (Fig. 4.14) can be described as an initial decrease in body temperature, the rate of which is determined by a combination of the local thermal environmental, the surface area and body mass of the animal. The nominal freezing should take place at the temperature equivalent to the osmotic concentration of the animal. In this example, and for amphibians in general, the actual freezing occurs at a lower temperature than is predicated from the colligative properties of its body fluids. This additional decrease in temperature before freezing occurs is known as supercooling. Eventually some of the animal's body water begins to freeze, and with this phase transition from liquid to solid, heat energy is released. This temperature is known as the crystallization temperature (when ice begins to crystallize). The heat energy released raises the body temperature of the animal, and this temperature increase is identified as the start of the exotherm. As more and more

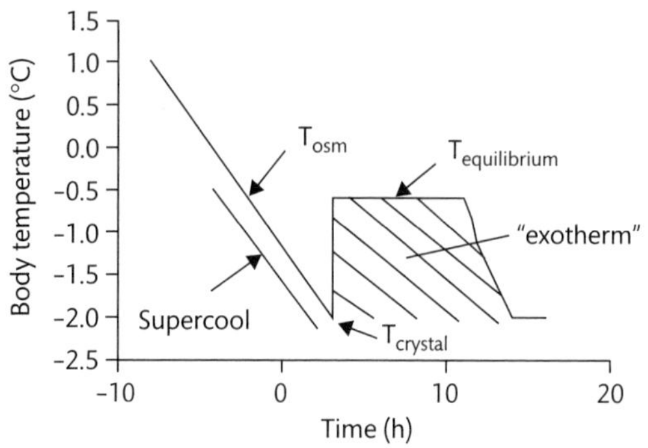

Fig. 4.14 A schematic of an amphibian's body temperature during a freezing episode; see text for explanation of the terms.

water is frozen there is continued release of heat and a continued elevated body temperature. Eventually, an equilibrium temperature is achieved when the water has stopped freezing. The integral of the exotherm represents a rough estimate of the total amount of body water that has frozen.

The cryopreservation literature gives some insight into the environmental and physical variables that increase cell survival from a freezing episode. As we saw above, osmotic volume stresses are placed on the cell both during the freezing phase (decrease in cell volume) and the thawing phase (increase in cell volume). This translates into mechanical stresses which can rupture the cell membrane (Pegg and Diaper 1988; Muldrew and McGann 1994). During freezing the concentration of the solutes increases as water is removed to form ice. The stress to red cell function was inferred to be disruption of protein structure and function as a consequence of the increased ionic strength of the concentrating solutes (Lovelock 1953a). The addition of glycerol ameliorated the effect of freezing damage, and the mechanism for this protection was interpreted as a colligative dilution of the ions (Lovelock 1953b). An alternative mechanism was proposed (Mazur *et al.* 1981) in which aqueous intercellular channels grow narrower as a consequence of being frozen, leaving less space for the cells and subjecting them to shear and compression forces. This is reflected functionally as the greater the fraction of unfrozen water the greater is cell survival (Mazur *et al.* 1981). Recent evidence seems to indicate that osmotic injury leads to cell membrane rupture which allows propagation of extracellular ice crystals to propagate into the intracellular space (Muldrew and McGann 1990, 1994). What is clear from empirical evidence is that freezing of the intracellular space compromises cell survival. The key to enhancing survival is to have freezing occur at a rate that allows the intracellular and extracellular environments to achieve an osmotic equilibrium as water is being removed during the formation of ice in the extracellular fluid. This obviously depends on the rate of water and solute flux across the cell membrane. If the rate of cooling is too fast, then osmotic equilibrium will not be achieved, the likelihood of membrane

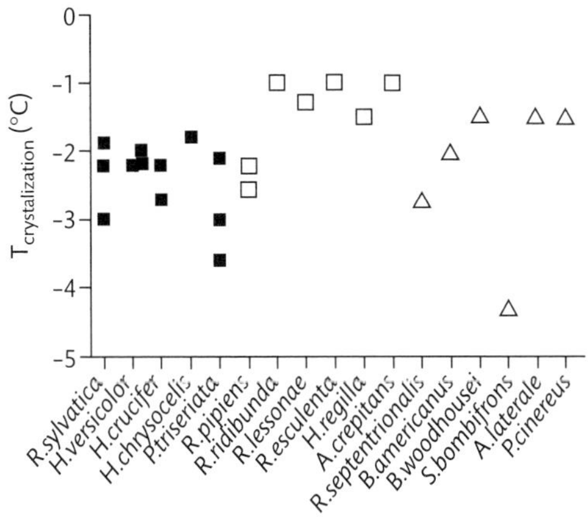

Fig. 4.15 The crystallization temperature when ice forms for a variety of amphibians. The filled squares are species with well developed freeze tolerance, the unfilled squares are species that will survive a short freezing exposure and the triangles are species that have not been demonstrated to withstand any freezing. Data from Schmid 1982; MacArthur and Dandy 1982; Storey and Storey 1986; Layne and Lee 1989; Constanzo *et al.* 1992; Layne 1995; Swanson and Graves 1995; Churchill and Storey 1996; Swanson *et al.* 1996; Irwin *et al.* 1999; Croes and Thomas 2000; Voituron *et al.* 2003; Voituron *et al.* 2005.

damage increases, and the probability of intracellular ice formation increases. At any non-equilibrium freezing rate, the degree of supercooling increases the probability of intracellular ice formation. Consequently, extensive supercooling then freezing increases the risk of intracellular ice formation and has a negative effect on cell survival.

Supercooling

The extent of supercooling and the tissue freezing (crystallization) temperature of amphibians varies from −1 to −4.3°C. It does not appear to differ between freeze tolerant and intolerant species (Fig. 4.15). The degree of supercooling is consistent with other vertebrates but less than for most insects (Lee and Constanzo 1998). Consequently, the extent of supercooling is modest and conveys limited protection to low environmental temperatures, but does not appear to contribute to the ability to tolerate freezing in amphibians.

The variation observed in crystallization temperature within and between species probably results from the experimental conditions under which the crystallization temperature was determined. Water will supercool but once an ice crystal begins to form it serves as a nucleation site to propagate further ice crystal formation. The initial formation of an ice crystal can result from spontaneous aggregation of water molecules or be initiated by another substance (Lee and Constanzo

1998). Amphibian plasma has agents that serve as ice-nucleating agents, which prevent extensive supercooling (Wolanczyk *et al.* 1990). The problem is that the crystallization temperature of the plasma and its ice-nucleating agents is around −8°C, obviously much lower than the experimentally determined T_C of all amphibians to date (Fig. 4.15). What then serves as the nucleator for ice crystal formation and prevents extensive supercooling down to the temperature for plasma ice-nucleation? There are a couple of nucleation site possibilities, one internal the other external. The intestinal tract and skin of amphibians contains bacteria, which nucleate ice crystal formation in supercooled water at −3°C (Lee *et al.* 1995). These are the two organs that supercool the least, and are closest to the whole animal crystallization temperature in comparison to the other organs of the body (Layne 1995). A possible external nucleation site is ice in the local environment with which the skin of the frog is in contact. To prevent dehydration, over-wintering amphibians will select moist substrates. Water in the interstices of soil and leaves has a lower osmotic concentration than an amphibian's body fluids, and hence freezes first. As frost forms on the surface of the hibernation site, these ice crystals could serve as nucleators across the permeable amphibian skin. External ice will trigger freezing in frogs at temperatures of −0.5 to −0.8°C, which is warmer than their normal supercooling temperature of −1.8°C (Layne 1991; Constanzo *et al.* 1999). Crystallization temperatures are consistently higher when determined on a wet substrate in comparison to a dry substrate (Swanson *et al.* 1996). It seems clear that the environmental conditions normally associated with a hibernation site can significantly raise the crystallization temperature; therefore the determination of crystallization temperature in a dry laboratory setting will have limited ecological relevance to an animal in a normally moist hibernation site.

The combination of a supercooling episode followed by a freezing episode can be lethal for amphibians. The western chorus frog (*Pseudacris triseriata*) shows 100% survival from freezing following supercooling down to −2°C, 60% survival from −2 to −3°C, 25% survival from −3 to −4°C and no survival if the freezing followed −4 to −6°C supercooling (Swanson *et al.* 1996). Clearly, natural selection should operate to favor behavioral selection of environments and/or physiological mechanisms that would raise crystallization temperatures and minimize the degree of supercooling for amphibians that tolerate freezing of their body fluids. The permeable integument of amphibians clearly is an advantage from the perspective of allowing external ice to serve as a nucleator and preventing extensive supercooling (Constanzo *et al.* 1999).

Pattern and Magnitude of Frozen Body Water

The sequence of ice propagation throughout the body has been followed during freezing and thawing with proton magnetic resonance imaging (MRI) of *Rana sylvatica* (Rubinsky *et al.* 1994). The first region to freeze was adjacent to the region of skin contacting the chamber, consistent with environmental nucleation. The progression of ice crystal formation was from the exterior to the interior, at first in the subcutaneous lymphatic sacs and pleuroperitoneal cavity. The extracellular fluids (lymph and interstitial) were the first to freeze, and the last to freeze were the major vessels and liver. Thawing was a more uniform process with the core organs melting first (Rubinsky *et al.*1994).

The pattern of organ dehydration during freezing for two freeze-tolerant species and four species with low freeze tolerance (Fig. 4.16), shows that the freeze tolerant species have greater water loss from the organs than freeze-intolerant species. This would suggest that the freeze tolerant species are more likely to equilibrate with the hyperosmolality induced by ice formation than the intolerant species. This loss of water and concentration of solute would decrease the likelihood of intracellular ice formation.

The amount of total body water that freezes seems to be an important aspect of surviving freezing. The fraction of body water frozen can be measured by calorimetry. The ice content is influenced by the plasma osmolality (Layne 1999), although the influence of osmolality is less than would be predicted from the colligative effects alone (Claussen and Constanzo 1990). Freeze tolerance in the three water frog species of the *Rana esculenta* complex is related to the ice content with freezing; above a threshold fraction of frozen body water (50–58%) there was no survival of freezing (Voituron *et al.* 2003, 2005). This is consistent of low

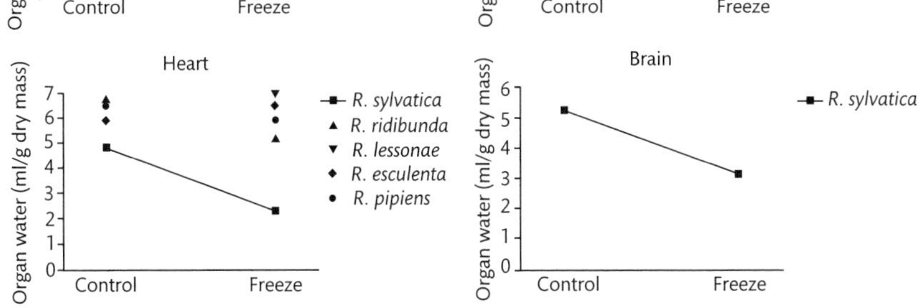

Fig. 4.16 The pattern of tissue water content upon freezing in two freeze-tolerant species (line connected) and four freeze-intolerant species: (A) muscle; (B) liver; (C) heart; (D) brain. Data are from Constanzo *et al.* 1993; Lee *et al.* 1992; Layne and Jones 2001; Irwin and Lee 2003; Voituron *et al.* 2003; Voituron *et al.* 2005.

fractions of frozen water (35–46%) in the treefrogs *Hyla versicolor* and *H. crucifer* (Schmid 1982; Storey and Storey 1985). However, the model study frog for freeze tolerance, *R. sylvatica*, seems to be able to tolerate a greater fraction (65%) of its body water being frozen (Layne and Lee 1987; Constanzo *et al.* 1993).

Cryoprotectants

A cryoprotectant is a chemical that enhances physiological function and/or increases survival during freezing. Cryoprotectants are usually divided into two broad categories, colligative and membrane stabilizers (see Storey 1997). Colligative cryoprotectants are those that reach high enough concentrations that they can significantly depress the freezing point and reduce the amount of ice formed during freezing. Membrane stabilizing cryoprotectants are generally in lower concentrations and prevent liquid–gel transitions of the cell membranes. The three chemicals invoked as being cryoprotectants in amphibians are glucose, glycerol and urea. All are important colligative cryoprotectants. Their role and the role of other possible membrane stabilizing cryoprotectants is essentially unstudied for amphibians.

Glucose is the most studied of amphibian cryoprotectants, especially for *R. sylvatica*. The freeze-tolerant *R. sylvatica* has six times the liver glycogen and 12 times the glycogen phosphylase activity compared to the non-freeze tolerant *R. pipiens* (Mommsen and Storey 1992). The synthesis of glucose has both a short and long-term regulatory component initiated at the start of freezing. At the initiation of ice formation an inactive form of the enzyme responsible for breaking down glycogen to form glucose-1-phosphate, glycogen phosphorylase a, is activated (Storey *et al.* 1996). At the same time there is a longer-term increased synthesis of glycogen phosphorylase enzymes (Storey and Storey 1988). This conversion to the active form (minutes) and upregulation of synthesis of phosphorylase (hours) leads to a sustained increase in the formation of glucose-1-phosphate. The glucose-1-phosphate is converted to glucose-6-phosphate, which has the potential to move down a variety of metabolic pathways. Inhibition of the glycolytic pathway is correlated with the inhibition of 6-phosphofructo-1-kinase and 6-phosphofructo-2-kinase (Storey 1987; Vasquez-Illanes and Storey 1993). There is also a decrease in activity of the enzyme glycogen synthetase, responsible for synthesis of glycogen from glucose (Russel and Storey 1995). This glucose mobilization process is not initiated until ice begins to form, so it is not activated in an anticipatory fashion, based on maintenance at low temperature (Storey and Storey 1985; Storey and Storey 1986a, b). How the ice starts this regulatory cascade is not understood but it seems to be mediated via the sympathetic nervous system since it can be stopped with β-blockers (Storey and Storey 1996). The mediation via the sympathetic nervous system is consistent with the more general amphibian stress response mediated by catecholamines released by the sympathetic nervous system. The glucose mobilized in the liver is distributed to the other organs of the body via the circulation (Storey 1987; Storey and Storey 1986). To escape the liver cells and enter the cells of other organs it has to traverse the cell membranes of the organs. This is achieved by an increase in the number and/or activity of membrane glucose transporters in the freeze tolerant *R. sylvatica* compared to the less tolerant *R. pipiens* (King *et al.* 1993, 1995). A summary of the various physiological and biochemical events associated with glucose mobilization during a freezing episode in *R. sylvatica* are delineated in Fig. 4.17.

The potential advantages of glucose as a "cryoprotectant" are numerous. Glycogen (a glucose polymer) is an important storage form of chemical energy, hence glucose is readily mobilized requiring no anabolic reactions. Glucose is the principal substrate that can be used to supply anaerobic metabolism in

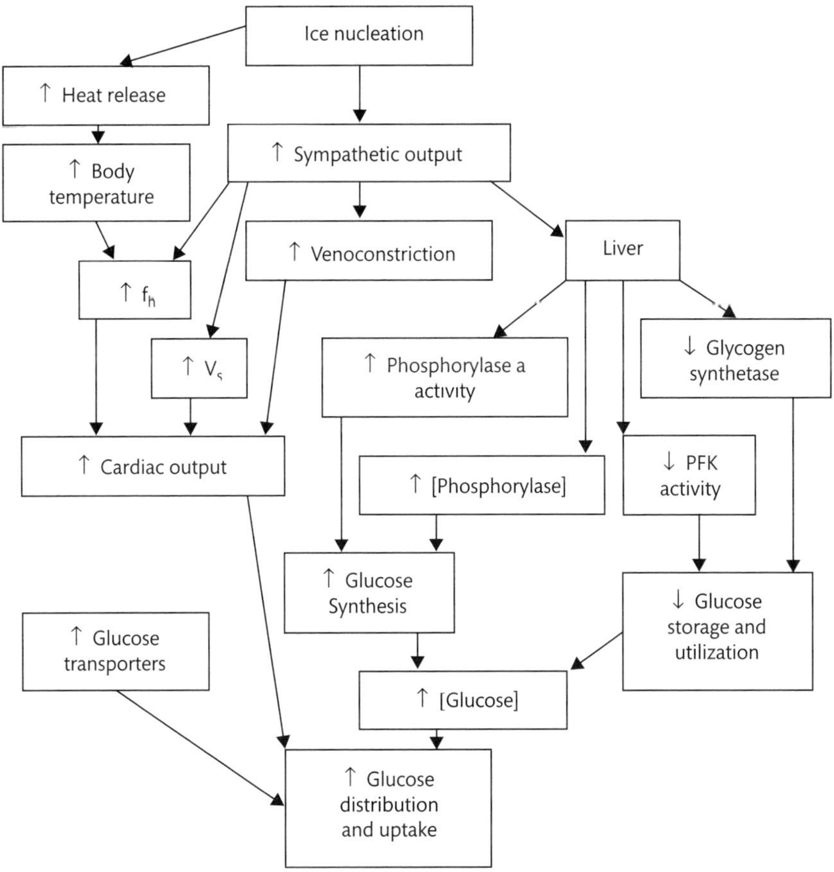

Fig. 4.17 A summary of the various physiological and biochemical events associated with glucose mobilization during a freezing event in *Rana sylvatica* (see text).

amphibians. As we will see below with the arrest of circulation upon freezing, the only way that O_2 can move in a frozen frog is by diffusion, hence aerobic metabolism is essentially stopped and cellular ATP needs have to be supplied anaerobically (Layne *et al.* 1989). Glucose will raise the osmotic concentration of the cell, limiting the decrease in cell volume associated with the removal of water from the intracellular fluid with ice formation. Finally there is some evidence that glucose can serve to stabilize cell and organ function during the increase in osmotic concentration associated with freezing. There is an inverse relationship between the severity of freezing injury and the quantity of exogenous glucose added in rapidly cooled frogs, a process where endogenous glucose mobilization is compromised (Constanzo *et al.* 1991). Exogenously administered glucose did not influence the degree of tissue water lost during freezing but did reduce the body ice content and red cell injury (Constanzo *et al.* 1993) as well as the length of time frogs could survive freezing (Layne *et al.* 1998). The magnitude of the hyperglycemia upon freezing in ranid frogs is correlated with the tolerance of freezing; *R. sylvatica* shows the greatest response and other ranids accumulate less than 20 mmol/l in their plasma. Exogenously administered glucose did not make a normally freeze intolerant species, *Bufo paracnemius*, cryoresistant (Steiner *et al.* 2000). It is clear that glucose is acting as a cryoprotectant from both a metabolic and colligative perspective.

Glycerol is another putative cryoprotectant produced by amphibians (Schmid 1982). The concentration of glycerol has been reported to increase in a variety of amphibians, but to the greatest degree and only significant colligative way in the gray treefrog, *Hyla versicolor* (Schmid 1982; Storey 1984; Storey and Storey 1985, 1986, 1987; Layne 1999; Irwin *et al.* 1999; Croes and Thomas 2000; Layne and Jones 2001). Glycerol concentrations of about 0.3 M have been measured for *H. versicolor* in both fall and late winter (Schmid 1982). The concentration increase upon freezing seems to be better developed in adult compared to immature *H. versicolor* (Storey and Storey 1985). The pattern of increase and removal suggests that it may be less mobile than glucose. Some data to suggest it is constitutively synthesized, as well as being stimulated during freezing (Schmid 1982; Irwin and Lee 2003; Zimmerman *et al.* 2007). The kidneys increase their fractional reabsorption of glycerol with cold acclimation (Zimmerman *et al.* 2007). There do not appear to be differences in glycerol synthetic capacity in colder climates for *H. versicolor* (Irwin and Lee 2003).

Urea has recently been described as a cryoprotectant for amphibians (Constanzo and Lee 2005). Amphibians commonly use urea to balance their internal osmotic concentration with that of the environment (see above). In this manner water can be osmotically taken up from the environment. The freeze-tolerant wood frog (*Rana sylvatica*) enclosed in outdoor enclosures accumulated up to 65 mmol/l urea in the fall, and accumulated up to 90 mmol/l in dry laboratory conditions (Constanzo and Lee 2005). During freezing, the concentrations of urea and glucose were similar, even though glucose was synthesized during the freezing, and urea was synthesized before the freeze. This suggested that the cryoprotectant properties of urea should be explored. Urea was comparable to glycerol, and as good or better than glucose, in preventing freeze–thaw damage to erythrocytes (Constanzo and Lee 2005). Urea is suited for a role in cryoprotection as it is the principal nitrogenous waste of terrestrial amphibians, equilibrates quickly between the intracellular and extracellular spaces, and is broadly used as an organic osmolyte during osmotic stress. The physiological role of urea as a solute to ameliorate osmotic stress in terrestrial hibernation sites has already been demonstrated in most amphibians studied to date. Given urea's osmotic role for amphibians, it is preadaptated to function as a colligative cryoprotectant. Urea may also be a contributor to metabolic depression (Constanzo and Lee 2005). The concentrations of urea, glycerol and glucose reported in the plasma of freeze-tolerant amphibians, *R. sylvatica* and *H. versicolor*, subjected to freezing during fall and winter are summarized in Fig. 4.18. The concentration of the principal cryoprotectant for each species is quite variable, probably reflective of synthetic rate variation, rate of freezing, glycogen storage and time of sampling.

Physiology of Freezing

The process of freezing has important implications for organ system function. The cardiovascular system plays a key role in the distribution of cryoprotectant synthesized and released from the liver in conjunction with its convective gas transport role. As we saw, with ice formation water is removed, thereby increasing the osmotic concentration of the remaining unfrozen extracellular fluid and creating a hyperosmotic environment for the cells. As ice forms in the extracellular space there is less lymph and plasma to fill the circulatory system, creating a hypovolemic stress on the cardiovascular system, which would decrease cardiac output via a Frank–Starling mechanism. As plasma volume is decreased, the red cell fraction (hematocrit) will increase, raising the blood viscosity. Combined with

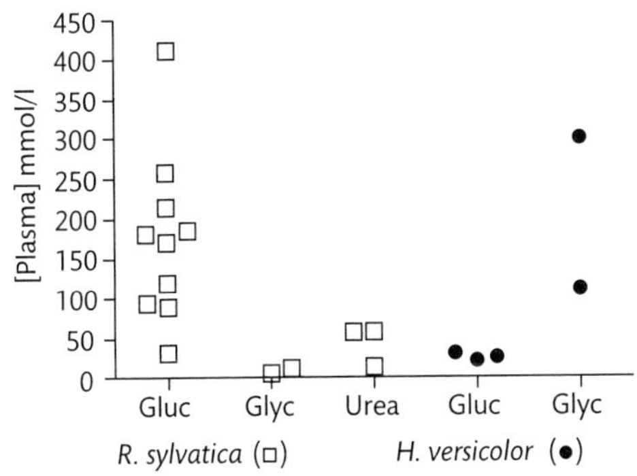

Fig. 4.18 The plasma concentration of glucose, glycogen and urea in *R. sylvatica* and *H. versicolor* during a freezing event. Data from summary of Constanzo and Lee 2005; Layne and Jones 2001; Storey and Storey 1985; Layne 1999.

the increase in viscosity generated by a decreased temperature and the presence of ice, the hyperviscous stress is magnified; the end result of these perturbations is a decrease in blood flow. Since glucose is not produced until ice has started to form, the cardiovascular system has to function under these hyperosmotic, hypovolemic and hyperviscous stresses. Couple this to a low heart rate of four beats per minute and the delivery of cryoprotectant to peripheral tissues will be a slow process. The release of the heat energy of fusion raising the body temperature as ice forms (exotherm) actually may assist the process of cryprotectant distribution via a Q_{10} contribution to an increase in heart rate to 8 beats per minute, doubling the blood flow as the body temperature increases from its supercooled point (Layne *et al.* 1989). Also contributing to the increased heart rate will be sympathetic stimulation. The synthesis of glucose is mediated via sympathetic stimulation (Storey and Storey 1996). Sympathetic stimulation would also increase heart rate, increase the strength of cardiac contraction and mobilize blood via venoconstriction. As freezing reaches its equilibrium point the heart stops beating (Layne *et al.* 1989). Cardiac contractility is the first physiological function to return upon thawing (Layne and First 1991) and is the least affected organ system by freeze duration (Layne *et al.* 1998). Interestingly, isolated frozen ventricular strips incubated in glucose from the freeze-tolerant *R. sylvatica* rapidly regain contractility after thawing (Canty *et al.* 1986). Ventricular strips from the mildly freeze-tolerant frog *R. pipiens* not only do no regain contractility after freezing but lose contractility when exposed to glucose equivalent in concentration to cryprotectant concentrations (Canty *et al.* 1986). The mechanism for this difference is currently not understood, but it suggests that the ability to function at low temperature and recover function upon thawing is more complicated than simply the synthesis and accumulation of cryoprotectant.

There are lower survival rates with rapid freezing in comparison to slow freezing (Constanzo *et al.* 1991). One of the dangers of a rapid freezing event is that it may not allow enough time for the synthesis and distribution of cryoprotectant at these low temperatures (Constanzo *et al.* 1992). Given a natural hibernation site with reasonable leaf litter and snow cover, the rate of cooling during a freezing event will probably be slow.

The metabolic effect of the freezing-induced circulatory arrest is that the cells have to operate anaerobically. Two things ameliorate the dangers of being forced to operate solely anaerobically: (1) metabolic rates are extremely low at these subzero temperatures; and (2) the principal substrate for anaerobic metabolism, glucose, has been distributed as a cryoprotectant so glucose exists in very high intracellular concentrations. Not all tissues accumulate glucose to the same degree: the organs accumulating the most are liver, heart, brain, kidney and leg muscles accumulate the least (Storey 1987). The energy charge within cells goes down during freezing and lactate begins to accumulate most dramatically in brain, heart and kidney (Storey and Storey 1984, 1985; Storey 1987). There is some evidence for alanine accumulation in leg muscles and skin, indicating that anaerobic breakdown of muscle amino acids (aspartate, glutamate and glutamine) during freezing may be more important than glucose breakdown in skeletal muscle during a freezing event (Storey 1987; Hemmings and Storey 2001).

The freeze-tolerant frog *R. sylvatica* has different responses of its neural and muscular function in comparison to other non-freeze tolerant amphibians. In particular, the structure and function of the Ca-ATPase from isolated skeletal muscle sarcoplasmic reticulum is temperature insensitive and has seven unique amino acid substitutions in comparison with a freeze intolerant congener (Dode *et al.* 2001; Hemmings and Storey 2001). This would allow normal rates of calcium cycling, and presumably muscle function at very low temperatures. The nervous system of *R. sylvatica* was also found to be less likely to show large, prolonged depolarization with accompanying uncontrolled nervous system activity in response to cold exposure when compared to freeze intolerant species (Dalo *et al.* 1995). The mechanisms of these adaptations remain to be fully delineated. It does suggest a suite of adaptations that would allow coordinated normal activity at low temperatures for *R. sylvatica*. It normally takes about 2 days post-thaw for *R. sylvatica* to return to normal behaviors and demonstrate normal behaviors (Constanzo *et al.* 1997; Layne *et al* 1998).

There has been much discussion of the similarities of freezing and dehydration tolerance. The corollaries are obvious in terms of hyperosmotic, hypovolemic and hyperviscous stresses, compromising cardiovascular function. Couple this to a stimulation of glucose mobilization that differs only in the magnitude of the increase in glucose concentration between dehydration and freeze stress, and the conclusion that freeze tolerance is an extension of adaptations involved in dehydration tolerance appears inevitable. Unfortunately, there are some strong distinctions in the behavior and osmotic response of organs between the two stresses. First, there are distinct differences in both the proteins synthesized and the rate of protein synthesis in response to either freezing or dehydration (Storey *et al.* 1997). Dehydration is a stress that can occur at any time of the year and is very rapid (hours to days). The key to escaping the environmental conditions leading to the negative water balance requires movement on the animal's part to find water. Amphibians increase their activity when they start to dehydrate, presumably trying to find water (Heatwole and Newby 1972; Putnam and Hillman 1977). This requires a mobilization of metabolic capacity, not the shutdown of metabolism one sees with surviving a freezing event during winter. Freeze tolerance is part of an environmentally predictable, long-term low energy and low temperature winter adaptation, rather than an acute episode that can occur in any season. The overall osmotic response of organs graphically demonstrates this dichotomy. If we compare the magnitude of water loss from muscle, heart, liver and brain during a freezing event and a dehydrational event we see completely different responses in the behavior of these organs. Freezing leads to substantial water losses from heart

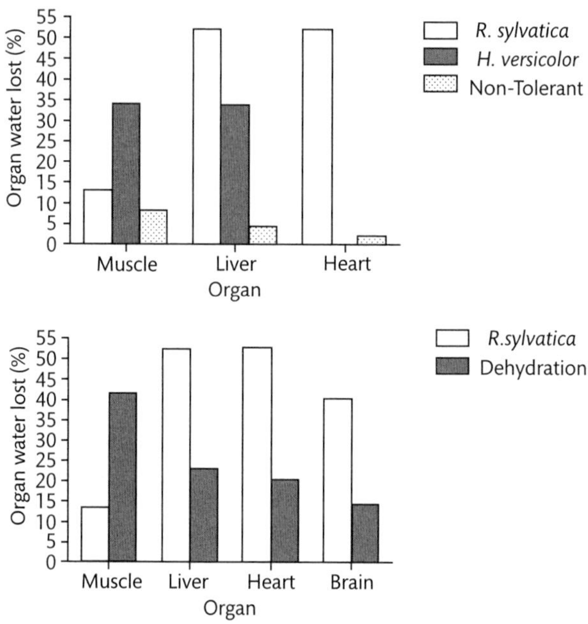

Fig. 4.19 (A) The amount of organ water lost in *R. sylvatica* and *H. versicolor* during a freezing event compared to freeze intolerant frogs and (B) a comparison of *R. sylvatica* to data from seven species of amphibians at their critical activity point following dehydration (data taken from Fig. 3.17).

liver and brain, but little lost from the skeletal muscle (Fig. 4.19). This decreases the likelihood of intracellular freezing occurring in the vital organs. Dehydration on the other hand results in a large amount of water being lost from the skeletal muscle (the largest organ system) and the least from the crucial organs (brain, liver and cardiac muscle) involved in a coordinated metabolic response to move and find water (Fig. 4.19). It is important to recognize not only the similarities but also the fundamental physiological differences between these stresses. As Swanson *et al.* (1996) point out, the anuran families with the greatest dehydration tolerance (Pelobatidae and Bufonidae) are not tolerant of freezing at all.

4.3.3 Deep Terrestrial Hibernation

The key to being a good deep, terrestrial hibernator is an ability stay below the frostline. The groups of amphibians that over-winter in this environment are the toads and many salamanders, which hibernate in deep hibernacula. There is strong evidence that toads will actively change depth in response to soil temperature changes (Breckenridge and Tester 1961; Tester and Breckenridge 1964a, b;

van Gelder *et al.* 1986), though not all stay below the frost line (Breckenridge and Tester 1961). Toads burrowed deeper to stay below the freezing point in a severe winter, while the toads were relatively inactive as the freezing point stayed well above their initial burrow depths in a milder winter. Salamanders will also respond to decreasing soil temperature by moving deeper (Vernberg 1953). There is some evidence that over-winter mortality is correlated with the depth of the soil frost line (Tester and Breckenridge 1964b; Kelleher and Tester 1969; Petranka 1979). Juvenile toads appear to be more susceptible to winter mortality, and this difference has been attributed to their hibernating at shallower depths, based on a reduced ability to dig compared to adults (Kelleher and Tester 1969). Populations of salamanders appear more vulnerable to winterkill in areas where their possible depth of soil penetration is limited during severe weather (Petranka 1979). Annual variation in the depth of the soil frost line is inversely related to the depth of snow cover (Tester and Breckenridge 1964b). Presumably the hypometabolic characteristics associated with aestivators are important for deep terrestrial hibernators as well (see above). Again, the environmental conditions of winter can act as a strong selective agent in deep terrestrial hibernators, just as for aquatic and shallow terrestrial hibernators.

4.4 Metabolic Depression

Metabolic depression is an adaptation whereby animals can survive with a metabolic rate below their normal minimal value (Hochachka and Guppy 1987; Storey 1988; Storey and Storey 1990; Guppy and Withers 1999; Withers and Cooper 2008). In ectotherms such as amphibians, metabolic depression is survival below the standard metabolic rate (SMR), which is the minimum resting rate of metabolism at any particular temperature (see Section 3.3). Metabolic depression occurs during periods of restricted resources, for example survival of short- or long-term anoxia (see Section 4.3), during extended periods of starvation such as hibernation (see Section 4.3) or aestivation (see Section 4.2). The adaptive significance of metabolic depression is to prolong survival until appropriate environmental conditions are restored. Metabolic depression has been reported in virtually all animal phyla, and the extent of depression can vary from moderate at about 60–80% of SMR, to commonly 20–25% of SMR, and in extreme examples to <1% of SMR (Guppy and Withers 1999). Metabolic depression in amphibians is generally to 20–25% of SMR.

Amphibians occur in a surprisingly wide range of habitats given their thin, permeable skin, high rates of evaporative water loss, poor tolerance of high temperatures, and susceptibility to freezing. Behavioral avoidance rather than physiological adaptation is often the first line of defense for amphibians against adverse environmental conditions. Dormancy, or inactivity, is a common response of amphibians to cold or desiccating conditions. It is sometimes termed hibernation or brumation in winter cold, and aestivation or retraherence in summer heat and dryness. Behavioral avoidance in particularly harsh environments is associated with long periods of inactivity interspersed with short periods of intense feeding and reproduction. For example, the annual temperature cycle of subpolar environments can restrict amphibians to a short spring/ summer activity period. In desert environments, activity may be restricted to short periods after rainfall, which in extreme cases is less frequent than every 12 months.

Metabolic depression to below standard metabolic rate is an important physiological adaptation for survival of such extended periods of inactivity, particularly in desert environments where the inactivity period can extend for a year or more, and ambient temperature is high so standard metabolic rate remains high. The adaptive significance of metabolic depression is self-evident. For every decrement in metabolic rate there is an equivalent (reciprocal) increment in survival time based on endogenous energy stores; a reduction to 90% of SMR results in a 111% survival time; reduction to 50% a 200% survival time; 20% of SMR a 500% survival time.

4.4.1 Aestivation and Hibernation

Many arid-zone amphibians burrow underground or seek refuge under bark or in tree hollows to escape the temperature and humidity extremes of an arid environment (Section 4.2.3). They remain dormant underground for extended periods of months to even years. The aestivation period is a well-defined physiological state of inactivity, especially for cocoon-forming species but also for non-cocooning species. For example, in natural conditions the water-holding frog *Cyclorana platycephala* can apparently survive cocooned underground for over 5 years if above-ground conditions are unfavorable, although an annual activity after heavy cyclonic rains is more typical (van Beurden 1980). For non-cocooning fossorial frogs such as *Scaphiopus* and *Notaden*, the period of survival underground is more likely less than annual, and even extended seasons of adverse environmental conditions might be interspersed with short periods of activity during favorable surface conditions (e.g. *Notaden*; Thompson *et al.* 2005). In the laboratory, *C. platycephala* can survive for over 18 months (Cartledge 2007), and 3 to 12 months is routine for many cocooning frogs and salamanders. Non-cocooning frogs survive in appropriate burrowed conditions for 3 to 6 months, for example *Scaphiopus* (Seymour 1973) and *Notaden* (McMaster 2007).

The steady-state metabolic rate of amphibians during aestivation is generally about 20–25% of the normal SMR (Withers 1993; Guppy and Withers 1999; Fig. 4.20). This level of depression is similar to that of aestivating lungfish (*Protopterus aethiopicus*; Delaney *et al.* 1974; Fishman *et al.* 1986), marbled swamp eel (*Synbranchus marmoratus*; Rosa 1977; Bicudo and Johansen 1979) and

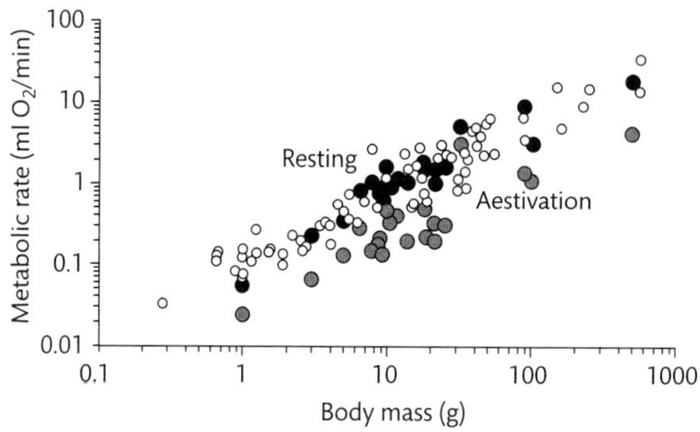

Fig. 4.20 Summary of resting metabolic rate (SMR; large black circles) and depressed metabolic rate during aestivation (large gray circles) for amphibians, with SMR for anuran amphibians (small open circles). Data from: Seymour (1973), Loveridge and Withers (1981), McClanahan *et al.* (1983), Etheridge (1990), Gatten *et al.* (1992), Withers (1993), Flanigan and Guppy (1997), Fuery *et al.* (1998), Flanigan *et al.* (1991), Withers and Thompson (2000), McMaster (2007).

salamander fish (*Lepidogalaxias salamandroides*; Pusey 1989), and some reptiles and invertebrates such as molluscs (Guppy and Withers 1999).

The metabolic rate of amphibians declines relatively rapidly from SMR to a depressed level, generally taking 2–4 weeks from the onset of aestivation to plateau at about 20–25% of SMR (Fig. 4.21). After 4 weeks or so, the metabolic rate is relatively constant for weeks to months. This pattern of metabolic decline during aestivation is very similar to that of lungfish (Delaney *et al.* 1974; Fishman *et al.* 1986) and the marbled swamp eel (Rosa 1977; Bicudo and Johansen 1979). In lungfish, the decline in metabolic rate with onset of aestivation is like that for the onset of starvation, so induction of metabolic depression seems to involve the same cues and mechanisms. In amphibians, starvation also depresses metabolism but apparently not so dramatically; starved *Scaphiopus* have about 50% of the normal metabolic rate (Seymour 1973a).

The adaptive value of metabolic depression during aestivation is to extend the survival time using endogenous energy stores. The primary energy store and hence metabolic fuel during aestivation is lipid, as indicated by the respiratory exchange ratio and depletion of body fat reserves during aestivation (Jones 1980a; Loveridge and Withers 1981; McClanahan 1967; Seymour 1973; van Beurden 1980). Lipid is stored primarily in the abdominal fat bodies, and sometimes axillary and inguinal fat pads, as well as in the liver and other organs. The fat bodies can comprise 3–5% of the body mass prior to aestivation, and are considerably reduced after aestivation. For the water-holding frog *Cyclorana platycephala*, both small and large individuals have about 5% fat body mass, which would sustain depressed metabolic rate for over a year (van Beurden 1980; Cartledge 2007).

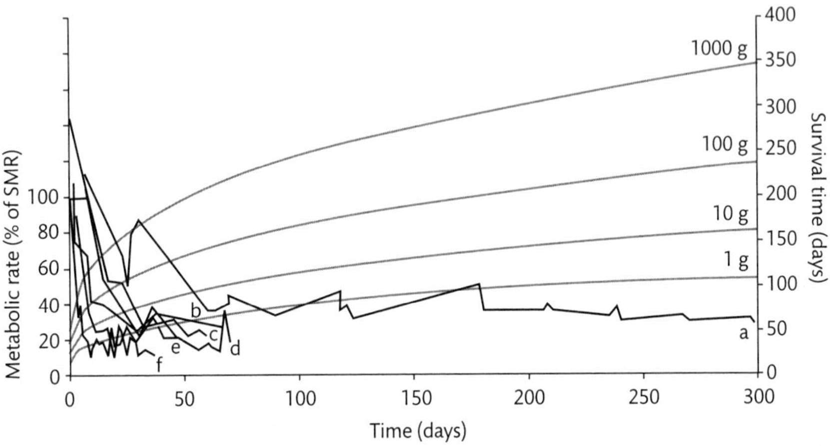

Fig. 4.21 Decline in metabolic rate from 100% of SMR during entry into metabolic depression for a salamander (a, *Siren intermedia*, Etheridge 1986) and anurans (b, *Neobatrachus centralis*, Withers 1993; c, *Scaphiopus couchii*, Seymour 1973; d, *Pyxicephalus adspersus*, Loveridge and Withers 1981; e, *Cyclorana maini*, Withers 1993; *Lepidobatrachus llanensis*; f, McClanahan *et al.* 1983). Approximate survival time based on 5% initial body mass as abdominal fat bodies and calculated from average metabolic rate during metabolic depression for these amphibians (assuming 38 kJ/g, 20.1 ml/kJ).

Loveridge and Withers (1981) estimated that 3% body fat stores would suffice for metabolic needs over 260 days of aestivation for the large (about 500 g) *Pyxicephalus adspersus*. For the smaller *Scaphiopus couchii*, lipid stores account for about 72% of their aestivation energy requirements and protein for about 23% (McClanahan 1967; Jones 1980). Large aestivating frogs are expected to have a longer survival time than smaller individuals/ species because of their lower mass-specific metabolic rate. van Beurden (1980) modeled energy store utilization of *Cyclorana platycephala* over extensive periods of aestivation. Some carbohydrate (e.g. liver glycogen) is used at the start of aestivation but the primary energy source during aestivation is the fat body, with more protein use towards the end of prolonged aestivation when the fat stores approach exhaustion. Survival is longer for large *C. platycephala* because of their lower mass-specific metabolic rate but equivalent mass-specific fat body stores (see Fig. 4.21).

Lipid metabolism during aestivation presumably follows the standard β-oxidation pathway of vertebrate cellular metabolism which requires additional input of C to prevent the depletion of TCA cycle intermediates. This additional C is generally provided by metabolism of pyruvate or phosphoenolpyruvate to oxaloacetate (Pinder *et al.* 1992). Although metabolism of lipid stores is often viewed as a concomitant "water store" because of the high metabolic water yield from fat metabolism (e.g. 146 moles H_2O from metabolism of 1 mole of palmitoyl-CoA; Pinder *et al.* 1992), the total water yield of fat metabolism is less than that of carbohydrate or protein metabolism because of its higher "bound" water content (see Chapter 2).

Protein metabolism provides some energy during aestivation, particularly towards the end of a long aestivation period (e.g. van Beurden 1980; Jones 1980) and results in the accumulation of often substantial concentrations of urea (see Section 4.2) particularly for cocoon-forming species that are sealed from the environment by their cocoon (except for gas exchange through their nares). Protein metabolism can even be accelerated during aestivation as the external soil water potential declines (McClanahan *et al.* 1972). During extended aestivation, urea accumulation can reflect 17–32% metabolism of body protein in *Scaphiopus* (McClanahan 1967; Jones 1980). This urea is apparently not accompanied by accumulation of counteracting solutes (Withers and Guppy 1996; see Section 4.2.2) and acts as a balancing osmolyte and ameliorates the adverse effects of high ionic concentration on enzyme function (Grundy and Storey 1994; Cowan and Storey 2002).

Behavior is the first key to surviving winter for amphibians (Section 4.3). For example, winter-acclimated *Rana temporaria* decreases its preferred temperature to 6.8°C in normoxia and 1.9°C in hypoxia (Tattersall and Boutilier 1997) compared to their normal thermal preference of about 29.6°C (Strübing 1954). Although hibernating amphibians are sometimes described as "torpid" (e.g. Holzapfel 1937; Willis *et al.* 1956; Pinder *et al.* 1992), it is not easy to determine whether this is simply a cold-induced lethargy or actually an altered metabolic state i.e. metabolic depression. Field and laboratory evidence suggests that overwintering frogs remain responsive and behavior is important in avoiding stresses such as hypoxia (see Tattersall and Boutilier 1997). Nevertheless, hibernating amphibians must cope with physiological effects of cold and starvation (Section 4.3). Similar strategies are used by both aquatic and terrestrial hibernators such as accumulation of lipid energy stores prior to hibernation, reliance on aerobic lipid metabolism (as in aestivating amphibians), and maintenance of cell membrane and protein function at low temperatures (Pinder *et al.* 1992).

In general, metabolic rate does not appear to decrease during hibernation other than from the standard physiological Q_{10} effect; the Q_{10} for biological processes such as metabolism is generally between about 2 and 3, and increases with decreasing temperature (Herbert and Jackson 1985; Ott *et al.* 1980; Withers 1992; Rome *et al.* 1992; see Section 3.5). The metabolic savings accruing from decreased temperature (e.g. from 22 to 2°C with Q_{10} decreasing from 4 to 2) is about 90% (Figure 4.22) which is similar to the savings accruing from metabolic depression during aestivation (about 80%) at a constant 22°C. Further metabolic depression after the thermal Q_{10} effect would provide a small, additional absolute energy savings but a substantial factorial reduction in MR, and hence an increase in survival time. However, the metabolic depression from decreased T_a and a Q_{10} effect results in >100 days survival time for even a 1-g frog, >200 days for a 10-g frog, and even longer survival times for larger frogs (Fig. 4.22). These survival

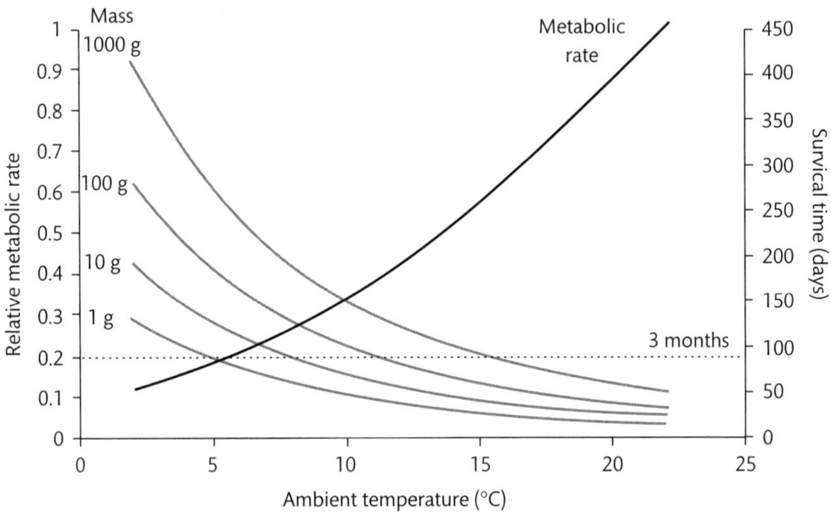

Fig. 4.22 Model of relative decline in metabolic rate of an amphibian with a decrease in body temperature from 22 to 2°C (assuming Q_{10} decreases linearly with T_a from 4 to 2; see text). Survival time is calculated from predicted standard metabolic rate (see Chapter 1.4) assuming 5% fat body stores and 38 kJ/g fat for frogs varying in mass from 1 to 1000 g.

times would seem more than adequate for survival of a cold winter period (e.g. 3 months), so further reduction in energy consumption by metabolic depression would appear unnecessary (and would result in relatively small absolute energy savings). This lack of necessity to further prolong survival time during hibernation is in contrast to the necessity for metabolic depression during aestivation, since survival times of even a year or more may be routinely required at often high ambient temperatures.

Nevertheless, the decrease in metabolic demand for O_2 during hibernation can be augmented by metabolic depression as well as an hypoxic induction of further hypometabolism beyond the Q_{10} effect (Donohoe *et al.* 1998; Section 4.3). For example, SMR of *R. temporaria* declined by 40% after 3 months of submergence at 3°C compared to prior to submersion at 3°C, and this hypometabolism was intensified with a 75% reduction from SMR if frogs were submerged in water made progressively hypoxic (Donohoe and Boutilier 1998). These 40 and 75% reductions in aerobic metabolism increase survival time by two and four fold.

4.4.2 Mechanisms of Metabolic Depression

It is not immediately clear how metabolic depression to below SMR is achieved. It could be caused by a reduced ATP supply that forces ATP-consuming processes to be reduced, or ATP-consuming processes could be turned down, causing a reduction in the demand for ATP. The mechanisms may or may not be similar for intrinsic metabolic depression during aestivation and hypoxia-induced metabolic depression. Metabolic depression that accompanies aestivation occurs in the absence of any obvious cell signals, that is there is no change in temperature, no obvious cause of cell hypoxia and no change in body fluid solute concentrations. In hibernation, the decrease in temperature reduces metabolic rate through a Q_{10} effect, and there may or may not be additional intrinsic metabolic depression. To understand possible mechanisms of metabolic depression, we first need to understand the processes that require O_2 and ATP.

The partitioning of O_2 consumption and ATP expenditure varies amongst organs, but skeletal muscle is a useful model for metabolic depression since it represents about 40% of the body mass in amphibians (Putnam 1979), and its metabolic rate conforms to tissue P_{O_2} (Boutilier 2001). In rat skeletal muscle, mitochondria consume most (about 86%) of the total O_2 used, with about 14% consumed by various peroxisomal oxidases associated with fatty acid oxidation, endoplasmic reticulum oxidases, and the production of superoxide and hydrogen peroxide free radicals (Rolfe and Brand 1996). About 52% of total O_2 consumption is associated with the "proton leak" across the inner mitochondrial membrane, and 34% is coupled to the production of ATP. With respect to ATP-consuming processes, about 40% of the ATP is utilized by Na-K-ATPases, Ca-ATPases and myosin ATPases to maintain ion concentration differences between the intracellular and extracellular spaces, and 60% is used in the synthesis of protein, mRNA, glycogen, urea and everything else. Skeletal muscle of resting *Neobatrachus centralis* has a low protein synthesis rate that is unaffected by aestivation (Fuery *et al.* 1998a). The liver has a different partitioning of O_2 and ATP with a larger synthetic role than skeletal muscle; protein synthesis is more than ten times that of skeletal muscle and accounts for about 38% of the resting cell O_2 consumption (Fuery *et al.* 1998a). In a whole rat, estimates suggest that SMR is partitioned into about 8% non-mitochondrial and 20% proton leak from the skeletal muscle and liver alone, with about 20% for ATPases, 20% for protein synthesis and 32% for "other" processes (Rolfe and Brand 1996; Fig. 4.23). Whether this partitioning in rats applies equally well to an ectothermic amphibian or not, it is clear that to achieve a metabolic depression to 20% of SMR there must be a substantial decrease in non-mitochondrial and mitochondrial O_2 consumption, the rate of ATP production and the extent of a proton leak, and that ATPase-dependent processes for both ion gradient maintenance and biosynthesis must be impacted. Adenylate energy charge is maintained during aestivation by *Neobatrachus kunapalari* (Flanigan and Guppy 1997), so ATP production has not been compromised relative to demand.

There is little scope for metabolic depression of the non-mitochondrial component of cellular O_2 consumption because of its small overall contribution to

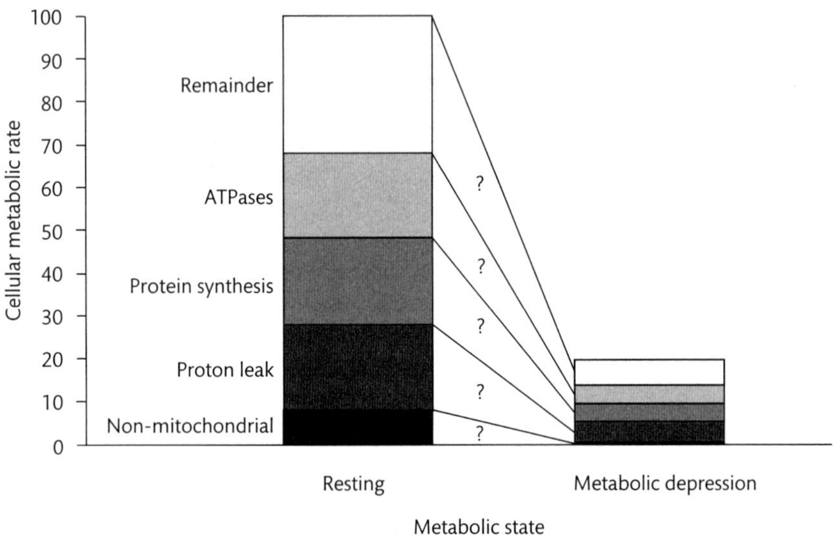

Fig. 4.23 Hypothetical partitioning of whole animal O_2 consumption into non-mitochondrial respiration and mitochondrial respiration, and mitochondrial respiration into the proton leak, ATPase activity, protein synthesis and "other" processes, at rest (based on data for rats; Rohlf and Brand 1996). The change in partitioning during metabolic depression is not known, but presumably involves reduction in all of the SMR components (see text).

cell metabolism. Nevertheless, there are decreases in lipid peroxidation and antioxidant defenses in aestivating spadefoot toads *Scaphiopus couchii* (Grundy and Storey 1998).

There are four general characteristics of metabolically-depressed cells, at the molecular level; a decrease in pH, latency of mRNA but reduced protein turnover, changed protein phosphorylation and maintenance of membrane ion gradients (Guppy and Withers 1999). Reduced intracellular pH is often associated with metabolic depression but whether it is necessary for metabolic depression is equivocal (Guppy *et al.* 1994). Decreased protein synthesis is at the level of translation, as mRNA levels are generally stable during metabolic depression. Liver protein synthesis varies from 12% of the O_2 consumption of liver in *Bufo marinus* to 38% in *Neobatrachus centralis* (Fuery *et al.* 1998a, b). Liver protein synthesis and O_2 consumption are reduced *in vitro* during aestivation by 55% and 67% respectively in *N. centralis*, but this only contributes about 5% of the metabolic depression of the whole frog (Fuery *et al.* 1998). No *in vitro* metabolic depression or decrease in protein synthesis was observed for skeletal muscle of *N. centralis*, and the low protein turnover of skeletal muscle precludes any significant contribution to overall metabolic depression. In contrast, in *Neobatrachus pelobatoides* there was a 70% decrease in skeletal muscle O_2 consumption during aestivation, which accounted for about 70% of the frog's metabolic depression (Flanigan *et al.* 1991), and in *N. wilsmorei* there was a 50% reduction in skeletal muscle *in vitro* (Flanigan *et al.* 1993). A mechanism for reduced protein synthesis during aestivation might involve EIF2α regulation of initiation of translation; the amount and extent of EIF2α phosphorylation increases during aestivation in *Neobatrachus sutor* (Pakay *et al.* 2003). Protein phosphorylation has been suggested to play a role in metabolic depression by reducing enzyme activity including protein synthesis. Phosphorylation of enzymes provides an important control of intermediary metabolism during aestivation in the spadefoot toad *Scaphiopus couchii* (Cowan and Storey 1999). There is a major reorganization of cell metabolism, with changes in activities of key enzymes and depression of signal transducing enzymes. However, there has been no direct quantification of the role of protein phosphorylation in metabolic depression. Ion gradients are consistently maintained during metabolic depression, even though the cost of maintaining ion gradients is thought to be high. Aestivating *N. wilsmorei* maintain ion gradients (Flanigan *et al.* 1993). In aestivating *Neobatrachus kunapalari*, ion gradients are also maintained, as are the levels of Na-K-ATPase (Flanigan and Guppy 1997). It seems that during metabolic depression "high priority" functions such as ion gradient maintenance are sustained (perhaps at a reduced cost?) whereas "low priority" functions such as protein synthesis are turned down.

In the aestivating frog *Neobatrachus kunapalari*, skeletal muscle represents 35% of its body mass and seems to account for 60% of its SMR and 85% of its depressed metabolic rate (Flanigan and Guppy 1997). Obviously, if 50% of this depression was involved with the futile proton leak, it would be hard to sustain the rest of the cell functions that require ATP, let alone maintain chemical energy reserves to extend survival time. Consequently, it is not surprising that the proton leak is decreased during hypoxic exposure by reducing state 4 mitochondrial respiration (St-Pierre *et al.* 2000).

We have already considered mechanisms for hypoxia-induced metabolic depression during hibernation with respect to the chemical energy budget at a cellular level (Fig. 4.12). There is a reduced activity of the electron transport chain (St-Pierre *et al.* 2000a; St-Pierre and Boutilier 2001), hence reduced aerobic ATP production by the mitochondria. Hypoxic submergence also decreases enzyme activities in both the TCA cycle and glycolysis (St-Pierre and Boutilier 2001). The activity of citrate synthase, a key regulatory enzyme of the TCA cycle, is decreased following hypoxic submergence. The activity of lactate dehydrogenase is also significantly lower in frog skeletal muscle during hypoxic submergence (St-Pierre and Boutilier 2001). Consequently, the rate of ATP synthesis via aerobic and anaerobic metabolism is reduced during hypoxia.

5

Approaches and Techniques

5.1 The Comparative Method Approach

The Lissamphibia is a monophyletic clade with an extensive evolutionary radiation and specialization of taxa (see Chapter 1). Closely related amphibian species are likely to be the most similar in all aspects of their biology, because of their recent phylogenetic relatedness. So, when we compare physiological (or other) measurements for various amphibian species it is important to appreciate that comparisons of taxa can be biased by the phylogenetic relatedness of the species. For example, in a comparative analysis of eight *Neobatrachus* species (that form a cocoon) with five other more-distantly related myobatrachid species (that do not form a cocoon), the eight closely-related *Neobatrachus* spp. are not necessarily separate and independent data points. Cocoon formation probably evolved in their common ancestor and is a derived trait for each species. Using n = 13 (eight for the *Neobatrachus* spp. and five for the other spp.) would over-estimate the number of independent species and therefore over-inflate the sample size and hence degrees of freedom and probability level for a statistical analysis. In fact, it might be more appropriate if the eight *Neobatrachus* spp. are averaged into a single datum for comparison with the other species, and n = 6.

Distantly related species can, of course, converge in aspects of their biology by adaptation to particular aspects of their environment. For example, Australia species of *Cyclorana* (Hylidae) are convergent with *Neobatrachus* (Myobatrachidae) in forming a cocoon during aestivation, because the nearest relatives of *Neobatrachus* (i.e. other myobatrachids) and of *Cyclorana* (i.e. other hylids) do not form a cocoon. Other genera of frogs and salamanders, from other parts of the world, have also convergently evolved cocoon formation. Probably, each genus of cocooning amphibians should be considered an independent datum in a comparative analysis of cocoon-formation, rather than individual species of each genus, as it is likely that all species in a genus will have derived the adaptation from a common ancestor rather than evolving it independently.

One, but not the only, role of biological (including physiological) study is to interpret findings as evidence of adaptation. For example, is cocoon formation by *Neobatrachus* an adaptation during aestivation that extends survival time? Such an adaptive significance of a cocoon, for example in reducing evaporative water loss (EWL), would seem obvious, but because all *Neobatrachus* species burrow and aestivate in a semiarid or arid environment it is possible (although unlikely) that cocoon-formation is not adaptive for aestivation but has some other role (e.g. nutrient storage) or possibly no adaptive advantage at all (e.g. recycling of shed skin nutrients is equally effective if a cocoon is formed then re-eaten, or if skin is re-eaten individually as it is shed). However, adaptive significance of cocoon-formation for aestivation can be inferred if it is apparent that it has independently evolved in different clades of frogs (and also salamanders) in response to the same stimulus, aestivation during extended dry periods. So, cocoon formation by eight species of *Neobatrachus* does not itself provide convincing evidence that a cocoon is adaptive for aestivation, but the independent evolution of cocoon formation by nine disparate amphibian genera that aestivate during extended dry periods does. Consequently, one task of "modern" comparative methods is to quantitatively analyze biological traits (e.g. cocoon formation) in a phylogenetic context to infer such adaptive significance.

5.1.1 Two-species Comparisons

An astute scientist might well guess the adaptive significance of a particular trait based on a comparison of two species, or genera. For example, cocoon formation by arid-zone *Neobatrachus* spp. that burrow in clay pans but not by the closely-related spadefoot toad *Notaden* spp. that burrow in sand dunes, might well suggest that cocoon formation is a derived adaptation for aestivation in clay soil. However, such a two-group comparison cannot provide scientific "proof" of an adaptive role of cocoon formation (Adolph and Garland 1994). In fact, the trait in question might be unrelated to the hypothetical adaptive role, and the observation of the hypothesized pattern might have been obtained by chance. For a two-species comparison, the probability of observing the hypothesis is 50:50, or P = 0.5, even if the hypothesis is incorrect.

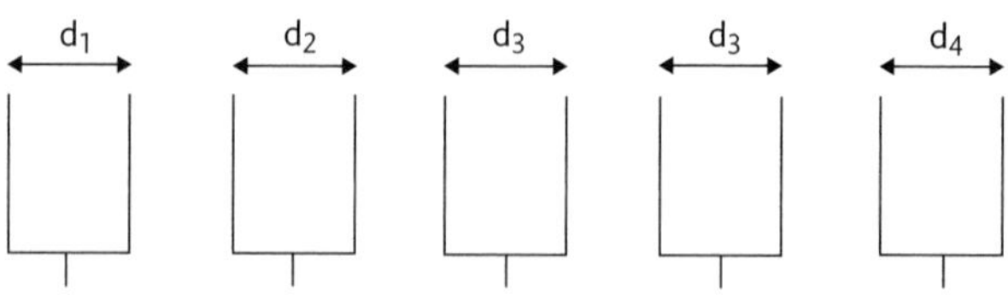

$P = (0.5)^2 = 0.03125$

Fig. 5.1 Schematic representation of species differences (d) derived from five pairs of two-species comparisons, which yield statistical significance ($P < 0.05$) if all five differences support the hypothesized difference.

To increase the veracity of our hypothesis for the adaptive value of a trait, we could examine more pairs of species. Observing the same putative adaptive pattern in a second pair of species would reinforce the value of the hypothesized adaptiveness. Now, the probability of observing the hypothesis by chance is $P = 0.25$ (0.5^2). Adding a third or fourth pair of species reduces the chance probability even further (0.125 or 0.0625 respectively). Adding a fifth species pair reduces the chance probability to 0.03125 (0.5^5), which is less than the scientifically accepted level of probability of $\alpha = 0.05$ for rejecting a null hypothesis. So, a five species-pair comparison can provide scientific credibility that the adaptive role of a trait is consistent with a hypothesized role (Fig. 5.1; Adolph and Garland 1994).

Although multiple two-species comparisons are a useful concept for comparative analysis, a more complete and rigorous comparative approach is provided by a phylogenetically-informed comparison.

5.1.2 Phylogenetic Methods

Accounting for phylogeny when analyzing biological data is not a particularly new concept. Comparative physiologists have routinely accounted for phylogeny when making species comparisons. For example, analyzing allometric data by ANCOVA (analysis of covariance) has been used to compare marsupial versus placental BMR (Dawson and Hulbert 1969), passerine versus "non-passerine" BMR (Lasiewski and Dawson 1967), and salamander versus anuran standard metabolic rate (SMR) (Gatten *et al.* 1992). A variety of other taxonomic (rather than phylogenetic) statistical approaches have also been used, for example nested ANOVA and partitioning of variability at various taxonomic levels (Harvey and Pagel 1991). However, modern phylogenetic methods provide a means of quantitatively accounting for the phylogeny of a group of species, which is particularly useful now that reasonable phylogenies for many taxa, including lissamphibians, are available (see Chapter 1). A variety of phylogenetic methods are now available for analyzing physiological data by determining whether there is a phylogenetic component to the trait, and if so quantitatively accounting for the effect of phylogeny on the trait.

5.1.3 Is There a Phylogenetic Signal?

The first basic question in a comparative analysis is whether there is a phylogenetic signal in a trait. Although many (or most) biological traits will show some phylogenetic pattern based on evolutionary inertia, traits do not necessarily have a phylogenetic signal. Some traits, especially more plastic ones related to behavior or physiology, do not necessarily have a strong phylogenetic signal. If the trait is sufficiently plastic that it does not have a significant phylogenetic signal, then it is not necessary to analyze it using a phylogenetic correction.

Tests for the presence of a phylogenetic signal include the C test (Abouheif 1999) and the k/k* test (Blomberg *et al.* 2003). These tests essentially randomly reallocate trait values to the tips of a phylogenetic tree, and determine how well these randomized trait values fit the tree compared to the actual arrangement. A similar test for categorical traits compares the number of nodes connecting species with the same category value for the actual tip categories with randomly reallocated tip categories (Vanhooydonck and Van Damme 1999). For example, there is no significant phylogenetic pattern of cocoon formation amongst amphibian families; each group of cocoon-forming amphibians appears to have independently derived the character.

5.1.4 Phylogenetic Correction

When there is a significant phylogenetic signal in a trait, then an important role of phylogenetic analysis is to "correct" for this phylogenetic relatedness in traits. For example, the allometric relationship for SMR of frogs compared to salamanders has the same slope, but a higher elevation (Gatten *et al.* 1992). The allometric relationships for SMR, based on $\log_{10}$ mass (M) and $\log_{10}$ SMR of genus averages (Fig 5.2), are:

salamanders: $\mathbf{\log_{10} SMR = 0.861 \log M - 1.197}$

frogs: $\mathbf{\log_{10} SMR = 0.834 \log_{10} M - 1.005}$

It is not reasonable to simply combine the SMR data for salamanders and frogs because their elevations differ (and the elevation of the pooled relationship would be unevenly weighted by the number of salamanders relative to frogs in the dataset). An appropriate phylogenetic method of analysis would account for the relatedness of salamanders and frogs, and "rescale" their SMR according to phylogeny. If we "rescale" SMR-based phylogeny using autocorrelation (one method for phylogenetic correction), then salamanders and frogs fit the same allometric relationship with respect to slope and elevation (Fig. 5.2 inset).

Independent Contrasts (IC; Felsenstein 1985) is a clever but conceptually simple approach to phylogenetic correction that superficially resembles the multiple two-species comparative approach. Consider a simple example (Fig. 5.3) of ten species arranged in five two-species pairs (cf Fig. 5.1 for five two-species comparisons). Each pair of closely related species allows the calculation of a difference (or contrast) between the species, for example d_1 for the two species of the left-hand pair, and d_2, d_3, d_4 and d_5 for the other species pairs. These five differences are essentially the five two-species pairs comparisons as described above, but they are not the only differences (contrasts) that can be calculated if we take into account the phylogenetic structure. Consider the two closely-related species on the left; we can estimate their common ancestral value, d_6, essentially as the average of the two species values. We can do the same for the next two most-closely-related

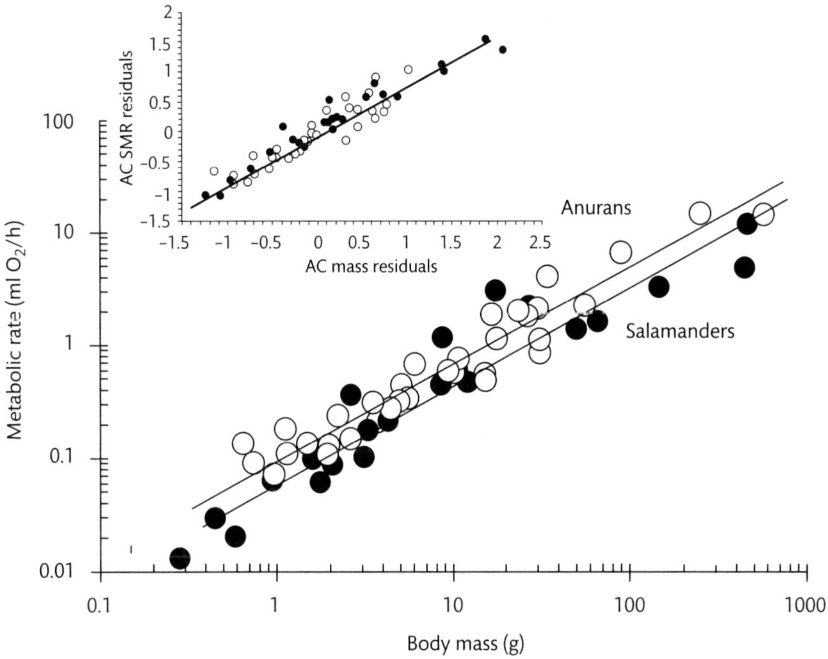

Fig. 5.2 Allometric relationship for standard metabolic rate of salamanders and frogs, based on genus averages calculated from data summary of Gatten *et al.* (1992) using the phylogenetic relationship from Wiens (2007) and Frost *et al.* (2006); the regression lines for anurans and salamanders do not differ in slope but differ significantly in elevation (see text). Inset shows phylogenetically-corrected mass and SMR residuals (using autocorrelation); there is no significant difference in slope or elevation after phylogenetic correction.

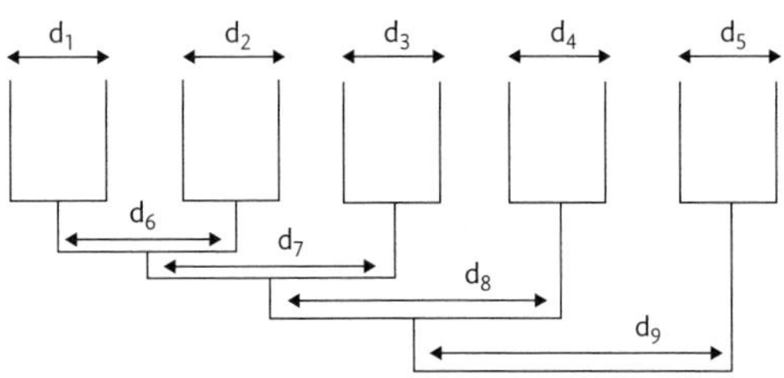

Fig. 5.3 Simple example of phylogenetic analysis by Independent Contrasts, showing the nine different contrasts that can be made between tips and nodes.

species, and calculate a difference between these two ancestral values (i.e. d_6). We can similarly continue to calculate differences between nodes, and obtain in total nine differences. These nine differences are not particularly useful in their own right, but can compared with the differences for a second trait, for example contrasts for SMR could be compared with contrasts for body mass to determine the phylogenetically-corrected allometry for SMR. The slope of this relationship for Y contrasts against X contrasts expresses the phylogenetically-independent relationship. The relationship between Y and X before phylogenetic correction might be similar to that after phylogenetic correction if phylogenetic patterns do not influence the overall relationship. Independent contrasts are differences between pairs of species (or ancestral nodes), so they cannot be examined for species-related patterns. However, it is possible to obtain species residuals and thereby compare species, for example salamanders versus frogs or aquatic versus terrestrial amphibians (Garland and Ives 2000; see below).

A number of other comparative methods are available for phylogenetic correction of traits and allometric analysis, for example autocorrelation, phylogenetic vector regression and phylogenetic least squares regression (Cheverud and Dow 1985; Diniz-Filho *et al.* 1998; Rohlf 2001; Bloomfield *et al.* 2003). Phylogenetic least squares regression (PGLS) simultaneously analyses an X and a Y trait by weighted regression, where the weighting matrix is the variance–covariance matrix derived for phylogenetic distances (Bloomfield *et al.* 2003). IC is a submodel of PGLS, and the two approaches produce essentially identical results (Garland and Ives 2000; Rohlf 2001).

Phylogenetic correction by autocorrelation (AC) is used to analyze a single trait (X or Y) by maximizing the correlation between that trait and the trait premultiplied by a constant (ρ) and a matrix representing the phylogenetic structure (W), with the residuals (ε) being the phylogenetically independent component of the traits (Cheverud and Dow 1985). The phylogenetic matrix W is based on the reciprocal of phylogenetic distances between species pairs; the diagonal (i.e. "similarities" for any species compared with itself) is set to 1. The value of ρ is calculated by an iterative procedure, using the maximum-likelihood function; ρ is constrained to the range $1/\lambda_{max}$ to $1/\lambda_{min}$, where λ_{min} and λ_{max} are the negative and positive Eigenvalues of W closest to 0 (Cheverud and Dow 1985; Rohlf 2001). The phylogenetically-independent Y-trait values are regressed against the phylogenetically-independent X-trait values to obtain the phylogenetically-corrected relationship between X and Y, and the residuals from the phylogenetically-corrected relationship.

Phylogenetic vector regression (PVR) summarizes a phylogenetic matrix as its significant Eigenvalues and Eigenvectors (Diniz-Filho *et al.* 1998). The phylogenetic matrix consists of distances between species pairs (D_{ij}) transformed to $-0.5D^2$. The D_{ij} matrix is double-centered (rows and columns are transformed to a mean of 0) before Eigenvectors and Eigenvalues are calculated. The most significant Eigenvectors are selected using a scree plot and weighted by the square root of the corresponding Eigenvalues. Multiple regression is used to determine the relationship between the significant eigenvectors and the Y trait. The multiple regression reflects the phylogentically-corrected relationship, and the residuals from this multiple regression can be further analyzed as phylogenetically-corrected trait values.

5.1.5 Ancestral Traits

It is possible to estimate the trait values for ancestral forms from the trait values of living species, based on phylogeny. In IC, the value of the ancestral node is calculated from daughter tip or more recent node values, essentially as averages of daughter values weighted for branch length. For the example described above (Fig. 5.2), the body mass and SMR of the ancestral salamander–frog are calculated to be 18.5 g and 0.97 ml O_2/h. For categorical traits, the ancestral state can be estimated from tips and nodes by determining the phylogenetic pattern that requires the least number of changes (e.g. MacClade; Maddison and Maddison 1992). For example, a phylogenetic analysis of amphibians for a categorical trait, cocooning or non-cocooning, would predict that the ancestral state was non-cocooning.

5.1.6 Phylogenetically-corrected Residuals

Phylogenetic analyses can provide residuals (ε) for species, which are differences for species between their actual trait value and their phylogenetically-corrected trait value. These residuals are especially useful for further statistical analysis of patterns amongst species, because they account for phylogeny (and other factors such as body mass). For example, SMR can be compared for salamanders versus frogs after mass and phylogenetic correction; a t-test of residuals indicates no significance difference ($P = 0.312$) in residuals from the regression of phylogeny-independent mass and SMR (from inset regression in Fig. 5.2) This is the same conclusion that we reached above by comparing the salamander and frog regressions after phylogenetic correction.

5.1.7 What is the Question?

Phylogenetic methodologies are essential tools for comparative biologists because they allow us to interpret patterns of biological traits in a phylogenetic context. However, the nature of the questions that we can ask after phylogenetic correction is different from the questions we can ask of data before phylogenetic correction.

After phylogenetic correction we can only make judgments pertaining to the phylogenetically corrected data; we have lost the direct connection to the real data. For example, does a frog have a lower-than-expected SMR. Before we account for phylogeny, this could be easily tested by determining if the SMR of the frog falls below the 95% prediction limits for the regression of SMR against mass (Cooper and Withers 2006). After we account for phylogeny, we can also determine whether the frog has a lower-than-expected residual SMR. However, if it does, then this means that the frog has a low SMR after we have accounted for any phylogenetic pattern; that is the frog has a lower than expected SMR based on its phylogenetic position. But, this does not mean that it has a low SMR. For example, it may well be in a clade that generally has a high SMR, and even though it has a low SMR relative to its nearest relatives it might still have a high SMR relative to other amphibians. Similarly, after phylogenetic correction it is apparent that frogs do not differ from salamanders. We do not conclude that frogs and salamanders do not differ in SMR; they actually do. Rather, we conclude that there is no difference in SMR between frogs and salamanders after accounting for their phylogenetic history.

Correcting for phylogeny can also remove other patterns. If a trait has a pattern that is correlated with phylogeny but has some other causality, then correcting for phylogeny will also remove the correlated causal factor. For example, if the rate of EWL of frogs is related to environmental aridity, and there is a phylogenetic pattern to how the species have radiated into arid environments, then "phylogenetically-correcting" the EWL data would simultaneously remove the environmental pattern. It would not be useful to conclude that there is no environmental influence on EWL. Rather, the environmental influence has been removed along with the phylogenetic pattern.

5.2 Techniques

5.2.1 Methods for Whole Animal Water and Ion Balance

Gravimetric Methods

As we saw in Chapter 3, the most common method for measuring evaporative water loss and rehydration rates is by changes in body mass. You will see two terms of expression for body mass, body mass and body weight. Weight is mass times gravity. As gravity will vary with elevation weight will vary with elevation, while mass is independent of elevation. Mass (kg) is the appropriate SI physical descriptor. Mass is determined by balancing the object in question with a known mass. Weight is measured as a force. Traditional-style scales were balances and measured mass. Most current weighing devices involve force-sensitive transducers and hence measure weight. The distinction is important when distance from the earth's core is significant; however even at an altitude of 5 km (15,840 ft) the difference between mass and weight is negligible. Since 1 g water = 1 ml of water, the mass change can be expressed as a volume flux. Defecation can influence measurement of weight loss or gain so animals are frequently fasted prior to water balance experiments. As metabolism is so low in amphibians, the time interval is short and the animal does not eat, a change in mass is assumed to result from the exchange of water and not carbon loss with metabolism (see water budgets, Chapter 2).

Body Water Compartmentalization

The total body water is partitioned into the intracellular and extracellular compartments. Water contained within the urinary bladder represents a storage compartment independent of total body water (see Chapter 2) and is determined as the body mass difference before and after draining the urinary bladder. The mass of a hydrated animal with an empty bladder has been termed the standard weight (Ruibal 1962).

Total body water is usually determined as the mass difference between the standard weight and the mass after drying the euthanized body at 60°C for some period of time until mass does not change. Amphibians are characteristically 75–79% water. Alternatively, tritiated water of known volume (V_i) and radioactive activity (C_i) can be injected into a living animal, which is given time to equilibrate, and then the final dilution volume (V_f) calculated from the plasma or lymphatic radioactivity (C_f) by the equation: $V_f = C_iV_i/C_f$.

Extracellular water is determined by measuring the concentration of a substance contained in the extracellular space (ECS) from the relationship $C_iV_i = C_fV_f$. As with tritiated water, a known volume (V_i) and concentration (C_i) is injected into the ECS, allowed to equilibrate and then a final concentration (C_f) is determined. The volume of the ECS $= C_iV_i/C_f$. The most common substance injected is inulin, a polymeric carbohydrate that is neither taken into cells nor broken down. The only complicating feature of inulin is that it will be filtered and excreted, which is why it is also used as a marker to determine glomerular filtration rate (GFR) (see Chapter 3). Consequently, if too much time is allowed for equilibration one is liable to overestimate the ECS as a result of this excretion.

The plasma volume is determined by injecting a chemical that is contained within the plasma space and again using the $C_iV_i = C_fV_f$ relationship to determine the final volume. It is difficult to find a marker that does not escape from the circulatory system. Red blood cells incubated and labeled with radioactive chromium is probably the marker with the least likelihood to escape, but it carries with it the attendant logistical difficulties of radioisotope use in a post 9/11 world. The most commonly used solute is plasma proteins labeled with Evans Blue. Because of the high permeability of amphibian capillaries to plasma proteins (Conklin 1930) they will escape into the lymphatic spaces with time, so a correction factor has to be employed to adjust for this loss. Generally, a series of samples are collected over time and the concentration of Evans Blue labeled protein within the circulation is plotted as the log of concentration over time to linearize the disappearance of the Evans Blue label. The linearization occurs if a constant fraction of the plasma turns over with time. This allows one to extrapolate the concentration at time 0 which will be used as the C_f. Anurans turn over plasma volume quickly, at 4% min, so there have to be numerous sample collections in the initial 10 min. Eventually, the Evans Blue label equilibrates throughout the total ECS, hence there are two components to the disappearance curve, loss from the plasma space and equilibration within the total extracellular space.

- Total blood volume is calculated as the plasma volume/(1 − fractional hematocrit).
- The interstitial volume is calculated as the difference between the extracellular space volume and plasma volume.
- The intracellular fluid volume is calculated as the difference between the total body water and extracellular fluid volume.

Evaporative Water Loss (EWL)

Typically, terrestrial species maintain a hydrated state when water is available *ad lib* and the standard weight serves as a starting point for measuring EWL. The decrease in an animal's mass over time in a specific environmental condition is historically the most commonly used method used to measure rates of EWL but requires handling the animals, which itself can affect water loss. To avoid handling, animals can be placed in a flow-through chamber and the decrease in animal mass correlated with an increase in the mass of a hygroscopic chemical (silica gel or drierite) in the excurrent air from the chamber. Both these methods are suitable for long-time average rates of EWL.

The use of humidity sensors has allowed a finer temporal resolution of rates of EWL and is the most common technique in use today. The degree of temporal resolution is really only limited by the turnover time of air in the animal chamber. By monitoring the rate of air flow and the relative humidity and temperature of incurrent and excurrent air from the animal chamber, EWL can be calculated from the equation:

$$\mathbf{EWL} = [(\mathbf{C_{wv}} \times \mathbf{RH\ of\ the\ incurrent\ air}) - (\mathbf{C_{wv}} \times \mathbf{RH\ of\ the\ excurrent\ air})] \times \mathbf{F}$$

In this equation, C_{wv} is the water vapor density of water-saturated air (mg/cm^3) and relative humidity is the fractional saturation (0–1) at the experimental temperature. C_{wv} values are generally obtained from reference sources many of which are available from internet sites. F is the flow rate over the animal. When F is expressed as cm^3/h, EWL is in units of mg/h. In some cases the area-specific rates of water loss are obtained to better characterize the properties of the skin. This can be accomplished by calculating the surface area from the body mass (Hutchison *et al.* 1968) or applying a small chamber of known area to the surface of the skin (Gomez *et al.* 2006). In this study a total resistance value, Rt, was calculated from the driving force for evaporation and the rate of water loss:

$$\mathbf{Rt} = (\mathbf{C_{wv}\ of\ the\ skin\ surface} - \mathbf{C_{wv}} \times \mathbf{RH\ of\ the\ excurrent\ air}) \times \mathbf{EWL^{-1}}$$

When EWL is expressed in area specific units (mg water/cm^2 × s), Rt values are in units of s/cm. In addition, boundary layers can affect evaporation so flow rates are critical and contribute to the resistance to water loss. Rt is then the sum of the skin resistance (Rs) and boundary layer resistance (Rb) and Rs = Rt − Rb. Rb can be estimated from the EWL of a saturated tissue of the same surface area as the animal under the same conditions of flow and temperature (Gomez *et al.* 2006). Lillywhite (2006) estimated Rs values from a large selection of literature values for EWL (which he termed transepithelial water loss, TEWL). They values range from 0 to 5 s/cm for semiaquatic ranids and bufonids to over 200 s/cm for cocoon forming anurans such as *Cyclorana austrailis* with a formed cocoon in place and as high as 400–900 s/cm for *Chiromantis* spp.

Rehydration

As noted in Chapter 3, terrestrial species are able to rehydrate rapidly because of the insertion of aquaporins into the apical membrane of granular cells in the skin. Rates of rehydration can be measured by weighing animals at different times after water is made available. Factors that can affect rehydration are the degree of dehydration and whether rehydration is obtained by placing the animals in a container of water or giving them access to a wetted surface. In the latter case, many anurans adopt the water absorption response posture (Fig. 3.3A) in which the pelvic "seat patch" region of the ventral skin is applied to the surface (Stille 1958), introducing a behavioral component to rehydration (Hillyard *et al.* 1998). Values for water absorption may be expressed as percent of the standard weight recovered over time or as volume per unit of surface area calculated from body mass. Smaller animals will have a greater surface area to volume ratio and thus absorb a greater percentage of the body mass for a given level of cutaneous water permeability. Because the seat patch accounts for most of the water absorption by many anuran species, calculations of water absorption by the calculated surface area will underestimate flow across the seat patch. This source of error was addressed by Marrero and Hillyard (1985) who compared rehydration rates of *Bufo woodhousei* with the seat patch covered with petroleum jelly with that of uncovered controls and found the seat patch contributed approximately 85% of the water absorbed. Another technique is to place the animal on a platform that only allows the seat patch contact with water in a reservoir connected to a pipette (McClanahan and Baldwin 1969; Parsons and Mobin 1991). Water absorption measured volumetrically, corresponds well with gravimetric measurements (Viborg *et al.* 2006).

The ability of AVT and β-adrenergic agents to stimulate cutaneous water absorption has been evaluated by injecting hydrated animals with these substances and measuring water absorption above the standard weight. As a general rule, urine formation by injected animals or dehydrated animals is relatively small and is deposited into the bladder so urine formation does not affect the body weight. Some studies apply a purse string ligature around the cloaca to prevent urinary water loss.

Whole Animal Solute Balance

The ability of amphibians to absorb NaCl from dilute solutions was documented by Krogh (1937, 1939). Frogs (*Rana esculenta*) were depleted of both NaCl by immersion in deionized water or by Na^+ or Cl^- separately by immersion in dilute choline chloride or sodium sulfate solutions. Salt-depleted frogs were then placed in dilute solutions and their ability to absorb salt was evaluated by the lowering of salt concentration in the solutions. Frogs depleted of Na^+ were able to absorb Na^+ from solutions as low as 0.01 mmol/l, while those depleted of Cl^- could absorb Cl^- from similarly dilute solutions, thus demonstrating independent transport mechanisms which Krogh suggested were accompanied by exchange with a counter ion. Garcia-Romeu *et al.* (1969) obtained similar results with the Leptodactylid frog *Caliptocephallela gayi*, cannulating the cloaca to insure bladder water was externalized (Fig. 5.4). By titrating the bathing solution they demonstrated that Na^+ absorption from dilute solutions was linearly correlated with acid secretion while Cl^- absorption was correlated with base secretion, thereby laying the ground work for studies on transport mechanisms in MR cells, described in Chapter 3.

Krogh also observed that frogs sprayed with deionized water for several weeks were able to maintain normal plasma Na^+ and Cl^- concentrations, as did McAfee and Kalckar (1972). Under most conditions, dietary intake is sufficient to offset cutaneous loss but requires renal reabsorption of salts. Analysis of urine and plasma samples is described below.

Catheterization of the Ureters and Urinary Bladder

The ureters of amphibians are relatively easy to catheterize, and an excellent description of the procedure is given in Middler *et al.* (1968). The urinary bladders can be easily catheterized via their ventral cloacal sphincter, and will seal naturally around small tubing inserted through the sphincter. With a catheter in place a mock urine containing a large, impermeable solute can be injected into the

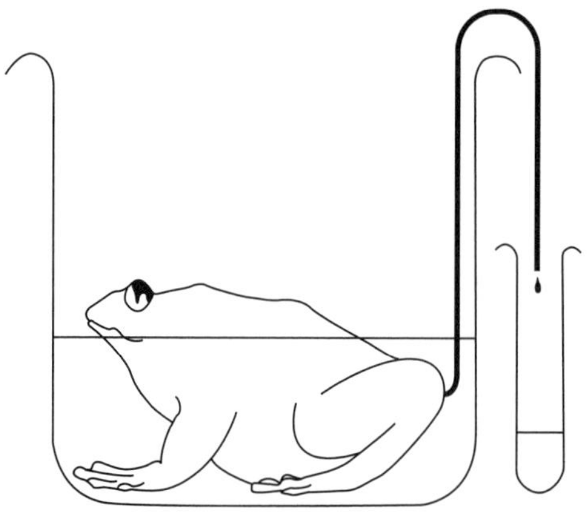

Fig. 5.4 The set up used by Garcia-Romeu *et al* (1969) to study salt absorption by *C. gayi*. Animals depleted of Na^+ or Cl^- were placed in dilute solutions of these ions and uptake was evaluated from the decrease in concentration of each ion in the bath. At the same time the change in bath pH was measured and showed Na^+ absorption was correlated with H^+ secretion while Cl^- absorption was correlated with base secretion.

bladder and reabsorption of bladder water calculated from the increase in concentration of that solute. This technique has been applied to study the stimulation of bladder water reabsorption during dehydration or hemorrhage and the role of arginine vasotocin in this response (Ewer 1952; Sawyer and Shisgall 1956; Kleeman *et al.* 1967).

Cannulation of Blood Vessels

Cannulation of blood vessels requires actual surgical intervention and the insertion of fine-bore polyethylene cannula filled with a heparinized saline solution to prevent blood clotting. The type of heparin utilized to prevent clotting is an important consideration when solutes are to be analyzed from the blood sample. Heparin comes as a salt of various cations. It is important to choose an appropriate cation that will not contaminate the type of ion being analyzed. For instance, if one is interested in N metabolism, ammonium heparin has the greatest likelihood of compromising the accuracy of ammonia and urea concentrations of the plasma. Once the blood sample is withdrawn via the cannula, the plasma is separated from the cells by centrifugation. The sample can be frozen as plasma and stored for future analysis. The freezing of whole blood will lead to lysis of the erythrocytes and contaminate the plasma for specific solute analysis.

5.2.2 Methods for Isolated Tissue

Because amphibian tissues are long lived when isolated from the animal and they have historically been readily available for investigators, the skin and bladder in particular have served as model tissues for the study of solute and water transport and have given rise to some of the fundamental concepts used by transport physiologists. Dutrochet (1839) developed a technique in which a tissue was placed as a diaphragm over the end of a tubular container. The inside of the container was filled with a known solution and the tube was placed in a container of water. Fluid movement into or out of the container, termed an osmometer, could be measured as volume change. Matteucci, (1845) adopted this technique with frog skin and generally found that water moved from more dilute to more concentrated solutions. More recently, this technique has been used to demonstrate the high water permeability of the seat patch region of terrestrial species (McClanahan and Baldwin 1969; Hillyard and Marrero 1985).

The force exerted by the rising fluid level in an osmometer was termed osmotic pressure, a concept that Pfeffer (1885) developed using artificial membranes. Van't Hoff (1887) showed that the ideal gas law could be applied to dilute solutions and related the pressure exerted by a solution relative to pure water (π) in terms of the ideal gas law:

$$\pi = CRT$$

where C is the concentration, R is the gas constant and T the absolute temperature. Historically, values for R have been used to calculate π in units of Atmospheres or Bars but more recently the SI unit of kPa has been used (1 Bar = 0.99 Atm = 101 kPa). Concentrations determined from osmotic pressure are expressed as (milli)osmolar or (milli)osmolal and refer to the ideal behavior of the solutes. When estimating (milli)osmolar concentrations from molar concentrations it is important to correct for non-ideal behavior. For example, the osmotic coefficient of NaCl is 0.93 so a 100 mmol/l solution will have an osmotic concentration of 186 mOsm/l. Conversely, to make a 200 mOsm/l solution requires 107.5 mmol/l NaCl per liter of solution. Another point to consider is the difference between osmolar and osmolal solutions. The thermodynamic term for the free energy of water is termed the water potential (ΔG_{H2O}) that is a function of the mole-fraction (X_A), which is the ratio of the number of water (nW) molecules to the total number of solute and water molecules (nS + nW) in a solution:

$$\Delta G_{H_2O} = RT \ln X_A = RT \ln nW/(nW + nS)$$

For a molal solution the amount of water is fixed at 1 kg and osmotic concentration is expressed as mOsmoles/kg while for a molar solution less water is needed to make more concentrated solutions up to a 1 liter volume and the osmotic concentration is expressed as mOsmoles/l. Thus, the mole fraction for pure water is one and the water potential (ln 1) is zero. As more solute is added the mole faction becomes less than 1 and the water potential is negative so that water moves passively down a free energy gradient. For dilute solutions, on the order of amphibian skin body fluids, the difference between osmolar and osmolal solutions is negligible. Instruments currently used to calculate osmotic concentration rely on the colligative properties of freezing point depression and reduced vapor pressure. However, the same considerations regarding osmolarity, osmolality and ideality apply to these techniques and may become important for studying osmotic balance of animals in brackish water habitats.

Isotopic Water Flux

The availability of isotopes made it possible to measure made it possible to measure the diffusion of salts and water across the isolated skin and bladder. Koefoed-Johnsen and Ussing (1953) compared the diffusion of deuterium (D_20) across isolated frog skin with osmotic flow measured volumetrically. They found osmotic flow was stimulated to a much greater degree than was the inward diffusion of D_20 which led them to hypothesize that water permeability was mediated by pores in the membrane. The logic for this argument was based on the mathematical formulas for diffusion versus pores. Specifically, diffusion follows Fick's law: flux = A × D × (Co − Ci) where A is the area of a pore, D the diffusion coefficient for water and Co and Ci the concentrations outside and inside the membrane. Increasing the radius of a pore by ADH action will increase the area by the square of the radius ($A = \pi r^2$). In contrast, flow through a pore will be described by the Poiseuille equation: flux $= \Delta P \times \pi r^4/8 \times l \times \eta$ where ΔP is the osmotic pressure gradient, l is the length of the pore and η the viscosity. Thus, flux will increase by the fourth power of the radius as a pore becomes opened by ADH. Hays and Leaf (1962) obtained similar results with isolated toad bladder (Fig. 5.5), and showed the pores to be located in the apical membrane.

Discovery of Aquaporins

The concept of pores was supported by electron microscopic observations of control and ADH-stimulated toad bladder that showed intramembranous particle aggregates in the apical membrane of granular cells following hormone treatment (Chevalier *et al.* 1974) along with traffic from subapical vesicles to the apical membrane (Humbert *et al.* 1977; Wade *et al.* 1981). The pores themselves were first identified in red blood cells as channel-forming intramembranous particles with a molecular weight of 28 kD (CHIP 28) (Preston and Agre 1991; Preston *et al.* 1992). This was accomplished by purifying the channel protein and obtaining a partial amino acid sequence. From this sequence a putative nucleotide sequence could be constructed to give a primer that allowed the polymerase chain reaction to amplify cDNAs from a human cDNA library. Once the cDNA was identified it was possible to identify the water-conducting

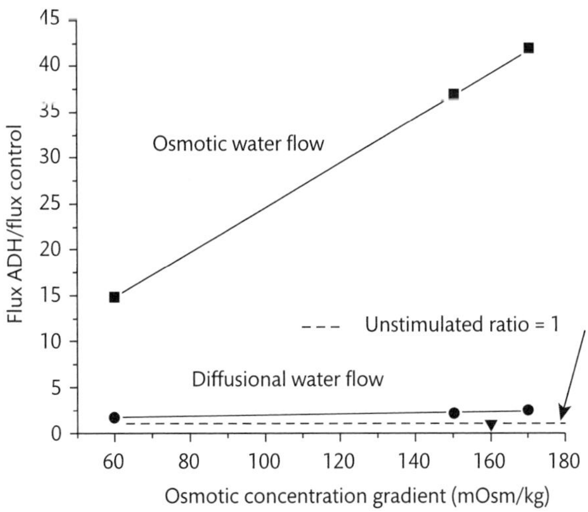

Fig. 5.5 The stimulation of osmotic versus diffusional water flow across isolated toad bladder expressed as the ratio of the stimulated values to the control. The control values were obtained with an osmotic gradient of 160 mOsm/kg prior to treatment with ADH, 1 unit/20 ml. Note the much greater increase in osmotic flow that is consistent with opening of a pore versus widening of a channel. Data plotted from Hays and Leaf (1962).

channels, termed aquaporins, from a large number of tissues (reviewed by Agre *et al.* 1993). Aquaporins serve a number of important physiological functions in human tissues so the focus on their identification has been largely oriented toward medial applications. Only recently has the detailed analysis of amphibian aquaporins, presented in Chapter 3, been undertaken (reviewed by Suzuki *et al.* 2007).

Salt Transport Across Isolated Tissues

As with water permeability, a variety of techniques were developed with amphibian skin and bladder to study ion transport across cell membranes. These are presented in an historical perspective so that the role of technology can be seen as this gives a more detailed understanding of electrolyte balance in amphibians. In a series of experiments, Emil du Bois Reymond studied electric currents in frog tissues, including the skin (du Bois Reymond 1848, 1884). He observed an inward-directed current that was ascribed to glandular secretion. Galaeotti (1904) later showed the current to be dependent on Na^+ in the external solution. During the same time period, Reid (1892) used volumetric measurements to observe that water could be absorbed by isolated frog skin when both sides were bathed with identical saline solutions. Maxwell (1913) developed a method of dissecting the hind leg skin as a tube of skin and ligated the tube to form a sac. The sac could be filled with one solution and immersed in an outer solution that could be the same or different from the inner solution. Movement of fluid into or out of the sac was evaluated by weighing the sac before and after a given time period or by emptying its contents into a calibrated vessel. This period was often 24 hours and no indication is given of providing nutrients or aeration. Nonetheless, inward fluid movement was observed when both sides of the skin were exposed to identical saline solutions. In some experiments, the sacs were everted and fluid moved out of the sac. Curiously, he thought that movement was due to a physical property of the skin and concluded: "The assumption of a vital activity on the part of the frog skin in transferring water through itself is shown to be unnecessary." In contrast, Huf (1935, 1951) used the same preparation and showed the uptake of fluid with identical saline solutions on either side of the tissue was accompanied by accumulation of salt and water. Salt and the resulting water transport could be blocked by metabolic inhibitors and he coined the familiar phrase "active transport" to describe the process.

This was confirmed by Ussing and Zerhan (1951) using an apparatus, the "Ussing Chamber", that has been and continues to be widely used to study transepithelial ion transport. In this apparatus, the isolated skin is clamped between two chamber halves that are filled with identical frog Ringer's solutions. A pair of voltage electrodes measures the voltage across the skin while a second pair of electrodes allows a current to be passed so the charge carried by the transported ions is short circuited, and the potential across the skin is regulated to zero. Thus, there is neither a voltage nor a concentration gradient across the skin and the current, termed the short circuit current (Isc, measured in amperes = coulombs × sec^{-1}), should provide a measure of active Na^+ transport by the following equation, (F is Faraday's constant = 10^5 Coulombs × $mole^{-1}$):

Na^+ transport (moles × sec^{-1}) = Isc (coulombs × sec^{-1})/ F (coulombs × $mole^{-1}$)

Isc is typically on the order of microamperes and usually normalized for surface area of skin.

Ussing and Zerhan (1951) also measured the rate of Na^+ transport as the difference between influx and efflux of isotopic Na^+ and found it to be equivalent to the Isc. They utilized the flux ratio equation (Ussing 1949) as the criterion for active transport:

$$Ji/Jo = (Co/Ci)\, e^{ZF\Delta V/RT}$$

Ji and Jo refer to the inward and outward flux, Co and Ci the ion concentration at the outer and inner surface of the tissue. Z is the charge on the ion, F is Faraday's constant, ΔV is the voltage across the tissue, R is the gas constant and T the absolute temperature. The exponential term describes the effect of a voltage gradient on the flux ratio for charged ions. Under short circuited conditions, Co and Ci are equal and ΔV is zero so a net inward flux represents active transport. Several terms have been used to describe the outer and inner surface of the skin and other epithelial tissues studied with the Ussing chamber. The outer surface represents the apical membrane of the epithelia which is also termed the mucosal surface. The inner surface is referred to as the serosal surface because it would normally receive blood supply. It may also be called the basolateral side since solutions in the inner facing chambers bathe the basolateral membranes.

By varying the ionic concentrations in the mucosal and serosal solutions, Ussing and Zerhan (1951) demonstrated that the apical membranes are permeable to Na^+ while the basolateral membranes are permeable to K^+. The mechanism for apical Na^+ transport as a selective ion channel was elucidated by the development of the voltage clamp which uses a feedback circuit to pass current across the tissue and the diuretic drug, amiloride, that selectively inhibits Na^+ transport when added to the mucosal solution. Van Driessche and Lindemann (1978) constructed a very sensitive voltage clamp that could measure fluctuations in Isc which arise from spontaneous opening and closing of subsets of the active channels and increased the fluctuations by blocking them with amiloride. The analysis of current fluctuations to study the properties of ion channels is summarized by Van Driessche (1994). Apical Na^+ entry is the regulated step in transepithelial transport and described by the formula:

$$Isc = i \times Po \times N$$

In this formula, Isc represents net Na^+ transport, i is the current through single channels, N the number of channels and Po the probability of a channel being open. Thus, the product i × Po represents the number of active channels. There continues to be debate as to the extent that Na^+ transported is regulated by an increase in the number of active channels by activation of channels resident in the membrane (i.e. via Po) or by insertion of new channels (via N). As noted in Chapter 3, noise analysis has also characterized a K^+-selective channel in the apical membrane that may serve for K^+ secretion in K^+-loaded frogs (Zeiske and Van Driessche 1980) and also a non-selective cation channel whose function remains unknown (Van Driessche and Zeiske 1985).

Identification of the Epithelial Na^+ Channel

As with aquaporins, the molecular identity of the epithelial Na^+ channel was established with a mammalian system. RNA from rat colon was injected into *Xenopus* oocytes (see below), and Na^+ transport was evaluated from voltage clamp measurements of amiloride-sensitive current across the membrane (Canessa *et al.* 1993). The amiloride-sensitive epithelial Na^+ channel, termed ENaC, was shown to be homologous with mechanosensitive channels in the body wall of the nematode *Caenorhabditis elegans* and to consist of three subunits, α, β and γ, that form a heterogenous multimer with the channel in the middle of the subunit aggregation. Canessa *et al.* (1994) proposed the native channel to consist of a 2 α, 1 β, 1 γ tetramer. A large family of ENaC-related transport proteins have since been identified in many tissues having medically important functions (reviewed by Kellenberger and Schild 2002). The expression and control of ENaCs in the skin, bladder, colon and kidneys is important in amphibians for overall electrolyte balance described in Chapters 3 and 4.

Amphibian Cell Cultures and Oocytes

A cellular model for epithelial Na^+ transport was established with cultured cells, termed the A6 cell line, from *Xenopus laevis* kidney that displays amiloride-sensitive Na^+ transport like that of the skin and bladder in a single cell layer (Perkins and Handler 1981). This development shifted the study of epithelial Na^+ transport from native epithelia to cell cultures and ENaCs studied in A6 cells are often referred to in the literature as "the" amphibian Na^+ channel (Shane *et al.* 2006). The challenge now is to use these same molecular techniques to identify transport mechanisms as they relate to osmotic and ionic homeostasis in the diversity of amphibian species over the range of habitats, aquatic to terrestrial, to which they are adapted.

The oocytes of *Xenopus laevis* have provided critical information regarding the transport function of membrane proteins, including the definition of the mammalian CHIP 28 as a water-conducting channel (Preston and Agre 1991), and the epithelial Na^+ channel. Oocytes, prior to fertilization, must transcribe messenger RNA for the synthesis of many proteins required to sustain the fertilized zygote during the initial cell divisions. The translational machinery will also synthesize proteins from heterologous mRNA (mRNA obtained from other species, both plant and animal) that is injected into oocytes harvested from the ovaries and, if they are membrane proteins, insert them into the oocyte membrane. This has been done for mRNAs from aquaporin-expressing tissues and injected oocytes swell and burst because of the increased water permeability conferred by aquaporins in the membrane.

5.2.3 Methods for Cardiovascular Dynamics

As we saw in Chapter 2, convective flow is driven by a pressure difference and opposed by the frictional resistance of the vascular system, flow = Δ P/R. Usually, the two variables measured are flow and the Δ P, the resistance is then calculated as Δ P/flow.

Blood Flow Rate

The rate of blood flow is the most difficult variable to measure. The most common techniques in use today involve positioning either Doppler flow probes or ultrasonic flow probes on the vessels of interest; the key variable is to maintain a consistent orientation of the probe and vessel. Doppler probes measure rate of flow, hence have to be individually calibrated *in situ* after the experiment in order to quantify volume flow. Doppler flow probes are also sensitive to hematocrit. Ultrasonic flow probes (Transonic Inc.) on the other hand are calibrated for volume flow, hence the need for *in situ* calibration is less necessary, though recommended. The drawback of ultrasonic probes in the past has been their bulk, though more recent, miniaturized versions developed for mice may work well with many amphibian species.

The gold standard for cardiac output for many anesthetized small mammal studies has been the use of conductance probes. These probes use impedance changes in a varying volume heart to quantify volume change. They were developed to monitor cardiac output in mice and rats, and we are unaware of their use in amphibians at this stage.

Pressure

Blood pressures are characteristically measured by cannulating the blood vessels of interest with inert polyethylene tubing and connecting these to sensitive force transducers that have been calibrated with static pressure heads. Cannulation requires surgical implantation and periodic flushing of anticoagulants to maintain patency of the cannula. The work of Jones (1970) gives the most detailed account of cannulation procedures in amphibians.

The use of these transducers has a variety of potential pitfalls that investigators must avoid. First, the connection from vessel to the face of the transducer has to be filled with fluid. The incompressibility of the water allows the recording of high frequency pressure changes. Any air bubbles in the lines or transducer housing essentially dampen pressure change events. The height difference between the transducer and vessel will be recorded. This means that animal movement can elicit a pressure difference with no actual change in vessel pressure. Finally, the diameter and length of the cannulation tubing will influence the frequency response of the pressures recorded. A small diameter or long tube tends to dampen high frequency components of the pressure trace. The investigator is left with a tradeoff between large bore to assure signal fidelity, balanced by the surgical difficulty of implanting large bore tubing in small-diameter blood vessels.

The most trouble-free approach for monitoring pressure is the use of cannulae that have the transducer built into to the tubing (see Mikro-Tip® pressure catheters, Millar Instruments). This sensor obviates the bubble, height difference and frequency issues, but is quite expensive.

Colloid Osmotic Pressure (COP)

The COP of plasma or lymph is the subset of total osmotic concentration that is contributed by plasma proteins. It can be estimated from protein concentration (Landis and Pappenheimer 1963), but each class of plasma proteins has a different osmotic effect. The most common technique is to separate with a dialysis membrane impermeable to high molecular weight molecules the solution to be measured from a saline filled solution attached to a pressure transducer (Kakiuchi *et al.* 1979). Water will move from the saline solution to the plasma or lymph creating a pressure difference, which is monitored with the pressure transducer.

5.2.4 Temperature

The measurement of temperature is essential for the study of environmental physiology of amphibians, as temperature is an all-pervasive aspect of the environment of amphibians, and body temperature has profound impacts on physiological function. To fully understand the thermal balance of amphibians, it is necessary to describe the integrated thermal environment (T_e) of the amphibian, its own body temperature (T_b), and its surface temperature (T_s) (e.g. Stanier *et al.* 1984; Porter and Gates 1969; Monteith 1990).

Various temperature measurement systems have been used to measure body and ambient temperature in amphibian laboratory and field studies (Brattstrom 1970; Hutchison and Dupré 1992). These include mercury thermometers, electronic thermometers (thermocouples, thermistors), temperature-sensitive radiotelemeters, passive implantable transponders (PITs), temperature loggers and infrared (IR) thermography (Fig. 5.6).

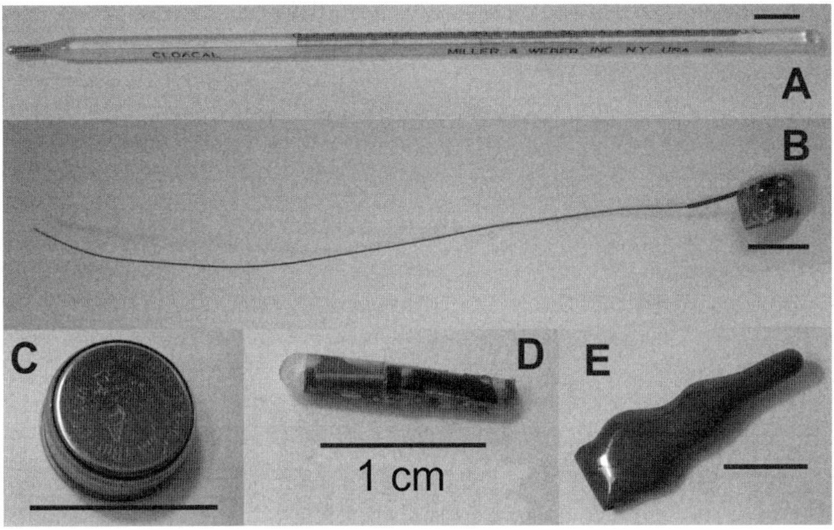

Fig. 5.6 Thermal measurement devices; (A) Schultheis quick-reading thermometer; (B) external-attachment temperature-sensitive radiotelemeter; (C) temperature logger; (D) passive implantable transponder (PIT); (E) implantable temperature-sensitive radiotelemeter. (Photo: P. C. Withers).

Environmental Temperature

Water temperature (T_{water}) is an appropriate measure of environmental temperature for aquatic amphibians. It can be measured simply and directly using any of a variety of temperature sensing elements (see below). Water temperature can be quickly measured; the high heat capacity of water quickly brings a temperature-sensing element to the water temperature. Air temperature requires longer for the probe to attain equilibrium with the air and should be measured in the shade to avoid radiation effects. A thermocouple painted with white enamel paint is a simple and inexpensive means of measuring air temperature with little error from the influence of incident radiation (Christian and Tracy 1985).

Air temperature (T_{air}) is a very common measure of environmental temperature for terrestrial amphibians. However, their thermal environment is determined by the complex interactions of conduction, convection, radiation and evaporation (see Chapter 1). Relative humidity of the ambient air influences the rate of evaporative cooling. Condensation is a theoretically possible means of heat and water gain (Lasiewski and Bartholomew 1969), although it is not clear if condensation has any ecologically relevant role. The integrated thermal environment can be expressed as a single temperature, the operative or effective environmental temperature (T_e; e.g. Bakken *et al.* 1985; Bakken 1992). It can be calculated from measurement of every one of the different modes of heat exchange or by measurement of the temperature of an integrating sphere (e.g. a black-body sphere; Hetem *et al.* 2007). If radiative heat exchange is negligible, then for most amphibians the wet-bulb air temperature (T_{wet}) is often a good estimate of skin and body temperature, as it includes the cooling effect of evaporation. For waterproof frogs, or frogs in humid environments, dry bulb temperature (T_{dry}) would be a better measure of body temperature. Taxidermic mounts or models can be used as "integrating thermometers" to estimate the equilibrium body temperature (e.g. Heath 1964; Boyer 1965; Bakken and Gates 1975; Bakken 1976; Walsberg 1986, 1996). For amphibians, sponges, agar casts, cloth-covered copper electroforms and plaster casts have been used as thermal models (Bradford 1984; Spotila and Berman 1976; Wygoda 1984; Hasegawa 2005).

Body Temperature

Amphibians are generally sufficiently small that internal thermal gradients are negligible to unmeasurable, with the exception that the high rates of evaporative water loss of most amphibians can substantially lower the surface (skin) temperature. Core body temperature is typically measured by inserting a fine temperature-measuring device into the hind gut via the cloaca ("grab-and-jab" or "noose-and-goose" school of thermal biology; Brattstrom 1963; Cabanac and Cabanac 2004; John-Alder *et al.* 1998; Passmore and Malherbe 1985; Thorson 1955; Vences *et al.* 2002). Sometimes internal temperature is measured orally (Brattstrom 1963; Lillywhite *et al.* 1973, 1998) or estimated from skin temperature (Brattstrom 1963; Heath 1975; Lillywhite 1970; Navas 1996; Tracy 1976). Thermally-sensitive radiotransmitters can be attached to the body surface or implanted internally for remote measurement of T_b (Bradford 1984; Heath 1975; Seebacher and Alford 2002). Physical models have also been used to estimate the equilibrium temperature of an amphibian in its thermal environment (Bartelt and Peterson 2005; Hasegawa *et al.* 2005; Navas and Araujo 2000; O'Connor and Tracy 1987; Seebacher and Alford 2002), and infrared thermometry has been recently used (Young *et al.* 2005; Rowley and Alford 2007; Hare *et al.* 2007).

Whatever the instrument and mode of measurement, there are some basic requirements for accurate measurement of T_b. First, the temperature of the amphibian must not be altered by the process of measurement. This is difficult and problematic for the smaller (majority) of amphibians because of their rapid thermal exchange with their environment (captor, holding bag or surface they are placed on for measurement) due to their low thermal inertia and high thermal conductance. Activity (e.g. calling, locomotion) can slightly elevate T_b due to internal heat production, particularly for larger species, and exposure to air movement can promote evaporative heat loss and lower T_s and T_b. For non-waterproof amphibians, evaporation has a substantial influence on T_s and T_b. For example, frogs and toads can be over 8 °C cooler than the air at low ambient humidity in moving air;

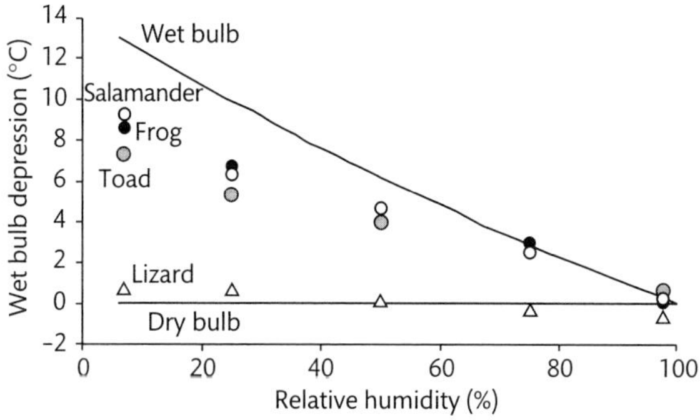

Fig. 5.7 Effect of ambient humidity on wet bulb temperature depression, and body temperature measurements for a salamander (*Plethodon glutinosus*), frog (*Rana pipiens*), toad (*Bufo fowleri*) and a lizard (*Sceloporus undulatus*). Amphibian and lizard data from Hall and Root (1930).

the thermal difference is essentially linearly related to relative humidity (e.g. Hall and Root 1930; Fig. 5.7).

Mercury Thermometers

Early mercury (or alcohol) thermometers were generally too large to be useful for T_b measurement of amphibians, but the small Schultheis "quick reading" mercury thermometer proved to be ideal for laboratory and field measurement of T_b (and T_a) of amphibians, and a large dataset was quickly amassed after its development (e.g. Brattstrom 1970; Hutchison and Dupré 1992). These are still available and used but have been largely supplanted by small electronic thermometers.

Electronic Thermometers

Thermocouples and thermistors have been commonly used for measurement of T_b since they are available as simple, portable and hand-held electronic thermometers. Both methodologies allow for very small sensing elements. Thermocouples rely on the principle that the contact zone of two dissimilar metals creates a voltage that is dependent on the temperature (Peltier effect). Commonly used pairs of dissimilar metals that are used for thermocouples are copper and constantan (type T) or chromel and alumel (type K). A major advantages of thermocouples, other than small size, is the constancy of the thermal effect, but a disadvantage is that the induced voltage is small and non-linear (e.g. about 40–50 μV per °C for T and K-type thermocouples). Modern electronic components make the manufacture of stable, precise thermocouple meters relatively inexpensive and simple. Thermistors are small temperature-sensitive resistive elements. They can be very small, and their reasonably large thermal sensitivity can provide a relatively large electrical signal through an appropriate electronic bridge. Thermistors tend to have quite non-linear thermal characteristics and can be unstable over time. Other thermal sensing elements have also been used to make electronic thermometers such as diodes, but these are generally larger and less suited for T_b measurement.

Radiotelemeters

Radiotelemetry has been commonly used for locating amphibians in the field (e.g. Richards *et al.* 1994) along with other techniques such as radioactive tags (Ashton 1994; Semlitsch 1981), thread bobbins (Heyer 1994), fluorescent yarn tags and luminescent capsules (Windmiller 1996). Temperature-sensitive radiotelemeters are generally relatively simple electrical circuits that periodically emit a radio-signal (generally AM but sometimes FM) whose pulse (or frequency) characteristics are thermally sensitive because of the inclusion of a temperature-sensitive electrical component (often a thermistor) in the circuit. A radiotelemeter can be quite small, particularly with the advent of modern microelectronic components, but the limit to small size is ultimately the size of the battery. Small battery size limits the life of the radiotelemeter and the range of the radio-signal. Temperature-sensitive radiotelemeters can have an external whip antenna, restricting it to external attachment but providing a greater range for signal reception, or a coiled internal antenna that makes it smaller and more suitable for implantation in the body cavity but reduces the range for reception of the signal. External transmitters can be attached with glue or using a belt or backpack harness (e.g. Rathbun 1996; Fisher and Muth 1995; Fig. 5.8).

Passive Implantable Transponders (PITs)

Passive implantable transponders avoid the requirement for a bulky battery power supply (that limits the small size of telemeters) by passively gaining the energy for transmission of a temperature signal from an external electro-magnetic field. Inexpensive diode tags can be used to locate animals over a relatively short range (1–10 m) by harmonic radar, but external attachment can influence behavior (e.g. Lovei *et al.* 1997; Langkilde and Alford 2002). Many types of implantable transponders are available for alphanumeric identification of an implanted chip (e.g. for checking the identity of individual animals) but some also have a temperature-sensitive circuit and include temperature in the signal output. The relatively short range (a few cm) limits the usefulness of PITs to identification and temperature measurement of animals in the hand or at least very close to the sensing unit. The small size of PITs enables their implantation in the body cavity of medium to large amphibians, for remote monitoring of core body temperature.

Temperature Loggers

Computerized data-loggers have been used for many years particularly for remote logging of ambient temperature and other environmental data. Modern

Fig. 5.8 Water-holding frog (*Cyclorana platycephala*) with temperature-sensitive radiotransmitter. (Photo: K. McMaster), University of Western Australia.

technology has enabled the miniaturization of data-loggers to the point where they can be implanted in large amphibians for long-term monitoring of core body temperature (e.g. Forsythe *et al.* 2004; Taylor 2004; Wernecke *et al.* 2007). The memory size of the logging unit determines the combination of sampling frequency and duration of monitoring.

Infrared Thermography

All objects at a temperature higher than 0°K emit radiation, and amphibians (and other animals) with a surface temperature between about 0 and 50°C emit in the infrared spectrum (e.g. Cena and Monteith 1975). Consequently, surface temperature of amphibians and objects in their environment can be measured by infrared thermography (e.g. Cena and Clark 1973; Spruyt *et al.* 1995; Speakman and Ward 1998). Early infrared (IR) thermographic equipment was large and cumbersome and not convenient for field work but was nevertheless very useful in the laboratory environment (Jones and Avery 1989). The advent of small portable IR cameras has revolutionized the applicability of IR thermography for animal studies.

Hand-held IR thermometers are now relatively inexpensive and readily available, particularly in the clinical temperature range (about 34–42°C) but wider temperature-range units are also available. IR thermography has been used to measure temperature of skeletal muscles of bullfrogs at rest and during isometric

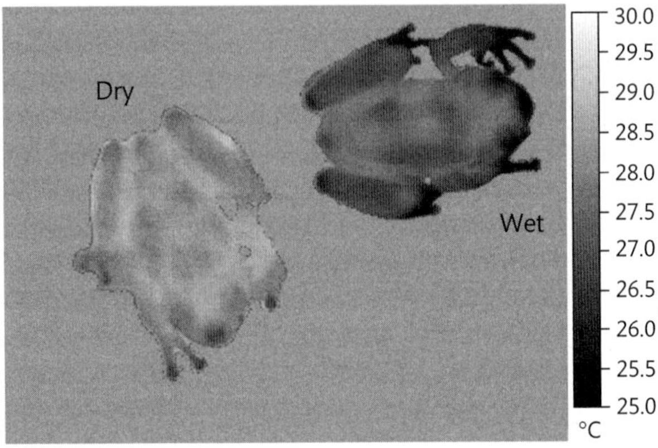

Fig. 5.9 Thermal image of a hylid frog. *Bokermannohyla alvarengai*, when dry and wet. (Photo: G. Tattersall).

tetanus (Kogayashi and Sugi 1980; Kobayashi *et al.* 1998) and body temperature of amphibians (Rowley and Alford 2007; Hare *et al.* 2007).

IR imagery has been used for many varied roles in animal research, husbandry and welfare, including detection signs of diseases such as rabies, effects of exercise on temperature of muscles, surface temperature of insects, birds and mammals, incubation temperature for geckos, and even plant–water relations and freezing. IR imagery has not been commonly used for amphibians. Their surface temperature is affected by ambient humidity and can therefore vary substantially between wet and dry conditions. For example, the moderately waterproof hylid frog, *Bokermannohyla alvarengai*, (Fig. 5.9) has lower surface temperatures when its skin is wet, reflecting evaporative cooling, than when it's skin is dry and hence close to ambient air temperature.

5.2.5 Energetics

Energy content of biological materials (food, feces, tissues) and rate of energy use are important measurements that allow calculation of metabolic rate and energy needs. Metabolic rate is the rate at which an amphibian uses energy to sustain a range of biological functions (cell maintenance, secretion, digestion, activity, movement, gamete production, etc.). Standard metabolic rate (SMR) is the lowest sustainable metabolic rate of an amphibian, when at rest, not digesting, and not stressed. It depends on body temperature (see Chapter 3), so SMR is lower at low T_a and higher at higher T_a. Amphibians typically have low SMRs, but during intense activities their metabolic rates can be an order of magnitude or more higher than SMR. Standard, resting and low activity rates of metabolism are supported by aerobic metabolic pathways. During hibernation or aestivation, the metabolic rate can be reduced to about 20–25% of SMR (see Chapter 4). At high levels of activity, the maximal sustainable aerobic metabolic rate is insufficient and metabolic energy production is assisted by anaerobic metabolism. For amphibians, the only significant end-product of anaerobic metabolism is lactic acid (Bennett 1978).

Energy Content

The energy content of biological materials is typically from about 17 kJ/g for carbohydrate to about 23 kJ/g for protein and 38 kJ/g for lipids (Brafield and Llewellyn 1982). The energy content of animals is typically about 19 to 30 kJ/g, depending primarily on their percent lipid content. The energy content of biological materials can be directly measured by the heat produced through complete oxidation of a sample in the presence of excess oxygen. A bomb calorimeter consists of a high-pressure chamber in which the sample is placed, at a high pressure of pure oxygen. The sample is ignited (usually by an electrical current that heats an ignition wire of high resistance) and the sample is completely combusted. The heat produced by complete combustion (corrected for the heat input of the ignition system) provides a direct measure of the total energy content of the sample.

Metabolic rate can be measured as the difference between energy consumption (i.e. food) and energy excretion (i.e. feces and to a minor extent urine), so long as growth (or weight loss) and gamete production are accounted for. Such a measure of metabolic rate is typically made over an extended period (e.g. days or weeks, rather than minutes or hours) and is generally converted to an average daily metabolic rate (ADMR).

Aerobic Metabolic Rate

Aerobic metabolism is sustained at the biochemical level by oxidative metabolism of substrates. For example, the aerobic oxidation of 1 mole of glucose metabolism consumes 6 moles of oxygen, produces 6 moles of carbon dioxide and 6 moles of water, and releases 2874 kJ of energy;

$$C_6H_{12}O_6 + 6\,O_2 \rightarrow 6\,CO_2 + 6\,H_2O + 2874\,\text{kJ/mole}$$

At metabolic levels slightly above SMR, O_2 delivery to tissues and CO_2 removal can be increased to match metabolic demand, and metabolism remains aerobic. There are a variety of possible measurements of aerobic metabolic rate. Considering the above equation illustrating aerobic metabolism of a carbohydrate substrate, the overall "metabolic rate" could in theory be measured as the rate of substrate utilization (e.g. $C_6H_{12}O_6$ consumption), O_2 consumption (V_{O_2}), CO_2 production (V_{CO_2}), metabolic H_2O production (MWP), or heat production (HP). In practice, heat production is the most direct measure of metabolic rate, but it is seldom measured. It is generally not possible to measure substrate utilization by a living amphibian, or metabolic water production (even with isotopic tracer techniques, as there is so much water already present in an amphibian and it turns over very quickly relative to the rate of metabolic water production). O_2 consumption and CO_2 production are the most common measures of metabolic rate, made by manometry (pressure change) or respirometry (measurement of O_2 or CO_2 concentration). The ratio of CO_2 produced per O_2 consumed is called the respiratory quotient (RQ) when measured at the biochemical level, or respiratory exchange ratio (RER) when measured for an animal as respiratory uptake of O_2 and CO_2 excretion. RQ varies from about 0.7 for protein and lipid metabolism to 1.0 for carbohydrate metabolism (Withers 1992) but can, under non-steady state conditions, be less than 0.7 (e.g. carbohydrate synthesis from lipid; Kleiber 1975) or greater than 1.0 (e.g. fat synthesis; Blaxter 1989). RER can differ from the actual RQ in non-steady state. For example, changes in O_2 levels stored on hemoglobin or myoglobin can result in minor discrepancies between RER and RQ. Changes in CO_2 levels in body fluids, for example during activity or temperature change, can result in more substantial discrepancies because of the high solubility of CO_2 in body fluids.

Manometry

Manometric measurement of the rate of O_2 consumption by an amphibian is accomplished by pressure/volume change in a closed metabolic chamber (Fig. 5.10). If the CO_2 that is produced is removed by chemical absorption (e.g. by KOH, Ascarite, soda lime, etc.) then either the volume (V) of gas is reduced if the pressure (P) is kept constant, or P is reduced if V is kept constant. This reflects Boyle's Law, which states that P and V are inversely related for a constant amount of gas (PV = constant). If O_2 is consumed and not replaced by CO_2 then PV is reduced. If V is constant then P is reduced, and if P is constant then V is reduced. There are many designs for manometers, derived from the classical Warburg manometer (Warburg 1923; Scholander and Edwards 1942).

Respirometry

Unlike manometry, respirometry directly measures the change in concentration of O_2 and sometimes CO_2 as a result of the animal's metabolism, rather than relying on pressure changes; the advantage of respirometry is the avoidance of thermal artifacts (Hillman *et al.* 1979). The O_2 level is measured directly by Clark-type O_2 electrodes in water, and various O_2 sensors for air (e.g. paramagnetic or fuel cell). The CO_2 level in water can be measured by CO_2 electrodes, and in air by infrared gas analyzers. Microgasometric determination of samples of water and air can be accomplished by chemical analysis (e.g. Scholander 1947).

Metabolic rate is calculated based on the percent depletion of O_2 or percent accumulation of CO_2, in conjunction with the volume of the metabolic chamber (closed system respirometry) or air flow rate through the metabolic system

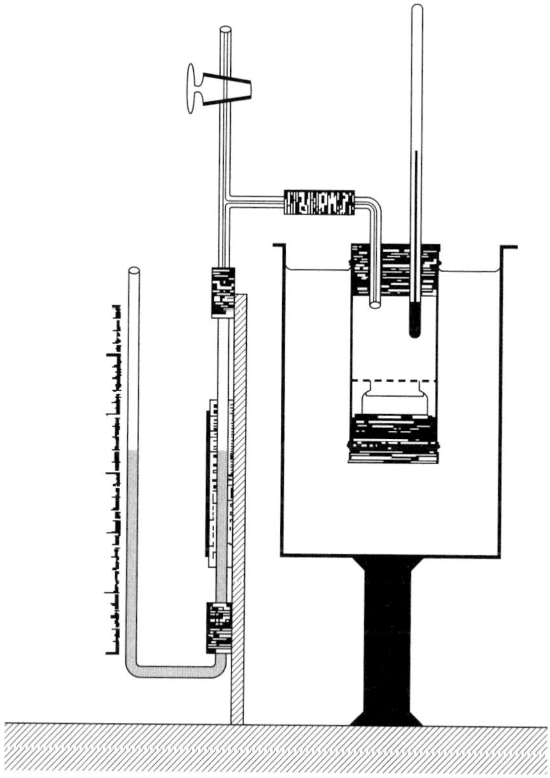

Fig. 5.10 Schematic of a classical Warburg manometric system for measuring oxygen consumption rate. Modified from Wells (1938).

(open-flow respirometry). In closed-system respirometry, the amphibian is sealed in a closed chamber, of known air volume, and an air sample is removed after a measured period of time for analysis of O_2 (and sometimes CO_2). In flow-through respirometry, the amphibian is placed in a sealed chamber that has a known flow rate of air passing through it, and the O_2 (and/or sometimes CO_2) level of the excurrent air is measured. The advantages of closed-system respirometry are that a number of animals can be measured simultaneously since the gas samples can be analyzed sequentially with a single analyzer, and the fractional depletion of O_2 can be relatively high (e.g. 0.005 to 0.010) depending on the size and metabolic rate of the amphibian, the volume of the chamber, and the elapsed time. The disadvantage of closed-system respirometry is that it determines the average metabolic rate over the entire period of measurement; any metabolic peaks (e.g. due to activity) are included in the average measurement. The advantage of flow-through respirometry is that changes in metabolic rate can be monitored over time and minimum rate (SMR) can be distinguished from active rates; virtually instantaneous values of metabolic rates can be calculated (Bartholomew *et al.* 1981). The disadvantage of flow-through respirometry is that the excurrent O_2 level is relatively low, often close to the maximum resolution of O_2 analyzers. Closed system respirometers have been used for aquatic and terrestrial amphibians (e.g. Preslar and Hutchison 1978; Withers and Hillman 1983; Walsberg *et al.* 1986; Seymour 1995). Ultsch and Dade (1990) and Abel *et al.* (1992) describe aquatic flow-through respirometry systems for salamanders and anuran tadpoles respectively, and Full (1986), Full *et al.* (1988), Walton and Anderson (1988) and Word (2008) describe an aerial flow-through system for aestivating, resting and active terrestrial anurans.

The measurement principle of respirometry is simple. The oxygen consumption rate (V_{O_2}) and carbon dioxide production rate (V_{CO_2}) are calculated from the mass balance of O_2 and CO_2 entering and leaving the metabolic chamber. The percent O_2 concentration entering the chamber is typically 20.95% and the fractional O_2 content (F_{IO_2}) is 0.2095 (standard atmospheric air, dried to remove water vapor; Walker 1977) and the fractional O_2 content leaving the chamber (F_{EO_2}) is measured. Similarly, the fractional CO_2 content entering the chamber (F_{ICO_2}) is typically about 0.0003 and the fractional content leaving the chamber (F_{ECO_2}) is measured.

Although the principle of respirometric measurement of metabolic rate by mass balance between air in and air out is simple, the actual calculation of V_{O_2} or V_{CO_2} is more complex because consumption of O_2 and production of CO_2 change the airflow out of the metabolic chamber. Mathematical derivations of the appropriate equations for calculation of V_{O_2} and V_{CO_2} have been presented in various forms, for closed, open-flow, and mask systems (e.g. Depocas and Hart 1957; Hill 1972; Withers 1977; Frappell *et al.* 1989; Withers 2001; Chaui-Berlinck and Bicudo 2006). The rate of air flow entering the metabolic chamber (V_I) is generally measured and the flow rate exiting the chamber (V_E) must be calculated; it is possible to measure the excurrent flow and calculate the incurrent flow. The mass inflow of nitrogen into the chamber is calculated from the V_I and F_{IO_2} and F_{ICO_2} (inert gasses other than N_2 are included as "N_2") as $F_{IN2} = V_I\,(1-F_{IO_2}-F_{ICO_2})$, and this must equal the mass outflow of N_2 which is calculated as $F_{EN2} = V_E\,(1-F_{EO_2}-F_{ECO_2})$. This equality allows the calculation of V_E from V_I (or vice versa) providing that water vapor is removed from the incurrent and excurrent airstreams (e.g. using Drierite or magnesium perchlorate), and that F_{IO_2}, F_{ICO_2}, F_{EO_2} and F_{ECO_2} are assumed or measured. If CO_2 is removed from the incurrent and excurrent air streams using a chemical absorber (e.g. Ascarite or Sodasorb) then F_{ICO_2} and F_{ECO_2} can be ignored and the commonly presented formula for V_{O_2} is derived as $V_{O_2}\,(\text{ml}\,O_2/\text{h}) = V_I\,(\text{ml/h})\,(F_{IO_2}-F_{EO_2})/(1-F_{EO_2})$. CO_2 can be left in the excurrent air stream (for convenience of not using a CO_2 absorbant) and V_{O_2} calculated with relatively minor error if an RQ is assumed (Koteja 1996). V_{O_2} is corrected to mass-specific units (ml O_2/gh) by dividing by body mass (g), and to standard temperature and pressure (STPD; 0°C and 1 atm). For a closed system, the equivalent formula is V_{O_2} (ml O_2/h) = V (ml) $(F_{IO_2}-F_{EO_2})/[(1-F_{EO_2})\ \Delta t]$ (h), where V is the chamber air volume and Δt is the time that the chamber is sealed.

Respirometric measurement of oxygen consumption provides equivalent values to values obtained by manometric techniques for resting amphibians; both techniques have contributed to the large amount of data for resting amphibians (see Gatten *et al.* 1992). It also provides equivalent measures to direct calorimetry for ectotherms (Walsberg and Hoffman 2006). For sustainable submaximal exercise, such as during sustained locomotion, flow-through respirometry has been used (e.g. Full 1986; Full *et al.* 1988; Walton and Anderson 1988). For non-sustainable maximal oxygen consumption rate, closed system respirometry with manually-induced exercise is preferable to electrical stimulation to elicit activity and manometric measurement of oxygen consumption rate (Hillman *et al.* 1979). The latter methods suggest that amphibians have little or no oxygen consumption during activity, but maximal uptake during recovery (e.g. Bennett and Licht 1973; Bennett and Wake 1974; Feder 1977) whereas manual prodding during swimming or rotation in a sealed metabolic chamber and removal of an air sample for direct measurement of O_2 (Seymour 1973; Hillman and Withers 1979; Walsberg 1986) indicate that maximal activity V_{O_2} of anurans and salamanders is high during activity, at up to ten times resting, and decreases during recovery (e.g. Withers 1980; Hillman and Withers 1979; Withers and Hillman 1981; Chapter 2). Activities such as calling by male anurans can also elicit high metabolic rates, even greater than for enforced activity (e.g. Taigen and Wells 1985; Taigen *et al.* 1985; Bucher *et al.* 1982; Prestwich *et al.* 1989; Grafe 1996, Chapter 3).

To convert measurements of oxygen consumption into heat production requires an oxycalorific quotient, the value of which depends on the substrate (or mix of substrates) being oxidized. The oxycalorific quotient varies from 21.1 kJ/l O_2 for carbohydrate to 19.6 kJ/l O_2 for lipid to 19.1 kJ/l O_2 for protein (Brafield and Llewellyn 1982). For protein metabolism, the oxycalorific coefficient varies depending on the nature of the nitrogenous waste product, from 19.1 kJ/l O_2 with ammonia excretion to 19.4 kJ/l O_2 for urea or uric acid excretion. It is more accurate to convert metabolic rate in units of ml O_2 to heat production, than units of CO_2, because the conversion coefficient varies less with RQ for O_2, from 18.8 to 21.4 kJ/ml O_2 than for CO_2, from 21.4 to 27.9 kJ/ml CO_2 (Withers 1992). The energy lost via the nitrogenous waste product is intermediate for ammonia (3.86 kJ/l O_2) compared to urea (3.51 kJ/l O_2) and uric acid (5.89 kJ/l O_2); RER is 0.96, 0.84 and 0.72 respectively (Brafield and Llewellyn 1982).

Anaerobic Metabolism

High and intense (maximal) rates of metabolism are supported by anaerobic metabolic pathways. For vertebrates, including lissamphibians, the primary anaerobic end product is lactic acid; intermediary metabolites such as pyruvate, alanine and succinate contribute negligibly to total ATP production (Bennett 1978; Withers 1992). The accumulated lactic acid yields 17 μmol of ATP per 1 mg lactate assuming glycogen is the anaerobic substrate (Bennett and Licht 1972) or 11 μmol of ATP per 1 mg lactate (from glucose; Withers 1992). Muscle stores of creatinine phosphate contribute to ATP production, and intracellular ATP levels also become depleted during activity; these contribute a minor fraction of total anaerobic ATP production (Fitts and Holloszy 1976; Gatz and Piiper 1979).

Biochemical assay of accumulated lactic acid is the general means of determining the rate of anaerobic metabolism, because of its predominance as the anaerobic end product. Measurement of whole-body lactate is accomplished by homogenizing the animal in perchloric acid then enzymatic assay of lactate in the supernatant (e.g. Hutchison and Miller 1979; Miller and Hutchison 1980). Particular tissue samples (e.g. blood, skeletal muscle) can be analyzed for lactic acid content, but it is difficult to determine the total anaerobic energy production from specific tissues because of the marked variability in lactic acid distribution in various body tissues. About 75% of total lactate after exercise is accounted for by muscle, blood and liver in *Xenopus laevis*, *Rana pipiens* and *Bufo boreas* (Putnam 1979). Species differences in lactate accumulation are explained by variation in the muscles involved in activity, by the percent muscle content of the body and by differences in lactate accumulation in different muscles (Putnam 1979). Most of the lactate accumulated during activity is resynthesized to glycogen for restorage, rather than aerobic oxidation (Withers *et al.* 1988).

6
Conclusions and Future Directions

6.1 What We Know Now

Living amphibians probably have all evolved from a terrestrial rather than an aquatic ancestor, but have subsequently radiated into a very wide range of environments and niches, including both some very terrestrial and hostile ones as well as aquatic environments. They represent a wide spectrum of evolutionary solutions to survival in these environments, and consequently are excellent model systems for studying adaptations associated with the aquatic terrestrial transition in vertebrates.

The basic terrestrial strategy is ectothermy (low energy exchange with high conversion efficiencies of ingested energy to body mass), nocturnality in favorable hydric environments with rapid cutaneous exchange of water with the environment and dilute urine and large bladder volumes for a substantial role of bladder storage and reabsorption of bladder urine. Behavioral responses to avoid desiccating conditions, and physiological tolerance to dehydration to cope with desiccation when it happens, are important correlates of habitation in more terrestrial environments. In conjunction, the physiological control of water uptake across the skin and urinary bladder via hormones and the sympathetic nervous system are more developed in more terrestrial species. Tolerance of long periods of inactivity/aphagia during hibernation/aestivation, in conjunction with metabolic depression to eke out energy stores and cocoons to eke out water stores, characterize dormant periods. The waterproof frogs, *Chiromantis* and *Phyllomedusa*, are probably the highlight of adaptive change in amphibians as a whole, for retaining activity/feeding in arid environments and being uricotelic.

Existence in very dilute aquatic environments entails skin adaptations to minimize diffusive water entry and the active transport of sodium and chloride from the environment. These adaptations prevent the body fluid osmotic concentration from decreasing during periods of aphagia.

Survival in drying soil or hypersaline environments entails the accumulation of solutes, in particular urea, to raise the body's osmotic concentration and allow the diffusive uptake of water from theses environments, very much like the ureo-osmoconforming strategy of sharks and rays. Unlike chondrichthean fishes, there does not seem to be an accumulation of solutes to counterbalance the effects of urea on biochemical processes.

Handling the extremes of winter entails either the tolerance of extracellular freezing with the accumulation of cryoprotectants or behavioral avoidance by selecting microenvironments that do not freeze. Selecting an aquatic environment decreases the likelihood of a freezing event but the amphibians become more subject to hypoxia. Consequently, selection for winter survival spans a spectrum of remarkable behavioral and physiological adaptations coupled to either freeze tolerance or hypoxia tolerance.

6.2 Phylogenetic Breadth

There is still much to be learned about both basic amphibian physiology as well as the remarkable interspecific variability contained within the Class. So much of what we currently understand about how amphibians work derives from such a narrow phylogenetic window that it is premature to describe our current knowledge as "typical" of amphibian function. Much of our functional understanding of amphibians was generated by investigators in general physiology whose interest was more in the organ than in the biology of the species from which the organ was extracted. There is a much smaller literature base derived from environmental or comparative physiology, where function and field biology received more emphasis. The point of comparative and environmental physiology is in exploring the use of shared problems or environmental stresses in multiple species to understand the principles governing physiological and evolutionary solutions. Consequently, we still are at a very rudimentary level in being able to analyze how environmental differences have shaped intra- and interspecific variation in amphibians' physiological performance.

Moreover, without a clear knowledge of evolutionary relationships it is impossible for us to predict or extrapolate to other amphibians. However, we can broaden the scope and meaning of the results of laboratory investigations by looking for corroboration in distantly related species. While this may never be feasible by direct physiological experimentation, it is certainly possible at the level of morphology. For instance, the observations summarized in Table 1.1 are based on representatives of 144 species within 39 of the 40 currently recognized monophyletic anuran families (Frost *et al.* 2006). Such a comparative examination of morphological characteristics is possible only because of the availability of large collections in natural history museums and universities, in this case the California Academy of Sciences in San Francisco.

In recent years, researchers have attempted to account for the potential of phylogeny in interpreting amphibian adaptations to environmental gradients. The Amphibian Tree of Life (Frost *et al.* 2006), the latest amphibian phylogenetic construct, is an estimate of evolutionary relationships based on analysis of morphological and molecular characteristics for over 500 amphibian taxa; while nowhere near complete, it is the most comprehensive study of its kind to date. Table 1.1 is simply a survey of a few representatives of each family; its results cannot be placed in any meaningful context until our knowledge of anuran relationships is much more finely tuned. Nevertheless, the data in Table 1.1 are preliminarily intriguing; they indicate that many of these monophyletic lineages contain taxa that represent a spectrum of the various adaptations we have included, ranging from complete absence to most derived.

For instance, ventral dermal ornamentation associated with rapid water uptake (character B, Table 1.1) is absent (B2) or most extensive (B4) in various taxa *within* 14 families: alytids, megophryids, myobatrachids, hylids, cyclorhamphids, microhylids, brevicipitids, hyperoliids, arthroleptids, ceratobatrachids, petropedetids, pyxicephalids, rhacophorids and ranids; in 14 families (not necessarily the same ones) there are some species in each family that have dorsal dermal ornamentation (C2) and still others that lack the character entirely (character C1, Table 1.1).

Species that we would characterize as semifossorial or fossorial/terrestrial (Chapter 1.5), based upon the presence of an enlarged spade-like metatarsal tubercle (Table 1.1, D3) are another example. Eight families contain species that possess the enlarged digging tubercle but additional species lack the structure entirely (Table 1.1, D1): these are Limnodynastidae, Hylidae, Ceratophryidae, Microhylidae, Brevicipitidae, Arthroleptidae, Ptychadenidae and the Pyxicephalidae. Similarly, all species we examined that possess enlarged digital tips indicating scansorial ability (Chapter 1.5, Table 1.1, Character E) belong to families that also contain species that wholly lack digital modifications (E2), including the megophryids, brachycephalids, hylids, dendrobatids, bufonids, microhylids, hyperoliids, arthroleptids, ceratobatrachids, petropedetids and the ranids.

The most interesting families, and perhaps those most in need of further, more comprehensive systematic analysis, are those that exhibit the greatest range of variation of all of the character states in Table 1.1; these include the Australian Limnodynastidae and Myobatrachidae, the New and Old World Hylidae and Microhylidae, and the African Pyxicephalidae, Brevicipitidae and Arthroleptidae. A single genus of the latter family, *Leptopelis*, is almost entirely comprised of scansorial species, but also contains two taxa that possess all of the characteristics of semifossorial species.

Why should anyone really care about this imbalance in our knowledge of basic systematics and amphibian physiological function? As the remaining amphibious descendents of the first terrestrial vertebrates to invade the land, they present unique evolutionary models for our understanding of the rigors and solutions of these "fishes" to the aquatic–terrestrial transition to more terrestrial tetrapod vertebrates.

6.3 Population Declines and Environmental Physiology

As a class, amphibians also represent the vertebrate group that appears most at risk to changing environmental conditions. A greater proportion of amphibians met the IUCN standards for threatened, endangered or extinct species than any other vertebrate group in the past 25 years (Stuart *et al.* 2004). In many ways amphibians have become the "canary in the mine" for indicating potential problems with environmental change. The current rate of decline exceeds historic background rates of extinction by 200 fold (McCallum 2007). The limitation with this "canary" approach is that if we do not understand how amphibians normally function and respond to environmental variables, then it is impossible to interpret what environmental variables are important to their evolutionary success and what is their physiological capacity/incapacity to handle these environmental challenges. This means that amphibian conservation biology has at its root understanding of comparative environmental physiology. It also means that if we cannot project or extrapolate this knowledge across evolutionary lineages, the cause may be lost.

The amphibian decline phenomenon has been the subject of much interest in the past three decades (see review of Beebe and Griffiths 2005). Much of the confusion as to the causes for this decline has arisen from an inability to attribute a cause to the "enigmatic" declines, which represents 47% of the total documented population declines (Stuart *et al.* 2004). Enigmatic declines are those of unknown cause or disease. The other 53% of population declines have been attributed to the more customary causes of "overexploitation" (11%) and "habitat loss" (42%). Environmental changes (rainfall, temperature, ultraviolet radiation and chemical pollutants) have been suggested to contribute to both the enigmatic and habitat loss declines, being involved in effectively 90% of all declines. Habitat loss by definition means that habitat alteration exceeds the physiologic capacity for a species to survive. It does not describe whether the decline is a consequence of an altered thermal, hydric or energetic environment.

Enigmatic declines are subtler than marked habitat alteration, and encompass both disease or unknown causes. An excellent review of the fungal, viral and bacterial agents involved in disease-based enigmatic declines can be found in Carey *et al.* (1999). Enigmatic declines are more common in species associated with riparian habitats and narrow, high-elevation distributions (Lips *et al.* 2003; Stuart *et al.* 2004). Clearly, water availability and temperature and their potential impacts on energy and immune competence are contributing to these enigmatic declines (Carey and Alexander 2003). This is consistent with hydro-periods being an important determinant of reproductive success. It is important to realize that it is not just the environments required for adult survival but the conditions that allow successful reproduction. Consequently, larval environments are also an important consideration in understanding population declines.

The degree of genetic diversity measured as heterozygosity is significantly correlated with life zone (greater in the tropics) and habitat type, decreasing from terrestrial and arboreal species, to aquatic and subterranean forms (Nevo and Bieles 1991). Therefore genetic diversity decreases with decreasing ecological heterogeneity (Nevo and Bieles 1991). Amphibians tend to have highly genetically substructured populations, with restricted gene flow (see Shaffer *et al.* 2000). This has led to high endemism (Williams *et al.* 2000) and a propensity to not be at climatic equilibrium as a consequence of historic climatic fluctuations (Araujo and Pearson 2005). These genetic and ecological consequences result from the propensity of amphibians to rapidly dehydrate, their limited tolerance of elevated temperatures and their limited mobility (Blaustein *et al.* 1994) preventing gene flow in isolated populations and limiting recolonization following local population extinctions. Essentially, their relatively fragile physiological capacities predispose them to extinction due to local environmental change or severe climatic events, independent of habitat loss and overexploitation. Amazingly little is known about the genetic bases for intraspecific and interspecific variation in physiological performance characteristics. What are the physiological and morphological characteristics that allow for increased mobility and thermal and dehydration tolerance allowing for movement of amphibians to recolonize environments or survive environmental stress imposed by changing environmental conditions? Obviously within the Class Amphibia there are behavioral, morphological and physiological capacities that allow habitation of extreme environmental conditions.

6.4 Genomics, Proteomics and Metabolomics

The reductionist biological paradigm for any physiological process starts with DNA, carries through the transcription of that DNA as messenger RNA, and the eventual translation of the messenger RNA into proteins. These proteins interacting with environmental exchanges functionally create the intracellular and extracellular environment present within the organism. The regulation of the production of mRNA (genomics), the synthesis of proteins (proteomics) and the effects of these changes on the concentrations of extra- and intracellular molecules (metabolomics) is essentially understanding organismal function from a physiological perspective. The past has entailed narrow glimpses of these processes at the level of a single gene or a few genes at a time. The times are changing to where it is now possible to monitor tens of thousands genes at a time with DNA microarrays, and advanced protein and solute analyses. A DNA microarray allows an investigator to monitor the expression of mRNA from thousands of genes present in the DNA library at a time. In the past investigators might have monitored the regulation of one or two genes at a time. The ability to now monitor a large fraction of the genome at a single time not only provides the basis for monitoring known genes that control enzyme concentration, but also allows the ability to "fish" for groups of genes that change in similar fashion with a change in environmental condition. This opens up the possibility of identifying which genes are grouped to a particular stress, as well as delineating unexpected genes that play a role in a physiological response to an environmental challenge. Though this ability to "fish" for genes up or down regulated is not hypothesis driven science up front, it opens unexpected links to be tested in an hypothesis driven fashion.

Not all mRNA necessarily leads to changes in protein expression. The field of proteomics, and its accompanying analytical technologies, allows for the quantitative determination of thousands of proteins simultaneously. By coupling this to the emerging field of metabolomics, where the detection and concentrations of thousands of small molecules are monitored simultaneously, it will be possible in the future to monitor how transcription, translation and their effects interact with the environment to determine the actual conditions cells face with changing environmental conditions. Right now our vision has been as narrow as one or two genes at a time and their effects on a few chemical species. Recent application of these technologies has broadened this to tens of genes (Cowan *et al.* 2000; Cai and Storey 1997a, b; Storey *et al.* 1997, 1999; Storey 1999, 2004). With the advent of new technologies this will soon be hundreds of thousands genes at a time. Couple this to phylogenetic diversity and two things become obvious. First the analyses appear daunting, but with their solution answers as to what are the integrated physiological responses to particular environmental stresses and why some species handle these better than others will finally become clear.

6.5 Levels of Organization

An amphibian can be studied from a holistic or reductionist perspective. The holistic perspective entails its interactions with the environment as well as its interactions with other organisms both intra- and interspecifically. A reductionist approach could entail studies at a progressively lower level, for example an organ system, organ, tissue, cell, organelle or molecular/biochemical level. Any organism can only be partially understood from studies at a given level of holism or reductionism. Real understanding of organismal function can only occur with the integration of all levels of study. Competitive science has tended to create specialists at a given level to ensure a flow of resources to support the laboratory. This creates great expertise at a particular level, but there is little integration of the information to other levels of study. This specialization tends to be reinforced by professional societies that tend to focus upon a limited breadth of organizational levels. No one level of organizational study is better or worse science, though reductionists tend to ignore holistic ecologists, and ecologists dismiss reductionists as students of plumbing. A successful future will entail scientists that not only tolerate colleagues operating at a different level of organization, but actually embrace the strengths different levels of study bring to our overall understanding of organismal function. This requires a degree of courage on the part of investigators to actually operate out of their "comfort zone" at different levels of organization occasionally and follow the problem at hand. The successful amphibian biologist of the future will have colleagues operating at levels of organization different from, yet complimentary to, their own levels. The reductionist will aspire to put their findings in context with the ecology of the species they are studying and the ecologist will strive to understand whether physiological limitations or specializations constrain the distribution of their species in space and time. The time for integrated systems biology has arrived. We now need the nerve and courage to exploit systems biology.

Bibliography

Abdulnour-Nakhoul, S., Khuri, R. N. and Nakhoul, N. L. (1994). Effect of norepinephrine on cellular sodium transport in *Ambystoma* kidney proximal tubule. *Am J Physiol* **267**, F725–F736.

Abe, A. S. (1991). Responses to temperature in the oxygen uptake of awake and dormant frogs (Amphibia, Leptodactylidae). *Studies Neotropical Fauna Environ* **26**, 135–141.

Abe, A. S. (1995). Estivation in South American amphibians and reptiles. *Brazil J Med Biol Res* **28**, 1241–1247.

Abe, A. S. and Bicudo, J. (1991). Adaptations to salinity and osmoregulation in the frog *Thoropa miliaris*. *Zool Anz* **227**, 313–318.

Abel, D., Searle, D. B. and Borass, M. E. (1992). Periodicities and transient shifts in anuran (*Xenopus laevis*, *Rana clamitans*) oxygen consumption revealed with flow-through respirometry. *Comp Biochem Physiol A* **101**, 425–432.

Abercrombie, R. F., Putnam, R. W. and Roos, A. (1983). The intracellular pH of frog skeletal regulation in isotonic solutions. *J Physiol* **345**, 175–187.

Abouheif, E. (1999). A method to test the assumption of phylogenetic independence in comparative data. *Evol Ecol Res* **1**, 895–909.

Acher, R., Chauvet, J. and Rouille, A. (1997). Adaptive evolution of water homeostasis regulation in amphibians: vasotocin and hydrins. *Biol Cell* **89**, 283–291.

Adolph, E. F. (1927). The excretion of water by the kidneys of frogs. *Am J Physiol* **81**, 315–324.

Adolph, E. F. (1932). The vapor tension relations of frogs. *Biol Bull* **62**, 112–125.

Adolph, E. F. (1936). Epinephrine and urine formation in the frog. *Am J Physiol* **115**, 200–209.

Agre, P., Preston, G. M., Smith, B. L., Jung, J. S., Raina, S., Moon, C., Guggino, W. B. and Nielsen, S. (1993). Aquaporin CHIP: the archetypal molecular water channel. *Am J Physiol* **265**, F463–F476.

Alcocer, I., Santacruz, X., Steinbeisser, H., Theirauch, K. H. and Del Pino, E. M. (1992). Ureotelism as the prevailing mode of nitrogen excretion in larvae of the marsupial frog *Gastrotheca riobambae* (Anura, Hylidae). *Comp Biochem Physiol* **101A**, 229–231.

Alexander, S. S. and Blerby, C. W. (1938). Experimental studies on the sexual cycle of the South African clawed toad (*Xenopus laevis*). *J Exp Biol* **15**, 74–81.

Altig, R. A. and McDearman, W. (1975). Percent assimilation and clearance times of five anuran tadpoles. *Herpetologica* **31**, 67–69.

Altig, R. A. and McDiarmid, R. W. (2007). Morphological diversity and evolution of eggs and clutch structure in Amphibians. *Herpetol Monographs.* 27, 1–32.

Altland, P. D. and Brace, K. C. (1962). Red cell life span in the turtle and toad. *Am J Physiol* **203**, 1188–1190.

Alvarado, R. H. (1967). The significance of grouping on water conservation in *Ambystoma*. *Copeia* 1967, 667–668.

Alvarado, R. H. (1972). The effects of dehydration on water and electrolytes in *Ambystoma tigrinum*. *Physiol Zool* **45**, 43–53.

Alvarado, R. H. (1979). Amphibians. In: *Comparative Physiology of Osmoregulation in Animals*, ed. G. M. O. Maloiy, Vol. 1, pp. 261–303. Academic Press.

Alvarado, R. H. and Moody, A. (1970). Sodium and chloride transport in tadpoles of the bullfrog *Rana catesbeiana*. *Am J Physiol* **218**, 1510–1516.

Alvarez, D. and Nicieza, A. G. (2002a). Effects of induced variation in anuran larval development on postmetamorphic energy reserves and locomotion. *Oecologia* **131**, 186–195.

Alvarez, D. and Nicieza, A. G. (2002b). Effects of temperature and food quality on anuran larval growth and metamorphosis. *Functional Ecology* **16**, 640–648.

Ambrosoli, G., Cananau, S. and Cerreteli, P. (1973). Energetics of twitch and tetanus in the frog gastrocnemius. *Arch Fisiol* **70**, 9–11.

Ambrosoli, G. and Cerreteli, P. (1973). The anaerobic recovery of frog muscle. *Pflugers Arch* **345**, 131–143.

Amey, A. P. and Grigg, G. C. (1995). Lipid-reduced evaporative water loss in two arboreal hylid frogs. *Comp Biochem Physiol A* **111**, 282–291.

Ammar, A., Rajerison, R. M., Roseau, S., Bloch-Faure, M. and Butlen, D. (1994). Frog glomerular vasotocin receptors resemble mammalian V1b receptors. *Am J Physiol* **267**, R1198–R1208.

Andersen, J. B., Hedrick, M. S. and Wang, T. (2003). Cardiovascular responses to hypoxia and anaemia in the toad *Bufo marinus*. *J Exp Biol* **206**, 857–865.

Andersen, J. B., Jensen, F. B. and Wang, T. (2001). Effects of temperature and oxygen availability on circulating catecholamines in the toad *Bufo marinus*. *Comp Biochem Physiol* **129A**, 473–486.

Andersen, J. B. and Wang, T. (2003). Cardiorespiratory effects of forced activity and digestion in toads. *Physiol Biochem Zool* **76**, 459.

Anderson, K. A. and Beitinger, T. L. (1979). Body heating and cooling in the mudpuppy, *Necturus maculosus*. *Herpetologica* **35**, 234–239.

Andrade, D. V. and Abe, A. S. (1997). Evaporative water loss and oxygen uptake in two casque-headed tree frogs, *Aparasphenodon brunoi* and *Corythomantis greeningi* (Anura, Hylidae). *Comp Biochem Physiol A* **118**, 685–689.

Angell, C. S., Hipona, F. A. and Novick, A. (1966). Angiocardiographic studies of the Anuran double circulation. *Anat Rec* **156**, 41–53.

Angilletta, M. J., Niewiarowski, P. H. and Navas, C. A. (2002). The evolution of thermal physiology in ectotherms. *J Thermal Biol* **27**, 249–268.

Anholt, B. R., Hotz, H., Guex, G. and Smelitsch, R. D. (2003). Overwinter survival of *Rana lessonae* and its hemiclonal associate *Rana esculenta*. *Ecology* **84**, 391–397.

Arak, A. (1983). Male-male competition and mate choice in anuran amphibians. In: *Mate Choice*, ed. P. Bateson, pp. 181–210. Cambridge University Press, New York.

Araujo, M. B. and Pearson, R. G. (2005). Equilibrium of species' distributions with climate. *Ecography* **28**, 693–695.

Arieff, A. I., Guisado, R. and Lazarowitz, V. C. (1977). Pathophysiology of hyperosmolar states. In: *Disturbances in Body Fluid Osmolality*, eds T. E. Andreoli, J. J. Grantham, and F. C. Rector, pp. 227–250 . Am. Physiol. Soc.: Bethesda, MD.

Ashley, H., Katti, P. and Frieden, E. (1968). Urea excretion in the bullfrog tadpole: effect of temperature, metamorphosis, and thyroid hormones. *Dev Biol* **17**, 293–307.

Ashton, R. E. (1994). Tracking with radioactive tags. In: *Measuring and Monitoring Biological Diversity: Standard Methods for Amphibians*, eds W. R. Heyer, M. A. Donnelly,

R. W. McDiarmid, L. C., Hayek and M. S. Foster, pp. 158–163. Sminthsonian Institution Press: Washington.
Atkinson, D. E. (1977). *Cellular Energy Metabolism and its Regulation*. Academic Press: New York.
Atkinson, D. E. (1992). Functional roles of urea synthesis in vertebrates. *Physiol Zool* **65**, 243–267.
Atkinson, D. E. and Bourke, E. (1984). The role of ureagenesis in pH homeostasis. *Trends Biochem Sci* **9**, 297–300.
Atkinson, D. E. and Camien, M. N. (1982). The role of urea synthesis in the removal of metabolic bicarbonate and and the regulation of blood pH. *Curr Top Cell Regul* **21**, 261–302.
Avery, R. A. (1982). Field studies in body temperatures and thermoregulation. In: *Biology of the Reptilia*, eds C. Gans and F. H. Pough, Vol. 12, pp. 93–166. Academic Press: New York.
Bagnara, J. T. (2003). Enigmas of pterhodin, a red melanosomal pigment of tree frogs. *Pigment Cell Res* **16**, 510–516.
Bagnara, J. T. and Hadley, M. E. (1973). *Chromatophores and Color Change. The Comparative Physiology of Animal Pigmentation*. Prentice-Hall: Englewood-Cliffs.
Baker, C. A. and Hillyard, S. D. (1992). Capacitance, short-circuit current and osmotic water flow across different regions of the isolated toad skin. *J Comp Physiol B* **162**, 707–713.
Bakken, G. S. (1976). A heat transfer analysis of animals: Unifying concepts and the application of metabolism chamber data to field ecology. *J Theor Biol* **60**, 337–384.
Bakken, G. S. (1981). Heated taxidermic mounts: A means of measuring the standard operative temperature affecting small animals. *Ecology* **62**, 311–318.
Bakken, G. S. (1992). Measurement and application of operative and standard operative temperatures in ecology. *Am Zoologist* **32**, 194–216.
Bakken, G. S. and Gates, D. M. (1975). Heat-transfer analysis of animals: Some implications for field ecology, physiology, and evolution. In: *Perspectives of Biophysical Ecology*, eds D. M. Gates and R. B.Schmerl, pp. 255–290. Springer-Verlag: New York.
Bakken, G. S., Santee, W. R. and Erskine, D. J. (1985). Operative and standard operative temperature: Tools for thermal energetics studies. *Am Zoologist* **25**, 933–943.
Baldwin, A. L., Ferrer, P., Rozum, J. S. and Gore, R. W. (1993). Regulation of water balance between blood and lymph in the frog *Rana pipiens*. *Lymphology* **26**, 4–18.
Baldwin, R. A. (1974). The water balance response of the "pelvic" patch of *Bufo punctatus* and *Bufo boreas*. *Comp Biochem Physiol A* **47**, 1285–1295.
Balinsky, J. B. (1981). Adaptation of nitrogen metabolism to hyperosmotic environment in amphibia. *J Exp Zool* **215**, 335–350.
Balinsky, J. B. and Baldwin, E. (1961). The mode of excretion of ammonia and urea in *Xenopus laevis*. *J Exp Biol* **38**, 695–705.
Balinsky, J. B., Choritz, E. L., Coe, C. G. and van der Schans, G. S. (1967). Amino acid metabolism and urea synthesis in naturally aestivating *Xenopus laevis*. *Comp Biochem Physiol* **22**, 59–68.
Balinsky, J. B., Cragg, M. and Baldwin, E. (1961). The adaptation of amphibian waste nitrogen excretion to dehydration. *Comp Biochem Physiol* **3**, 236–244.
Bannikov, A. G. (1948). On the fluctuation of anuran populations. *Dokl Acad Nauk SSSR* **61**, 131–134.
Barch, S. H., Shaver, J. R. and Wilson, G. B. (1966). An electron microscopic study of the nephric unit in the frog. *Trans Amer Microsc Soc* **85**, 350–359.
Barica, J. and Mathias, J. A. (1979). Oxygen depletion and winter-kill risk in small prarie lakes under extended ice cover. *J Fish Res Board Can* **36**, 980–986.

Bartelt, P. E. and Peterson, C. R (2005). Physically modelling operative temperatures and evaporation rates in amphibians. *J Thermal Biol* **30**, 93–102.

Bartholomew, G. A., Vleck, D.and Vleck, C. A. (1981). Instantaneous measurement of oxygen consumption during pre-flight warm-up and psot-flight cooling in sphingid moths. *J Exp Biol* **90**, 17–32.

Baudinette, R. V., Miller, A. M. and Sarre, M. P. (2000). Aquatic and terrestrial locomotory energetics in a toad and a turtle: a search for generalizations among ectotherms. *Physiol Biochem Zool* **73**, 672–682.

Baustian, M. (1988). The contribution of lymphatic pathways during recovery from hemorrhage in the toad *Bufo marinus*. *Physiol Zool* **61**, 555–563.

Baxter, C. F., Baldwin, R. E., Lu, P., Imaki, H. and Sturman, J. A. (1993). Taurine in toad brain and blood under different conditions of osmolality: an immunohistochemical study. *Neurochem Res* **18**, 425–435.

Baxter, C. F. and Ortiz, C. L. (1966). Amino acids and the maintenance of osmotic equilibrium in brain tissue. *Life Sciences* **5**, 2321–2329.

Beachy, C. K. and Bruce, R. C. (1992). Lunglessness in plethodontid salamanders is consistent with the hypothesis of a mountain stream origin: A response to Ruben and Boucot. *Am Nat* **139**, 839–847.

Beard, K. H., Vogt, K. A. and Kulmatiski, A. (2002). Top-down effects of a terrestrial frog on forest nutrient dynamics. *Oecologia* **133**, 583–593.

Beaupre, S. J. (2005). Ratio representations of specific dynamic action (mas-specific SDA and SDA coefficient) do not standardize for body mass and meal size. *Physiol Biochem Zool* **78**, 126–131.

Beck, C. W. and Congdon, J. D. (2003). Energetics of metamorphic climax in the southern toad (*Bufo terrestris*). *Oecologia* **137**, 344–351.

Becker, B., Rathscheck, W., Rathscheck, H., Muller, H. and Schroeder, W. (1970). Oxygen tension, oxygen consumption and blood flow in the extremity muscles of the unanesthetized frog. *Pflugers Arch* **321**, 15–33.

Beebe, T. J. C. and Griffiths, R. A. (2005). The amphibian decline crisis:: A watershed for conservation biology? *Biolog Conserv* **125**, 271–285.

Beebee, T. J. C. (1995). Amphibian breeding and climate. *Nature* **374**, 219–220.

Beitinger, T. L. and Anderson, K. (1979). Internal convective and conductive heat loss in the mudpuppy, *Necturus maculosus*. *Comp Biochem Physiol A* **62**, 1035–1039.

Bennett, A. F. (1974). Enzymatic correlates of activity metabolism in anuran amphibians. *Am J Physiol* **226**, 1149–1151.

Bennett, A. F. (1978). Activity metabolism of the lower vertebrates. *Annu Rev Physiol* **40**, 447–469.

Bennett, A. F. (1984). Thermal dependence of muscle function. *Am J Physiol* **247**, R217–229.

Bennett, A. F. and Licht, P. (1972). Anaerobic metabolism during activity in lizards. *J Comp Physiol* **81**, 277–288.

Bennett, A. F. and Licht, P. (1973). Relative contributions of anaerobic and aerobic energy production during activity in Amphibia. *J Comp Physiol B* **87**, 351–360.

Bennett, A. F. and Wake, M. H. (1974). Metabolic correlates of activity in the caecilian *Geotrypetes seraphini*. *Copeia* 1974, 764–769.

Bennett, W. O., Summers, A. P. and Brainerd, E. L. (1999). Confirmation of the passive exhalation hypothesis of a terrestrial caecilian, *Dermophis mexicanus*. *Copeia* 1999, 206–209.

Bentley, P. J. (1958). The effects of neurohypophyseal extracts on water transfer across the wall of the isolated urinary bladder of the toad *Bufo marinus*. *J Endocrinology* **17**, 201–209.

Bentley, P. J. (1966). Adaptations of Amphibia to arid environments. *Science* **152**, 619–623.

Bentley, P. J. (1969). Neurohypophyseal hormones in Amphibia: A comparison of their actions and storage. *General and Comparative Endocrinology* **13**, 39–44.

Bentley, P. J. (1971). *Endocrines and Osmoregulation. ZooPhysiology and Ecology.* Springer Verlag.

Bentley, P. J. (1973). Role of the skin in amphibian sodium metabolism. *Science* **181**, 686–687.

Bentley, P. J. (1975). The electrical P.D. across the integument of some neotenous urodele amphibians. *Comp Biochem Physiol* **50A**, 639–643.

Bentley, P. J. and Baldwin, G. F. (1980). Comparison of transcutaneous permeability in skins from larval and adult samamanders (*Ambystoma tigrinum*). *Am J Physiol* **239**, R505–R508.

Bentley, P. J. and Heller, H. (1964). The action of neurohypophysial hormones on the water and sodium metabolism of urodele amphibians. *J Physiol* **171**, 434–453.

Bentley, P. J. and Heller, H. (1965). The water retaining action of vasotocin on the fire salamander (*Salamandra maculosa*): The role of the urinary bladder. *J Physiol* **181**, 124–129.

Bentley, P. J., Lee, A. K. and Main, A. R. (1958). Comparison of dehydration and hydration of two genera of frogs (*Heleioporus* and *Neobatrachus*) that live in areas of varying aridity. *J Exp Biol* **35**, 677–684.

Bentley, P. J. and Main, A. R. (1972). Zonal differences in permeability of the skin of some anuran Amphibia. *Am J Physiol* **223**, 361–363.

Bentley, P. J. and Schmidt-Nielsen, K. (1971). Acute effects of sea water on frogs (*Rana pipiens*). *Comp Biochem Physiol A* **40**, 547–548.

Bentley, P. J. and Shield, J. W. (1973). Respiration of some urodele and anuran amphibia. II. In air, role of the skin and lungs. *Comp Biochem Physiol A* **46**, 29–38.

Bentley, P. J. and Yorio, T. (1977). The permeability of the skin of a neotenous urodele amphibian, the mudpuppy *Necturus maculosus*. *J Physiol* **265**, 537–547.

Bentley, P. J. and Yorio, T. (1979). Do frogs drink? *J Exp Biol* **79**, 41–46.

Benton, M. J. (1990). Phylogeny of the major tetrapod groups: morphological data and divergence dates. *J Mol Evol* **30**, 409–424.

Bentzel, C. J., Davies, M., Scott, W. N., Zatzman M. and Soloman, A. K. (1968). Osmotic volume flow in the proximal tubule of *Necturus* kidney. *J Gen Physiol* **51**, 517–533.

Bentzel, C. J., Parsa, B. and Hare, D. K. (1969). Osmotic flow across proximal tubule of *Necturus*: correlation of physiologic and anatomic studies. *Am J Physiol* **217**, 570–580.

Bentzel, C. J. and Reczek, P. R. (1978). Permeability changes in *Necturus* proximal tubule during volume expansion. *Am J Physiol* **234**, F225–234.

Berger, L. (1982). Hibernation of the European water frogs (*Rana esculenta* complex). *Zool Pol* **29**, 57–72.

Berman, D. I., Leirkh, A. N. and Mikhailova, E. I. (1984). Winter hibernation of the Siberian salamander, *Hynobius keyserlingi*. *Zhurnal Evolyutsionnoi Biokihimii i Fiziologii (J Evol Biochem Physiol)* **20**, 323–327.

Berman, D. M., Soria, M. O., Proto de Grifasi, M. and Coviello, A. (1994). Atrial natriuretic peptide inhibited the natriferic and hydrosmotic effects of arginine vasotocin on toad skin. *Comp Biochem Physiol C Pharmacol Toxicol Endocrinol* **109**, 277–282.

Bernheim, H. A. and Kluger, M. J. (1976). Fever and antipyresis in the lizard *Dipsosaurus dorsalis*. *Am J Physiol* **231**, 198–203.

Berven, K. A., Gill, D. E. and Smith-Gill, S. J. (1979). Counter-gradient selection in the green frog, *Rana clamitans*. *Evolution* **33**, 609–623.

Beuchat, C. A., Pough, F. H. and Stewart, M. M. (1984). Response to simultaneous dehydration and thermal stress in three species of Puerto rican frogs. *J Comp Physiol* **154B**, 579–585.

Bevier, C. R. (1997). Utilization of energy substrates during calling activity in tropical frogs. *Behav Ecol Sociobiol* **41**, 343–352.

Bicego, K. C., Barros, R. C. H. and Branco, L. G. S. (2007). Physiology of temperature regulation: Comparative aspects. *Comp Biochem Physiol A* **147**, 616–639.

Bicego, K. C., Steiner, A. A. Antunes-Rodrigues, J. and Branco, L. G. S. (2002a). Indomethacin impairs LPS-induced behavioral fever in toads. *J Applied Physiol* **93**, 512–516.

Bicego, K. C., Steiner, A. A., Gargaglioni, L. H. and Branco, L. G. S. (2002b). Is lactate a mediator of hypoxia-induced anapyrexia? *Pflugers Archiv* **444**, 810–815.

Bicego-Nahas, K. C., Steiner, A. A., Carnio, E. C., Antunes-Rodrigues, J. and Branco, L. G. S. (2000). Antipyretic effect of arginine vasotocin in toads. *Am J Physiol* **278**, R1408–1414.

Bicudo, J. E. P. and Johansen, K. (1979). Respiratory gas exchange in the airbreathing fish, *Synbranchus marmoratus*. *Environ Fish Biol* **4**, 55–64.

Bieter, R. N. (1930). The effect of the splanchnics upon glomerular blood flow in the frog kidney. *Am J Physiol* **91**, 436–460.

Biewald, V. G. A. (1980). Responses of cardiac vagal receptors to pressoric and osmotic stimulation in the frog. *Zool J Physiol* **84**, 383–395.

Binzoni, T., Ferretti, G., Barbalat, F. and Cerreteli, P. (1990). Energetics of resting anaerobic frog gastrocnemius at different temperatures by 31P-NMR. *Respir Physiol* **82**, 137–148.

Bishop, W. R., Mumbach, M. W. and Scheer, B. T. (1961). Interrenal control of sodium transport across frog skin. *Am J Physiol* **200**, 451–453.

Blanchard, E. M. and Solaro, R. J. (1984). Inhibition of activation and troponin calcium binding of dog cardiac myofibrils by acidic pH. *Circ Res* **55**, 382–391.

Blaustein, A. R., Wake, D. B. and Sousa, W. P. (1994). Amphibian declines: judging stability, persistence, and susceptibility of populations to local and global extinctions. *Conserv Biol* **8**, 60–71.

Blaxter, K. (1989). *Energy Metabolism in Animals and Man*. Cambridge University Press: Cambridge.

Blaylock, L. A., Ruibal, R. and Platt-Aloai, K. (1976). Skin structure and wiping behavior of phyllomedusine frogs. *Copeia* 1976, 283–295.

Blazer-Yost, B. L., Liu, X. and Helman, S. I. (1998). Hormonal regulation of ENaCs: insulin and aldosterone. *Am J Physiol* **274**, C1373–C1379.

Blomberg, S. P., Garland, T. and Ives, A. R. (2003). Testing for phylogenetic signal in comparative data: behavioral traits are more labile. *Evolution* **57**, 717–745.

Boas, E. V. (1882). Ueber den vonus arteriosus und die Arterienbogen der Amphibien. *Morph Jb* 7, 488–572.

Bohlin, R. G. and Beitinger, T. L. (1979). Heat exchange in the aquatic salamander, *Amphiuma means*. *J Thermal Biol* **4**, 63–67.

Bokma, F. (2004). Evidence against universal metabolic allometry. *Functional Ecology* **18**, 184–187.

Bolt, J. R. (1977). Dissorophoid relationships and ontogeny, and the orogin of the Lissamphibia. *J Paleontology* **51**, 235–249.

Bolton, P. M. and Beuchat, C. A. (1991). Cilia in the urinary bladder of reptiles and amphibians: a correlate of urate production? *Copeia* 1991, 711–717.

Bond, A. N. (1960). An analysis of the response of salamander gills to changes in the oxygen concentration of the medium. *Dev Biol* 2, 1–20.

Bonetto, R., del Castillo, E. and Uranga, J. (1981). Effect of glomerulopressin on the lymphatic hearts of the toad, probably mediated by prostaglandins. *Cell Mol Life Sci* 37, 209–210.

Booth, D. T. (1994). Effects of unstirred layers on cutaneous gas exchange. In: *Amphibian Biology*, eds H. Heatwole and G.T. Barthalmus, pp. 168–177. Surrey Beatty and Sons: Australia.

Booth, D. T. (2006). Effect of soil type on burrowing behaviour and cocoon formation in the green-striped burrowing frog, *Cyclorana alboguttata*. *Can J Zool* **84**, 832–838.

Borkin, L. J. (1999). Distribution of amphibians in North Africa, Europe, Western Asia, and the Former Soviet Union. In: *Patterns of Distribution of Amphibians A Global Perspective*, ed. W. E. Duellman, pp. 329–420. Johns Hopkins University Press: Baltimore and London.

Borsook, H. (1936). The specific dynamic action of protein and amino acids in animals. *Biol Rev* **11**, 147–180.

Bott, P. A. (1962). Micropuncture study of renal excretion of water, K, Na, and Cl in *Necturus*. *Am J Physiol* **203**, 662–666.

Boutilier, R. G. (2001). Mechanisms of metabolic defense against hypoxia in hibernating frogs. *Respir Physiol* **128**, 365–377.

Boutilier, R. G. (2001). Mechanisms of cell survival in hypoxia and hypothermia. *J Exp Biol* **204**, 3171–3181.

Boutilier, R. G., Donohoe, P. H., Tattersall, G. J. and West, T. G. (1997). Hypometabolic homeostasis in overwintering aquatic amphibians. *J Exp Biol* **200**, 387–400.

Boutilier, R. G., Emilio, M. G. and Shelton, G. (1986). Aerobic and anaerobic correlates of mechanical work by gastrocnemius muscles of the aquatic amphibian *Xenopus laevis*. *J Exp Biol* **122**, 223–235.

Boutilier, R. G., Emilio, M. G. and Shelton, G. (1986). The effects of mechanical work on electrolyte and water distribution in amphibian skeletal muscle. *J Exp Biol* **120**, 333–350.

Boutilier, R. G., Glass, M. L. and Heisler, N. (1986). The relative distribution of pulmocutaneous blood flow in *Rana catesbeiana*: effects of pulmonary or cutaneous hypoxia. *J Exp Biol* **126**, 33–39.

Boutilier, R. G., Glass, M. L. and Heisler, N. (1987). Blood gases, and extracellular/intracellular acid-base status as a function of temperature in the anuran amphibians *Xenopus laevis* and *Bufo marinus*. *J Exp Biol* **130**, 13–25.

Boutilier, R. G. and Heisler, N. (1988). Acid-base regulation and blood gases in the anuran amphibian, *Bufo marinus*, during environmental hypercapnia. *J Exp Biol* **134**, 79–98.

Boutilier, R. G., McDonald, D. G. and Toews, D. P. (1980). The effects of enforced activity on ventilation, circulation and blood acid-base balance in the aquatic gill-less urodele, *Cryptobranchus alleganiensis*; a comparison with the semi-terrestrial anuran, *Bufo marinus*. *J Exp Biol* **84**, 289–302.

Boutilier, R. G., Randall, D. J., Shelton, G. and Toews, D. P. (1979). Acid-base relationships in the blood of the toad, *Bufo marinus*. I. The effects of environmental CO_2. *J Exp Biol* **82**, 331–344.

Boutilier, R. G., Randall, D. J., Shelton, G. and Toews, D. P. (1979). Acid-base relationships in the blood of the toad, *Bufo marinus*. II. The effects of dehydration. *J Exp Biol* **82**, 345–355.

Boutilier, R. G., Randall, D. J., Shelton, G. and Toews, D. P. (1979). Acid-base relationships in the blood of the toad, *Bufo marinus*. III. The effects of burrowing. *J Exp Biol* **82**, 357–365.

Boutilier, R. G. and Shelton, G. (1986). Gas exchange, storage and transport in voluntarily diving *Xenopus laevis*. *J Exp Biol* **126**, 133–155.

Boutilier, R. G. and St-Pierre, J. (2000). Surviving hypoxia without really dying. *Comp Biochem Physiol A Mol Integr Physiol* **126**, 481–490.

Boutilier, R. G. and St-Pierre, J. (2002). Adaptive plasticity of skeletal muscle energetics in hibernating frogs: mitochondrial proton leak during metabolic depression. *J Exp Biol* **205**, 2287–2296.

Boutilier, R. G. and Toews, D. P. (1977). The effect of progressive hypoxia on respiration in the toad *Bufo marinus*. *J Exp Biol* **68**, 99–107.

Boutilier, R. G. and Toews, D. P. (1981). Respiratory, circulatory and acid-base adjustments to hypercapnia in a strictly aquatic and predominantly skin-breathing urodele, *Cryptobranchus alleganiensis*. *Respir Physiol* **46**, 177–192.

Boutilier, R. G. and Toews, D. P. (1981). Respiratory properties of blood in a strictly aquatic and predominantly skin-breathing urodele, *Cryptobranchus alleganiensis*. *Respir Physiol* **46**, 161–176.

Boyd, S. K. and Moore, F. L. (1990). Autoradiographic localization of putative arginine vasotocin receptors in the urodele amphibian. *General Comp Endocrinology* **78**, 344–350.

Boyer, D. R. (1965). Ecology of the basking habit in turtles. *Ecology* **46**, 99–118.

Bradford, D. F. (1983). Winterkill, oxygen relations, and energy metabolism of a submerged dormant amphibian, *Rana muscosa*. *Ecology* **64**, 1171–1183.

Bradford, D. F. (1984). Temperature modulation in a high-elevation amphibian, *Rana muscosa*. *Copeia* 1984, 966–976.

Brafield, A. E. and Llewellyn, M. J. (1982). *Animal Energetics*. Blackie: Glasgow and London.

Brainerd, E. L. (1999). New perspectives on the evolution of lung ventilation mechanisms in vertebrates. *Exp Biol Online* **4:2**, 11–28.

Branco, L. G. S. and Glass, M. L. (1995). Ventilatory responses to carboxyhaemoglobinanemia and hypoxic hypoxia in *Bufo paracnemias*. *J Exp Biol* **198**, 1417–1421.

Branco, L. G. S., Glass, M. L. and Hoffmann, A. (1992). Central chemoreceptor drive to breathing in unanesthetized toads, *Bufo paracnemis*. *Respir Physiol* **87**, 195–204.

Branco, L. G. S., Glass, M. L., Wang, T. and Hoffmann, A. (1993). Temperature and central chemoreceptor drive to ventilation in toad (*Bufo paracnemis*). *Respir Physiol* **93**, 337–346.

Brattstrom, B. H. (1963). A preliminary review of the thermal requirements of amphibians. *Ecology* **44**, 238–255.

Brattstrom, B. H. (1968). Thermal acclimation in anuran amphibians as a function of latitude and altitude. *Comp Biochem Physiol* **24**, 93–111.

Brattstrom, B. H. (1970). Amphibia. In: *Comparative Physiology of Thermoregulation: Invertebrates and Nonmammalian Vertebrates*, ed. G. C. Whittow, Vol. 1, pp.135–166. Academic Press: New York.

Brattstrom, B. H. (1970). Thermal acclimation in Australian amphibians. *Comp Biochem Physiol* **35**, 69–103.

Brattstrom, B. H. (1979). Amphibian temperature regulation studies in the field and laboratory. *Am Zoologist* **19**, 345–356.

Brauer, A. (1902). Beitrage zur kenntniss der entwicklung und anatomie der gymnophionen. III. Die entwicklung der excretionsorgane. *Zool Jahrb Abt Anat Ontog Tiere* **16**, 1–176.

Breckenridge, W. J. and Tester, J. R. (1961). Growth, local movements and hibernation of the Manitoba toad, *Bufo hemiophrys*. *Ecology* **42**, 637–646.

Brekke, D. R., Hillyard, S. D. and Winokur, R. M. (1991). Behavior associated with the water absorption response by the toad, *Bufo punctatus*. Copeia 1991, 393–401.

Brooks, G. A. (2000). Intra- and extra-cellular lactate shuttles. *Med Sci Sports Exer* **32**, 790–799.

Brown, A. V. and Fitzpatrick, L. C. (1981). Metabolic acclimation to temperature in the Ozark salamander *Plethodon dorsalis augusticlavius*. *Comp Biochem Physiol* **69A**, 499–503.

Brown, A. V. and Fitzpatrick, L. C. (1981). Thermal acclimation and metabolism in the gray-bellied salamander, *Eurycea multiplicata griesogaster*. *Comp Biochem Physiol* **69A**, 505–509.

Brown, G. W., Brown, W. R. and Cohen, P. P. (1959). Comparative biochemistry of urea synthesis. II. Levels of urea cycle enzymes in metamorphosing *Rana catesbeiana* tadpoles. *J Biol Chem* **234**, 1775–1780.

Brown, J. H., West, G. B. and Enquist, B. J. (2005). Yes, West, Brown and Enquist's model of allometric scaling is both mathematically correct and biologically relevant. *Functional Ecology* **19**, 735–738.

Brown, P. S. and Brown, S. C. (1973). Prolactin and thyroid hormone interactions in salt and water balance in the newt *Notophthalmus viridescens*. *General Comp Endocrinol* **20**, 456–466.

Brown, P. S. and Brown, S. C. (1977). Water balance responses to dehydration and neurohypophyseal peptides in the newt, *Notophthalmus viridescens*. *General Comp Endocrinol* **31**, 189–201.

Brown, P. S. and Brown, S. C. (1982). Effects of hypophysectomy and prolactin on the water balance response of the newt, *Taricha torosa*. *General Comp Endocrinol* **46**, 7–12.

Brown, P. S. and Frye, B. E. (1969). Effects of prolactin and growth hormone on growth and metamorphosis of tadpoles of the frog, *Rana pipiens*. *General Comp Endocrinol* **13**, 126–138.

Brown, S. C. and Brown, P. S. (1980). Water balance in the California newt, *Taricha torosa*. *Am J Physiol* **238**, R113–R118.

Brown, S. C., Savage, L. M. and Brown, P. S. (1988). Electrophysiology of newt skin: responses to salt-depletion, aldosterone, arginine vasotocin and epinephrine. *Comp Biochem Physiol A* **91**, 715–719.

Bruce, R. C. (2007). Out of the frying pan into the fire: an ecological perspective on evolutionary reversal in life history in plethodontid salamanders (Amphibia: Plethodontidae). *Evolution Ecol,* **21**, 703–726.

Bruce, R. C., Beachy, C. K., Lenzo, P. G., Proynch, S. P. and Wassersug, R. J. (1994). Effects of lung reduction on rheotactic performance in amphibian larvae. *J Exp Zool* **268**, 377–380.

Brunn, F. (1921). Beitrag zur Kenntnis der Wirkung von Hypophysenextracten auf den Wasserhaushalt des Frosches. *Zeitschrift fur die Gesammte Experimentelle Medizin* **25**, 120–125.

Bruno, M. P. and Coviello, A. (1996). Effects of angiotensin II antagonists on the contractile and hydrosmotic effect of AT II and AT III in the toad (*Bufo arenarum*). *J Comp Physiol* **165B**, 565–570.

Bucher, T. L., Ryan, M. J. and Bartholomew, G. A. (1982). Oxygen consumption during resting, calling, and nest building in the frog *Physalaemus pustulosus*. *Physiol Zool* **55**, 10–22.

Buckley, L. B. and Jetz, W. (2007). Environmental and historical constraints on global patterns of amphibian richness. *Proc R Soc B* **274**, 1167–1173.

Budtz, P. E. (1979). Epidermal structure and dynamics of the toad *Bufo bufo* deprived of the pars distalis of the pituitary gland. *J Zool* **189**, 57–92.

Budtz, P. E., Christoffersen, B. C., Johansen, J. S., Spies, I. and Willumsen, N. J. (1995). Tissue kinetics, ion transport and recruitment of mitochondria-rich cells in the skin of the toad (*Bufo bufo*) in response to exposure to distilled water. *Cell Tissue Res* **280**, 65–75.

Bungaard, M. and Forjaer-Jensen, J. (1982). Functional aspects of the ultrastructire of terminal blood vessels; a qualitative study on consecutive segments of the frog microvasculature. *Microvasc Res* **23**, 1–30.

Burggren, W. and Doyle, M. (1986). Ontogeny of regulation of gill and lung ventilation in the bullfrog, *Rana catesbeiana. Respir Physiol* **66**, 279–291.

Burggren, W. and Doyle, M. (1986). The action of acetylcholine upon heart rate changes markedly with development in bullfrogs. *J Exp Zool* **240**, 137–140.

Burggren, W. and Doyle, M. (1986). Ontogeny of heart rate regulation in the bullfrog, *Rana catesbeiana. Am J Physiol* **251**, R231–239.

Burggren, W. and Moalli, R. (1984). 'Active' regulation of cutaneous gas exchange by capillary recruitment in amphibians: experimental evidence and a revised model for skin respiration. *Respir Physiol* **55**, 379–392.

Burggren, W. and Mwalukoma, A. (1983). Respiration during chronic hypoxia and hyperoxia in larval and adult bullfrogs (*Rana catesbeiana*). I. Morphological responses of lungs, skin and gills. *J Exp Biol* **105**, 191–203.

Burggren, W. W., Bicudo, J. E., Glass, M. L. and Abe, A. S. (1992). Development of blood pressure and cardiac reflexes in the frog *Pseudis paradoxsus. Am J Physiol* **263**, R602–608.

Burggren, W. W. and Feder, M. E. (1986). Effect of experimental ventilation of the skin on cutaneous gas exchange in the bullfrog. *J Exp Biol* **121**, 445–449.

Burggren, W. W., Infantino, R. L. and Townsend, D. S. (1990). Developmental changes in cardiac and metabolic physiology of the direct-developing tropical frog *Eleuthrodactylus coqui. J Exp Biol* **152**, 129–147.

Burggren, W. W. and Johansen, K. (1986). Circulation and respiration in lungfishes (Dipnoi). *J Morph Suppl* **1**, 217–236.

Burggren, W. W. and Pinder, A. W. (1991). Ontogeny of cardiovascular and respiratory physiology in lower vertebrates. *Annu Rev Physiol* **53**, 107–135.

Burggren, W. W. and West, N. H. (1982). Changing respiratory importance of gills, lungs and skin during metamorphosis in the bullfrog, *Rana catesbeiana. Resp Physiol* **47**, 151–164.

Burton, T. C. (1986). A reassessment of the Papuan subfamily Asterophyryinae (Anura: Microhylidae). *Records of the South Australian Museum* **19**, 405–450.

Burton, T. M. and Likens, G. E. (1975). Energy flow and nutrient cycling in salamander populations in the Hubbard Brook experimental forest, New Hampshire. *Ecology* **56**, 1068–1080.

Burton, T. M. and Likens, G. E. (1975). Salamander populations and biomass in the Hubbard Brook Experimental Forest, New Hampshire. *Copeia* 1975, 541–546.

Bush, F. M. (1963). Effects of light and temperature on the gross composition of the toad, *Bufo fowleri. J Exp Zool* **153**, 1–13.

Buttemer, W. A. (1990). Effect of temperature on evaporative water loss in the Australian tree frogs *Litoria caerulea* and *Litoria chloris. Physiol Zool* **63**, 1043–1057.

Buttemer, W. A. and Thomas, C. (2003). Influence of temperature on evaporative water loss and cutaneous resistance to water vapour diffusion in the orange-thighed frog (*Litoria xanthomera*). *Aust J Zool* **51**, 111–118.

Cabanac, A. J. and Cabanac, M. (2004). No emotional fever in toads. *J Thermal Biol* **29**, 669–673.

Cabanac, M. (1999). Emotion and phylogeny. *Jap J Physiol* **49**, 1–10.

Cai, Q. and Storey, K. B. (1997). Upregulation of a novel gene by freezing exposure in the freeze-tolerant wood frog (*Rana sylvatica*). *Gene* **198**, 305–312.

Cai, Q. and Storey, K. B. (1997). Freezing-induced genes in wood frog (*Rana sylvatica*): fibrinogen upregulation by freezing and dehydration. *Am J Physiol* **272**, R1480–1492.

Campbell, J. P., Aiyawar, R. M., Berry, E. R. and Huf, E. G. (1967). Electrolytes in frog skin secretions. *Comp Biochem Physiol* **23**, 213–223.

Campbell, J. W. (1991). Excretory nitrogen metabolism. In: *Experimental and Metabolic Animal Physiology, Comparative Animal Physiology*, ed. C. L. Prosser, pp. 277–324, 4th edn. Wiley-Interscience: New York.

Campbell, J. W. (1997). Mitochondrial ammonia metabolism and the proton-neutral theory of hepatic ammonia detoxification. *J Exp Zool* **278**, 308–321.

Campbell, J. W., Vorhaben, J. E. and Smith, D. D., Jr. (1984). Hepatic ammonia metabolism in a uricotelic treefrog *Phyllomedusa sauvagei*. *Am J Physiol* **246**, R805–810.

Canessa, C. M., Horisberger, J. D. and Rossier, B. C. (1993). Epithelial sodium channel related to proteins involved in neurodegeneration. *Nature* **361**, 467–470.

Canessa, C. M., Schild, L., Buell, G., Thorens, B., Gautschi, I., Horisberger, J. D. and Rossier, B. C. (1994). Amiloride-sensitive Na^+ channel is made of three homologous subunits. *Nature* **367**, 463–467.

Canty, A., Driedzic, W. R. and Storey, K. B. (1986). Freeze tolerance of isolated ventricle strips of the wood frog, *Rana sylvatica*. *Cryo-Letters* **7**, 81–86.

Carey, C. (1978). Factors affecting body temperature of toads. *Oecologia* **35**, 197–219.

Carey, C. (1979). Effect of constant and fluctuating temperatures on resting and active oxygen consumption of toads, *Bufo boreas*. *Oecologia* **39**, 201–212.

Carey, C. and Alexander, M. A. (2003). Climate change and amphibian declines: is there a link? *Diversity Distributions* **9**, 111–121.

Carey, C., Cohen, N. and Rollins-Smith, L. (1999). Amphibian declines: an immunological perspective. *Develop Comp Immunol* **23**, 459–472.

Carlstrom, D. (1963). A crystallographic study of vertebrate otoliths. *Biol Bull Mar Biol Lab Woods Hole* **125**, 441–463.

Carrier, D. R. and Wake, M. H. (1995). Mechanism of lung ventilation in the caecilian *Dermophis mexicanus*. *J Morph* **226**, 289–295.

Carrier, O. (1965). The local control of blood flow: an illustration of homeostasis. *Bioscience* **15**, 665–668.

Carrol, R. L. (2001). The origin and early radiation of terrestrial vertebrates. *J Paleontology* 75, 1202–1213.

Carroll, R., Boisvert, C., Bolt, J., Green, M., Philip, N., Rolian, C., Schoch, R., et al. (2004). Changing patterns of ontogeny from osteolepiform fish through Permian tetrapods as a guide to the early evolution of land vertebrates. In: *Recent Advances in the Origin and Early Radiation of Vertebrates*, eds G. Arratia, M. V. H. Wilson and R. Cloutier, pp. 321–343. Pfeil: Munich.

Carroll, R. L., Irwin, J. T. and Green, D. M. (2005). Thermal physiology and the origin of terrestriality in vertebrates. *Zool J Linn Soc* **143**, 345–358.

Carter, D. B. (1979). Structure and function of the subcutaneous lymph sacs in anura (Amphibia). *J Herpetol* **13**, 321–327.

Cartledge, V. A. (2006). *Water Balance in Australian Cocooning and Non-cocooning Desert Burrowing Frogs*. PhD thesis. School of Animal Biology, University of Western Australia, Perth.

Cartledge, V. A., Withers, P. C., McMaster, K. A., Thompson, G. G. and Bradshaw, S. D. (2006). Water balance of field-excavated aestivating Australian desert frogs, the cocoon-forming

Neobatrachus aquilonius and the non-cocooning *Notaden nichollsi* (Amphibia: Myobatrachidae). *J Exp Biol* **209**, 3309–3321.

Cartledge, V. A., Withers, P. C., Thompson, G. G. and McMaster, K. A. (2005). Water relations of the burrowing sandhill frog, *Arenophryne rotunda* (Myobatrachidae). *J Comp Physiol B* **176**, 295–302.

Carvahlo, E. T. C. and Junqueira, L. C. U. (1999). Histology of the kidney and urinary bladder of *Siphonops annulatus* (Amphibia-Gymnophiona). *Arch Histol Cytol* **62**, 39–45.

Casterlin, M. E. and Reynolds, W. W. (1977). Behavioral fever in anuran amphibian larvae. *Life Sciences* **20**, 593–596.

Cei, J. M. (1969). The Patagonian telomatobiid fauna of the volcanic Somuncurá Plateau. *J Herpetology* **3**, 1–18.

Cena, K. and Clark, J. A. (1973). Thermographic measurements of the surface temperatures of animals. *J Mammalogy* **54**, 1003–1007.

Cena, K. and Monteith, J. L. (1975). Transfer process in animal coats. I. Radiative transfer. *J Royal Soc London B* **188**, 377–393.

Cerra, M. C., Amelio, D., Tavolaro, P., Palma, A. and Marciano, V. (2003). Pericardium of the frog, *Rana esculenta*, is morphologically designed as a lymphatic space. *J Morph* **257**, 72–77.

Cerreteli, P., di Prampero, P. E. and Ambrosoli, G. (1972). High-energy phosphate resynthesis from anaerobic glycolysis in frog gastrocnemius muscle. *Am J Physiol* **222**, 1021–1026.

Cerreteli, P. and Grassi, B. (2001). Gas exchange, MRS, and NIRS assessment of metabolic transients in skeletal muscle. *Amer Zool* **41**, 229–246.

Chan, H. Y., O'Shea, J. E., Withers, P. C. and Stewart, T. (2002). Podocyte complexity and terrestriality in frogs. *J Royal Soc W Aust* **85**, 181–182.

Chapman, B. M. and Chapman, R. F. (1958). A field study of a population of leopard toads (*Bufo regularis regularis*). *J Animal Ecol* **27**, 265–286.

Chaui-Berlinck, J. G. and Bicudo, J. E. P. W. (2006). *Respirometrica—a Técnica*. Santos: São Paulo.

Chauvet, J., Rouille, Y., Chauvet, M. T. and Acher, R. (1990). Occurrence of hydrin 2 (vasotocinyl-Gly), a new hydroosmotic neurohypophyseal peptide, in secretory granules isolated from the frog (*Rana esculenta*) neurointermediate pituitary. *Neuroendocrinology* **51**, 233–236.

Chauvet, J., Rouille, A., Ouedraogo, Y. and Acher, R. (1991). Adaptive differential processing of provasotocin in amphibians: occurrence of hydrin 2 (vasotocinyl-glycine) in anura but not in urodela. *CR Acad Sci Paris* **313**, 351–358.

Chevalier, J., Bourguet, J. and Hugon, J. S. (1974). Membrane associated particles: distribution in frog urinary bladder epithelium at rest and after oxytocin treatment. *Cell Tissue Res* **152**, 129–140.

Cheverud, J. M. and Dow, M. M. (1985). An autocorrelation analysis of genetic variation due to lineal fission in social groups of Rhesus macaques. *Am J Physiol Anthropol* **67**, 113–121.

Chew, M. M., Elliot, H. B. and Wong, H.Y. (1972). Permeability of urinary bladder of *Rana cancrivora* to urea in the presence of oxytocin. *J Physiol* **223**, 757–772.

Chieffi, G., Rastogi, R. K., Iela, L. and Milone, M. (1975). The function of fat bodies in relation to the hypothalamo-hypophysial-gonadal axis in the frog *Rana esculenta*. *Cell Tissue Res* **161**, 157–165.

Christensen, C. U. (1974a). Effect of arterial perfusion on net water flux and active sodium transport across the isolated skin of *Bufo bufo bufo*. *J Comp Physiol* **93**, 93–104.

Christensen, C. U. (1974b). Adaptations in the water economy of some anuran amphibians. *Comp Biochem Physiol A* **47**, 1035–1049.

Christian, K., Parry, D. and Green, B. (1988). Water loss and an extraepidermal lipid barrier in the Australian treefrog *Litoria caerulea*. *Am Zoologist* **28**, 17A.

Christian, K. and Parry, D. (1997). Reduced rates of evaporative water loss and chemical properties of skin secretions of the frogs *Litoria caerulea* and *Cyclorana australis*. *Aust J Zool* **45**, 13–20.

Christian, K. A. and Tracy, C. R. (1985). Measuring air temperature in field studies. *J Thermal Biol* **10**, 55–56.

Christiansen, J. and Penney, D. (1973). Anaerobic glycolysis and lactic acid accumulation in cold submerged *Rana pipiens*. *J Comp Physiol* **87A**, 237–245.

Churchill, T. A. and Storey, K. B. (1993). Dehydration tolerance in wood frogs: a new perspective on development of amphibian freeze tolerance. *Am J Physiol* **265**, R1324–1332.

Churchill, T. A. and Storey, K. B. (1996). Organ metabolism and cryoprotectant synthesis during freezing in spring peepers *Pseudacris crucifer*. *Copeia* 1996, 517–525.

Claussen, D. L. (1973). The water relations of the tailed frog, *Ascaphus truei*, and the Pacific treefrog *Hyla regilla*. *Comp Biochem Physiol* **44A**, 155–171.

Claussen, D. L. (1977). Thermal acclimation in ambystomatid salamanders. *Comp Biochem Physiol A* **58**, 333–340.

Claussen, D. L. and Constanzo, J. P. (1990). A simple model for estimating the ice content of freezing ectotherms. *J Therm Biol* **15**, 223–231.

Claussen, D. L. and Layne, J. R., Jr. (1983). Growth and survival of juvenile toads, *Bufo woodhousei*, maintained on four different diets. *J Herpetol* **17**, 107–112.

Cleeman, L., Pizarro, G. and Morad, M. (1984). Optical measurements of extracellular calcium depletion during a single heartbeat. *Science* **226**, 174–177.

Clothier, R. H., Worley, R. T. S. and Balls, M. (1978). The structure and ultrastructure of the renal tubule of the urodele amphibian, *Amphiuma means*. *J Anat* **127**, 491–504.

Cogger, H. G. (2000). *Reptiles and Amphibians of Australia*. Reed New Holland: Sydney.

Cohen, N. W. (1952). Comparative rates of dehydration and hydration in some California salamanders. *Ecology* **33**, 462–479.

Cole, D. F. (1968). The aqueous and vitreous humors of the giant toad *Bufo marinus*. *Comp Biochem Physiol* **26**, 353–357.

Conklin, R. E. (1930). The formation and circulation of lymph in the frog: I. The rate of lymph production. *Am J Physiol* **95**, 79–90.

Connell, J. M. C. and Davies, E. (2005). The new biology of aldosterone. *J Endocrinology* **186**, 1–20.

Connett, R. J. (1979). Glyconeogenesis from lactate in frog striated muscle. *Am J Physiol* **237**, C231–C236.

Conway, E. J. (1945). The physiological significance of inorganic levels in the internal medium of animals. *Biol Rev Camb Philos Soc* **20**, 56–72.

Conway, E. J. (1957). The nature and significance of concentration relations of potassium and sodium in skeletal muscle. *Physiol Rev* **37**, 84–132.

Cooper, C. A. and Withers, P. C. (2006). Numbats and aardwolves—how low is low? A re-affirmation of the need for statistical rigour in evaluating regression predictions. *J Comp Physiol B* **76**, 623–629.

Cope, E. D. (1865). Sketch of the primary groups of batrachia salientia. *Natural History Review* **5**, 97–120.

Coppo, J. A., Mussart, N. B. and Fioranelli, S. A. (2005). Blood and urine physiological variables in farm-cultured *Rana catesbeiana* (Anura: Ranidae) in Argentina. *Rev Biol Trop* **53**, 545–559.

Costanzo, J. P., Bayuk, J. M. and Lee, R. E., Jr. (1999). Inoculative freezing by environmental ice nuclei in the freeze-tolerant wood frog, *Rana sylvatica*. *J Exp Zool* **284**, 7–14.

Costanzo, J. P. and Lee, R. E., Jr. (2005). Cryoprotection by urea in a terrestrially hibernating frog. *J Exp Biol* **208**, 4079–4089.

Costanzo, J.P., Irwin, J.T. and Lee, R.E. (1997). Freezing Impairment of male reproductive behaviors of the freeze-tolerant Wood frog, *Rana sylvatica*. *Physiol. Zool.* **70**, 158–166.

Costanzo, J. P., Lee, R. E., Jr. and Lortz, P. H. (1993). Physiological responses of freeze-tolerant and -intolerant frogs: clues to evolution of anuran freeze tolerance. *Am J Physiol* **265**, R721–725.

Costanzo, J. P., Lee, R. E., Jr. and Lortz, P. H. (1993). Glucose concentration regulates freeze tolerance in the wood frog *Rana sylvatica*. *J Exp Biol* **181**, 245–255.

Costanzo, J. P., Lee, R. E., Jr. and Wright, M. F. (1991). Glucose loading prevents freezing injury in rapidly cooled wood frogs. *Am J Physiol* **261**, R1549–1553.

Costanzo, J. P., Lee, R. E., Jr. and Wright, M. F. (1991). Effect of cooling rate on the survival of frozen wood frogs, *Rana sylvatica*. *J Comp Physiol B* **161**, 225–229.

Costanzo, J. P., Lee, R. E., Jr. and Wright, M. F. (1992). Cooling rate influences cryoprotectant distribution and organ dehydration in freezing wood frogs. *J Exp Zool* **261**, 373–378.

Costanzo, J. P., Wright, M. F. and Lee, R. E., Jr. (1992). Freeze tolerance as an overwintering adaptation in Cope's grey treefrog (*Hyla chyrsocelis*). *Copeia* 1992, 565–569.

Coviello, A., Orce, G. and Causarano, J. (1974). Effect of a competetive antagonist of angiotensin II on sodium and water transport in toad skin. *Acta Phyiol Lat Am* **24**, 409–413.

Cowan, K. J., MacDonald, J. A., Storey, J. M. and Storey, K. B. (2000). Metabolic reorganization and signal transduction during estivation in the spadefoot toad. *Exp Biol Online* **5**, 1.

Cowan, K. J. and Storey, K. B. (1999). Reversible phosphorylation control of skeletal muscle pyruvate kinase and phosphofructokinase during estivation in the spadefoot toad, *Scaphiopus couchii*. *Mol Cell Biochem* **195**, 173–181.

Cowan, K. J. and Storey, K. B. (2002). Urea and KCl have differential effects on enzyme activities in liver and muscle of estivating versus nonestivating species. *Biochem Cell Biol* **80**, 745–755.

Cowles, R. B. and Bogert, C. M. (1944). A preliminary study of the thermal requirements of desert reptiles. *Bull Am Mus Nat Hist* **83**, 261–296.

Cox, M. and Singer, I. (1977). Insulin-mediated Na^+ transport in the toad urinary bladder. *Am J Physiol* **232**, F270–F277.

Crabbe, J. (1961). Stimulation of active sodium transport by the isolated toad bladder with aldosterone in vitro. *J Clin Invest* **40**, 2103–2110.

Cragg, M., Balinsky, J. B. and Baldwin, E. (1961). A comparative study of nitrogen excretion in some amphibia and reptiles. *Comp Biochem Physiol* **3**, 227–235.

Cramp, R. L. and Franklin, C. E. (2003). Is re-feeding efficiency compromised by prolonged starvation during aestivation in the green striped burrowing frog, *Cyclorana alboguttata*. *J Exp Zool* **300A**, 126–132.

Cramp, R. L., Franklin, C. E. and Meyer, E. A. (2005). The impact of prolonged fasting during aestivation on the structure of the small intestine in the green-striped burrowing frog, *Cyclorana alboguttata*. *Acta Zoologica* **86**, 13–24.

Cree, A. (1985). Water balance and nitrogen excretion of two introduced frogs (*Litoria raniformis* and *L. ewingi*). *N Z J Zool* **12**, 341–348.

Cree, A. (1989). Relationship between environmental conditions and nocturnal activity of the terrestrial frog, *Leiopelma archeyi*. *J Herpetol* **23**, 61–68.

Crim, J. W. (1972). Studies on the possible regulation of plasma sodium by prolactin in amphibia. *Comp Biochem Physiol* **43A**, 349–357.

Croes, S. A. and Thomas, R. E. (2000). Freeze tolerance and cryoprotectant synthesis of the Pacific tree frog *Hyla regilla*. *Copeia* 2000, 863–868.

Crossley, D. A., 2nd and Hillman, S. S. (1999). The role of pulmocutaneous baroreceptors in the control of lymphatic heart rate in the toad *Bufo marinus*. *Physiol Biochem Zool* **72**, 109–115.

Crump, M. L. and Kaplan, R. H. (1979). Clutch energy partitioning of tropical tree frogs (Hylidae). *Copeia* 1979, 626–635.

Cummins, C. P. (1986). Temporal and spatial variation in egg size and fecundity in *Rana temporaria*. *J Animal Ecol* **55**, 303–316.

Currie, S. and Boutilier, R. G. (2001). Strategies of hypoxia and anoxia tolerance in cardiomyocytes from the overwintering common frog, *Rana temporaria*. *Physiol Biochem Zool* **74**, 420–428.

Curtin, N. A. (1986). Buffer power and intracellular pH of frog sartorius muscle. *Biophys J* **50**, 837–841.

Curtin, N. A. and Edman, K. A. P. (1994). Force-velocity relation for frog muscle fibres: effects of moderate fatigue and of intracellular acidification. *J Physiol* **475**, 483–494.

Czopek, J. (1958). The vascularization of the respiratory surfaces in *Ambystoma mexicanum* in ontogeny. *Zool Polon* **8**, 131–149.

Czopek, J. (1959). Vascularization of respiratory surfaces in *Salamandra salamandra* in ontogeny. *Bull Acad Polon Sci Ser Sci Biol* **7**, 473–478.

Czopek, J. (1965). Quantitative studies on the morphology of respiratory surfaces in amphibians. *Acta Anat* **62**, 296–323.

Dalo, N.L., Hackman,J.C., Storey, K. and Davidoff, R.A. (1995). Changes in motorneuron membrane potential and reflex activity induced by sudden cooling of isolated spinal cords: differences Among cold-sensitive, cold-resistant and freeze-tolerant amphibian Species. *J Exp Biol* **198**, 1765–1774.

Damstra, K. S. J. (1986). A kinetic analysis of evaporative water loss barriers. *J Theor Biol* **121**, 439–448.

Daniels, C. B. and Orgeig, S. (2003). Pulmonary surfactant: the key to the evolution of air breathing. *News Physiol Sci* **18**, 151–157.

Daniels, C. B., Orgeig, S. and Smits, A. W. (1995). The evolution of the vertebrate pulmonary surfactant system. *Physiol Zool* **68**, 539–566.

Daniels, C. B., Orgeig, S., Wilsen, J. and Nicholas, T. E. (1994). Pulmonary type surfactants in lungs of terrestrial and aquatic amphibians. *Respir Physiol* **95**, 249–258.

Darveau, C. A., Suarez, R. K., Andrews, R. D. and Hochachka, P. W. (2002). Alometric cascade as a unifying principle of body mass effects on metabolism. *Nature* **417**, 166–170.

Dash, M. C. and Hota, A. K. (1980). Density effects on the survival, growth rate, and metamorphosis of *Rana tigrina* tadpoles. *Ecology* **61**, 1025–1028.

Daszak, P., Scott, D. E., Kilpatrick, A. M., Faggioni, C., Gibbons, J. W. and Porter, D. (2005). Amphibian population declines at Savannah River site are linked to climate, not chytridiomycosis. *Ecology* **86**, 3232–3237.

Davies, M. (1984). Osteology of the myobatrachine frog *Arenophryne rotunda* Tyler (Anura: Leptodactylidae) and comparisons with other myobatrachine genera. *Aust J Zool* **32**, 789–802.

Davies, M. and Withers, P. C. (1993). Morphology and physiology of the Anura. In: *Fauna of Australia*, Vol. 2A. Amphibia and Reptilia, eds C. Glasby, G. Ross and P. Beesley, pp. 15–27. Australian Government Printing Service: Canberra.

Davis, D. D. (1965). Wallace's flying frog. *Malayan Nature Journal* **19**, 149–151.

Dawson, A. B. (1924). Glomerular versus tubular activity in the mesonephros of *Necturus*: Elimination of iron salts. *Am J Physiol* **71**, 679–687.

Dawson, A. B. (1951). Functional, and degenerate or rudimentary glomeruli in the kidney of two species of Australian frog, *Cyclorana platycephalus* and *alboguttatus* (Gunther). *Anatomical Record* **109**, 417–429.

Dawson, T. J. and Hulbert, A. J. (1969). Standard energy metabolism of marsupials. *Nature* **221**, 383.

de Graaf, A. R. (1957). Investigations into the distribution of blood in the heart and aortic arches of *Xenopus laevis*. *J Exp Biol* **34**, 143–172.

De Ruyter, M. L. and Stiffler, D. F. (1985). Interrenal function in larval *Ambystoma tigrinum*. II. Control of aldosterone secretion and electrolyte balance by ACTH. *General Comp Endocrinology* **62**, 298–305.

De Vlaming, V. L. and Bury, R. B. (1970). Thermal selection in tadpoles of the tailed frog, *Ascaphus truei*. *J Herpetology* **4**, 179–189.

Deeds, D. G., Sullivan, L. P., Fento, R. A., Tucker, J. M. and Cuppage, F. E. (1977). Function and structure of perfused bullfrog kidney. *Am J Physiol* **233**, F481–F490.

DeGrauw, E. A. (1998). *A Comparative Approach to Blood Volume Control in Anurans*. PhD Thesis, Portland State University.

DeGrauw, E. A. and Hillman, S. S. (2004). General function and endocrine control of the posterior lymph hearts in *Bufo marinus* and *Rana catesbeiana*. *Physiol Biochem Zool* 77, 594–600.

Dejours, P. (1975). *Principles of Comparative Respiratory Physiology*. North-Holland Publishing Co: Amsterdam.

Delaney, R. G., Lahiri, S. and Fishman, A. P. (1974). Aestivation of the African lungfish *Protopterus aethiopicus*: Cardiovascular and respiratory functions. *J Exp Biol* **61**, 111–128.

DeLong, K. T. (1962). Quantitative analysis of blood circulation through the frog heart. *Science* **138**, 693–694.

del Pino, E. M., Alcocer, I. and Grunz, H. (1994). Urea is necessary for the culture of embryos of the marsupial frog *Gastrotheca riobambae*, and is tolerated by embryos of the aquatic frog *Xenopus laevis*. *Develop Growth Differ* **36**, 73–80.

Delson, J. and Whitford, W. W. (1973). Adaptation of the tiger salamander, *Ambystoma tigrinum*, to arid habitats. *Comp Biochem Physiol A* **46**, 631–638.

Dempster, W. T. (1930). The morphology of the amphibian endolymphatic system. *J Morph* **50**, 7–126.

Depocas, F. and Hart, J. S. (1957). Use of the Pauling oxygen analyzer for measurement of oxygen consumption of animals in open-circuit systyems and in short lag, closed-circuit apparatus. *J Applied Physiol* **10**, 388–392.

DeSousa, R. C. and Grosso, A. (1982). Osmotic water flow across the abdominal skin of the toad, *Bufo marinus*: effect of vasopressin and isoprenaline. *J Physiol* **329**, 281–296.

Devuyst, O., Beaujean, V. and Crabbi, J. (1991). Effect of environmental conditions on mitochondria-rich cell density and chloride transport in toad skin. *Pflgers Archiv* **417**, 577–581.

Dicker, S. E. and A. B. Elliott. (1967). Water uptake by *Bufo melanostictus*, as affected by osmotic gradients, vasopressin and temperature. *J Physiol* 190, 359–370.

Dicker, S. E. and Elliott, A. B. (1970). Effect of neurohypophysial hormones on fluid movement across isolated bladder of *Rana cancrivora*, *Rana temporaria* and *Bufo melanostictus*. *J Physiol* **210**, 137–149.

Dickson, K. A. and Graham, J. B. (2004). Evolution and consequences of endothermy in fishes. *Physiol Biochem Zool* 77, 998–1018.

DiGirolamo, M. and Owens, J. L. (1976). Water content of rat adipose tissue and isolated adipocytes in relation to cell size. *Am J Physiol* **231**, 1568–1572.

Dimmit, M. A. and Ruibal, R. (1980). Exploitation of food resources by spadefoot toads (*Scaphiopus*). *Copeia* 1980, 854–862.

Diniz-Filho, J. A. F., de Sant'Ana, C. E. R. and Bini, L. M (1998). An eigenvector method for estimating phylogenetic inertia. *Evolution* **52**, 1247–1262.

Dixon, M. (1952). Manometric Methods as Applied to the Mmeasurement of Cell Respiration and Other Processes. Cambridge University Press, Cambridge.

Dobkin, D. S. and Gettinger, R. D. (1985). Thermal aspects of anuran foam nests. *J Herpetology* **19**, 271–275.

Dobkin, D. S., Gettinger, R. D. and O'Connor, M. P. (1989). Importance of retreat site for hydration state in *Bufo marinus* during a dry rainy season *Am Zoologist* **29**, 88A.

Dodd, M. H. I. and Dodd, J. M. (1976). The biology of metamorphosis. In: *Physiology of Amphibia*, ed. B. Lofts, Vol. 3, pp. 447–599. Academic Press: New York.

Dode, L., Van Baelen, K., Wuytack, F. and Dean, W. L. (2001). Low temperature molecular adaptation of the skeletal muscle sarco(endo)plasmic reticulum Ca^{2+}-ATPase 1 (SERCA 1) in the wood frog (*Rana sylvatica*). *J Biol Chem* **276**, 3911–3919.

Dole, J. W. (1967). The role of substrate moisture and dew in the water economy of leopard frogs, *Rana pipiens*. *Copeia* 1967, 141–149.

Dolphin, J. L. and Frieden, E. (1955). Biochemistry of amphibian metamprphosis. II. Arginase activity. *J Biol Chem* **217**, 735–743.

Donald, J. A., Meier, S. K. and Riddell, S. (2002). Toad atrial natriuretic peptide: cDNA cloning and functional analysis in isolated perfused kidneys. *Physiol Biochem Zool* **75**, 617–626.

Donald, J. A. and Trajanovska, S. (2006). A perspective on the role of natriuretic peptides in amphibian osmoregulation. *General Comp Endocrinology* **147**, 47–53.

Donohoe, P. H. and Boutilier, R. G. (1998). The protective effects of metabolic rate depression in hypoxic cold submerged frogs. *Respiration Physiol* **111**, 325–336.

Donohoe, P. H., West, T. G. and Boutilier, R. G. (1998). Respiratory, metabolic, and acid-base correlates of aerobic metabolic rate reduction in overwintering frogs. *Am J Physiol* **274**, R704–710.

Donohoe, P. H., West, T. G. and Boutilier, R. G. (2000). Factors affecting membrane permeability and ionic homeostasis in the cold-submerged frog. *J Exp Biol* **203**, 405–414.

Dorlochter, M., Astrow, S. H. and Herrera, A. A. (1994). Effects of testosterone on a sexually dimorphic frog muscle: repeated in vivo observations and androgen receptor distribution. *J Neurobiol* **25**, 897–916.

Drewes, R. C., Hedrick, M. S., Hillman, S. S. and Withers, P. (2007). Unique role of skeletal muscle contraction in vertical lymph movement in anurans. *J Exp Biol* **210**, 3931–3939.

Drewes, R. C., Hillman, S. S., Putnam, R. B. and Sokol, O. M. (1977). Water, nitrogen and ion balance in the African treefrog *Chiromantis petersii* Boulenger (Anura: Rhacophoridae), with comments on the structure of the integument. *J Comp Physiol* **116**, 257–267.

Dubois, A. (1983). Classification et nomenclature supragenerique des amphibians anoures. *Bull Menseul de la Societe Linneenne de Lyon* **52**, 270–276.

Du Bois-Reymond, E. (1848). *Undersuchungen ber Thierische Electricitdt* (Book 1). Georg Reimer: Berlin.

Du Bois-Reymond, E. (1884). *Untersuchungen ber Thierische Electricitdt* (Book 2). Georg Reimer: Berlin.

Duellman, W. E. (1979). On the classification of frogs. *Occ Papers Museuem Nat Hist Univ Kansas* **42**, 1–14.

Duellman, W. E. (1999). *Patterns of Distribution of Amphibians. A Global Perspective.* Johns Hopkins University Press: Baltimore and London.

Duellman, W. E. and Trueb, L. (1986). *Biology of Amphibians.* Johns Hopkins University Press: Baltimore.

Duellman, W. E. and Trueb, L. (1994). *Biology of Amphibians.* Johns Hopkins University Press: Baltimore and London.

Duke, J. T. and Ultsch, G. R. (1990). Metabolic oxygen regulation and conformity during submergence in the salamanders *Siren lacertina, Amphiuma means* and *Amphiuma tridactylum*, and a comparison with other giant salamanders. *Oecologia* **84**, 16–23.

Dunlap, D. G. (1971). Acutely measured metabolic rate-temperature curves in the cricket frog, *Acris crepitans. Comp Biochem Physiol* **38A**, 1–16.

Dunn, E. R. (1928). The habits of Plethodontidae. *Am Nat* **62**, 236–248.

Dupré, R. K., Taylor, R. F. and Frazier, D. T. (1985). Static lung compliance during development of the bullfrog, *Rana catesbeiana. Respir Physiol* **59**, 231–238.

Dupré, R. K. and Wood, S. C. (1988). Behavioural temperature regulation by aquatic ectotherms during hypoxia. *Can J Zool* **66**, 2649–2652.

Dutrochet, H. (1839). Endosmosis. In: *The Cyclopaedia of Anatomy and Physiology.* Sherwood, Gilbert and Piper: London.

Dyson, M. L., Henzi, S. P., Halliday, T. R. and Barrett, L. (1998). Success breeds success in mating reed frogs (*Hyperolius marmoratus*). *Proc R Soc Lond B Biol Sci* **265**, 1417–1421.

Ecker, A. (1864). *Die Anatomie des Frosches.* Braunschweig.

Edman, K. A. P., Reggiani, C., Schiaffino, S. and Kronnie, G. (1988). Maximum velocity of shortening related to myosin isoform composition in frog skeletal muscle fibres. *J Physiol* **395**, 679–694.

Elias, H. and Shapiro, J. (1957). Histology of the skin of some toads and frogs. *Am Mus Nov* **1819**, 1–27.

Elkan, E. (1968). Mucopolysaccharides in the anuran defence against desiccation. *J Zool* **155**, 19–53.

Ellsworth, M. L., Forrester, T., Ellis, C. G. and Dietrich, H. H. (1995). The erythrocte as a regulator of vascular tone. *Am J Physiol* **269**, H2155–H2161.

Else, P. L. and Bennett, A. F. (1988). The thermal dependence of locomotor performance and muscle contractile function in the salamander *Ambystoma tigrinum nebulosum. J Exp Biol* **128**, 219–233.

Emerson, S. B. (1976). Burrowing in frogs. *J Morph* **149**, 437–458.

Emilio, M. G. (1974). Gas exchange and blood gas concentrations in the frog. *J Exp Biol* **60**, 901–908.

Emilio, M. G. and Shelton, G. (1972). Factors affecting blood flow to the lungs in the amphibian, *Xenopus laevis. J Exp Biol* **56**, 67–77.

Emilio, M. G. and Shelton, G. (1974). Gas exchange and its effect on blood gas concentrations in the amphibian, *Xenopus laevis. J Exp Biol* **60**, 567–579.

Emilio, M. G. and Shelton, G. (1980). Carbon dioxide exchange and its effects on pH and bicarbonate equilibria in the blood of the amphibian, *Xenopus laevis. J Exp Biol* **85**, 253–262.

Estes, R. and Wake, M. H. (1972). The first fossil record of caecilian amphibians. *Nature* **239**, 228–231.

Etheridge, K. (1990). The energetics of aestivating sirenid salamanders (*Siren lacertina* and *Pseudobranchus striatus*). *Herpetologica* **46,** 407–414.

Etkin, W. and Gona, A. G. (1967). Antagonism between prolactin and thyroid hormone in amphibian development. *J Exp Zool* **165,** 249–268.

Evans, S. E., Milner, A. R. and Werner, C. (1996). Sirenid salamanders and a gymnophionan amphibian from the Cretaceous of the Sudan. *Palaeontology* **39,** 77–95.

Evans, S. E. and Borsuk-Bialynicka, M. (1998). A stem-group frog from the Early Triassic of Poland. *Acta Palaeontologica Polonica* **43,** 573–580.

Ewer, R. F. (1952). The effect of pituitrin on fluid distribution in *Bufo regularis* Reuss. *J Exp Biol* **29,** 173–177.

Ewer, R. F. (1952). The effects of posterior pituitary extracts on water balance in *Bufo carens* and *Xenopus laevis,* together with some general considerations of anuran water economy. *J Exp Biol* **29,** 429–439.

Fain, W. and Rosen, S. (1973). Carbonic anhydrase activity in amphibian and reptilian lungs: a histochemical and biochemical analysis. *Histochem Jour* **5,** 519–528.

Fanelli, G. M. and Goldstein, L. (1964). Ammonia excretion in the neotenous newt, *Necturus maculosus. Comp Biochem Physiol* **13,** 193–204.

Farias, A., Fiorito, L. E. and Hermida, G. N. (1998). Structure of the *Bufo arenarum* kidney: renal corpuscle, neck segment and proximal tubule. *Biocell* **22,** 187–196.

Farmer, S. W., Licht, P. and Papkoff, H. (1977). Biological activity of bullfrog growth hormone in the rat and the bullfrog (*Rana catesbeiana*). *Endocrinology* **101,** 1145–1150.

Farquhar, M. G. and Palade, G. E. (1964). Cell junctions in amphibian skin. *J Cell Biol* **26,** 263–291.

Farrell, M. P. (1971). Effect of temperature and photoperiod on the water economy of *Hyla crucifer. Herpetologica* **27,** 41–48.

Farrell, M. P. and MacMahon, J. A. (1969). An eco-physiological study of water economy in eight species of tree frogs (Hylidae). *Herpetologica* **25,** 279–294.

Feder, M. E. (1976a). Oxygen consumption and body temperature in Neotropical and temperate zone lungless amphibians (Amphibia, Plethodontidae). *J Comp Physiol* **110B,** 197–208.

Feder, M. E. (1976b). Lunglessness, body size, and metabolic rate in salamanders. *Physiol Zool* **49,** 398–406.

Feder, M. E. (1977). Oxygen consumption and activity in salamanders: Effects of body size and lunglessness. *J Exp Zool* **202,** 403–414.

Feder, M. E. (1982a). Environmental variability and thermal acclimation of metabolism in tropical anurans. *J Thermal Biol* **7,** 23–28.

Feder, M. E. (1982b). Thermal ecology of neotropical lungless salamanders (Amphibia: Plethodontidae): Environmental temperatures and behavioural responses. *Ecology* **63,** 1665–1674.

Feder, M. E. (1983). Integrating the ecology and physiology of plethodontid salamanders. *Herpetologica* **39,** 291–310.

Feder, M. E. (1992). A perspective on environmental physiology of the amphibians. In: *Environmenal Physiology of the Amphibia*, eds M. E. Feder and W. W. Burggren, pp. 1–6. University of Chicago Press: Chicago.

Feder, M. E. and Arnold, S. J. (1982). Anaerobic metabolism and behavior during predatory encounters between snakes (*Thamnophis elegans*) and salamanders (*Plethodon jordani*). *Oecologia* **53,** 93–97.

Feder, M. E. and Burggren, W. W. (1985). Cutaneous gas exchange in vertebrates: design, patterns, control and implications. *Biol Rev Camb Philos Soc* **60,** 1–45.

Feder, M. E. and Burrgren, W. W. (1992). *Environmental Physiology of the Amphibians*. University of Chicago Press: Chicago and London.

Feder, M. E. and Lynch, J. F. (1982). Effects of altitude, elevation, and microclimate on field body temperature of neotropical and temperate zone salamanders. *Ecology* **63**, 1657–1664.

Feder, M. E., Lynch, J. F., Shaffer, H. B. and Wake, D. B. (1982). Field body temperatures of tropical and temperate zone salamanders. *Smithsonian Herpetological Information Service* **52**, 1–23.

Felsenstein, J. (1985). Phylogenies and the comparative method. *Am Naturalist* **125**, 1–15.

Fenoglio, C., Visai, L., Addario, C., Gerzeli, G., Milanesi, G., Vaccarone, R. and Barni, S. (2004). Expression of natriuretic peptides, nitric oxide synthase, and guanylate cyclase activity in frog mesonephros during the annual cycle. *General Comp Endocrinology* **137**, 166–176.

Fernandes, M. S., Giusti, H. and Glass, M. L. (2005). An assessment of dead space in pulmonary ventilation of the toad *Bufo schneideri*. *Comp Biochem Physiol A Mol Integr Physiol* **142**, 446–450.

Ferriera, H. G. and Jesus, C. H. (1973). Salt adaptation in *Bufo bufo*. *J Physiol* **228**, 583–600.

Figge, F. H. J. (1936). The differential reaction of the blood vessels of a brancial arch of *Ambystoma tigrinum* (Colorado axolotl) I. The reaction to adrenalin, oxygen, and carbon dioxide. *Physiol Zool* **9**, 79–101.

Fisher, M. and Muth, A. (1995). A backpack method for mounting radio transmitetrs to small lizards. *Herpetological Review* **26**, 139–140.

Fishman, A. F., Pack, A. I., Delaney, R. G. and Galante, R. J. (1986). Estivation in *Protopterus*. *J Morphology* Suppl. **1**, 237–248.

Fitch, H. S. (1956). Temperature responses in free-living amphibians and reptiles of northeastern Kansas. *University of Kansas Publications, Museum of Natural History* **8**, 417–476.

Fitts, R. H. and Holloszy, J. O. (1976). Lactate and contractile force in frog muscle during development of fatigue and recovery. *Am J Physiol* **231**, 430–433.

Fitzpatrick, L. C. (1973). Influence of seasonal temperature on the energy budget and metabolic rates of the northern two-lined salamander *Eurycea bislineata bislineata*. *Comp Biochem Physiol* **45A**, 807–818.

Fitzpatrick, L. C. (1973). Energy allocation in the Allegheny mountain salamander, *Desmognathus ochrophaeus*. *Ecol Monogr* **43**, 43–58.

Fitzpatrick, L. C. and Atebara, M. Y. (1974). Effects of acclimation to seasonal temperatures on energy metabolism in the toad, *Bufo woodhousei*. *Physiol Zool* **47**, 119–129.

Fitzsimons, J. T. (1998). Angiotensin, thirst and sodium appetite. *Physiological Reviews* **78**, 583–686.

Flanigin, J. E. and Guppy, M. (1997). Metabolic depression and sodium-potassium ATPase in the aestivating frog, *Neobatrachus kunapalari*. *J Comp Physiol* **167B**, 135–145.

Flanigan, J. E., Withers, P. C., Fuery, C. J. and Guppy, M. (1993). Metabolic depression and Na^+/K^+ gradients in the aestivating Australian goldfields frog, *Neobatrachus wilsmorei*. *J Comp Physiol B* **163**, 587–593.

Flanigan, J. E., Withers, P. C. and Guppy, M. (1991). *In vitro* metabolic depression in tissues from the aestivating frog, *Neobatrachus pelobatoides*. *J Exp Biol* **161**, 273–283.

Flindt, R. (1966). Relative coordination der lympherzbewegungen bei Anuen. *Pflugers Arch* **290**, 28–37.

Ford, L. S. and Canatella, D. C. (1993). The major clades of Frogs. *Herpetological Monographs* **7**, 94–117.

Forsman, A. and Hagman, M. (2006). Calling is an honest indicator of paternal genetic quality in poison frogs. *Evolution* **60**, 2148–2157.

Forster, R. P. (1938). The use of inulin and creatinine as glomerular filtrate measuring substances in the frog. *J Cell Comp Physiol* **12**, 213–222.

Forster, R. P. (1942). The nature of the glucose reabsorptive process in the frog renal tubule. Evidence for intermittency of glomerular function in the intact animal. *J Cell Comp Physiol* **20**, 55–68.

Foster, R. W. (1974). Adrenoceptors and salt and water movement in epithelia *Naunyn-Schmiedeberg's Arch Pharmacol* **281**, 315–326.

Fournier, P. A. and Guderly, H. (1992). Metabolic fate of lactate after vigorous activity in the leopard frog, *Rana pipiens*. *Anat Embryol* **262**, R245–R254.

Fournier, P. A. and Guderly, H. (1993). Muscle: the predominant glucose producing organ in the leopard frog during exercise. *Am J Physiol* **264**, R239–R243.

Fox, H. (1963). The amphibian pronephros. *Quart Rev Biol* **38**, 1–25.

Fox, H. (1984). *Amphibian Morphogenesis*. Humana Press: New Jersey.

Foxon, G. E. H. (1951). A radiographic study of the passage of blood through the heart in the frog and in the toad. *Proc Zool Soc (London)* **121**, 529–538.

Frangioni, G. and Borgioli, G. (1978). The fundamental role of lactate in erythropoiesis: Experimental studies on larvae of Amphibia Anura. *Monitore Zool Ital* **12**, 181–197.

Frangioni, G. and Borgioli, G. (1980). Lactate sensitive cells in newt kidneys. *Cell Tissue Res* **207**, 233–239.

Frangioni, G. and Borgioli, G. (1989). Relationships between spleen and respiration in the newt. *J Exp Zool* **252**, 118–25.

Frangioni, G. and Borgioli, G. (1991). Effect of spleen congestion and decongestion on newt blood. *J Zool Lond* **223**, 15–25.

Frangioni, G. and Borgioli, G. (1993). Splenic respiratory compensation and blood volume in the newt. *J Zool Lond* **229**, 505–514.

Frank, O. [translators Chapman, C. B. and. Waterman., E.] (1960). On the dynamics of cardiac muscle. *Am Heart J* **58**, 282–317, 467–478.

Frappell, P. B., Blevin, H. A. and Baudinette, R. V. (1989). Understanding respirometry chambers: What goes in must come out. *J Theor Biol* **138**, 479–494.

Frazier, L. W. and Vanatta, J. C. (1981). Excretion of K^+ by frog with rate varying with K^+ load. *Comp Biochem Physiol A* **69**, 157–160.

Freed, A. N. (1980). An adaptive advantage of basking behavior in an anuran amphibian. *Physiol Zool* **54**, 433–444.

Frieden, E. and Just, J. J. (1970). Hormone responses in amphibian metamorphosis. In: *Mechanisms of Hormone Action,* ed. G. Litwack, pp. 1–52. Academic Press: New York.

Fritsch, B., Drewes, R. C. and Ruibal, R. (1987). The retention of the lateral-line nucleus in adult anurans. *Copeia* 1987, 127–135.

Fritsche, R. (1997). Ontogeny of cardiovascular control in amphibians. *Amer Zool* **37**, 23–30.

Frost, D., Grant, T., Faivovich, J., Bain, R. H., Haas, A., Haddad, H. F. E., De Sa, R. O., Channing, A., Wilkinson, M., Donellan, S. C. *et al.* (2006). The amphibian tree of life. *Bull Am Mus Nat Hist* **297**, 1–370.

Fuery, C. J., Attwood, P. V., Withers, P. C., Yancey, P. H., Baldwin, J. and Guppy, M. (1997). Effects of urea on M4-lactate dehydrogenase from elamobranchs and urea-accumulating Australian desert frogs. *Comp Biochem Physiol B* **117**, 143–150.

Fuery, C. J., Withers, P. C. and Guppy, M. (1998). Protein synthesis in the liver of *Bufo marinus*: cost and contribution to oxygen consumption. *Comp Biochem Physiol A* **119**, 459–467.

Fuery, C. J., Withers, P. C., Hobbs, A. A. and Guppy, M. (1998). The role of protein synthesis during metabolic depression in the Australian desert frog *Neobatrachus centralis. Comp Biochem Physiol A* **119**, 469–476.

Full, R. J. (1986). Locomotion without lungs: Energetics and performance of a lungless salamander *Am J Physiol* **130**, 137–154.

Full, R. J., Anderson, B. D., Finnerty, C. M. and Feder, M. E. (1988). Exercising with and without lungs: 1. The effects of metabolic cost, maximal oxygen transport, and body size on terrestrial locomotion in salamander species. *J Exp Biol* **138**, 471–485.

Funkhouser, D. and Goldstein, L. (1973). Urea response to pure osmotic stress in the aquatic toad *Xenopus laevis. Am J Physiol* **224**, 524–529.

Gahlenbeck, H. and Bartels, H. (1968). Temperaturadaptation der Sauerstoffaffinitat des blutes von *Rana esculenta. Z Vergl Physiol* **59**, 232–240.

Gahlenbeck, H. and Bartels, H. (1970). Blood gas transport properties in gill and lung forms of the axolotl (*Ambystoma mexicanum*). *Respir Physiol* **9**, 175–182.

Galeotti, G. (1904). Concerning the EMF which is generated at the surface of animal membranes in contact with different electrolytes. *Zeit Physikalische Chemie* **49**, 542–562.

Gallardo, R., Pang, P. K. T. and Sawyer, W. H. (1980). Neural influences on bullfrog renal function. *Proc Soc Exp Biol Med* **165**, 233–240.

Gamperl, A. K., Milsom, W. K., Farrell, A. P. and Wang, T. (1999). Cardiorespiratory responses of the toad (*Bufo marinus*) to hypoxia at two different temperatures. *J Exp Biol* **202**, 3647–3658.

Gans, C. (1970). Respiration in early tetrapods: The frog is a red herring. *Evolution* **24**, 723–734.

Gans, C., de Jongh, H. J. and Farber, J. (1969). Bullfrog (*Rana catesbeiana*) ventilation: How does the frog breathe? *Science* **163**, 1223–1225.

Garcia Romeu, F., Masoni, A., and Isaia, J. (1981). Active urea transport through isolated skins of frog and toad. *Am J Physiol* **241**, R114–R123.

Garcia Romeu, F., Salibian, A. and Pezzani-Hernandez, S. (1969). The nature of in vivo sodium and chloride uptake mechanisms through the epithelium of the Chilean frog, *Calyptocephallela gayi* (Dumeril and Bibron, 1841). *J Gen Physiol* **53**, 616–835.

Gardner, M. N., Smits, A. W. and Smatresk, N. J. (2000). The ventilatory responses of the caecilian *Typhlonectes natans* to hypoxia and hypercapnia. *Physiol Biochem Zool* **73**, 23–29.

Garland, H. O. and Henderson, I. W. (1975). Influence of environmental salinity on renal and adrenocortical function in the toad *Bufo marinus. Gen Comp Endocrin* **27**, 136–143.

Garland, T. and Adolph, S. C. (1994). Why not to do two-species comparative studies: Limitations on inferring adaptation. *Physiol Zool* **67**, 797–828.

Garland, T. and Ives, A.R. (2000). Using the past to predict the present: confidence intervals for regression equations in phylogenetic comparative methods. *Am Naturalist* **155**, 346–364.

Garlick, R. L., Davis, B. J., Farmer, M., Fyhn, H. J., Fyhn, U. E., Nobil, R. W., Powers, D. A., Riggs, A. F. and Weber, R. E. (1979). A fetal-maternal shift in the oxygen equilibrium of hemoglobin from the viviparous caecilian, *Typhlonectes compressicauda. Comp Biochem Physiol* **62A**, 239–244.

Garty, H. and Palmer, L. G. (1987). Epithelial sodium channels: stucture, function, regulation. *Physiol Rev* **77**, 359–396.

Gates, D. M. (1980). *Biophysical Ecology*. Springer-Verlag: New York.

Gatten, R. E. (1987). Activity metabolism of anuran amphibians: tolerance to dehydration. *Physiol Zool* **60**, 576–585.

Gatten, R. R., Miller, K. and Full, R. J. (1992). Energetics at rest and during locomotion. In: *Environmental Physiology of the Amphibians*, eds M. E. Feder and W. W. Burrgren, pp. 314–377. University of Chicago Press: Chicago and London.

Gatz, R. N., Crawford, E. C. and Piiper, J. (1974). Metabolic and heart rate response of the plethodontid salamander, *Desmognathus fuscus* to hypoxia. *Respir Physiol* **20**, 43–49.

Gatz, R. N., Crawford, E. C., Jr. and Piiper, J. (1975). Kinetics of inert gas equilibration in an exclusively skin-breathing salamander, *Desmognathus fuscus*. *Respir Physiol* **24**, 15–29.

Gatz, R. N. and Piiper, J. (1979). Anaerobic energy metabolism during severe hypoxia in the lungless salamander *Desmognathus fuscus* (Plethodontidae). *Respiration Physiol* **38**, 377–384.

Gaupp, E. (1904). *A. Ecker und R. Wiedersheim's Anatomie des Frosches*. pp. 443–961. Braunschweig.

Gealekman, O. and Warburg, M. R. (2000). Changes in the numbers and dimensions of glomeruli during metamorphosis of *Pelobates syriacus* (Anura:Pelobatidae). *Eur J Morphol* **38**, 80–87.

Gehlbach, F. R., Gordon, R. and Jordan, J. B. (1973). Aestivation of the salamander, *Siren intermedia*. *Am Midland Naturalist* **89**, 455–463.

Gehlbach, F. R., Kimmell, J. R. and Weems, W. A. (1969). Aggregations and body water relations in tiger salamanders (*Ambystoma tigrinum*) from the Grand Canyon rims, Arizona. *Physiol Zool* **42**, 173–182.

Gehr, P., Mwangi, D. K., Ammann, A., Maloiy, G. M. O. and Taylor, C. R. (1981). Design of the mammalian respiratory system. V. Scaling morphometric diffusing capacity to body mass: wild and domestic animals. *Respir Physiol* **44**, 61–86.

Geise, W. and Linsenmair, K. E. (1986). Adaptations of the reed frog *Hyperolius viridiflavus* (Amphibia, Anura, Hyperoliidae)to its arid environment: 2. Some aspects of the water economy of *Hyperolius viridiflavus nitidulus* under wet and dry season conditions. *Oecologia* **68**, 542–548.

Gentry, R. M. and Johnson, P. C. (1972). Reactive hyperemia in arterioles and capillaries of frog skeletal muscle following microocclusion. *Circ Res* **31**, 953–965.

Gerhardt, H. C., Daniel, R. E., Perrill, S. A. and Schramm, S. (1987). Mating behavior and success in the green treefrog. *Anim Behav* **35**, 1490–1503.

Giardina, B., Mosca, D. and De Rosa, M. C. (2004). The Bohr effect of haemoglobin in vertebrates: an example of molecular adaptation to different physiological requirements. *Acta Physiol Scand* **182**, 229–244.

Gibbons, C. A. and Shadwick, R. E. (1991). Circulatory mechanics in the toad *Bufo marinus* I. Structure and mechanical design of the aorta. *J Exp Biol* **158**, 275–289.

Gibbons, C. A. and Shadwick, R. E. (1991). Circulatory mechanics in the toad *Bufo marinus* II. Haemodynamics of the arterial Windkessel. *J Exp Biol* **158**, 291–306.

Giese, W. and Linsenmair, K. E. (1988). Adaptations of the reed frog *Hyperolius viridiflavus* (Amphibia, Anura, Hyperoliidae) to its arid environment: 4. Ecological significance of water economy with comments on thermoregulation and energy allocation. *Oecologia* 77, 327–338.

Gil, N. and Katz, U. (1996). Oxygen consumption, heart rate and respiratory movements are maintained almost unchanged in toads (*Bufo viridis*) on soil without access to free water. *J Arid Environ* **33**, 237–245.

Gillis, J. E. (1979). Adaptive differences in the water economies of two species of leopard frogs from eastern Colorado (Amphibia, Anura, Ranidae). *J Herpetology* **13**, 445–450.

Gillis, J. E. and Breuer, W. J. (1984). A comparison of rates of evaporative water loss and tolerance to dehydration in te red-eft and newt *Notopthalmus viridescens*. *J Herpetology* **18**, 81–82.

Girgenrath, M. and Marsh, R. L. (2003). Season and testosterone affect contractile properties of fast calling muscles in the gray tree frog *Hyla chrysocelis*. *Am J Physiol* **284**, R1513–R1520.

Girish, S. and Saidapur, S. K. (2000). Interrelationship between food availability, fat body, and ovarian cycles in the frog, *Rana tigrina*, with a discussion of the role of the fat body in anuran reproduction. *J Exp Zool* **286**, 487–493.

Giselen, T. and Kauri, H. (1959). Zoogeography of the Swedish amphibians and reptiles. *Acta Vertebratica* **1**, 197–397.

Glass, M. L., Branco, L. G. S., Antunes-Rodrigues, J. and Gutkowska, J. (1996). Characterization of the atrial natriuretic factor system in the lungs of the toad *Bufo paracnemis*. *J Exp Biol* **199**, 1493–1499.

Glass, M. L., Burggren, W. W. and Johansen, K. (1981). Pulmonary diffusing capacity of the bullfrog (*Rana catesbeiana*). *Acta Physiol Scand* **113**, 485–490.

Gleeson, T. T. (1991). Patterns of metabolic recovery from exercise in amphibians and reptiles. *J Exp Biol* **160**, 187–207.

Gnionsahe, A., Claire, M., Koechlin, N., Bonvalet, J. P. and Farman, N. (1989). Aldosterone binding sites along nephron of *Xenopus* and rabbit. *Am J Physiol* **257**, R87–R95.

Godt, R. E. and Maughan, D. W. (1988). On the composition of the cytosol of relaxed skeletal muscle of the frog. *Am J Physiol* **254**, C591–C604.

Goldenberg, S. and M. R. Warburg (1977). Changes in the effect of vasotocin on water balance of *Rana ridibunda* during ontogenesis. *Comparative Biochemistry and Physiology* **57A**, 451–455.

Gomez, N. A., Acosta, M.A., Zaidan III, F. and Lillywhite, H.B. (2006). Wiping behaviour, skin resistance, and metabolic response to dehydration in the arboreal frog *Phyllomedusa hypochondrialis*. *Physiol Biochem Zool* **79**, 1058–1068.

Goniakowska-Witainska, L. (1978). Ultrastructure and morphometric study of the lung of the European salamander, *Salamandra salamandra*. *Cell Tissue Res* **191**, 343–356.

Goniakowska-Witainska, L. (1986). Lung of the tree frog, *Hyla arborea*: a scanning and transmission electron microscope study. *Anat Embryol* **174**, 379–389.

Gordon, M. S. (1962). Osmotic regulation in the green toad (*Bufo viridis*). *J Exp Biol* **39**, 261–270.

Gordon, M. S., Schmidt-Nielsen, K., and Kelly, H. M. (1961). Osmotic regulation in the crab-eating frog (*Rana cancrivora*). *J Exp Biol* **38**, 659–678.

Gottlieb, G. and Jackson, D. C. (1976). Importance of pulmonary ventilation in respiratory control in the bullfrog. *Am J Physiol* **230**, 608–613.

Gower, D. J., Loader, S. P., Montcrieff, C. B. and Wilkinson, M. (2004) Niche separation and comparative abundance of *Boulengerula boulengeri* and *Scolecomorphus vittatus* (Amphibia: Gymnophiona) in an East Usambara forest, Tanzania. *African J Herpetol* **53**, 183–190.

Grafe, T. U. (1996). Energetics of vocalization in the African reed frog (*Hyperolius marmoratus*). *Comp Biochem Physiol* **114A**, 235–243.

Grafe, T. U. (2000). *Leptopelis viridis* (West African tree frog). Cocoon formation. *Herpetolog Rev* **31**, 100–101.

Grafe, T. U., Kaminsky, S. K. and Linsenmair, K. E. (2005). Terrestrial larval development and nitrogen excretion in the adro-tropical pig-nosed frog, *Hemisus marmoratus*. *J Trop Ecol* **21**, 219–222.

Grafe, T. U., Schmuck, R. and Linsenmair, K. E. (1992). Reproductive energetics of the African reed frogs, *Hyperolius viridiflavius* and *Hyperolius marmoratus*. *Physiol Zool* **65**, 153–171.

Grafe, T. U. and Thein, J. (2001). Energetics of calling and metabolic substrate use during prolonged exercise in the European treefrog *Hyla arborea*. *J Comp Physiol* **171B**, 69–76.

Grandchamp, A. and Boulpaep, E. L. (1974). Pressure control of sodium reabsorption and intercellular backflux across the proximal kidney tubule. *J Clinical Investigation* **54**, 69–82.

Grayson, K. L., Cook, L. W., Todd, M. J., Pierce, D., Hopkins, W. A., Gatten, R. E. and Dorcas, M. E. (2005). Effects of prey type on specific dynamic action, growth, and mass conversion efficiencies in the horned frog, *Ceratophrys cranwelli*. *Comp Biochem Physiol* **141A**, 298–304.

Greber, K. and Schipp, R. (1986). Besonderheiten dur struktur und innervation des lympherz-myocards vo Anuren-eine vergleichende elektronenmikroskopische, histochemische und fluoreszenmikroskopische untersuchung. *Verh Dtsch Zool Ges* **79**, 75–82.

Greber, K. and Schipp, R. (1990). Early development and myogenesis of the posterior anuran lymph hearts. *Anat Embryol* **181**, 75–82.

Greenbank, J. (1945). Limnological conditions in ice-covered lakes, especially as related to winter-kill of fish. *Ecol Monogr* **15**, 343–392.

Greenwald, L. (1972). Sodium balance in amphibians from different habitats. *Physiol Zool* **45**, 229–237.

Grill, G., Granger, P. and Thurau, K. (1972). The renin-angiotensin system of amphibians. I. Determination of the renin content of amphibian kidneys. *Pflugers Archiv* **331**, 1–12.

Grover, M. (2000). Determinants of salamander distributions along moisture gradients. *Copeia* 2000, 156–168.

Grundy, J. E. and Storey, K. B. (1994). Urea and salt effects on enzymes from estivating and non-estivating amphibians. *Mol Cell Biochem* **131**, 9–17.

Grundy, J. E. and Storey, K. B. (1998). Antioxidant defenses and lipid peroxidation damage in estivating toads, *Scaphiopus couchii*. *J Comp Physiol B* **168**, 132–142.

Guggino, W. B., Oberleithner, H. and Giebisch, G. (1988). The amphibian diluting segment. *Am J Physiol* **254**, F615–F627.

Guha, K., Jorgensen, C. B. and Larsen, L. O. (1980). Relationship between nutritional state and testes function, together with observations on patterns of feeding in the toad, *Bufo bufo bufo*. *J Zool Lond* **192**, 147–155.

Guimond, R. W. and Hutchison, V. H. (1972). Pulmonary, branchial, and cutaneous gas exchange in the mudpuppy, *Necturus maculosus maculosus*. *Comp Biochem Physiol* **42A**, 367–392.

Guimond, R. W. and Hutchison, V. H. (1973a). Aquatic respiration: an unusual strategy in the hellbender, *Cryptobranchus alleganiensis alleganiensis*. *Science* **182**, 1263–1265.

Guimond, R. W. and Hutchison, V. H. (1973b). Trimodal gas exchange in the large aquatic salamander, *Siren lacertina*. *Comp Biochem Physiol* **46A**, 249–268.

Guimond, R. W. and Hutchison, V. H. (1974). Aerial and aquatic respiration in the congo eel *Amphiuma means means*. *Respir Physiol* **20**, 147–159.

Guimond, R. W. and Hutchison, V. H. (1976). Gas exchange of the giant salamanders of North America. In: *Respiration of Amphibious Vertebrates*, ed. G. M. Hughes, pp. 313–338. Academic Press: New York.

Guirguis, S. M., Yee, J. C. and Stiffler, D. F. (1997). Active calcium ion transport *Xenopus laevis* skin. *J Comp Physiol B* **167**, 328–334.

Guppy, M., Fuery, C. J. and Flanigan, J. E. (1994). Biochemical principles of metabolic depression. *Comp Biochem Physiol B* **109**, 175–189.

Guppy, M., Fuery, C. J. and Flanigan, J. E. (2004). The biochemistry of metabolic depression: A history of perceptions. *Comp Biochem Physiol B* **139**, 435–442.

Guppy, M. and Withers, P. C. (1999). Metabolic depression in animals: physiological perspectives and biochemical generalizations. *Biolog Rev* **74**, 1–40.

Guyton, A. C., and Hall, J. E. (2006). *Textbook of Medical Physiology*, 11th edn. Elsevier Saunders: Philadelphia.

Haas, A. (2003). Phylogeny of frogs as inferred from primarily larval characters (Amphibia: Anura). *Cladistics* **19**, 23–90.

Haberich, F. J. (1965). The functional separation of venous and arterial blood in the univentricular frog heart. *Ann NY Acad Sci* **127**, 459–476.

Hall, F. G. and Root, R. W. (1930). The influence of humidity on the body temperature of certain poikilotherms. *Biological Bulletin* **58**, 52–58.

Hancock, T. V., Hoagland, T. M. and Hillman, S. S. (2000). Whole-body systemic transcapillary filtration rates, coefficients, and isogravimetric capillary pressures in *Bufo marinus* and *Rana catesbeiana*. *Physiol Biochem Zool* **73**, 161–168.

Hanke, W., Kaltenhauser, U., Maser-Gluth, C. and Pathak, V. (1990). Regulation of interrenal activity by electrolytes and peptide hormones. In: *Biology and Physiology of Amphibians*, ed. W. Hanke, 38, pp. 241–255. Fortschritte der Zoologie.

Hare, J. R., Whitworth, E. and Cree, A. (2007). Correct orientation of a hand-held infrared thermometer is important for accurate measurement of body temperatures in small lizards and tuatara. *Herpetological Review* **38**, 311–315.

Harlan, R. A. and Wilkinson, R. F. (1981). The effects of progressive hypoxia and rocking activity on blood oxygen tensions for hellbenders, *Cryptobranchus allaganiensis*. *J Herpetol* **15**, 383–387.

Harri, M. N. E. and Talo, A. (1975a). Effect of season and temperature acclimation on the heart rate-temperature relationship in the frog, *Rana temporaria*. *Comp Biochem Physiol A* **50**, 469–472.

Harri, M. N. E. and Talo, A. (1975b). Effect of season and temperature acclimation on the heart rate-temperature relationship in the isolated frog's heart, *Rana temporaria*. *Comp Biochem Physiol A* **50**, 409–412.

Hartzell, H. C. (1979). Adenosine receptors in frog sinus venosus: slow inhibitory potentials produced by adenosine compounds and acetylcholine. *J Physiol* **293**, 23–49.

Harvey, P. H. and Pagel, M.P. (1991). *The Comparative Method in Evolutionary Biology*. Oxford University Press: Oxford.

Hasegawa, M., Suzuki, Y. and Wada, S. (2005). Design and performance of a wet sponge model for amphibian thermal biology. *Current Herpetology* **24**, 27–32.

Hasegawa, T., Tanii, H., Suzuki, M. and Tanaka, S. (2003). Regulation of water absorption in the frog skins by two vasotocin-dependent water-channel aquaporins, AQP-h2 and AQP-h3. *Endocrinology* **144**, 4087–4096.

Hasegawa, Y., Watanabe, T. X., Sokabe, H., and Nakajima, T. (1983). Chemical structure of angiotensin in the bullfrog *Rana catesbeiana*. Gen Comp Endocrinol **50**, 75–90.

Hayes, J. P. (2001). Mass-specific and whole-animal metabolism are not the same concept. *Physiol Biochem Zool* **74**, 147–150.

Hayman, J. M. (1927). Estimations of afferent arteriole and glomerular capillary pressures in the frog kidney. *Am J Physiol* **79**, 389–409.

Hayman, J. M. (1928). Notes on the arrangement of blood vessels within the frog's kidney together with some measurements of blood pressure in the renal portal and renal veins. *Am J Physiol* **86**, 331–339.

Hazard, E. S. and Hutchison, V. H. (1978). Ontogenetic changes in erythrocytic organic phosphates in the bullfrog, *Rana catesbeiana*. *J Exp Zool* **206**, 109–118.

Hazard, E. S. and Hutchison, V. H. (1982). Distribution of acid-soluble phosphates in the erythrocytes of selected species of amphibians. *Comp Biochem Physiol* **73A**, 111–124.

Heath, J. E. (1964). Reptilian thermoregulation: Evaluation of field studies. *Science* **146**, 784–785.

Heath, J. E. (1975). Behavioral thermoregulation in high altitude tiger salamanders, *Ambystoma tigrinimum*. *Herpetologica* **31**, 84–93.

Heatwole, H. (1984). Adaptations of amphibians to aridity. In: *Arid Australia*, eds H. G. Cogger and E. E. Cameron, pp. 177–222. Sydney Australian Museum.

Heatwole, H., Cameron, E. and Webb, G. J. W. (1971). Studies on anuran water balance. II. Evaporative water loss, vital limit, and behavioral responses to desiccation in *Notaden bennetti*. *Herpetologica* 27, 365–378.

Heatwole, H. and Newby, R. C. (1972). Interaction of internal rhythm and loss of body water in influencing activity levels of amphibians. *Herpetologica* **28**, 156–162.

Hedges, S. B., Nussbaum, R. A. and Maxson, L. R. (1993). Caecilian phylogeny and biogeography inferred from mitochondrial DNA sequences of the 12s rRNA and 12s rRNA genes (Amphibia: Gymnophiona). *Herpetolog Mono* 7, 64–67.

Hedrick, M. S. (1985). *Aspects of Cardiovascular Oxygen Transport in Vertebrates*. MS Thesis, Portland State University, Portland, OR.

Hedrick, M. S., Drewes, R. C., Hillman, S. S. and Withers, P. (2007). Lung ventilation contributes to vertical lymph movement in anurans. *J Exp Biol* **210**, 3940–3945.

Hedrick, M. S., Fahlman, C. S. and Bickler, P. E. (2005). Intracellular calcium and survival of tadpole forebrain cells in anoxia. *J Exp Biol* **208**, 681–686.

Hedrick, M. S., Palioca, W. B. and Hillman, S. S. (1999). Effects of temperature and physical activity on blood flow shunts and intracardiac mixing in the toad *Bufo marinus*. *Physiol Biochem Zool* 72, 509–19.

Heller, H. (1974). History of neurohypophyseal research. In: *Handbook of Physiology*, Section 7: *Endocrinology*, Vol. IV *The Pituitary Gland and its Neuroendocrine Control*, eds E. Knobil and W. H. Sawyer, pp. 103–117. American Physiological Society: Washington D.C.

Heller, H. and Bentley, P. J. (1965). Phylogenetic distribution of the effects of neurohypophysial hormones on water and sodium metabolism. *Gen Comp Endocrinol* **56**, 96–108.

Helman, S. I., Cox, T. C. and Van Driessche, W. (1983). Hormonal control of apical membrane Na transport in epithelia. Studies with fluctuation analysis. *J General Physiol* **82**, 201–220.

Hemmings, S. J. and Storey, K. B. (2001). Characterization of sarcolemma and sarcoplasmic reticulum isolated from skeletal muscle of the freeze tolerant wood frog, *Rana sylvatica*: the beta(2)-adrenergic receptor and calcium transport systems in control, frozen and thawed states. *Cell Biochem Funct* **19**, 143–152.

Henderson, I. W., Edwards, B. R., Garland, H. O. and Chester Jones, I. (1972). Renal function in two toads, *Xenopus laevis* and *Bufo marinus*. *General Comp Endocrinol Suppl.* **3**, 350–359.

Henderson, K. K., Wagner, H., Favret, F., Britton, S. L., Koch, L. G., Wagner, P. D. and Gonzales, N. C. (2002). Determinants of maximal O_2 uptake in rats selectively bred for endurance running capacity. *J Appl Physiol* **93**, 1265–1274.

Hepple, R. T. (2000). Skeletal muscle: microcirculatory adaptation to metabolic demand. *Med Sci Sports Exer* **32**, 117–123.

Herbert, C. V. and Jackson, D. C. (1985). Temperature effects on the responses to prolonged submergence in the turtle *Chrysemys picta belli*: 2. Metabolic rate, blood acid-base, and ionic changes, and cardiovascular function in aerated and anoxic water. *Physiol Zool* **58**, 670–681.

Herman, C. A. (1977). Comparative effects of epinephrine and norepinephrine on plasma glucose and hematocrit levels in the American bullgrog (*Rana catesbeiana*). *Gen Comp Endocrinol* **32**, 321–329.

Hermes-Lima, M. and Storey, K. B. (1996). Relationship between anoxia exposure and antioxidant status of the frog *Rana pipiens*. *Am J Physiol* **271**, R918–R925.

Hermes-Lima, M. and Storey, K. B. (1998). Role of antioxidant defenses in the tolerance of severe dehydration by anurans. The case of the leopard frog *Rana pipiens*. *Mol Cell Biochem* **189**, 79–89.

Herrera, F. C. (1965). Effect of insulin on short-circuit current across toad urinary bladder. *Am J Physiol* **209**, 819–824.

Hetem, R. S., Maloney, S. K., Fuller, A., Meyer, L. C. R. and Mitchell, D. (2007). Validation of a biotelemetric technique, using ambulatory miniature black globe thermometers, to quantify thermoregulatory behaviour in ungulates. *J Exp Zool* **307A**, 343–356.

Heyer, W. R. (1994). Thread bobbins. In: *Measuring and Monitoring Biological Diversity: Standard Methods for Amphibians*, eds W. R. Heyer, Donnelly, M. A., McDiarmid, R. W., Hayek, L. C., and Foster, M. S., pp. 153–155. Smithsonian Institution Press: Washington.

Hicks, J. W. and Wood, S. C. (1985). Temperature regulation in lizards: Effects of hypoxia. *Am J Physiol* **248**, R595–600.

Hill, A. V. (1938). The heat of shortening and the dynamic constants of muscle. *Proc Royal Soc London B* **126**, 136–195.

Hill, A. V. (1970). *First and Last Experiments in Muscle Mechanics*. Cambridge University Press: Cambridge.

Hill, R. W. (1972). Determination of oxygen consumption by use of the paramagnetic oxygen analyser. *J Applied Physiol* **33**, 261–263.

Hillis, D. M. (1991). The phylogeny of amphibians: current knowledge and the role of cytogenetics. In: *Amphibian Cytogenetics and Evolution,* eds D. M. Green and S. K. Sessions, pp. 7–31. Academic Press, San Diego.

Hillman, S. S. (1974). The effect of arginine vasopressin on water and sodium balance in the urodele amphibian *Aniedes lugubris*. *Gen Comp Endocrinol* **24**, 74–82.

Hillman, S. S. (1976). Cardiovascular correlates of maximal oxygen consumption rates in anuran amphibians. *J Comp Physiol B* **109**, 199–207.

Hillman, S. S. (1978a). The roles of oxygen delivery and electrolyte levels in the dehydrational death of *Xenopus laevis*. *J Comp Physiol* **128B**, 169–175.

Hillman, S. S. (1978b). Some effects of dehydration on internal distributions of water and solutes in *Xenopus laevis*. *Comp Biochem Physiol* **61A**, 303–307.

Hillman, S. S. (1980a). Physiological correlates of differential dehydration tolerance in anuran amphibians. *Copeia* 1980, 125–129.

Hillman, S. S. (1980b). The effect of anemia on metabolic performance in the frog, *Rana pipiens*. *J Exp Zool* **211**, 107–111.

Hillman, S. S. (1982). Effects of DL-propranolol on exercise heart rate and maximal rates of oxygen consumption in *Scaphiopus intermontanus*. *Experientia* **38**, 940–941.

Hillman, S. S. (1984). Inotropic influence of dehydration and hyperosmolal solutions on amphibian cardiac muscle. *J Comp Physiol* **154B**, 325–328.
Hillman, S. S. (1987). Dehydrational effects on cardiovascular and metabolic capacity in two amphibians. *Physiol Zool* **60**, 608–613.
Hillman, S. S. (1988). Dehydrational effects on brain and cerebrospinal fluid electrolytes in two amphibians. *Physiol Zool* **61**, 254–259.
Hillman, S. S. (1991). Cardiac scope in amphibians: transition to terrestrial life. *Can J Zool* **69**, 2010–2013.
Hillman, S. S., Hedrick, M. S., Withers, P. C. and Drewes, R. C. (2004). Lymph pools in the basement, sump pumps in the attic: the anuran dilemma for lymph movement. *Physiol Biochem Zool* 77, 161–73.
Hillman, S. S. and Schimpf, B. (1994). Renal responses to blood volume expansion of varying osmotic concentration in two species of amphibian, *Bufo marinus* and *Rana catesbeiana*. *Physiol Zool* **67**, 995–1005.
Hillman, S. S., Shoemaker, V. H., Putnam, R. and Withers, P. C. (1979). Reassessment of aerobic metabolism in amphibians during activity. *J Comp Physiol B* **129**, 309–313.
Hillman, S. S. and Withers, P. C. (1979). An analysis of respiratory surface area as a limit to activity metabolism in anurans. *Can J Zool* 57, 2100–2105.
Hillman, S. S. and Withers, P. C. (1988). The hemodynamic consequences of hemorrhage and hypernatremia in two amphibians. *J Comp Physiol B* **157**, 807–812.
Hillman, S. S., Withers, P. C. and Drewes, R. C. (2000). Correlation of ventricle mass and dehydration tolerance in amphibians. *Herpetologica* **56**, 413–420.
Hillman, S. S., Withers, P. C., Hedrick, M. S. and Drewes, R. C. (2005). Functional roles for the compartmentalization of the subcutaneous lymphatic sacs in anuran amphibians. *Physiol Biochem Zool* **78**, 515–523.
Hillman, S. S., Withers, P. C., Hedrick, M. S. and Kimmel, P. B. (1985). The effects of erythrocythemia on blood viscosity, maximal systemic oxygen transport capacity and maximal rates of oxygen consumption in an amphibian. *J Comp Physiol B* **155**, 577–581.
Hillman, S. S., Withers, P. and Kimmel, P. B. (1998). Plasma catecholamines with hemorrhage in the bullfrog, *Rana catesbeiana*. *J Exp Zool* **280**, 174–181.
Hillman, S. S., Withers, P. C., Palioca, W. B. and Ruben, J. A. (1987). Cardiovascular consequences of hypercalcemia during activity in two species of amphibian. *J Exp Zool* **242**, 303–308.
Hillman, S. S., Zygmunt, A. and Baustian, M. (1987). Transcapillary fluid forces during dehydration in two amphibians. *Physiol Zool* **60**, 339–345.
Hillyard, S. D. (1976). The movement of soil water across the isolated amphibian skin. *Copeia* 1976, 314–320.
Hillyard, S. D. (1979). The effect of isoproterenol on the amphibian water balance response. *Comp Biochem Physiol* **62C**, 93–95.
Hillyard, S. D. (1999). Behavioral, molecular and integrative mechanisms of amphibian osmoregulation. *J Exp Zool* **283**, 662–674.
Hillyard, S. D., Hoff, K. and Propper, C. (1998). The water absorption response: a behavioral assay for physiological processes in terrestrial amphibians. *Physiol Zool* **71**, 127–138.
Hillyard, S. D., Viborg, A., Nagai, T. and van Hoff, K. (2007). Chemosensory function of salt and water transport by the amphibian skin. *Comp Biochem Physiol* **148A**, 44–54.
Hillyard, S. D., Zeiske, W. and Van Driessche, W. (1982). Poorly selective cation channels in the skin of the larval frog (stage < XIX). *Pflugers Archiv* **394**, 287–293.

Hirano, M. and Rome, L. C. (1984). Jumping performance of frogs (*Rana pipiens*) as a function of muscle temperature. *J Exp Biol* **108**, 429–439.

Hoagland, T. M. (1997). Whole body in vivo compliance in two amphibians, *Bufo marinus* and *Rana catesbeiana*. MS Thesis, Portland State University.

Hochachka, P. W., Darveau, C. A., Andrews, R. D. and Suarez, R. K. (2003). Allometric cascade: a model for resolving body mass effects on metabolism. *Comp Biochem Physiol* **134A**, 675–691.

Hochachka, P. W. and Guppy, M. (1987). *Metabolic Arrest and the Control of Biological Time.* Harvard University Press: Cambridge.

Hochachka, P. W. and Somero, G. N. (1984). *Biochemical Adaptation*. Princeton University Press: New Jersey.

Hoff, K. and Hillyard, S. D. (1991). Angiotensin II stimulates cutaneous drinking in the toad, *Bufo punctatus*. *Physiol Zool* **64**, 1165–1172.

Hoff, K. and Hillyard, S. D. (1993). Inhibition of cutaneous water absorption in dehydrated toads by saralasin is associated with changes in barometric pressure. *Physiol Zool* **66**, 89–98.

Hoffman, J., U. Katz. (1997). Salt and water balance in the toad *Bufo viridis* during recovery from two different osmotically stressful conditions. *Comp Biochem Physiol A* **117**, 147–154.

Hoffman, J. and Katz, U. (1999). Elevated plasma osmotic concentration stimulates water absorption response in a toad. *J Exp Zool* **284**, 168–173.

Holenweg, A. K. and Reyer, H. U. (2000). Hibernation behavior of *Rana lessonae* and *R. esculenta* in their natural habitat. *Oecologia* **123**, 41–47.

Holenweg Peter, A. K. (2001). Survival in adults of the water frog *Rana lessonae* and its hybridogenetic associate *Rana esculenta*. *Can J Zool* **79**, 652–661.

Holroyd, S. M. and Gibbs, C. L. (1993). The energetics of shortening amphibian cardia muscle. *Pflugers Arch* **424**, 84–90.

Holzapfel, R. A. (1937). The cyclic character of hibernation in frogs. *Quart Rev Biol* **12**, 65–84.

Hou, P. C. and Burggren, W. W. (1995). Cardiac output and peripheral resistance during larval development in the anuran amphibian *Xenopus laevis*. *Am J Physiol* **269**, R1126–1132.

Houck, M. A. and Bellis, E. D. (1972). Comparative tolerance to desiccation in the salamanders *Desmognathus f. fuscus* and *Desmognathus o. ochrophaeus*. *J Herpetology* **6**, 209–215.

Howell, B. J., Baumgardner, F. W., Bondi, K. and Rahn, H. (1970). Acid-base balance in cold-blooded vertebrates as a function of body temperature. *Am J Physiol* **218**, 600–606.

Howland, R. B. (1916). On the effect of removal of the pronephros of the amphibian embryo. *PNAS* **2**, 231–232.

Hoyt, R. W., Eldridge, M. and Wood, S. C. (1984). Noninvasive pulsed doppler determination of cardiac output in an unanesthetized neotenic salamander, *Ambystoma tigrinum*. *J Exp Zool* **230**, 491–493.

Hsu, A. C. and Dawson, M. J. (2003). Muscle glycogenolysis is not activated by changes in cytosolic P-metabolites: α^{31}P and ^{1}H MRS demonstration. *Magnetic Resonance Medicine* **49**, 626–631.

Huang, H. and Brown, D. D. (2000a). Prolactin is not a juvenile hormone in *Xenopus laevis* metamorphosis. *PNAS* **97**, 195–199.

Huang, H. and Brown, D. D. (2000b). Overexpression of *Xenopus laevis* growth hormone stimulates growth of tadpoles and frogs. *PNAS* **97**, 190–194.

Huber, G. C. (1917). On the morphology of the renal tubules of vertebrates. *Anat Rec* **18**, 305–339.

Huey, R. B. and Slatkin, M. (1976). Costs and benefits of lizard thermoregulation. *Quart Rev Biol* **51**, 363–384.

Huey, R. B. and Stevenson, R. D. (1979). Integrating thermal physiology and ecology of ectotherms: A discussion of approaches. *Am Zoolog* **19**, 357–366.

Huf, E. (1935). Ber den anteil vitaler krdfte bei der resorption von flussigkeit durch die froschhaut. *Pflugers Archiv* **236**, 1–19.

Huf, E. G., Parrish, J. and Weatherford, C. (1951). Active salt and water uptake by isolated frog skin. *Am J Physiol* **167**, 137–142.

Hughes, G. M. and Vergara, G. A. (1978). Static pressure-volume curves for the lung of the frog (*Rana pipiens*). *J Exp Biol* **76**, 149–165.

Humbert, F., Montesano, R., Grosso, A., de Sousa, R. C. and Orci, L. (1977). Particle aggregates in plasma and intracellular membranes of toad bladder (granular cell). *Cell Mol Life Sci* **33**, 1364–1367.

Hurst, A. M. and Hunter, M. (1990). Acute changes in channel density of amphibian diluting segment. *Am J Physiol* **259**, C1005–C1009.

Hutchison, V. H. (1958). The distribution and the ecology of the cave salamander, *Eurycea lucifuga*. *Ecolog Mono* **28**, 1–20.

Hutchison, V. H. and Erskine, D. J. (1981). Thermal selection and prostaglandin E1 fever in the salamander *Necturus maculosus*. *Herpetologica* **37**, 195–198.

Hutchison, V. H. and Dupré, R. K. (1992). Thermoregulation. In: *Environmental Physiology of the Amphibia*, eds M. E. Feder and W. W. Burrgren, pp. 314–377. University of Chicago Press: Chicago.

Hutchison, V. H., Haines, H. B. and Engbretson, G. (1976). Aquatic life at high altitude: respiratory adaptations in the Lake Titicaca frog, *Telamatobius culeus*. *Respir Physiol* **27**, 115–129.

Hutchison, V. H. and Hazard, E. S. (1984). Erythrocytic organic phosphates: diel and seasonal cycles in the frog, *Rana berlanderi*. *Comp Biochem Physiol* **79A**, 533–538.

Hutchison, V. H. and Miller, K. (1979). Anaerobic capacity of amphibians. *Comp Biochem Physiol A* **63**, 213–216.

Hutchison, V. H., Whitford, W. G. and Kohl, M. A. (1968). The relation of body size and surface area to gas exchange in anurans. *Physiol Zool* **41**, 65–85.

Iga, T. and Bagnara, J. T. (1967). An analysis of color change in the leaf frog, *Agalychnis dacnicolor*. *J Zool* **192**, 331–341.

Inger, R. F. (1980). Densities of floor-dwelling frogs and lizards in lowland forests of Southeast Asia and Central America. *Am Nat* **115**, 761–770.

Ireland, P. H. (1976). Reproduction and larval development of the gray-bellied salamander *Eurycea multiplicata grisegaster*. *Herpetologica* **32**, 233–238.

Irisawa, H. (1978). Comparative physiology of the cardiac pacemaker mechanism. *Physiol Rev* **58**, 461–498.

Irwin, J. T., Costanzo, J. P. and Lee, R. E., Jr. (1999). Terrestrial hibernation in the northern cricket frog, *Acris crepitans*. *Can J Zool* 77, 1240–1246.

Irwin, J. T. and Lee, R. E., Jr. (2003). Geographic variation in energy storage and physiological responses to freezing in the gray treefrogs *Hyla versicolor* and *H. chrysoscelis*. *J Exp Biol* **206**, 2859–2867.

Jackson, D. C. (1978). Respiratory control and CO2 conductance: temperature effects in a turtle and a frog. *Respir Physiol* **33**, 103–114.

Jacobsen, A. and Fritsche, R. (1999). Development of adrenergic and cholinergic cardiac control in larvae of the African clawed frog, *Xenopus laevis*. *Physiol Biochem Zool* **72**, 328–338.

Jacobsen, S. K. (1985). Reproductive behavior and male mating success in two species of glass frogs (Centrolenidae). *Herpetologica* **41**, 396–404.

Jagger, J. E., Bateman, R. M., Ellsworth, M. L. and Ellis, C. G. (2001). Role of erythrocyte in regulating local O_2 delivery mediated by hemoglobin oxygenation. *Am J Physiol* **280**, H2833–H2839.

Jia, X. and Burggren, W. (1997). Developmental changes in chemoreceptive control of gill ventilation in larval bullfrogs (*Rana catesbeiana*). II. Sites of O_2-sensitive chemoreceptors. *J Exp Biol* **200**, 2237–2248.

Johansen, K. (1962). Double circulation in the amphibian *Amphiuma tridactylum. Nature* **194**, 991–992.

Johansen, K. (1963). Cardiovascular dynamics in the amphibian *Amphiuma tridactylum. Acta Physiol Scand Suppl* **217**, 1–82.

Johansen, K. and Ditadi, A. S. F. (1966). Double circulation in the giant toad, *Bufo paracnemis. Physiol Zool* **39**, 140–150.

Johansen, K. and Hanson, D. (1968). Functional anatomy of the hearts of lungfishes and amphibians. *Amer Zool* **8**, 191–210.

Johansen, K. and Lenfant, C. (1972). A comparative approach to the adaptability of O_2-Hb affinity. *Proc A Benzon Symp* **4**, 750–780.

Johansen, K., Lykkeboe, G., Kornerup, S. and Maloiy, G. M. O. (1980). Temperature insensitive O_2 binding in blood of the tree frog *Chiromantis petersi. J Comp Physiol* **136**, 71–76.

John-Alder, H. B., Barnhart, M. C. and Bennett, A. F. (1989). Thermal sensitivity of swimming performance and muscle contraction in northern and southern populations of tree frogs (*Hyla crucifer*). *J Exp Biol* **142**, 357–372.

John-Alder, H. B., Morins, P. J., Lawler, S. (1988). Thermal physiology, phenology, and distribution of tree frogs. *Am Naturalist* **132**, 506–520.

Johnsen, A. and Nielsen, R. (1984). Correlation between C-AMP in isolated frog skin epithelium and stimulation of sodium transport and osmotic water flow by antidiuretic hormone and phosphodiesterase inhibitors. *General Comp Endocrinology* **54**, 144–153.

Johnson, C. R. (1969). Aggregation as a means of water conservation in juvenile *Limnodynastes* from Australia. *Herpetologica* **25**, 275–276.

Johnson, C. R. (1971). Thermal relations and water balance in the day frog, *Taudactylus diurnis*, from an Australian rainforest. *Aust J Zool* **19**, 35–39.

Johnson, C. R. (1972). Diel variation in the thermal tolerance of *Litoria gracilenta* (Anura: Hylidae). *Comp Biochem Physiol A* **41**, 727–730.

Johnson, W. E. and Propper, C. R. (2000). Effects of dehydration on plasma osmolality, thirst-related behavior, and plasma and brain angiotensin concentrations in Couch's spadefoot toad, *Scaphiopus couchii. J Exp Zool* **286**, 572–584.

Jokumsen, A. and Weber, R. E. (1980). Haemoglobin-oxygen binding properties in the blood of *Xenopus laevis*, with special reference to the influences of aestivation and of temperature and salinity acclimation. *J Exp Biol* **86**, 19–37.

Jolivet-Jaudet, G., Inoue, M., Takada, K., and Ishii, S. (1984). Circannual changes in plasma aldosterone levels in *Bufo japonicus formosus. Gen Comp Endocrinol* **53**, 163–167.

Jolly, J. (1946). Recherches sur le systeme lymphatique des batrachiens. *Arch Anat, Microscop Morphol Exp* **36**, 3–44.

Jones, C. (1971). Notes on hairy frogs (*Trichobatrachus robustus*) collected in Rio Muni, West Africa. *Herpetologica* **27**, 51–54.

Jones, D. R. (1970). Experiments on amphibian respiratory and circulatory systems. *Exp Physiol Biochem* **3**, 233–293.

Jones, D. R. and Shelton, G. (1972). Factors affecting diastolic blood pressures in the systemic and pulmocutaneous arches of anuran amphibia. *J Exp Biol* **57**, 789–803.

Jones, J. M., Gamperl, A. K., Farrell, A. P. and Toews, D. P. (1997). Direct measurement of flow from the posterior lymph hearts of hydrated and dehydrated toads (*Bufo marinus*). *J Exp Biol* **200**, 1695–1702.

Jones, J. M. and Lindstedt, S. L. (1993). Limits to maximal performance. *Annu Rev Physiol* **55**, 547–569.

Jones, R. M. (1980a). Nitrogen excretion by *Scaphiopus* tadpoles in ephemeral ponds. *Physiol Zool* **53**, 26–31.

Jones, R. M. (1980b). Metabolic consequences of accelerated urea synthesis during seasonal dormancy of spadefoot toads, *Scaphiopus couchii* and *Scaphiopus multiplicatus*. *J Exp Zool* **212**, 255–267.

Jones, R. M. (1982). How toads breathe: control of air flow to and from lungs by the nares in *Bufo marinus*. *Respir Physiol* **49**, 251–265.

Jones, R. M. and Hillman, S. S. (1978). Salinity adaptation in the salamander, *Batrachoseps*. *J Exp Biol* **76**, 1–10.

Jones, S. M. and Avery, R. A. (1989). The use of pyroelectric vidicon infra-red camera to monitor the body temperatures of small terrestrial vertebrates. *Functional Ecol* **3**, 373–377.

Jorgensen, C. B. (1950). Osmotic regulation in the frog, *Rana esculenta* at low temperatures. *Acta Physiol Scand* **20**, 46–55.

Jorgensen, C. B. (1983). Pattern of growth in a temperate zone anuran (*Bufo viridis*). *J Exp Zool* **227**, 433–439.

Jorgensen, C. B. (1988). Metabolic costs of growth and maintenance in the toad, *Bufo bufo*. *J Exp Biol* **138**, 319–331.

Jorgensen, C. B. (1989). Patterns of growth and fattening in young toads, *Bufo bufo*, fed mealworms: effects of growth hormone and feeding regimen. *Copeia* 1989, 124–128.

Jorgensen, C. B. (1991a). Water and salt balance at low temperature in a cold temperate zone anuran, the toad, *Bufo bufo*. *Comp Biochem Physiol* **100A**, 377–384.

Jorgensen, C. B. (1991b). Water economy in the life of a terrestrial anuran, the toad *Bufo bufo*. *Biol Skr Dan Vid Selsk* **39**, 1–30.

Jorgensen, C. B. (1992). Relationships between feeding, digestion and water balance in a carnivorous vertebrate, the toad *Bufo bufo* L. *Comp Biochem Physiol A* **101**, 157–160.

Jorgensen, C. B. (1993). Effects of feeding on water balance and cutaneous drinking in the toad (*Bufo bufo* L.). *Comp Biochem Physiol Comp Physiol* **106**, 793–798.

Jorgensen, C. B. (1994a). Robert Towson's observations on amphibian water economy revived. *Comp Biochem Physiol* **109A**, 325–334.

Jorgensen, C. B. (1994b). Water economy in a terrestrial toad (*Bufo bufo*) with special reference to cutaneous drinking and urinary bladder function. *Comp Biochem Physiol* **109A**, 311–324.

Jorgensen, C. B. (1997a). Urea and amphibian water economy. *Comp Biochem Physiol A Physiol* **117**, 161–170.

Jorgensen, C. B. (1997b). 200 years of amphibian water economy: from Robert Townson to the present. *Biol Rev Camb Philos Soc* **72**, 153–237.

Jorgensen, C. B. (2000). Amphibian respiration and olfaction and their relationships: from Robert Townson (1794) to the present. *Biol Rev Camb Philos Soc* **75**, 297–345.

Jorgensen, C. B., Brems, K. and Geckler, P. (1978). Volume and osmotic regulation in the toad *Bufo bufo bufo* at low temperature, with special reference to amphibian hibernation. In : *Osmotic and Volume Regulation,* eds C. B. Jorgensen and E. Skadhauge, pp. 62–79. Munksgaard: Copenhagen.

Jorgensen, C. B., Larsen, L. O. and Lofts, B. (1979). Annual cycles of fat bodies and gonads in the toad *Bufo bufo bufo*. *Biol Skr Dan Vid Selsk* **22**, 1–37.

Jorgensen, C. B. and Wind-Larsen, H. (1987). Energetics of growth in a temperate zone toad, *Bufo bufo*: effects of growth hormone. *Acta Zoologica* **68**, 107–113.

Ju, Y. and Allen, D. G. (1999). How does B-adrenergic stimulation increase the heart rate? The role of intracellular Ca^{2+} release in amphibian pacemaker cells. *J Physiol* **516**, 793–804.

Jungreis, A. M. and Johnston, J. W. (1979). Aldosterone levels in *Rana pipiens*: Hormone titers do not always reflect rate of synthesis or target tissue responsiveness. *Comp Biochem Physiol* **63A**, 307–311.

Just, J. J. and Atkinson, B. G. (1972). Hemoglobin transitions in the bullfrog, *Rana catesbeiana*, during spontaneous and induced metamorphosis. *J Exp Zool* **182**, 271–280.

Juszczyk, W., Obrzut, K. and Zamachowski, W. (1966). Morphological changes in the alimentary canal of the common frog (*Rana temporaria)* in the annual cycle. *Acta Biologica Cracoviensia: Zoologia* **9**, 239–246.

Kabbara, A. A., Nguyen, L. T., Stephenson, G. M. M. and Allen, D. G. (2000). Intracellular calcium during fatigue of cane toad skeletal muscle in the absence of glucose. *J Musc Res Cell Motil* **21**, 481–489.

Kakiuchi, Y., Arai, T., Horimoto, M., Kikuchi, Y. and Koyama, T. (1979). A new needle-type colloid osmometer for continuous determination of blood oncotic pressure. *Am J Physiol* **236**, F419–F422.

Kampmeier, O. F. (1969). *Evolution and Comparative Morphology of the Lymphatic System.* Thomas, Springfield.

Karlstrom, E. L. (1962). The toad genus *Bufo* in the Sierra Nevada of California. *University of California Publications in Zoology* **61**, 1–104.

Katchburian, E., Antoniazzi, M. M., Jared, C., Faria, F. P., Souza Santos, H. and Freymüller, E. (2001). Mineralized dermal layer of the Brazilian tree-frog *Corythomantis greeningi. J Morphol* **248**, 56–63.

Katsoyannis, P. G. and du Vigneaud, V. (1959). Active principles of the neurohypophysis of cold-blooded vertebrates. Arginine Vasotocin. *Nature,* 1465.

Katz, U. (1973). Studies on the adaptation of the toad *Bufo viridis* to high salinities: oxygen consumption, plasma concentration and water content of the tissues. *J Exp Biol* **58**, 785–796.

Katz, U. (1992). Environmental challenge and the role of urea in amphibian osmoregulation. *Verhandlungen der Deutsch Zoologischen Gesellschaft* **85**, 247–257.

Katz, U., Garcia Romeu, F., Masoni, A. and Isaia, J. (1981). Active transport of urea across the skin of the euryhaline toad, *Bufo viridis*. *Pflugers Archiv* **390**, 299–300.

Katz, U. and Hoffman, J. (1990). Changing plasma osmolality—a strategy of adaptation of anuran amphibia to water scarcity under burrowing conditions. *Fortschritte Zoologie* **38**, 351–356.

Katz, U., Pagi, D., Hayat, S. and Degani, G. (1986). Plasma osmolality, urine composition and tissue water content of the toad *Bufo viridis* in nature and under controlled laboratory conditions. *Comp Biochem Physiol* **85A**, 703–713.

Kaul, R. and Shoemaker, V. H. (1989). Control of thermoregulatory evaporation in the waterproof treefrog *Chiromantis xerampelina*. *J Comp Physiol B* **158**, 643–649.

Kawamata, S. (1990). Fine structural changes in the endolymphatic sac induced by calcium loading in the tree frog, *Hyla arborea japonica*. *Arch Histol Cytol* **53**, 397–404.

KC, P., Bernard, D. G., Gardner, M. N. and Smatresk, N. J. (2000). Ventilatory mechanics and the effects of water depth on breathing pattern in the aquatic caecilian *Typhlonectes natans*. *J Exp Biol* **203**, 263–272.

Kelleher, K. E. and Tester, J. R. (1969). Homing and survival in the Manitoba toad, *Bufo hemiophrys*, in Minnesota. *Ecology* **50**, 1040–1048.

Kellenberger, S. and Schild, L. (2002). Epithelial sodium channels/degenerin family of ion channels:a variety of functions for a shared structure. *Physiological Reviews* **82**, 735–767.

Kiesecker, J. M., Blaustein, A. R. and Belden, L. K. (2001). Complex causes of amphibian population declines. *Nature* **410**, 681–684.

Killorn, E. E. and Toews, D. P. (2001). The dynamics of venous return and response to hypervolemia in the toad, *Bufo marinus* (L.). *BMC Physiol* **1**, 13.

Kimmel, P. B. (1988). Plasma cathepsin activity and the role of hypovolemic shock in dehydrational death in the toad, *Bufo marinus*. *Comp Biochem Physiol* **90A**, 127–133.

King, P. A., Rosholt, M. N. and Storey, K. B. (1993). Adaptations of plasma membrane glucose transport facilitate cryoprotectant distribution in freeze-tolerant frogs. *Am J Physiol* **265**, R1036–1042.

King, P. A., Rosholt, M. N. and Storey, K. B. (1995). Seasonal changes in plasma membrane glucose transport in freeze tolerant wood frogs. *Can J Zool* **73**, 1–9.

King, R. B., Huff, S. and Phillips, J. B. (1994). Physiological color change in the green treefrog. Responses to background brightness and temperature. *Copeia* 1994, 422–432.

Kingsbury, D. L. and Fenwick, J. C. (1989). The effect of eel calcitonin on calcium influx an plasma ion levels in axolotls, *Ambystoma mexicanum*. *Gen Comp Endocrin* **75**, 135–138.

Kinkead, R. and Milsom, W. K. (1994). Chemoreceptors and control of episodic breathing in the bullfrog (*Rana catesbeiana*). *Respir Physiol* **95**, 81–98.

Kleiber, M. (1961). *The Fire of Life*. John Wiley and Sons: New York.

Klein, U., Timme, M., Zeiske, W. and Ehrenfeld, J. (1997). The H^+ pump in frog skin (*Rana esculenta*): identification and localization of a V-ATPase. *J Membrane Biol* **157**, 117–126.

Kloas, W. and Hanke, W. (1993). Receptors for atrial natriuretic factor (ANF) in kidney and adrenal tissue of urodeles-lack of angiotensin II receptors in these tissues. *Gen Comp Endocrinol* **91**, 235–249.

Kluger, M. J. (1977). Fever in the frog *Hyla cinerea*. *J Thermal Biol* **2**, 79–88.

Knickerbocker, D. L. and Lutz, P. L. (2001). Slow ATP loss and the defense of ion homeostasis in the anoxic frog brain. *J Exp Biol* **204**, 3547–3551.

Kobayashi, T., Shimo, M. and Sugi, H. (1998). Infrared thermography of bullfrog skeletal muscle at rest and during isometric tetanus. *Jap J Physiol* **48**, 477–482.

Kobayashi, T. and Sugi, H. (1980). Measurement of heat production during the summation of isometric contraction in frog skeletal muscle with an infrared thermometer. *Jap J Physiol* **30**, 617–629.

Kobelt, F. and Linsenmair, K. E. (1986). Adaptation of the reed frog *Hyperolius viridiflavus* (Amphibia: Anura: Hyperoliidae) to its arid environment. *Oecologia* **68**, 533–541.

Kobelt, F. and Linsenmair, K. E. (1992). Adaptations of the reed frog *Hyperolius viridiflavus* (Amphibia Anura Hyperoliidae) to its arid environment. VI. The iridophores in the skin as radiation reflectors. *J Comp Physiol B* **162**, 314–326.

Koefoed-Johnsen, V. and Ussing, H. H. (1953). The contribution of diffusion and flow to the passage of D_2O through living membranes. Effect of neurohypophyseal hormone on isolated anuran skin. *Acta Physiol Scand* **28**, 60–76.

Kogo, N., Perry, S. F. and Remmers, J. E. (1994). Neural organization of the ventilatory activity in the frog *Rana catesbeiana*. *J Neurobiol* **25**, 1067–1079.

Konno, N., Hyodo, S., Matsuda, K. and Uchiyama, M. (2006). Effect of osmotic stress on expression of putative facilitative urea transporter in the kidney and urinary bladder of the marine toad, *Bufo marinus*. *J Exp Biol* **209**, 1207–1216.

Konno, N., Hyodo, S., Takei, Y., Matsuda, K. and Uchiyama, M. (2005). Plasma aldosterone, angiotensin II and arginine vasotocin concentrations in the toad, *Bufo marinus*, following osmotic treatments. *General Comp Endocrinol* **140**, 86–93.

Koteja, P. (1996). Measuring energy metabolism with open-flow respirometric systems: which design to choose? *Func Ecol* **10**, 675–677.

Kottek, M., Grieser, J., Beck, C., Rudolf, B. and Rubel, F. (2006). World map of the Köppen-Geiger climate classification updated. *Meteorologische Zeitschrift* **15**, 259–263.

Koyama, T. and Horimoto, M. (1982). Pulmonary microcirculatory response to localized hypercapnia. *Am J Physiol* **53**, 1556–1564.

Koyama, T. and Horimoto, M. (1983). Blood flow reduction in local poulmonary microvessels during acute hypoxia imposed on a small fraction of the lung. *Respir Physiol* **52**, 181–189.

Koyama, T. and Horimoto, M. (1988). Lymphatic microcirculation in frog lung. *Biorheology* **25**, 219–226.

Krakauer, T. (1970). Tolerance limits of the toad, *Bufo marinus*, in south Florida. *Comp Biochem Physiol* **33**, 15–26.

Krause, U. and Wegener, G. (1996). Exercise and recovery in frog muscle: metabolism of PCr, adenine nucleotides, and related compounds. *Am J Physiol* **270**, R811–R820.

Kreuzer, F. (1970). Facilitated diffusion of oxygen and its possible significance; a review. *Respir Physiol* **9**, 1–30.

Krishnamoorthy, R. V. and Shakunthala, N. (1974). Increased RBC count and pulmonary respiration in cold-adapted frogs. *J Exp Biol* **61**, 285–290.

Krogh, A. (1919). The supply of oxygen to the tissues and the regulation of the capillary circulation. *J Physiol* **52**, 457–474.

Krogh, A. (1937). Osmotic regulation in the frog (*R. esculenta*) by active absorption of chloride ions. *Skand Archiv Physiol* **76**, 60–64.

Krogh, A. (1939). *Osmotic Regulation in Aquatic Animals*. Cambridge University Press: Cambridge.

Kruhoffer, M., Glass, M. L., Abe, A. S. and Johansen, K. (1987). Control of breathing in an amphibian *Bufo paracnemis*: effects of temperature and hypoxia. *Respir Physiol* **69**, 267–275.

Kuhlmann, W. D. and Fedde, M. R. (1979). Intrapulmonary receptors in the bullfrog: sensitivity to CO_2. *J Comp Physiol* **132**, 69–75.

Kupferberg, S. J., Marks, J. C. and Power, M. E. (1994). Effects of variation in natural and algal detrital diets on larval anuran (*Hyla regilla*) life history traits. *Copeia* 1994, 446–457.

Kuramoto, M. (1981). Relationships between number, size and shape of red blood cells in amphibians. *Comp Biochem Physiol* **69A**, 771–775.

Kusano, T., Sakai, A. and Hatanaka, S. (2006). Ecological functions of the foam nests of the Japanese treefrog, *Rhacophorus arboreus* (Amphibia, Rhacophoridae). *Herpetological Journal* **16**, 163–169.

Laitinen, M. and Pasanen, S. (1998). Wintering selection by the common frog (*Rana temporaria*) and common toad (*Bufo bufo*) in Finland: a behavioural experiment. *Ann Zool Fennici* **35**, 59–62.

Lamoureux, V. S. and Madison, D. M. (1999). Overwintering habitats of radio-implanted green frogs, *Rana clamitans*. *J Herpetol* **33**, 430–435.

Landis, E. M. and Pappenheimer, J. R. (1963). Exchange of substances through capillary walls. In: *Handbook of Physiology Circulation,* eds W. F. Hamilton and P. Dow, pp. 961–1033. Am Physiol Soc, Washington, DC.

Lang, F., Busch, G. L., Ritter, M., Vokl, H., Waldegger, S., Gulbins, E. and Haussinger, D. (1998). Functional significance of cell volume regulatory mechanisms. *Physiol Rev* **78**, 247–306.

Lang, F., Tschernko, E. and Haussinger, D. (1992). Hepatic regulation of renal function. *Exp Physiol* 77, 663–673.

Langille, B. L. and Jones, D. R. (1977). Dynamics of blood flow through the hearts and arterial systems of anuran amphibians. *J Exp Biol* **68**, 1–17.

Langkilde, T. and Alford, R. A. (2002). The tail wags the frog: Harmonic radar transponders affect movement behavior in *Litoria lesueri*. *J Herpetology* **36**, 711–715.

Lannergren, J. and Hoh, J. F. Y. (1984). Myosin isoenzymes in single muscle fibers of *Xenopus laevis*: analysis of five different functional types. *Proc R Soc Lond* B **222**, 401–408.

Larsen, E. H. (1991). Chloride transport by high-resistance heterocellular epithelia. *Physiolog Rev* **71**, 235–283.

Larsen, E. H., Willumsen, N. J. and Christofferson, B. C. (1992). Role of proton pump of mitochondria rich cells for active transport of chloride ions in toad skin epithelium. *J Physiol* **450**, 203–216.

Larsen, L. O. (1984). Feeding in adult toads: physiology, behaviour, ecology. *Vidensk Meddr dansk Naturh Foren* **145**, 97–116.

Larsen, L. O. (1992). Feeding and digestion. In: *Environmental Physiology of the Amphibians*, eds W. Burggren and M. Feder, pp. 378–394. University of Chicago Press.

Larson, A. (1996a). Ambystomatidae. Mole salamanders. *Tree of Life Web Project*. http://www.tolweb.org.

Larson, A. (1996b). Amphiumidae. Amphiumas. *Tree of Life Web Project*. http://www.tolweb.org.

Larson, A. (1996c). Cryptobranchidae. Hellbenders. *Tree of Life Web Project*. http://www.tolweb.org.

Larson, A. (1996d). Rhyacotritonidae. Torrent Salamanders. *Tree of Life Web Project*. http://www.tolweg.org.

Larson, A. (1996e). Sirenidae. Sirens. *Tree of Life Web Project*. http://www.tolweb.org.

Larson, A. (2006). Proteidae. Mudpuppies, waterdogs. *Tree of Life Web Project*. http://www.tolweb.org.

Larson, A., Wake, D. B. and Devitt, T. (2006a). Hynobidae. *Tree of Life Web Project*. http://www.tolweb.org.

Larson, A., Wake, D. B. and Devitt, T. (2006b). Plethodontidae. Lungless salamanders. *Tree of Life Web Project*. http://www.tolweb.org.

Larson, A., Wake, D. B. and Devitt, T. (2007). Salamandridae. Newts and "true salamanders". *Tree of Life Web Project*. http://www.tolweb.org.

Larson, D. W. (1971). Desiccation tolerance of three races of *Ambystoma tigrinum* in North Dakota. *Copeia* 1971, 743–747.

Lasiewski, R. C. and Bartholomew, G. A. (1969). Condensation as a mechanism for water gain in nocturnal desert poikilotherms. *Copeia* 1969, 405–407.

Lasiewski, R. C. and Dawson, W. R. (1969). Calculation and miscalculation of the equations relating avian standard metabolism to body weight. *Condor* **71,** 335–336.

Laurent, R. F. (1979). Esquisse d'une phylogenese des anoures. *Bull Soc Zool France* **104,** 397–422.

Lawson, R. (1966). The anatomy of the heart of *Hypogeophysis rostratus* (Amphibia: Apoda) and possible mode of action. *J Zool, London* **149,** 320–336.

Layne, J. R., Jr. (1991). External ice triggers freezing in freeze-tolerant frogs at temperatures above their supercooling point. *J Herpetol* **25,** 129–130.

Layne, J. R., Jr. (1995). Crystalization temperatures of frogs and their individual organs. *J Herpetol* **29,** 296–298.

Layne, J. R., Jr. (1999). Freeze tolerance and cryoprotectant mobilization in the gray treefrog (*Hyla versicolor*). *J Exp Zool* **283,** 221–225.

Layne, J. R. and Claussen, D. L. (1982). Seasonal variation in the thermal acclimation of critical thermal maxima (CT_{max}) and minima (CT_{min}) in the salamander *Eurycea bislineata. J Thermal Biol* 7, 29–33.

Layne, J. R., Jr., Costanzo, J. P. and Lee, R. E., Jr. (1998). Freeze duration influences postfreeze survival in the frog *Rana sylvatica*. *J Exp Zool* **280,** 197–201.

Layne, J. R., Jr. and First, M. C. (1991). Resumption of physiological functions in the wood frog (*Rana sylvatica*) after freezing. *Am J Physiol* **261,** R134–137.

Layne, J. R., Jr. and Jones, A. L. (2001). Freeze tolerance in the gray treefrog: cryoprotectant mobilization and organ dehydration. *J Exp Zool* **290,** 1–5.

Layne, J.R. and Lee R.E. (1987). Freeze tolerance and the dynamics of ice formation in wood frogs (*Rana sylvatica*) from southern Ohio. Can J Zool 65:2062–2065.

Layne, J. R., Jr., Lee, R. E., Jr. and Heil, T. L. (1989). Freezing-induced changes in the heart rate of wood frogs (*Rana sylvatica*). *Am J Physiol* **257,** R1046–1049.

Layne, J. R., Romano, M. A. and Guttman, S. I. (1989). Responses to desiccation of the treefrogs *Hyla cinerea* and *H. gratiosa* and their natural hybrids. *Am Midland Naturalist* **121,** 61–67.

Lazure, C., Ong, H., McNicoll, N., Nechitalo, P., Chretian, M., De Lean, A. and Vaudry, H. (1988). The amino acid sequence of frog heart natriuretic-like peptide and mammalian ANF are closely related. *FEBS Lett* **238,** 300–306.

Leaf, A., Anderson, J. and Page, L. B. (1958). Active sodium transport by the isolated bladder. *J General Physiol* **41,** 657–688.

Lee, A. K. (1968). Water economy of the burrowing frog, *Heleioporus eyrei* (Gray). *Copeia* 1968, 741–745.

Lee, A. K. and Mercer, E. H. (1967). Cocoon surrounding desert dwelling frogs. *Science* **157,** 87–88.

Lee, M. R., Lee, R. E., Strong-Gunderson, J. M. and Minges, S. R. (1995). Isplation of ice-nucleating active bacteria from the freeze tolerant frog, *Rana sylvatica*. *Cryobiology* **32,** 358–365.

Lee, R. E. and Costanzo, J. P. (1998). Biological ice nucleation and ice distribution in cold-hardy ectothermic animals. *Annu Rev Physiol* **60,** 55–72.

Lee, R. E., Jr., Constanzo, J. P., Davidson, E. C. and Layne, J. R., Jr. (1992). Dynamics of body water during freezing and thawing in a freeze-tolerant frog (*Rana sylvatica*). *J Therm Biol* 17, 263–266.

Lenfant, C. and Johansen, K. (1967). Respiratory adaptations in selected amphibians. *Respir Physiol* **2**, 247–260.

Liberthson, R. R., Szidon, J. P., Bharati, S., Lev, M. and Fishman, A. P. (1975). Hemodynamic consequences of delayed venticuloconal conductionin the frog *Rana catesbeiana*. *Am J Physiol* **229**, 1085–1093.

Licht, L. E. (1991). Habitat selection of *Rana pipiens* and *Rana sylvatica* during exposure to warm and cold temperatures. *Am Midl Nat* **125**, 259–268.

Liggins, G. W. and Grigg, G. C. (1985). Osmoregulation of the cane toad, *Bufo marinus*, in salt water. *Comp Biochem Physiol A* **82**, 613–619.

Lihrmann, I., Netchitailo, P., Feuilloley, M., Cantin, M., Delarue, C., Leboulenger, F., DeLean A. and Vaudry, H. (1988). Effects of atrial natriuretic factor on corticosteroid production of perifused frog interrenal slices. *Gen Com Endocrinol* **71**, 55–62.

Lillywhite, H. B. (1970). Behavioral thermoregulation in the bullfrog, *Rana catesbeiana*. *Copeia* 1970, 158–168.

Lillywhite, H. B. (1971a). Thermal modulation of cutaneous mucus discharge as a determinant of evaporative water loss in the frog, *Rana catesbeiana*. *Zeit Vergleichende Physiologie* **73**, 84–104.

Lillywhite, H. B. (1971b). Temperature selection by the bullfrog, *Rana catesbeiana*. *Comp Biochem Physiol A* **40**, 213–227.

Lillywhite, H. B. (2006). Water relations of tetrapod integument. *J Exp Biol* **209**, 202–226.

Lillywhite, H. B. and Licht, P. (1974). Movement of water over toad skin: Functional role of epidermal sculpturing. *Copeia* 1974, 165–171.

Lillywhite, H. B. and Licht, P. (1975). A comparative study of integumentary mucous secretions in amphibians. *Comp Biochem Physiol A* **51**, 937–941.

Lillywhite, H. B., Licht, P., Chelgren, P. (1973). The role of behavioral thermoregulation in the growth energetics of the toad, *Bufo boreas*. *Ecology* **54**, 375–383.

Lillywhite, H. B. and Mittal, A.K. (1999). Amphibian skin and the aquatic-terrestrial transition: Constraints and compromises related to water exchange. In: *Water/Air transition*, eds A. K. Mittal, F. B. Eddy and J. S. Datta Munshi, pp. 131–144. Oxford and IBH Publishing: New Delhi and Calcutta.

Lillywhite, H. B., Mittal, A. K., Garg, T. K. and Agrawal, N. (1997a). Wiping behaviour and its ecophysiological significance in the Indian tree frog *Polypedates maculatus*. *Copeia* 1997, 88–100.

Lillywhite, H. B., Mittal, A. K., Garg, T. K. and Agrawal, N. (1997b). Integumentary structure and its relationship to wiping behaviour in the common Indian tree frog *Polypedates maculatus*. *J Zool* **243**, 675–687.

Lillywhite, H. B., Mittal, A. K., Garg, T. K., and Das, I. (1998). Basking behaviour, sweating and thermal ecology of the Indian tree frog, *Polypedates maculatus*. *J Herpetology* **32**, 169–175.

Lips, K. R., Reeve, J. D. and Witters, L. R. (2003). Ecological traits predicting amphibian poulation declines in Central America. *Conserv Biol* **17**, 1078–1088.

Littleford, R. A., Keller, W. F. and Philips, N. E. (1947). Studies on the vital limits of water loss in the plethodontid salamanders. *Ecology* **28**, 440–447.

Lombard, R. E. and Wake, D. B. (1977). Tongue evolution in the lungless salamanders, Family Plethodontidae IV. Phylogeny of plethodontid salamanders and the evolution of feeding dynamics. *Syst Zool* **35**, 532–551.

Lombard, R. E. and Wake, D. B. (1986). Tongue evolution in the lungless salamanders, Family Plethodontidae IV. Phylogeny of plethodontid salamanders and the evolution of feeding dynamics. *System Zool* **35**, 532–551.

Londos, P. L. and Brooks, R. J. (1988). Effect of temperature acclimation on locomotory performance curves in the toad, *Bufo woodhousii woodhousii*. *Copeia* 1988, 26–32.

Longphre, M. and Gatten, R. E. (1994). Individual variability in sustained performance, aerobic metabolism, oxygen transport, and enzyme activity in toads (*Bufo marinus*). *Copeia* 1994, 887–896.

Loong, A. M., Chew, S. F. and Ip, Y. K. (2002). Excretory nitrogen metabolism in the juvenile axolotl *Ambystoma mexicanum*: differences in aquatic and terrestrial environments. *Physiol Biochem Zool* **75**, 459–468.

Loument, C. (1992). Les amphibians de Sao Tome et Principe: revision systematique, cris nupitaux et cryotypes. *Alytes* **10**, 37–62.

Lövei, G. L., Stringer, I. A. N., Devine, C. D. and Cartellieri, M. (1997). Harmonic radar—a method using inexpensive tags to study invertebrate movement on land. *N Z J Ecol* **21**, 187–193.

Lovelock, J. E. (1953a). The haemolysis of human red blood cells by freezing and thawing. *Biochim Biophys Acta* **10**, 414–426.

Lovelock, J. E. (1953b). The mechanism for the protective action of glycerol against haemolysis by freezing and thawing. *Biochim Biophys Acta* **11**, 28–36.

Loveridge, J. P. (1970). Observations on nitrogenous excretion and water relations of *Chiromantis xerampelina* (Amphibia, Anura). *Arnoldia* **5**, 1–6.

Loveridge, J. P. (1976). Strategies of water conservation in southern African frogs. *Zoologica Africana* **11**, 319–333.

Loveridge, J. P. and Crayé, G. (1979). Cocoon formation in two species of southern African frogs. *South African J Sci* **75**, 18–20.

Loveridge, J. P. and Withers, P. C. (1981). Metabolism and water balance of active and cocooned African bullfrogs *Pyxicephalus adspersus*. *Physiol Zool* **54**, 203–214.

Lucas, E. A. and Reynolds, A. W. (1967). Temperature selection by amphibian larvae. *Physiol Zool* **40**, 159–171.

Lutterschmidt, W. I. and Hutchison, V. H. (1997). The critical thermal maximum: history and critique. *Can J Zool* **75**, 1561–1574.

Lutz, G. J., Bremner, S., Lajevardi, N., Lieber, R. L. and Rome, L. C. (1998). Quantitative analysis of muscle fibre type and myosin heavy chain distribution in the frog hindlimb: implications for locomotory design. *Jour Musc Res Cell Mot* **19**, 717–731.

Lutz, G. J., Cuizon, D. B., Ryan, A. F. and Lieber, R. L. (1998). Four novel myosin heavy chain transcripts define a molecular basis for muscle fibre types in *Rana pipiens*. *J Physiol* **508**, 667–680.

Lutz, G. J., Sirsi, S. R., Shaphard-Palmer, A. A., Bremner, S. and Lieber, R. L. (2002). Influence of myosin isoforms on contractile properties of intact muscle fibers from *Rana pipiens*. *Am J Physiol* **282**, C835–C844.

Lutz, P. L. and Reiners, R. (1997). Survival of energy failure in the anoxic frog brain: delayed release of glutamate. *J Exp Biol* **200**, 2913–2917.

Lykkeboe, G. and Johansen, K. (1978). An O_2-Hb "paradox" in frog blood? *Respir Physiol* **35**, 119–122.

MacArthur, D. L. and Dandy, J. W. T. (1982). Physiological aspects of overwintering in the boreal chorus frog (*Pseudacris triseriata maculata*). *Comp Biochem Physiol* **72A**, 137–141.

MacKenzie, J. A. and Jackson, D. C. (1978). The effect of temperature on cutaneous CO_2 loss and conductance in the bullfrog. *Respir Physiol* **32**, 313–323.

MacMahon, J. A. (1965). Factors influencing the rate of water loss in salamanders. *Am Zoologist* **4**, 304.

MacNalley, R. C. (1981). On the repeoductive energetics of chorusing males: Energy depletion, restoration, and growth for two sympatric species of *Ranidella* (Anura). *Oecologia* **51**, 181–188.

Maddison. W. P. and Maddison, D. R. (1992). *MacClade. Version 3.0. Analysis of Phylogeny and Character Evolution*. Sinauer Associates: Sunderland, MA.

Maetz, J., Sawyer, W. H., Pickford, G. E. and Mayer, N. (1967). Evolution de la balance du sodium chez *Fundulus heteroclitus* au cours du transfert d'eay de mer en d'eau douce: effets de l'hypophysectomie et de la prolactine. *General Comp Endocrinology* **8**, 169–176.

Maginniss, L. A., Song, Y., K. and Reeves, R. B. (1980). Oxygen equilibria of ectotherm blood containing multiple hemoglobins. *Respir Physiol* **42**, 329–343.

Main, A. R. (1968). Ecology, systematics, and ecology of Australian frogs. In: *Advances in Ecological Research*, Vol. 5, ed. J. B. Cragg, pp. 37–86. Academic Press: New York.

Main, A. R. and Bentley, P. J. (1964). Water relations of Australian burrowing frogs and tree frogs. *Ecology* **45**, 379–382.

Main, A. R., Littlejohn, M. J. and Lee, A. K. (1959). Ecology of Australian frogs. In: *Biogeography and Ecology in Australia Monographs in Biology 8*, eds A. Keast, R. L. Crocker and C. S. Christian, pp. 396–411. Junk: The Hague.

Maina, J. N. (1989). The morphology of the lung of the East African tree frog *Chiromantis petersii* with observations on the skin and the buccal cavity as secondary gas exchange organs; A TEM and SEM study. *J Anat* **165**, 29–43.

Maina, J. N. (2002). Structure, function and evolution of the gas exchangers: comparative perspectives. *J Anat* **201**, 281–304.

Maina, J. N. and Maloiy, G. M. O. (1988). A scanning and transmission electron microscopic study of the lung of a caecilian *Boulengerula taitanus*. *J Zool, London* **215**, 739–751.

Maina, J. N. and West, J. B. (2005). Thin and strong! The bioengineering dilemma in the structural and functional design of the blood-gas barrier. *Physiol Rev* **85**, 811–844.

Maleek, R., Sullivan, P., Hoff, K., Baula, V. and Hillyard, S. D. (1999). Salt sensitivity and hydration behavior of the toad, *Bufo marinus*. *Physiol Behav* **67**, 739–745.

Malvin, G. M. (1988). Microvascular regulation of cutaneous gas exchange in amphibians. *Am Zool* **28**, 999–1007.

Malvin, G. M. (1993). Vascular effects of arginine vasotocin in toad skin. *Am J Physiol* **265**, R426–432.

Malvin, G. M. and Hlastala, M. P. (1986a). Effects of lung volume and O_2 and CO_2 content on cutaneous gas exchange in frogs. *Am J Physiol* **251**, R941–946.

Malvin, G. M. and Hlastala, M. P. (1986b). Regulation of cutaneous gas exchange by environmental O_2 and CO_2 in the frog. *Respir Physiol* **65**, 99–111.

Malvin, G. M. and Hlastala, M. P. (1989). Effects of environmental O_2 on blood flow and diffusing capacity in amphibian skin. *Respir Physiol* **76**, 229–41.

Malvin, G. M., Macias, S., Sanchez, M., Dasalla, R., Park, A. and Duran, M. (1995). Lymphatic regulation of hematocrit during hypoxia in the toad *Bufo woodhousei*. *Am J Physiol* **269**, R814–821.

Malvin, G. M. and Wood, S. C. (1991). Behavioral thermoregulation of the toad, *Bufo marinus*: Effects of air humidity. *J Exp Zool* **258**, 322–326.

Malvin, G. M., Wood, S. C. and Riedel, C. (1989). Behavioral hypothermia in dehydrated toads. *FASEB J* **3**, A234.

Manion, J. J. and Cory, L. (1952). Winter kill of *Rana pipiens* in shallow ponds. *Herpetologica* **8**, 32.

Marangio, M. S. and Anderson, J. D. (1977). Soil moisture preference and water relations of the marbled salamander, *Ambystoma opacum* (Amphibia, Urodela, Ambystomatidae). *J Herpetol* **11**, 169–176.

Marcus, H. (1935). Zur stammensgeschichte der Herzens. *Morph Jb* **76**, 92–103.

Marker, G. M. and Gatten, R. E. (1993). Individual variability in sprint performance, lactate production, and enzyme activity in frogs (*Rana pipiens*). *J Herpetol* **27**, 294–299.

Marrero, M. B. and Hillyard, S. D. (1985). Differences in c-AMP levels in epithelial cells from pelvic and pectoral regions of the toad skin. *Comp Biochem Physiol C* **82**, 69–73.

Marsh, R. L. and Ellerby, D. J. (2006). Partitioning locomotor energy use among and within muscles muscle blood flow as a measure of muscle oxygen consumption. *J Exp Biol* **209**, 2385–2394.

Marsh, R. L. and Taigen, T. L. (1987). Properties enhancing aerobic capacity of calling muscles in gray tree frogs *Hyla versicolor*. *Am J Physiol* **252**, R786–R793.

Marshall, A. G. (1978). *Biophysical Chemistry. Principles, Techniques and Applications*. John Wiley and Sons: New York.

Marshall, E. K. (1932). The secretion of urea in the frog. *J Cell Comp Physiol* **2**, 349–353.

Marshall, E. K. and Smith, H. W. (1930). The glomerular development of the vertebrate kidneys in relation to habitat. *Biol Bull* **59**, 135–153.

Martin, A. A. and Cooper, A. K. (1972). The ecology of terrestrial anuran eggs, genus *Crinia* (Leptodactylidae). *Copeia* 1972, 163–168.

Martin, J. and Hillman, S. S. (2009). The physical movement of urine from the kidneys to rge urinary bladder and bladder compliance in two anurans. *Physiol Biochem Zool*, in press.

Martin, K. M. and Hutchison, V. H. (1979). Ventilatory activity in *Amphiuma tridactylum* and *Siren lacertina* (Amphibia:, Caudata). *J Herpetol* **13**, 427–434.

Mason, M. J. and Thomas, R. C. (1988). A microelectrodestudy of the mechnisms of L-lactate entry into and release from frog sartorius muscle. *J Physiol* **400**, 459–479.

Mathias, J. A. (1971). *The Comparative Ecologies of Two Species of Amphibia (Bufo bufo and Bufo calamita) on the Ainsdale Sand Dunes Natural Nature Reserve*. Ph.D., University of Manchester.

Matteucci, C. and Cima, A. (1845). Memoire sur l'endosmos. *Annals de Chimie et de Physique* **13**, 63–86.

Matthews, K. R. and Pope, K. L. (1999). A telemetric study of the movement pattern and habitat use of *Rana muscosa*, the mountain yellow-legged frog, in a high elevation basin in Kings Canyon National Park, California. *J Herpetol* **33**, 615–624.

Maunsbach, A. B. and Boulpaep, E. L. (1984). Quantitative ultrastruce and functional correlates in proximal tubule of *Ambystoma* and *Necturus*. *Am J Physiol* **246**, F710–F724.

Maxwell, S. S. (1913). On the absorption of water by the skin of the frog. *Am J Physiol* **32**, 286–294.

Mayhew, W. W. (1965). Adaptations of the amphibian, *Scaphiopus couchii*, to desert conditions. *Am Midland Naturalist* **74**, 95–109.

Mayhew, W. W. (1968). Biology of desert amphibians and reptiles. In: *Desert Biology*, ed. G. W. Brown, pp. 195–356. Academic Press: New York.

Mazur, P., Rall, W. F. and Rigopolous, N. (1981). Relative contributions of the fraction of unfrozen water and of salt concentration to the survival of slowly frozen human erythrocytes. *Biophys J* **36**, 653–675.

Mbangkollo, D. and de Roos, R. (1983). Comparative effects of epinephrine, norepinephrine and gentle handling stress on plasma lactate, glucose and hematocrit levels in the American bullfrog (*Rana catesbeiana*). *Gen Comp Endocrinol* **49**, 167–175.

McAfee, R. D. and Kalckar, H. M. (1972). Survival of *Rana pipiens* in deionized water. *Science* **178**, 183–185.

McBean, R. L. and Goldstein, L. (1970). Renal function during osmotic stress in the aquatic toad *Xenopus laevis*. *Am J Physiol* **219**, 1115–1123.

McCallum, M. L. (2007). Amphibian decline or extinction? Current declines dwarf background extinction rate. *J Herpetol* **41**, 483–491.

McClanahan, L. L. (1967). Adaptations of the spadefoot toad, *Scaphiopus couchi*, to desert environments. *Comp Biochem Physiol* **20**, 73–99.

McClanahan, L. L. (1972). Changes in body fluids of burrowed spadefoot toads as a function of soil water potential. *Copeia* 1972, 209–216.

McClanahan, L. L. and Baldwin, R. (1969). Rate of water uptake through the integument of the desert toad, *Bufo punctatus*. *Comp Biochem Physiol* **28**, 381–389.

McClanahan, L. L. and Shoemaker, V. H. (1987). Behavior and thermal relations of the arboreal frog *Phyllomedusa sauvagei*. *Nat Geogr Res* **3**, 11–21.

McClanahan, L. L., Shoemaker, V. H. and Ruibal, R. (1976). Structure and function of the cocoon of a ceratophyrid frog. *Copeia* 1976, 179–185.

McClanahan, L. L., Stinner, J. N. and Shoemaker, V. H. (1978). Skin lipids, water loss, and energy metabolism in a South American tree frog (*Phyllomedusa sauvagei*). *Physiol Zool* **51**, 179–187.

McClanahan, L. L., Ruibal, R. and Shoemaker, V. H. (1983). Rate of cocoon formation and physiological its correlates in a ceratophryid frog. *Physiol Zool* **56**, 430–435.

McClanahan, L. L., Ruibal, R. and Shoemaker, V. H. (1994). Frogs and toads in deserts. *Scientific Am* March, 64–70.

McClister, J. D. (2003). The metabolic cost of amplexus in the gray tree frog (*Hyla versicolor*): assessing the energetics of male mating success. *Can J Zool* **81**, 388–394.

McDiarmid, R. W. and Altig, R. A. (1999). *Tadpoles. The Biology of Anuran Larvae*. University of Chicago Press: Chicago and London.

McDiarmid, R. W. and Foster, M. S. (1987). Cocoon formation in another hylid frog, *Smilisca baudinii*. *J Herpetology* **21**, 352–355.

McKean, T., Scherzer, A. and Park, H. (1997). Hypoxia and ischaemia in buffer-perfused toad hearts. *J Exp Biol* **200**, 2575–2581.

McLister, J. D., Stevens, E. D. and Bogart, J. P. (1995). Comparative contractile dynamics of calling and locomotor muscles in three hylid frogs. *J Exp Biol* **198**, 1527–1538.

McMaster, K. A. (2006). *Ecophysiology of Cocooning and Non-cocooning Burrowing Desert Frogs*. School of Animal Biology, PhD, Perth, University of Western Australia.

McWhinnie, D. J. and Scopelliti, T. G. (1978). A comparative study of the influence of calcitonin on plasma mineral and volume dynamics in the amphibians, *Rana pipiens* and *Xenopus laevis* and in the rat. *Comp Biochem Physiol* **61A**, 487–495.

Measey, G. J. and Barot, S. (2006). Evidence of seasonal migration in a tropical subterranean vertebrate. *J Zool* **269**, 29–37.

Measey, G. J. and Van Dongen, S. (2006). Bergmann's rule and the terrestrial caecilian *Schistometopum thomense* (Amphibia: Gymniophiona: Caeciliidae). *Evol Ecol Res* **8**, 1049–1059.

Measey, G. J., Vences, M., Drewes, R. C., Chiari, Y., Melo, M. and Bourles, B. (2007). Freshwater paths across the ocean: molecular phylogeny of the frog *Ptychadena newtoni* gives insights into amphibian colonization of oceanic islands. *J Biogeography* **34**, 7–20.

Meban, C. (1980). Thicknesses of the air-blood barriers in vertebrate lungs. *J Anat* **131**, 299–307.

Meier, S. K., Toop, T. and Donald, J. A. (1999). Distribution and characterization of natriuretic peptide receptors in the kidney of the toad, *Bufo marinus*. *Gen Comp Endocrinol* **115**, 244–253.

Melichna, J., Gutmann, E., Herbrychova, A. and Stichova, J. (1972). Sexual dimorphism in contraction properties and fibre pattern of the flexor carpi radialis muscle of the frog (*Rana temporaria*). *Experientia* **28**, 89–91.

Mendiola, P. and De Costa, J. (1990). The effects of pH and temperature on the kinetic properties of skeletal muscle lactate dehydrogenase from anuran amphibians. *J Comp Physiol* **160B**, 105–111.

Meyer, A. and Zardoya, R. (2003). Recent advances in the (molecular) phylogeny of vertebrates. *Annu Rev Ecol Evol System* **34**, 311–338.

Meyer, D. J. and Huxley, V. H. (1990). Differential sensitivity of exchange vessel hydraulic conductivity to atrial natriuretic peptide. *Am J Physiol* **258**, H521–H528.

Meyer, D. J., Huxley, V. H. and McKay, M. K. (1992). Volume status influences atrial-peptide induved water conductivity changes in leopard frog mesenteric capillaries. *J Physiol* **447**, 33–47.

Meyers, R. S., Moalli, R., Jackson, D. C. and Millard, R. W. (1979). Microsphere studies of bullfrog central vascular shunts during diving and breathing in air. *J Exp Zool* **208**, 423–430.

Michel, C. C. (1984). Fluid movements through capillary walls. In: *Handbook of Physiology. The Cardiovascular System, Microcirculation*, ed. E. M. Renkin, pp. 375–409. Am Physiol Soc: Washington DC.

Michel, G., Ouedraogo, Y., Chauvet, J., Katz, U. and Acher, R. (1993). Differential processing of provasotocin: relative increase of hydrin 2 (vasotocinyl-Gly) in amphibians able to adapt to an arid environment. *Neuropeptides* **25**, 139–143.

Middler, S. A., Kleeman, C. R. and Edwards, E. (1967). Neurohypopysial function in the toad *Bufo marinus*. *Gen Comp Endocrinol* **9**, 38–48.

Middler, S. A., Kleeman, C. R. and Edwards, E. (1968a). The role of the urinary bladder in salt and water metabolism of the toad, *Bufo marinus*. *Comp Biochem Physiol* **26**, 57–68.

Middler, S. A., Kleeman, C. R. and Edwards, E. (1968b). Lymph mobilization following acute blood loss in the toads, *Bufo marinus*. *Comp Biochem Physiol* **24**, 343–353.

Miller, D. A., Standish, M. L. and Thurman, A. E. (1968). Effects of temperature on water and electrolyte balance in the frog. *Physiol Zool* **41**, 500–506.

Miller, K. (1982). Effect of temperature on sprint performance in the frog *Xenopus laevis* and the salamander *Necturus maculosus*. *Copeia* 1982, 695–698.

Miller, K. (1983). The role of lactate production in the metabolic support of locomotion by clawed frogs *Xenopus laevis*. *Physiol Zool* **56**, 580–584.

Miller, K. and Hutchison, V. H. (1980). Aerobic and anaerobic scope for activity in the giant toad, *Bufo marinus*. *Physiol Zool* **53**, 170–175.

Miller, K. and Packard, G. C. (1977). An altitudinal cline in critical thermal maxima of chorus frogs (*Pseudacris triseriata*). *Am Naturalist* **111**, 267–277.

Miller, K. and Sabol, J. L. (1989). The role of phosphocreatine breakdown in the metabolic support of locomotion by clawed frogs, *Xenopus laevis*. *Comp Biochem Physiol* **93B**, 251–254.

Mills, J. W., Ernst, S. A. and DiBona, D. R. (1977). Localization of Na^+ pump sites in the frog skin. *J Cell Biol* 73, 88–110.

Milner, A. R. (1993). The Paleozoic relatives of lissamphibians. *Herpetological Monographs* 7, 8–27.

Milsom, W. K. and Jones, D. R. (1977). Carbon dioxide sensitivity of pulmonary receptors in the frog. *Experientia* **33**, 1167–1168.

Milton, S. L., Manuel, L. and Lutz, P. L. (2003). Slow death in the leopard frog *Rana pipiens*: neurotransmitters and anoxia tolerance. *J Exp Biol* **206**, 4021–4028.

Mitchell, J. W. (1976). Heat transfer from spheres and other animal forms. *Biophys J* **16**, 561–569.

Mitchell, N. J. (2001). The energetics of endotrophic development in the frog *Geocrinia vittel lina* (Anura: Myobatrachidae). *Physiol Biochem Zool* 74, 832–842.

Mitchell, N. J. and Seymour, R. S. (2000). Effects of temperature on energy cost and timing of embryonic larval development of the terrestrially breeding moss frog, *Bryobatrachus nimbus*. *Physiol Biochem Zool* 73, 829–840.

Moalli, R., Meyers, R. S., Jackson, D. C. and Millard, R. W. (1980). Skin circulation of the frog, *Rana catesbeiana*: distribution and dynamics. *Respir Physiol* **40**, 137–48.

Mobjerg, N., Jesperson, A. and Wilkinson, M. (2004a). Morphology of the kidney in the West African caecilian, *Geotrypetes seraphini* (Amphibia, Gynophiona, Caeciliidae). *J Morphol* 262, 583–607.

Mobjerg, N., Larsen, E. H. and Novak, I. (2004b). Ion transport mechanisms in the mesonephric collecting duct system of the toad *Bufo bufo*: microelectrode recordings from isolated and perfused tubules. *Comp Biochem Physiol A Mol Integr Physiol* **137**, 585–595.

Mommsen, T. P. and Storey, K. B. (1992). Hormonal effects on glycogen metabolism in isolated hepatocytes of a freeze-tolerant frog. *Gen Comp Endocrinol* 87, 44–53.

Monteith, J. L. (1990). *Principles of Environmental Physics*. Edward Arnold: London.

Monteith, J. L. and Campbell, G. S. (1980). Diffusion of water vapour through integuments—potential confusion. *J Thermal Biol* 5, 7–9.

Moore, F. R. and Gatten, R. E. (1989). Locomotor performance of hydrated, dehydrated, and osmotically stressed anuran amphibians. *Herpetologica* **45**, 101–110.

Morad, M. and Goldman, Y. (1973). Excitation-contraction coupling in heart muscle: membrane control of development of tension. *Prog Biophys Mol Biol* 27, 257–313.

Morgunov, N. S. (1987). Electrophysiological analysis of B-receptor stimulation in salamander proximal tubules. *Am J Physiol* **253**, F636–F641.

Morgunov, N. S., Hirsch, D. J. and Mobbs, I. G. (1989). Low-sodium environment induces adaptation in salamander diluting segments. *Am J Physiol* **25**, F229–F238.

Morris, J. L. (1981). Structure and function of ciliated peritoneal funnels in the toad kidney (*Bufo marinus*). *Cell Tissue Res* **217**, 599–610.

Morris, J. L. and Nilsson, S. (1994). The circulatory system. In: *Comparative Physiology and Evolution of the Autonomic Nervous System*, eds S. Nilsson and S. Holmgren, pp. 193–246. Harwood Academic Publishers: Switzerland.

Morris, R. W. (1974). Function of the anuran conus arteriosus. *J Exp Biol* **61**, 503–520.

Moudgil, R., Michelakis, E. D. and Archer, S. L. (2005). Hypoxic pulmonary vasoconstriction. *J Appl Physiol* **98**, 390–403.

Muldrew, K. and McGann, L. E. (1990). Mechanisms of intracellular ice formation. *Biophys J* 57, 525–532.

Muldrew, K. and McGann, L. E. (1994). The osmotic rupture hypothesis of intracellular freezing injury. *Biophys J* **66**, 532–541.

Mullen, T. L. and Alvarado, R. H. (1976). Osmotic and ionic regulation in amphibians. *Physiol Zool* **49**, 11–23.

Murphy, C. G. (1994). Determinants of chorus tenure in barking treefrogs (*Hyla gratiosa*). *Behav Ecol Sociobiol* **34**, 285–294.

Murrish, D. E. and Schmidt-Nielsen, K. (1970). Water transport in the cloaca of lizards: active or passive? *Science* **170**, 324–326.

Muth, A. (1980). Physiological ecology of desert iguana (*Dipsosaurus dorsalis*) eggs: temperature and water relations. *Ecology* **61**, 1335–1343.

Muths, E. and Corn, P. S. (1997). Basking by adult arboreal toads (*Bufo boreas boreas*) during the breeding season. *J Herpetology* **31**, 426–428.

Myhre, K., Cabanac, M. and Myhre, G. (1977). Fever and behavioural temperature regulation in the frog, *Rana esculenta*. *Acta Physiologica Scandinavica* **101**, 219–229.

Nagai, T., Koyama, H., Hoff, K. and Hillyard, S. D. (1999). Desert toads discriminate salt taste with chemosensory function of their ventral skin. *J Comp Neurol* **408**, 125–136.

Nagel, W. and Crabbe, J. (1980). Mechanism of action of aldosterone on active sodium transport across toad skin. *Pflugers Archiv* **385**, 1432–2013.

Nagy, K. A. (2005). Review: Field metabolic rate and body size. *J Exp Biol* **208**, 1621–1625.

Nagy, K. A., Girard, I. A. and Brown, T. K. (1999). Energetics of free-ranging mammals, reptiles, and birds. *Annu Rev Nutr* **19**, 247–277.

Nagy, K. A. and Peterson, C. C. (1988). *Scaling of Water Flux Rates in Animals*. University of California Press.

Nash, G. and Fankhauser, G. (1959). Changes in the pattern of nitrogen excretion during the life cycle of the newt. *Science* **130**, 714–716.

Navas, C. A. (1996). Implications of microhabitat selection and patterns of activity on the thermal ecology of high-elevation neotropical anurans. *Oecologia* **108**, 617–626.

Navas, C. A. and Araujo, C. (2000). Misconceptions about colour, infrared radiation, and energy exchange between animals and their environments. *Herpetological Journal* **10**, 119–122.

Navas, C. A., Jared, C. and Antoniazzi, M. M. (2002). Water economy in the casque-headed tree-frog *Corythomantis greeningi* (Hylidae): Role of behaviour, skin, and skull skin co-ossification. *J Zool* **257**, 525–532.

Neill, W. T. (1958). The occurrence of amphians and reptiles in saltwater areas, and a bibliography. *Bulletin of Marine Science of the Gulf and Caribbean* **8**, 1–97.

Nevo, E. and Beiles, A. (1991). Genetic diversity and ecological heterogeneity in amphibian evolution. *Copeia* 1991, 565–592.

Newman, R. A. and Dunham, A. E. (1994). Size at metamorphosis and water loss in a desert anuran (*Scaphiopus couchii*). *Copeia* 1994, 372–381.

Nielsen, H. I. (1978). The effect of stress and adrenaline on the color of *Hyla cinerea* and *Hyla arborea*. *General Comp Endocrinology* **36**, 543–552.

Nietfelt, J. W., Jones, S. M.,. Droge, D. L. and Ballinger, R. E. (1980). Rate of thermal acclimation in larval *Ambystoma tigrinum*. *J Herpetology* **14**, 209–211.

Nikinmaa, M. (1990). *Vertebrate Red Blood Cells: Adaptations of function to respiratory requirements*. Springer-Verlag: New York.

Nikinmaa, M. (1997). Oxygen and carbon dioxide transport in vertebrate erythrocytes: an evolutionary change in the role of membrane transport. *J Exp Biol* **200**, 369–380.

Nobel, P. S. (1999). *Physicochemical and Environmental Plant Physiology*. Academic Press: San Diego.

Noble, G. K. (1925). The integumentary, pulmonary and cardiac modifications correlated ith increased cutaneous respiration in the Amphibia: a solution to the "hairy-frog" problem. *J Morph Physiol* **40**, 341–416.

Nolly, A. and Fasciolo, J. C. (1971). The renin-angiotensin system in *Bufo arenarum* and *Bufo paracnemis. Comp Biochem Physiol* **39A**, 823–831.

Nouwen, E. J. and Kuhn, E. R. (1985). Volumetric control of arginine vasotocin and mesotocin release in the frog (*Rana ridibunda*). *J Endocrinology* **105**, 371–377.

Nouwen, E. K. and Kuhn, E. R. (1983). Radioimmunoassay of arginine vasotocin and mesotocin in serum of the frog *Rana ridibunda. J Comp Endocrinology* **50**, 242–251.

Nussbaum, R. A. and Wilkinson, M. (1989). On the classification and phylogeny of caecilians (Amphibia: Gymnophonia) a critical review. *Herpetological Monographs* **3**, 1–42.

Nussbaum, R. A. and Wilkinson, M. (1995). A new genus of lungless tetrapod: a radically divergent caecilian (Amphibia: Gymnophiona). *Proc R Soc Lond B Biol Sci* **261**, 331–335.

Nyboe, O., Rosenkilde, P. and Ryttersgaard, L. (1985). Effects of hypophysectomy and substitution with growth hormone, prolactin, and thyroxin on the growth and deposition in juvenile frogs, *Xenopus laevis. Gen Comp Endocrinol* 57, 257–265.

O'Connor, M. P. and Tracy, C. R. (1987). Thermal and hydric relations of leopard frogs in the field. *Am Zoologist* **17**, 118A.

Oguro, C., Nagai, K., Tarui, H. and Sasayama, Y. (1981). Hypocalcemic factor in the ultimobranchial gland of the frog, *Rana rugosa. Comp Biochem Physiol* **68A**, 95–97.

Oguro, C. and Sasayama, Y. (1978). Function of the parathyroid gland in serum calcium regulation in the newt *Tylototriton andersoni. Gen Comp Endocrin* **35**, 10–15.

Oguro, C., Sasayama, Y., Ikadai, H. and Yoshizawa, H. (1975). Changes in calcium concentrations of serum and coelomic fluid of the bullfrog, *Rana catesbeiana*, during larval development and metamorphosis. *Dev Growth Diff* **17**, 71–75.

Oka, Y., Ohtani, R., Satou, M. and Ueda, K. (1984). Sexually dimorphic muscles in the forelimb of the Japanese toad, *Bufo japonicus. J Morph* **180**, 297–308.

Okada, Y., Miyamoto, T. and Sato, T. (1993) The ionic basis for the receptor potential of frog taste cells induced by water stimuli. *J Exp Biol* **174**, 1–17.

Olmo, E. and Morescalchi, A. (1975). Evolution of the genome sizes in salamanders. *Experientia* **31**, 804–806.

Olmo, E. and Morescalchi, A. (1978). Genome and cell size in frogs: a comparison with salamanders. *Experientia* **34**, 44–46.

O'Reilly, J. C., Summers, A. P. and Ritter, D. A. (2000). The evolution of the functional role of trunk muscles during locomotion in adult amphibians. *Am Zoologist* **40**, 123–135.

Orton, G. (1953). The systematics of vertebrate larvae. *Syst Zool* **2**, 63–75.

Orton, G. (1957). The bearing of larval evolution on some problems in frog classification. *Syst Zool* **6**, 79–86.

Oshima, N. (2001). Direct reception of light by chromatophores of lower vertebrates. *Pigment Cell Res* **14**, 312–319.

Ott, M., Heisler, N. and Ultsch, G. R. (1980). A re-evaluation of the relationship between temperature and the critical oxygen tension in freshwater fishes. *Comp Biochem Physiol A* **67**, 337–340.

Overton, E. (1904). Neunundereissing Thesen uber Wasserokonomie der Amphibien und die osmotischen Eigenshaften der Amphibienhaut. *Verhandlungen der Physikalish Medicinischen Gesellschaft in Wurtzberg* **36**, 277–296.

Owerkowicz, T., Farmer, C., Hicks, J. W. and Brainerd, E. L. (1999). Contributions of gular pumping to lung ventilation in monitor lizards. *Science* **284**, 1661–1663.

Ozorio de Almeida, M. (1923). Sur la separation des ondes arterielle et veineuse dans l'appareil circulation de la grenouille. *C R Soc Biol* **89**, 1019–1021.

Packard, G. C. (1971). Oxygen consumption of montane and piedmont chorus frogs (*Pseudacris triseriata*): a study of evolutionary temperature compensation. *Physiol Zool* **44**, 90–97.

Packard, G. C. and Boardman, T. J. (1988). The misuse of ratios, indices, and percentages in ecophysiological research. *Physiol Zool* **61**, 1–9.

Packard, G. C. and Boardman, T. J. (1999). The use of percentages and size-specific indices to normalize physiological data for variation in body size; wasted time, wasted effort? *Comp Biochem Physiol* **122A**, 37–44.

Packard, G. C. and Stiverson, R. K. (1976). Blood hemoglobin concentration in chorus frogs (*Pseudacris triseriata*): relations to body size and altitude. *Am Midl Nat* **96**, 482–487.

Paik, W. K. and Cohen, P. P. (1960). Biochemical studies on amphibian metamorphosis. I. The effect of thyroxine on protein synthesis in the tadpole. *J Gen Physiol* **43**, 683–696.

Pakay, J. L., Hobbs, A. H., Kimball, S. R. and Guppy, M. (2003). The role of eukaryotic initiation factor 2α during the metabolic depression associated with aestivation. *J Exp Biol* **206**, 2363–2371.

Pak Poy, R. K. F. (1957). Electron microscopy of the amphibian renal glomerulus. *Aust J Exp Biol* **35**, 583–594.

Paladino, F. V., O'Connor, M. P. and Spotila, J. R. (1990). Metabolism of leatherback turtles, gigantothermy, and thermoregulation of dinosaurs. *Nature* **344**, 858–860.

Pandian, T. J. and Marian, M. P. (1985). Time and energy costs of metamorphosis in the Indian bullfrog *Rana tigrina*. *Copeia* 1985, 653–662.

Pang, P., Galli-Gallardo, S. M., Collie, N. and Sawyer, W. H. (1980). Renal and peripheral vascular responsiveness to arginine vasotocin in bullfrog, *Rana catesbeiana*. *Am J Physiol* **239**, 156–160.

Pang, P. K. T. and Sawyer, W. H. (1974). Effects of prolactin on hypophysectomized mud puppies, *Necturus maculosus*. *Am J Physiol* **226**, 458–462.

Pang, P. K. T. and Sawyer, W. H. (1978). Renal and vascular responses of the bullfrog (*Rana catesbeiana*) to mesotocin. *Am J Physiol* **235**, F151–F155.

Pang, P. K., Uchiyama, M. and Sawyer, W. H. (1982). Endocrine and neural control of amphibian renal functions. *Fed Proc* **41**, 2365–2370.

Parry, C. R. and Cavill, R. (1979). A note on cocoon formation and structure in *Pyxicephalus adspersus* Tschudi (Anura: Ranidae). *Tran Rhodesia Sci Assoc* **58**, 55–58.

Parsons, R. H., McDevitt, V., Aggerwal, V., Le Blang, T., Manley, K., Kim, N., Lopez, J. and Kenedy, A. A. (1993). Regulation of pelvic patch water flow in *Bufo marinus*: role of bladder volume and ANG II. *Am J Physiol* **264**, R1260–1265.

Parsons, R. H. and Mobin, F. (1991). Water flow across the pectoral and ventral pelvic patch in *Rana catesbeiana*. *Physiol Zool* **64**, 812–822.

Parsons, R. H. and Schwartz, R. (1991). Role of circulation in maintaining Na^+ and K^+ concentration in the pelvic patch in *Rana catesbeiana*. *Am J Physiol* **261**, R686–R689.

Parsons, R. H., Thorn, L. M., Salerno, S. M. and Petronis, S. (1990). Salt load effects on osmotic water flow, tritiated water diffusion, and the cardiovascular system in *Rana pipiens* and *Rana catesbeiana*. *Physiol Zool* **63**, 571–586.

Parsons, T. S. and Williams, E. E. (1962). The teeth of amphibia and their relation to amphibian phylogeny. *J Morph* **110**, 375–390.

Pasanen, S. and Karhapaa, M. (1997). Can boreal common frog (*Rana temporaria*) survive in frost? *Ann Zool Fennici* **34**, 247–250.

Pasanen, S. and Koskela, P. (1974). Seasonal and age variation in the metabolism of the common frog, *Rana temporaria* in northern Finland. *Comp Biochem Physiol* **47A**, 635–654.

Pasanen, S. and Sorjonen, J. (1994). Partial terrestrial wintering in a northern common frog population (*Rana temporaria*). *Ann Zool Fennici* **31**, 275–278.

Pascual-Le Tallec, L. and Lombes, M. (2005). The mineralocorticoid receptor. A journey exporing its diversity and specificity of action. *Mol Endocrinol* **19**, 2211–2221.

Passmore, N. I. and Carruthers, V. C. (1979). *South African Frogs*. Witwatersrand University Press: Johannesburg.

Passmore, N. I. and Malherbe, M. (1985). High nocturnal body temperature in the painted reed frog (*Hyperolius marmoratus*). *Herpetologica* **41**, 212–215.

Peachey, L. D. and Rasmussen, H. (1961). The structure of the toad's urinary bladder as related to its physiology. *J Biophys Biochem Cytol* **10**, 529–553.

Pearson, O. P. and Bradford, D. F. (1976). Thermoregulation of lizards and toads at high altitudes in Peru. *Copeia* 1976, 155–170.

Pechmann, J. H. K., Scott, D. E., Gibbons, J. W. and Smelitsch, R. D. (1989). Influence of wetland hydroperiod on diversity and abundance of metamorphosing juvenile amphibians. *Wetlands Ecol Management* **1**, 3–11.

Pegg, D. E. and Diaper, M. P. (1988). On the mechanism of injury to slowly frozen erythrocytes. *Biophys J* **54**, 471–488.

Pelster, B. and Burggren, W. W. (1991). Central arterial hemodynamics in larval bullfrogs (*Rana catesbeiana*): developmental and seasonal influences. *Am J Physiol* **260**, R240–246.

Perkins, F. M and Handler, J. S. (1981). Transport properties of toad kidney epithelia in culture. *Am J Physiol* **241**, C154–C159.

Persson, B., Sakai, T. and Marsh, D. J. (1988). Juxtaglomerular interstitial hypertonicity in *Amphiuma*: tubular origin-TGF signal. *Am J Physiol* **254**, F445–F449.

Persson, B. E. (1981). Dynamics of glomerular ultrafiltration in *Amphiuma means*. *Pflugers Arch* **391**, 135–140.

Persson, B. E. and Persson, A. E. (1981). The existence of a tubulo-glomerular feedback mechanism in the *Amphiuma* nephron. *Pflugers Arch* **391**, 129–134.

Peters, S. E. and Aulner, D. A. (2000). Sexual dimorphism in forelimb muscles of the bullfrog, *Rana catesbeiana*: A functional analysis of isometric contractile properties. *J Exp Biol* **203**, 3639–3654.

Petranka, J. W. (1979). The effects of severe winter weather on *Plethodon dorsalis* and *Plethodon richmondi* populations in central Kentucky. *J Herpetol* **13**, 369–371.

Petriella, S., Reboreda, J. C., Otero, M. and Segura, E. (1989). Antidiuretic responses to osmotic cutaneous stimulation in the toad, *Bufo arenarum*. A possible adaptive control mechanism for urine production. *J Comp Physiol B* **159**, 91–95.

Pfeffer, W. (1877). *Osmotische Untersuchungen*. Engelmann: Leipzig.

Pianka, E. R. and Schall, J. J. (1981). Species densities of terrestrial vertebrates in Australia. In: *Ecological Biogeography of Terrestrial Vertebrates in Australia*, ed. A. Keast, pp. 1375–1392. DW Junk: The Hague, Netherlands.

Pickering, B. T. and Heller, H. (1959). Active principles of the neurohypophysis of cold-blooded vertebrates. Chromatographic and biological characteristics of fish and frog neurohypophyseal extracts. *Nature* **184**, 1463–1464.

Piiper, J., Gatz, R. N. and Crawford, E. C. (1976). Gas transport characteristics in an exclusively skin-breathing salamander, *Desmognathus fuscus* (Plethodontidae). In: *Respiration of Amphibious Vertebrates*, ed. G. M. Hughes, pp. 339–356. Academic Press: New York.

Piiper, J. and Scheid, P. (1977). Comparative physiology of respiration: functional analysis of gas exchange organs in vertebrates. In: *International Review of Physiology, Respiratory Physiology II*, ed. J. G. Widdicombe, pp. 219–253. University Park Press: Baltimore.

Pilkington, J. B. and Simkiss, K. (1966). The mobilization of the calcium carbonate deposits in the endolymphatic sacs of metamorphosing frogs. *J Exp Biol* **45**, 329–341.

Pinder, A. and Burggren, W. (1983). Respiration during chronic hypoxia and hyperoxia in larval and adult bullfrogs (*Rana catesbeiana*). II. Changes in respiratory properties of whole blood. *J Exp Biol* **105**, 205–213.

Pinder, A. W. (1987). Cutaneous diffusing capacity increases during hypoxia in cold submerged bullfrogs (*Rana catesbeiana*). *Respir Physiol* **70**, 85–95.

Pinder, A. W. and Burggren, W. W. (1986). Ventilation and partitioning of oxygen uptake in the frog *Rana pipiens*: effects of hypoxia and activity. *J Exp Biol* **126**, 453–468.

Pinder, A. W., Clemens, D. and Feder, M. E. (1991). Gas exchange in isolated perfused frog skin as a function of perfusion rate. *Respir Physiol* **85**, 1–14.

Pinder, A. W. and Feder, M. E. (1990). Effect of boundary layers on cutaneous gas exchange. *J Exp Biol* **154**, 67–80.

Pinder, A. W., Storey, K. B. and Ultsch, G. R. (1992). Estivation and hibernation. In: *Environmental Physiology of the Amphibians*, eds M. E. Feder and W. W. Burrgren, pp. 250–274. University of Chicago Press: Chicago and London.

Pinz, I. and Portner, H. O. (2003). Metabolic costs induced by lactate in the toad *Bufo marinus*: new mechanism behind oxygen debt? *J Appl Physiol* **94**, 1177–1185.

Pitkin, R. B. (1977). Effects of temperature on respiration of *Notopthalmus viridescens*, the red-spotted newt. *Comp Biochem Physiol* **57A**, 413–416.

Pitman, W. C. I., Cande, S., LaBrecque, J. and Pindell, J. (1993). Fragmentation of Gondwana: the separation of Africa from South America. In: *Biological Relationships between Africa and South America*, ed. P. Goldblatt, pp. 15–34. Yale University Press: New Haven, Connecticut.

Podrabsky, J. (2001). Survival of water stress in annual fish embryos: dehydration avoidance and egg envelope amyloid fibers. *Am J Physiol* **280**, R123–131.

Polos, J. L. C. and Hedrick, M. S. (2006). Hypaxial musculature is recruited during increased respiratory drive in anuran amphibians. *FASEB J* **20** A822.

Porter, W. P., Gates, D. M. (1969). Thermodynamic equilibria of animals with environment. *Ecolog Mono* **39**, 227–244.

Portner, H. O., Maclatchey, L. M. and Toews, D. P. (1991). Metabolic responses of the toad *Bufo marinus* to environmental hypoxia: an analysis of the critical P_{O2}. *Physiol Zool* **64**, 836–849.

Pote, K. G., Hauer, C. R. I., Michel, H., Shabanowitz, J., Hunt, D. F. and Kretsinger, R. H. (1993). Otoconin-22 the major protein of aragonitic frog otoconia, is a homolog of phospholipase A_2. *Biochemistry* **32**, 5017–5024.

Pough, F. H. (1980). The advantages of ectothermy for tetrapods. *Am Naturalist* **115**, 92–112.

Pough, F. H. (1983). Amphibians and reptiles as low-energy systems. In: *Behavioral Energetics*, eds W. P. Aspey and S. I. Lustick, pp. 141–188. Ohio State University Press: Columbus.

Pough, F. H., Andrews, R. D., Cadle, J. E., Crump, M. L., Savitsky, A. H. and Wells, K. D. (1998). *Herpetology*. Prentice-Hall: New Jersey.

Pough, F. H. and Gans, C. (1982). The vocabulary of reptilian thermoregulation. In: *Biology of the Reptilia*, Vol. 12, eds C. Gans and F. H. Pough, pp. 17–24. Academic Press: New York.

Pough, F. H. and Gatten, R. E. (1984). The use of anaerobic metabolism by frogs in a breeding chorus. *Comp Biochem Physiol* **78A**, 337–340.

Pough, F. H., Magnusson, W. E., Ryan, M. J., Wells, K. D. and Taigen, T. L. (1992). Behavioral energetics. In: *Environmental Physiology of the Amphibians*, eds W. Burggren and M. Feder, pp. 395–436. University of Chicago Press.

Pough, F. H., Taigen, T. L., Stewart, M. M. and Brussard, P. F. (1983). Behavioral modification of evaporative water loss by a Puerto Rican frog. *Ecology* **64**, 244–252.

Pounds, J. A., Fogden, M. P. L. and Campbell, J. H. (1999). Biological responses to climate change on a tropical mountain. *Nature* **398**, 611–615.

Poupa, O. and Ostadal, B. (1969). Experimental cardiomegalies and "cardiomegalies" in free-living animals. *Ann N Y Acad Sci* **156**, 445–468.

Powell, M. K., Mansfield-Jones, J. and Gatten, R. E. (1999). Specific dynamic effect in the horned frog *Ceratophyrys cranwelli*. *Copeia* 1999, 710–717.

Poynton, J. C. (1964). The Amphibia of Southern Africa. *Ann Natal Museum* 17, 1–334.

Poynton, J. C. (1999). Distribution of amphibians in sub-Saharan Africa, Madagascar, and Seychelles. In: *Patterns of Distribution of Amphibians A Global Perspective*, ed. W. E. Duellman, pp. 483–539. Johns Hopkins University Press: Baltimore and London.

Prasadmurthy, Y. S. and Saidapur, S. K. (1987). Role of fat bodies in oocyte growth and recruitment in the frog *Rana cyanophlyctis*. *J Exp Zool* **243**, 153–162.

Preest, M. R., Brust, D. G. and Wygoda, M. L. (1992). Cutaneous water loss and the effects of temperature and hydration state on aerobic metabolism of canyon tree frogs, *Hyla arenicolor*. *Herpetologica* **48**, 210–219.

Preest, M. R. and Pough, F. H. (1989). Interactive effects of body temperature and hydration state on locomotor performance of the toad *Bufo americanus*. *Func Ecol* **3**, 693–699.

Preslar, A. J. and Hutchison, V. H. (1978). Energetics for activity in the salamander *Amphiuma tridactylum*. *J Comp Physiol* **128**, 139–146.

Preston, G. M. and Agre, P. (1991). Isolation of the cDNA for erythrocyte integral membrane protein of 28 kilodaltons: member of an ancient channel family. *PNAS* **88**, 11110–11114.

Preston, G. M., Carroll, T. P., Guggino, W. B. and Agre, P. (1992). Appearance of water channels in oocytes expressing red cell CHIP 28 protein. *Science* **256**, 385–387.

Prestwich, K. N., Brugger, K. E. and Topping, M. (1989). Energy and communication in three species of hylid frogs: power input, power output and efficiency. *J Exp Biol* **144**, 53–80.

Priestley, J. (1878). An account of the anatomy and physiology of batrachian lymph-hearts. *J Physiol* **1**, 1–18.

Propper, C. R., Hillyard, S. D. and Johnson, W. E. (1995). Central angiotensin II induces thirst-related responses in an amphibian. *Hormones Behavior* **29**, 74–84.

Propper, C. R. and Johnson, W. E. (1994). Angiotensin II induces water absorption behavior in two species of desert anurans. *Hormones Behavior* **28**, 41–52.

Pruett, S. J., Hoyt, D. F. and Stiffler, D. F. (1991). The allometry of osmotic and ionic regulation in amphibia with emphasis on intraspecific scaling in larval *Ambystoma tigrinum*. *Physiol Zool* **64**, 1173–1199.

Pusey, B. J. (1986). The effect of starvation on oxygen consumption and nitrogen excretion in *Lepidogalaxias salamandroides* (Mees). *J Comp Physiol B* **156**, 701–705.

Pusey, B. J. (1989). Aestivation in the teleost fish *Lepidogalaxias salamandroides* Mees. *Comp Biochem Physiol A* **92**, 137–138.

Pusey, B. J. (1990). Seasonality, aestivation and the life history of the salamanderfish *Lepidogalaxias salamandroides* (Pisces, Lepidogalaxiidae). *Environ Biol Fishes* **29**, 15–26.

Putnam, R. W. (1979a). The basis for differences in lactic acid content after activity in different species of anuran amphibians. *Physiol Zool* **52**, 509–519.

Putnam, R. W. (1979b). The role of lactic acid accumulation in muscle fatigue of two species of anurans, *Xenopus laevis* and *Rana pipiens*. *J Exp Biol* **82**, 35–51.

Putnam, R. W. and Bennett, A. F. (1981). Thermal dependence of behavioral performance of anuran amphibians. *Anim Behav* **29**, 502–509.

Putnam, R. W. and Bennett, A. F. (1983). Histochemical, enzymatic, and contractile properties of skeletal muscles of three anuran amphibians. *Am J Physiol* **244**, R558–R567.

Putnam, R. W. and Hillman, S. S. (1977). Activity responses of anurans to dehydration. *Copeia* 1977, 746–749.

Putnam, R. W. and Roos, A. (1986). Effect of calcium and other divalent cations on intracellular pH regulation of frog muscle. *J Physiol* **381**, 221–239.

Putnam, R. W., Roos, A. and Wilding, T. J. (1986). Properties of the intracellular pH-regulating systems of frog skeletal muscle. *J Physiol* **381**, 205–219.

Rage, J. and Z. Rocek. (1989). Redescription of *Triadobatrachus masinoti* (Piveteau, 1936) an anuran amphibian from the early Triassic. *Palenotographica Abteilung A Palaeozoologie-Stratigraphie* **206**, 1–16.

Rajerison, R. M., Montegut, M., Jard, S. and Morel, F. (1972a). The isolated frog skin epithelium: presence of α and β adrenergic receptors regulating active sodium transport and water permeability. *Pfleugers Archiv* **332**, 313–331.

Rajerison, R. M., Montegut, M., Jard, S. and Morel, F. (1972b). The isolated frog skin epithelium: permeability characteristics and responsiveness to oxytocin, cyclic AMP and theophylline. *Pflugers Archiv* **332**, 302–312.

Ralin, D. B. and Rogers, J. S. (1972). Aspects of tolerance to desiccation in *Acris crepitans* and *Pseudacris streckeri*. *Copeia* 1972, 519–524.

Ramaswami, L. S. (1944). An account of the heart and associated vessels in some genera Apoda. *Proc Zool Soc, Lond* **114**, 117–139.

Rathbun, G. B. (1996). Evaluation of a radio-belt for ranid frogs. *Herpetological Review* **27**, 187–189.

Ray, C. (1958). Vital limits and rates of desiccation in salamanders. *Ecology* **39**, 75–83.

Raynor, G. S. (1971). Wind and temperature structure in a coniferous forest and a contiguous field. *Forest Sci* **17**, 351–363.

Reagan, N. L. and Verrell, P. A. (1991). The evolution of plethodontid salamanders: Did terrestrial mating facilitate lunglessness. *Am Nat* **138**, 1307–1313.

Reboreda, J. C. and Segura, E. T. (1984). Electrolyte sensitivity of the skin in the toad *Bufo arenarum* Hensel. *Acta Physiologica et Pharmacologica Latinoamericana* **34**, 43–49.

Regester, K. J., Lips, K. R. and Whiles, M. R. (2006). Energy flow and subsidies associated with the complex life cycle of ambystomatid salamanders in ponds and adjacent forest in southern Illinois. *Oecologia* **147**, 303–314.

Regester, K. J. and Whiles, M. R. (2006). Decomposition rates of salamander (*Ambystoma maculatum*) life stages and associated energy and nutrient fluxes in ponds and adjacent forests in southern Illinois. *Copeia* 2006, 640–649.

Reichle, D. E., Shanks, M. H. and Crossley, D. A. (1969). Calcium, potassium, an sodium content of forest floor arthropods. *Ann Entomolog Soc Am* **62**, 57–62.

Reid, E. W. (1892). Experiments upon "absorption without osmosis". *BMJ* **1**, 323–326.

Reig, O. (1958). Proposiciones par una nueva macrosistematica de los anuros. *Physis* **21**, 109–118.

Reiss, J. O. (2002). The phylogeny of amphibian metamorphosis. *Zoology* **105**, 85–96.

Renaud, J. M. (1989). The effect of lactate on intracellular pH and force recovery of fatigued sartorius muscle of the frog, *Rana pipiens*. *J Physiol* **416**, 31–47.

Renaud, J. M. and Stevens, E. D. (1983). The extent of long-term temperature compensation for jumping distance in the frog, *Rana pipiens*, and the toad, *Bufo Amus*. *Can J Zool* **61**, 1284–1287.

Reno, H. W., Gehlbach, F. R. and Turner, R. A. (1972). Skin and aestivational cocoon of the aquatic amphibian, *Siren intermedia* Le Conte. *Copeia* 1972, 625–631.

Ressel, S. J. (2001). Ultrastructural design of anuran muscles used for call production in relation to the thermal environment of a species. *J Exp Biol* **204**, 1445–1457.

Rey, P. (1937). *Recherches Expérimentelles sur l'Economie de l'Eau chez les Batrachiens*. Lons-le-Saunier: Declume.

Richards, A. N. and Schmidt, C. F. (1924). A description of the glomerular circulation in the frog's kidney and observations concerning the action of adrenaline and various other substances upon it. *Am J Physiol* **71**, 178–208.

Richards, S. R., Sinsch, U. and Alford, R. A. (1994). Radio tracking. In: *Measuring and Monitoring Biological Diversity: Standard Methods for Amphibians*, eds W. R. Heyer, M. A. Donnelly, R. W. McDiarmid, L. C. Hayek and M. S. Foster, pp. 158–163. Smithsonian Institution Publications: Washington.

Richter, S. and Splechtna, H. (1996). The structure of anuran podocyte-determined by ecology? *Acta Zoologica* 77, 335–348.

Rick, R., Dorge, A., Arnim, E. V. and Thurau, K. (1978). Electron microprobe analysis of frog skin: evidence for a syncytial sodium transport compartment. *J Membrane Biol* **39**, 313–331.

Riedel, C. and Wood, S. (1988). Effects of hypercapnia and hypoxia on temperature of the toad, *Bufo marinus*. *FASEB J* **2**, A500.

Riggs, A. F. (1998). Self-association, cooperativity and supercooperativity of oxygen binding by hemoglobins. *J Exp Biol* **201**, 1073–1084.

Roberts, J. D. (1981). Terrestrial breeding in the Australian leptodactylid frog *Myobatrachus gouldii*. *Aust Wildlife Res* **8**, 451–462.

Roberts, J. D. (1984). Terrestrial egg deposition and direct development in *Arenophryne rotunda* Tyler, a myobatrachid frog from coastal sand dunes at Shark Bay, W. A. *Aust Wildlife Res* **11**, 191–200.

Roberts, J. D. (1990). The biology of *Arenophryne rotunda* (Anura: Myobatrachidae): a burrowing frog from Shark Bay, Western Australia. In: *Research in Shark Bay Report of the France-Australe Bicentennial Expeditionary Committee*, pp. 287–297. Western Australian Museum: Perth.

Robertson, D. R. (1969). The ultimobranchial body of *Rana pipiens*. X. Effects of glandular extirpation on fracture healing. *J Exp Zool* **172**, 425–442.

Robertson, D. R. (1977). The annual pattern of plasma calcium in the frog and the seasonal effect of ultimobrachialectomy and parathroidectomy. *Gen Comp Endocrinol* **33**, 336–343.

Robertson, D. R. (1985). Annual and diurnal patterns of plasma calcium fractions in anuran *Rana pipiens*. *Am J Physiol* **249**, R290–R295.

Robinson, D. H. and Heinzelman, M. B. (1987). Morphology of ventral epidermis of *Rana catesbeiana* during metamorphosis. *Anatomical Record* **217**, 305–317.

Robinson, D. H. and Mills, J. W. (1987a). Ouabain binding in tadpole vental skin I. Kinetics and effects of intracellular ions. *Am J Physiol* **253**, R402–R409.

Robinson, D. H. and Mills, J. W. (1987b). Ouabain binding in tadpole ventral skin II. Localization of Na pump sites. *Am J Physiol* **253**, R410–R417.

Rocha, P. L. and Branco, L. G. S. (1997). Cardiovascular, respiratory and metabolic responses to temperature and hypoxia of the winter frog *Rana catesbeiana. Braz J Med Biol Res* **30**, 125–131.

Rocha, P. L. and Branco, L. G. S. (1998). Seasonal changes in the cardiovascular, respiratory and metabolic responses to temperature and hypoxia in the bullfrog *Rana catesbeiana. J Exp Biol* **201**, 761–768.

Rogowitz, G. L., Cortes-Rivera, M. and Nieves-Puigdoller, K. (1999). Water loss, cutaneous resistance, and the effects of dehydration on locomotion of *Eleutherodactylus* frogs. *J Comp Physiol* **169B**, 179–186.

Rogowitz, G. L. and Sanchez-Rivoleda, J. (1999). Locomotor performance and aerobic capacity of the cave coqui, *Eleutherodactylus cooki. Copeia* 1991, 40–48.

Rohlf, D. F. S. and Brand, M. D. (1996). Contribution of mitochondrial proton leak to skeletal muscle respiration and to standard metabolic rate. *Am J Physiol* **271**, C1380–1389.

Rohlf, F. J. (2001). Comparative methods for the analysis of continuous variables: geometric interpretations. *Evolution* **55**, 2143–2160.

Rohr, J. R. and Madison, D. M. (2003). Dryness increases predation risk in efts: support for an amphibian decline hypothesis. *Oecologia* **135**, 657–664.

Rolfe, D. F. S. and Brown, G. C. (1997). Cellular energy utilization and molecular origin of standard metabolic rate in mammals. *Physiol Rev* **77**, 731–758.

Rome, L. C., Stevens, E. D. and John-Alder, H. B. (1992). The influence of temperature and thermal acclimation on physiological function. In: *Environmental Physiology of the Amphibians*, eds M. E. Feder and W. W. Burrgren, pp. 183–205. University of Chicago Press: Chicago.

Romspert, A. P. (1976). Osmoregulation of the African clawed frog, *Xenopus laevis*, in hypersaline media. *Comp Biochem Physiol A* **54**, 207–210.

Romspert, A. L. and McClanahan, L. L. (1981). Osmoregulation in the terrestrial salamander, *Ambystoma tigrinum*, in hypersaline media. Copeia 1981, 400–405.

Rosa, L. L. L. (1977). Respiração de *Synbranchus marmoratus* Pisces—Teleostei—Na ãgua e no ar. *Boletin de Fisiologia Animal, Universidade São Paulo* **1**, 39–70.

Rose, F. L., Zambernard, J. and Pogany, G. S. (1965). Hepatic glycogen depletion in *Amphiuma* during induced anoxia. *Science* **147**, 1467–1468.

Rosen, S. and Friedley, N. J. (1973). Carbonic anhydrase activity in *Rana pipiens* skin: biochemical and histochemical analysis. *Histochem Cell Biol* **36**, 1–4.

Rosen, S., Oliver, J. A. and Steinmetz, P. R. (1974). Urinary acidification and carbonic anhydrase distribution in bladders of Dominican and Columbian toads. *J Memb Biol* **15**, 193–205.

Rosoff, C. J., Baldwin, G. F. and Bentley, P. J. (1983). Active transport of calcium across the urinary bladder and colon of the toad *Bufo marinus. Am J Physiol* **245**, R91–94.

Roth, J. J. (1973). Vascular supply to the ventral pelvic region of anurans as related to water balance. *J Morphology* **140**, 443–460.

Roughton, F. J. W. and Forster, R. E. (1957). Relative importance of diffusion and chemical reaction rates in determining rate of exchange of gases in human lung, with special reference to true diffusion capacity of pulmonary membrane and volume of blood in the capillaries. *J Appl Physiol* **11**, 290–302.

Rovedatti, M. G., Castane, P. M., Saliban, A. and Espina, A. (1988). Studies on the urinary nitrogen waste products in different South American anurans from different habitats. *Comp Biochem Physiol* **90A**, 249–252.

Rowlerson, A. M. and Spurway, N. C. (1988). Histochemical and immunohistochemical properties of skeletal muscle fibers from *Rana* and *Xenopus*. *Histochem Jour* **20**, 657–673.

Rowley, J. J. L. and Alford, A. R. (2007). Non-contact infrared thermometers can acccurately measure amphibian body temperatures. *Herpetological Rev* **38**, 308–311.

Ruben, J. A. and Boucot, A. J. (1989). The origin of the lungless salamanders (Amphibia: Plethodontidae). *Am Nat* **134**, 161–169.

Ruben, J. A., Reagan, N. L., Verrell, P. A. and Boucot, A. J. (1993). Plethodontid salamander origins: A response to Beachy and Bruce. *Am Nat* **142**, 1038–1051.

Rubinsky, B., Wong, S. T., Hong, J. S., Gilbert, J., Roos, M. and Storey, K. B. (1994). 1H magnetic resonance imaging of freezing and thawing in freeze-tolerant frogs. *Am J Physiol* **266**, R1771–1777.

Ruegg, J. C. (1987). Dependence of cardiac contractility on myofirillar calcium sensitivity. *News Physiol Sci* **2**, 179–182.

Rugh, R. (1938). Structure and function of the peritoneal funnels of the frog, *Rana pipiens*. *Proc Soc Exptl Biol Med* **37**, 717–721.

Ruibal, R. (1962). The adaptive value of bladder water in the toad, *Bufo cognatus*. *Physiol Zool* **35**, 218–223.

Ruibal, R. and Hillman, S. S. (1981). Cocoon structure and function in the burrowing hylid frog, *Pternohyla fodiens*. *J Herpetology* **15**, 403–408.

Ruibal, R. and Shoemaker, V. H. (1984). Osteoderms in anurans. *J Herpetol* **18**, 313–328.

Ruibal, R., Tevis, L. and Roig, V. (1969). The terrestrial ecology of the spadefoot toad *Scaphiopus hammondii*. *Copeia* 1969, 1571–1584.

Rumyantsev, P. and Shmantzar, I. A. (1967). Ultrastructure of muscle fibers of the frog lymph heart. *Citologiya* **9**, 1129–1136.

Rupley, J. A. and Careri, G. (1991). Protein hydration and function. *Adv Protein Chem* **41**, 37–172.

Russel, E. L. and Storey, K. B. (1995). Glycogen synthetase and the control of cryoprotectant clearance after thawing in the freeze-tolerant wood frog. *Cryo-Letters* **16**, 263–266.

Ryan, M. J., Bartholomew, G. A. and Rand, A. S. (1983). Energetics of reproduction in a neotropical frog, *Physalaemus pustulosus*. *Ecology* **64**, 1456–1462.

Sackin, H. and Boulpaep, E. L. (1981). Isolated perfused salamander proximal tubule: methods, electrophysiology, and transport. *Am J Physiol* **241**, F39–F52.

Saidapur, S. K. and Hoque, B. (1995). Effect of photoperiod and temperature on ovarian cycle of the frog *Rana tigrina*. *J Biosci* **20**, 445–452.

Saint-Aubain, M. L. (1981). Shunts in the gill filament in tadpoles of *Rana temporaria* and *Bufo bufo* (Amphibia, Anura). *J Exp Zool* **217**, 143–145.

Sakai, T. and Kawahara, K. (1983). The structure of the kidney of Japanese newts, *Triturus (Cynops) pyrrhogaster*. *Anat Embryol* **166**, 31–52.

Sakakibara, Y. (1984). The pattern of respiratory nerve activity in the bullfrog. *Jap J Physiol* **34**, 269–282.

Sakata, J., Kangawa, K. and Matsuo, H. (1988). Identification of new atrial natriuretic peptides in frog heart. *Biochem Biophs Res Comm* **155**, 1338–1345.

Saltys, H. A., Jonz, M. G. and Nurse, C. A. (2006). Comparative study of gill neuroepithelial cells and their innervation in teleosts and *Xenopus* tadpoles. *Cell Tissue Res* **323**, 1–10.

Sanders, J. and Davies, M. (1984). Burrowing behaviour and associated hindlimb morphology of some Australian hylid and leptodactylid frogs. *Aust Zoolog* **21**, 123–142.

San Mauro, D., Vences, M., Alcobendas, M., Zardoya, R. and Meyer, A. (2006). Initial diversification of living amphibians predated the break-up of Pangaea. *Am Nat* **165**, 590–599.

Sasayama, Y. and Clark, N. B. (1984). Renal handling of phosphate, calcium, sodium and potassium in intact and parathyroidectomized *Rana pipiens. J Exp Zool* **299**, 197–203.

Sato, T. and Beidler, L. M. (1975). Membrane resistance change of the frog taste cells in response to water and NaCl. *J Gen Physiol* **66**, 735–763.

Satlin, L. M., Carattino, M. D., Liu, W. and Kleyman, T. R. (2006). Regulation of cation transport in the distal nephron by mechanical forces. *Am J Physiol* **291**, F923–F931.

Sawant, V. A. and Varute, A. T. (1973). Lipid changes in the tadpoles of *Rana tigrina* during growth and metamorphosis. *Comp Biochem Physiol* **44B**, 729–750.

Sawyer, W. H. (1957). Increased reabsorption of osmotically free water by the toad (*Bufo marinus*) in response to neurohypophysial hormones. *Am J Physiol* **189**, 564–568.

Sawyer, W. H., Munsick, R. A. and van Dyke, H. B. (1959). Active principles of the neurohypophysis of cold-blooded vertebrates. Pharmacologic evidence for the presence of arginine vasotocin and oxytocin in neurohypophyseal extracts of cold blooded vertebrates. *Nature* **184**, 1464–1465.

Sawyer, W. H. and Schisgall, R. M. (1956). Increased permeability of the frog bladder to water in response to dehydration and neurohypophyseal extracts. *Am J Physiol* **187**, 312–314.

Schaffner, A. and Rodewald, R. (1978). Glomerular permeability in the bullfrog *Rana catesbeiana. J Cell Biol* **79**, 314–328.

Scharrer, E. (1958). Moulting cycles in the toad, *Bufo marinus. Anatomical Record* **130**, 369–370.

Scheer, B. T. and Markel, R. P. (1962). The effect of osmotic stress and hypophysectomy on blood and urine urea levels in frogs. *Comp Biochem Physiol* **7**, 289–297.

Scheuplein, R. (1978a). The skin as a barrier. In: *The Physiology and Pathophysiology of the Skin*, ed. A. Jarrett, pp. 1669–1692. Academic Press: London.

Scheuplein, R. (1978b). Skin permeation. In: *The Physiology and Pathophysiology of the Skin*, ed. A. Jarrett, pp. 1693–1730. Academic Press: London.

Scheuplein, R. (1978c). Site variations in diffusion and permeability. In: *The Physiology and Pathophysiology of the Skin*, ed. A. Jarrett, pp. 1731–1752. Academic Press: London.

Schilling, C. (1935). Das Herz von *Hypogeophysis und seine Entwicklund. Morph Jb* **76**, 52–91.

Schindelmeiser, J. and Greven, H. (1981). Nitrogen excretion in intra- and extrauterine larvae of the ovoviviparous salamander, *Salamandra salamandra* (Amphibia, Urodela). *Comp Biochem Physiol* **70A**, 563–565.

Schipp, R. and Flindt, R. (1968). Zur feinstruktur und innervation der lympherzmuskulatur der Amphibien. *Z Anat Entwicki Gesch* **127**, 232–253.

Schlaghecke, R. and Blum, V. (1978). Seasonal variation in fat body metabolism of the green frog, *Rana esculenta. Experientia* **34**, 1019–1020.

Schlumberger, H. G. and Burk, D. H. (1953). Comparative study of the reaction to injury. II. Hypervitaminosis D in the frog with special reference to the lime sacs. *AMA Arch Pathol* **56**, 103–124.

Schmajuk, N. A. and Segura, E. T. (1982). Behavioral regulation of water balance in the toad *Bufo arenarum. Herpetologica* **38**, 296–301.

Schmid, W. D. (1965a). Some aspects of the water economies of nine species of amphibians. *Ecology* **46**, 261–269.

Schmid, W. D. (1965b). High temperature tolerances of *Bufo hemiophrys* and *Bufo cognatus*. *Ecology* **46**, 559–560.
Schmid, W. D. (1968). Natural variations in nitrogen excretion of amphibians from different habitats. *Ecology* **49**, 180–185.
Schmid, W. D. (1982). Survival of frogs in low temperature. *Science* **215**, 697–698.
Schmid, W. D. and Underhill, J. C. (1964). Sodium transport by skin of amphibian species from different habitats. *Ecology* **45**, 864–865.
Schmidt, T. J., Husted, R. F. and Stokes, R. B. (1993). Steroid stimulation of Na^+ transport in A6 cells is mediated via glucocorticoid receptors. *Am J Physiol* **264**, C875–C884.
Schmidt, W. D. (1982). Survival of frogs in low temperature. *Science* **215**, 697–698.
Schmidt, W. D. and Underhill, J. C. (1964). Sodium transport by skin of amphibian species from different habitats. *Ecology* **45**, 864–865.
Schmidt-Nielsen, B. and Forster, R. P. (1954). The effect of dehydration and low temperature on renal function in the bullfrog. *J Cellular Comp Physiol* **44**, 233–246.
Schmidt-Nielsen, K. and Lee, P. (1962). Kidney function of the crab eating frog (*Rana cancrivora*). *J Exp Biol* **39**, 167–177.
Schmuck, R. and Linsenmair, K. E. (1988). Adaptations of the reed frog *Hyperolius viridiflavus* (Amphibia: Anura: Hyperoliidae) to its arid environment: 3. Aspects of nitrogen metabolism and osmoregulation in the reed frog *Hyperolius viridiflavus taeniatus*, with special reference to the role of iridophores. *Oecologia* **75**, 354–361.
Schoen, H. F. and Erlij, D. (1987). Insulin action on electrophysiological properties of apical and basolateral membranes of frog skin. *Am J Physiol* **252**, CC411–C417.
Scholander, P. F. (1947). Analyzer for accurate estimation of respiratory gases on one-half cubic centimeter samples. *J Biolog Chem* **167**, 235–250.
Scholander, P. F. (1960). Oxygen transport through hemoglobin solutions. *Science* **131**, 585–590.
Scholander, P. F., G. A. Edwards. (1942). Volumetric microrespirometer for aquatic organisms. Rev Scientific Instruments **13**, 292–295.
Schuytema, G. S. and Nebeker, A. V. (1999). Comparative toxicity of ammonium and nitrate compounds to Pacific treefrog and African clawed frog tadpoles. *Emnviron Toxicol Chem* **18**, 2251–2257.
Schwalm, P. A., Starrett, P. H. and McDiarmid, R. W. (1977). Infrared reflectance in leaf-sitting Neotropical frogs. *Science* **196**, 1225–1227.
Schwartz, J. J., Resel, S. and Bevier, C. R. (1995). Carbohydrate and calling: depletion of muscle glycogen and the chorusing dynamics of the Neotropical frog *Hyla microcephala*. *Behav Ecol Sociobiol* **37**, 125–135.
Schwarzopf, L. and Alford, R. A. (1996). Desiccation and shelter-site use in a tropical amphibian: comparing toads with physical models. *Functional Ecology* **10**, 193–200.
Scott, N. J. (1976). The abundance and diversity of the herpetofaunas of tropical forest litter. *Biotropica* **8**, 41–58.
Scott, W. N., Saperstein, V. S. and Yoder, J. (1974). Partition of tissue functions an epithelial cell localization of enzymes in "mitochondria-rich" cells of toad urinary bladder. *Science* **184**, 797–800.
Secor, S. M. (2001). Regulation of digestive performance: a proposed adaptive response. *Comp Biochem Physiol* **128A**, 565–577.
Secor, S. M. (2005). Physiological responses to feeding, fasting and estivation for anurans. *J Exp Biol* **208**, 2595–2608.

Secor, S. M. and Boehm, M. (2006). Specific dynamic action of ambystomatid salamanders and the effects of meal size, meal type, and body temperature. *Physiol Biochem Zool* **79**, 720–735.

Secor, S. M. and Faulkner, A. C. (2002). Effects of meal size, meal type, body temperature, and body size on the specific dynamic action of the marine toad, *Bufo marinus*. *Physiol Biochem Zool* **75**, 557–571.

Secor, S. M., Wooten, J. A. and Cox, C. L. (2007). Effects of meal size, meal type, and body temperature on the specific dynamic action of anurans. *J Comp Physiol* **177B**, 165–182.

Sedmera, D., Reckova, M., deAlmeida, A., Sedmerova, M., Biermann, M., Volejnik, J., Sarre, A., Raddatz, E., McCarthy, R. A., Gourdlie, R. G. *et al.* (2003). Functional and morphological evidence for a ventricular conduction system in zebrafish and *Xenopus* hearts. *Am J Physiol* **284**, H1152–H1160.

Seebacher, F. and Salford, R. A. (2002). Shelter microhabits determine body temperature and dehydration rates of a terrestrial amphibian (*Bufo marinus*). *J Herpetol* **36**, 69–75.

Segura, E. T., Badsholm, U. C. and Bronstein, A. (1982). Role of the nervous system in the control of the water economy of the toad *Bufo arenarum* Hensel. II Adrenergic controlof water uptake across the skin. *J Comp Physiol* **146**, 101–106.

Segura, E. T. and D'Agostino, S. A. (1964). Seasonal variations of blood pressure, vasomotor reactivity and plasmatic catecholamines in the toad. *Acta Physiol Lat Am* **14**, 231–237.

Seibel, R. V. (1970). Variables affecting the critical thermal maximum of the leopard frog, *Rana pipiens*. *Herpetologica* **26**, 208–213.

Seiyama, A. (2006). Virtual cooperativity in myoglobin oxygen saturation curve in skeletal muscle in vivo. *Dynamic Medicine* 5:3.

Semlitsch, R. D. (1981). Terrestrial activity and summer home range of the mole salamander (*Ambystoma talpoideum*). *Can J Zool* **59**, 315–322.

Seymour, R. S. (1972). Behavioral thermoregulation by juvenile green toads, *Bufo debilis*. *Copeia* 1972, 572–575.

Seymour, R. S. (1973a). Physiological correlates of forced activity and burrowing in the spadefoot toad, *Scaphiopus hammondii*. *Copeia* 1973, 103–115.

Seymour, R. S. (1973b). Energy metabolism of dormant sadefoot toads (*Scaphiopus*). *Copeia* 1973, 435–445.

Seymour, R. S. (1995). Oxygen uptake by embryos in gelatinous egg masses of *Rana sylvatica*: The roles of diffusion and convection *Copeia* 1995, 626–635.

Seymour, R. S. and Lee, A. K. (1974). Physiological adaptations to anuran amphibians to aridity: Australian prospects. *Aust Zoolog* **18**, 53–65.

Shaffer, H. B., Fellers, G. M., Magee, A. and Voss, R. (2000). The genetics of amphibian declines: population substructure and molecular differentiation in the Yosemite toad, *Bufo canorus* (Anura, Bufonidae) based on single-strand conformation polymorphism analysis (SSCP) and mitochondrial DNA sequence data. *Molecular Ecology* **9**, 245–257.

Sham, J. S., Sawyer, W. H. and Pang, P. K. (1989). Direct cardiac stimulation by arginine vasotocin in bullfrogs (*Rana catesbeiana*). *Am J Physiol* **256**, R187–192.

Shane, M. A., Nofziger, C. and Blazer-Yost, B. (2006). Hormonal regulation of the epithelial Na^+ channel: from amphibians to mammals. *Gen Comp Endocrinol* **147**, 85–92.

Sharma, H. L. (1957). The anatomy and mode of action of the heart of the frog, *Rana tigrina*. *J Morph* **100**, 313–343.

Sharma, H. L. (1961). The circulatory mechanisms and anatomy of the heart of the frog, *Rana pipiens*. *J Morph* **109**, 323–349.

Shcherbak, N. N. and Kovalyukh, N. N. (1973). Age of a living amphibian *Hynobius keyserlingi* dug out from ice. *Doklady Akademia Nauk SSSR* **211**, 359–360.
Shelton, G. (1970). The effect of lung ventilation on blood flow to the lungs and body of the amphibian, *Xenopus laevis*. *Respir Physiol* **9**, 183–195.
Shelton, G. and Jones, D. R. (1965). Central blood pressure and heart output in surface and submerged frogs. *J Exp Biol* **42**, 339–357.
Shelton, G. and Jones, D. R. (1968). A comparative study of central blood pressures in five amphibians. *J Exp Biol* **49**, 631–643.
Shelton, G., Jones, D. R. and Milsom, W. K. (1986). Control of breathing in ectothermic vertebrates. In: *Handbook of Physiology*, eds N. S. Cherniak and J. D. Widdicombe, pp. 857–909. Am Physiol Soc: Bethesda, MD.
Sherman, W. M., Plyley, M. J., Sharp, R. L., Van Handel, P. J., McAllister, R. M., Fink, W. J. and Costill, D. L. (1982). Muscle glycogen storage and its relationship with water. *Int J Sports Med* **3**, 22–24.
Shibata, E. F. and Giles, W. R. (1985). Ionic currents that generate the spontaneous diastolic depolarization in individual cardiac pacemaker cells. *PNAS* **82**, 7796–7800.
Shoemaker, V. H. (1964). The effects of dehydration on electrolyte concentrations in a toad, *Bufo marinus*. *Comp Biochem Physiol* **13**, 261–271.
Shoemaker, V. H. (1965). The stimulus for the water-balance response to dehydration in toads. *Comp Biochem Physiol* **15**, 81–88.
Shoemaker, V. H. (1987). Osmoregulation in amphibians. In: *Comparative Physiology: Life in Water and on Land*. eds P.Dejours, L. Bolis, C.R. Taylor and E.R. Weibel, pp. 109–120. Fidia Research Series IX-Liviana Press: Padova.
Shoemaker, V. H., Baker, M. A. and Loveridge, J. P. (1989). Effect of water balance on thermoregulation in waterproof frogs (*Chiromantis* and *Phyllomedusa*). *Physiol Zool* **62**, 133–146.
Shoemaker, V. H., Balding, R., Ruibal, R. and McClanahan, L. L. (1972). Uricotelism and low evaporative water loss in a South American frog. *Science* **175**, 1018–1020.
Shoemaker, V. H. and Bickler, P. E. (1979). Kidney and bladder function in a uricotelic treefrog (*Phyllomedusa sauvagei*). *J Comp Physiol* **133**, 211–218.
Shoemaker, V. H., Hillman, S. S., Hillyard, S. D., Jackson, D. C., McClanahan, L. L., Withers, P. C. and Wygoda, M. L. (1992). Exchange of water, ions, and respiratory gases in terrestrial amphibians. In: *Physiological Ecology of the Amphibians*, eds M. E. Feder and W. W. Burrgren, pp. 125–150. University of Chicago Press: Chicago and London.
Shoemaker, V. H. and McClanahan, L. L. (1973). Nitrogen excretion in the larvae of a land-nesting frog (*Leptodactylus bufonius*). *Comp Biochem Physiol A* **44**, 1149–1156.
Shoemaker, V. H. and McClanahan, L. L. (1975). Evaporative water loss, nitrogen excretion, and osmoregulation in phyllomedusine frogs. *J Comp Physiol* **100**, 331–345.
Shoemaker, V. H. and McClanahan, L. L. (1980). Nitrogen excretion and water balance in amphibians of Borneo. *Copeia* 1980, 446–451.
Shoemaker, V. H. and McClanahan, L. L. (1982). Enzymatic correlates and ontogeny of uricotelism in tree frogs of the genus *Phyllomedusa*. *J Exp Zool* **220**, 163–169.
Shoemaker, V. H., McClanahan, L. L. and Ruibal, R. (1969). Seasonal changes in body fluids in a field population of spadefoot toads. *Copeia* 1969, 585–591.
Shoemaker, V. H., McClanahan, L. L., Withers, P. C., Hillman, S. S. and Drewes, R. C. (1987). Thermoregulatory response to heat in the waterproof frogs *Phyllomedusa* and *Chiromantis*. *Physiol Zool* **60**, 365–372.

Shoemaker, V. H. and Nagy, K. A. (1977). Osmoregulation in amphibians and reptiles. *Annu Rev Physiol* **39**, 449–471.

Shoemaker, V. H. and Waring, H. (1968). Effect of hypothalamic lesions on the water-balance response of a toad (*Bufo marinus*). *Comp Biochem Physiol* **24**, 47–54.

Shpun, S. and Katz, U. (1995). Renal function at steady state in a toad (*Bufo viridis*) acclimated in hyperosmotic NaCl and urea solutions. *J Comp Physiol B* **164**, 646–652.

Sick, T. J. and Kreisman, N. R. (1981). Potassium ion homeostasis in amphibian brain: Contribution of active transport and oxidative metabolism. *J Neurophysiol* **45**, 998–1012.

Siebert, E. A., Lillywhite, H. B. and Wassersug, R. J. (1974). Cranial coossification in frogs: relationships to rates of evaporative water loss. *Physiol Zool* **47**, 261–265.

Sievert, L. M. and Bailey, J. K. (2000). Specific dynamic action in the toad, *Bufo woodhouseii*. *Copeia* 2000, 1076–1078.

Simkiss, K. (1968). Calcium and carbonate metabolism in the frog (*Rana temporaria*) during respiratory acidosis. *Am J Physiol* **214**, 627–634.

Simons, J. R. (1959). The distribution of blood from the heart in some amphibia. *Proc Zool Soc (London)* **132**, 51–64.

Simons, R. S., Bennett, W. O. and Brainerd, E. L. (2000). Mechanics of lung ventilation in a post-metamorphic salamander, *Ambystoma tigrinum*. *J Exp Biol* **203**, 1081–1092.

Sinsch, U. (1987). Modifikation im wasserhaushalt von drei *Rana*-arten (Amphibia: Anura) als anpassungen an ihre spezifischen habitate. *Verh Gesellsch Okol* **15**, 365–372.

Sinsch, U. (1989). Behavioral thermoregulation of the Andean toad (*Bufo spinulosus*) at high altitude. *Oecologia* **80**, 32–38.

Sinsch, U. (1991). Reabsorption of water and electrolytes in the urinary bladder of intact frogs (Genus *Rana*). *Comp Biochem Physiol* **99A**, 559–565.

Sinsch, U., Seine, R., and Sherif, N. (1992). Seasonal changes in the tolerance of osmoic stress in natterjack toads (*Bufo calamita*). *Comp Biochem Physiol A* **101**, 353–360.

Sinsch, U. and Sherif, N. (1989). Osmoregulation of natterjack toads (*Bufo calamita*) in tapwater, urea and NaCl solutions. *Verh Dtsch Ges* Dusseldorf **82**, 307–308.

Skelly, D. K. and Golon, J. (2003). Assimilation of natural benthic substrates by two species of tadpoles. *Herpetologica* **59**, 37–42.

Slivkoff, M. D. and Warburton, S. J. (2001). Angiotensin II alters blood flow distribution in amphibians. *Physiol Biochem Zool* **74**, 576–583.

Smatresk, N. J. and Smits, A. W. (1991). Effects of central and peripheral chemoreceptor stimulation on ventilation in the marine toad, *Bufo marinus*. *Respir Physiol* **83**, 223–238.

Smith, D. G. and Rapson, L. (1977). Differences in pulmonary microvascular anatomy between *Bufo marinus* and *Xenopus laevis*. *Cell Tissue Res* **178**, 1–15.

Smith, G. C. (1976). Ecological energetics of three species of ectothermic vertebrates. *Ecology* **57**, 252–264.

Smith, V. D. E. and Jackson, C. M. (1931). The changes during desiccation and rehydration in the body organs of the leopard frog (*Rana pipiens*). *Biolog Bull* **60**, 80–93.

Smits, A. W. (1989). Fluid balance in vertebrate lungs. In: *Comparative Pulmonary Physiology: Current Concepts*, ed. S.C. Wood, pp. 503–537. Marcel-Dekker.

Smits, A. W. (1994). Lack of edema in toad lungs after pulmonary hypertension. *Am J Physiol* **266**, R1338–1344.

Smits, A. W. (1998). Fluid balance in vertebrate lungs: Are all lungs "dry"? In: *Pulmonary Edema,* eds E. K. Weir and J. T. Reeves, pp. 3–16. Futira Publishing: Armonk, New York.

Smits, A. W. and Flanagin, J. I. (1994). Bimodal respiration in aquatic and terrestrial apodan amphibians. *Amer Zool* **34**, 247–263.

Smits, A. W., West, N. H. and Burggren, W. W. (1986). Pulmonary fluid balance following pulmocutaneous baroreceptor denervation in the toad. *J Appl Physiol* **61**, 331–337.

Smyth, D. H. (1939). The central and reflex control of respiration in the frog. *J Physiol* **95**, 305–327.

Snyder, G. K. and Hammerson, G. A. (1993). Interrelations between water economy and thermoregulation in the canyon tree-frog *Hyla arenicolor*. *J Arid Environ* **25**, 321–329.

Snyder, G. K. and Sheafor, B. A. (1999). Red blood cells: centerpiece in the evolution of the vertebrate circulatory system. *Amer Zool* **39**, 189–198.

Soderstrom-Lauritzsen, V., Nilsson, G. E. and Lutz, P. L. (2001). Effect of anoxia and adenosine on cerebral blood flow in the leopard frog (*Rana pipiens*). *Neurosci Lett* **311**, 85–88.

Sokabe, H., Nishimura, H., Kawabe, K., Tenmoku, S. and Arai, T. (1972). Plasma renin activity in varying hydrated states in the bullfrog. *Am J Physiol* **222**, 142–146.

Somero, G. N. (1986). From dogfish to dogs: Trimethylamines protect proteins from urea. *News Physiol Sci* **1**, 9–12.

Speakman, J. R. and Ward, S. (1998). Infrared thermography: Principles and applications. *Zoology* **101**, 224–232.

Spight, T. M. (1967). The water economy of salamanders: water uptake after dehydration. *Comp Biochem Physiol* **20**, 767–771.

Spight, T. M. (1968). The water economy of salamanders: Evaporative water loss. *Physiol Zool* **41**, 195–203.

Spotila, J. R. (1972). Role of temperature and water in the ecology of lungless salamanders. *Ecolog Mono* **42**, 95–125.

Spotila, J. R. and Berman, E. N. (1976). Determination of skin resistance and the role of the skin in controlling water loss in amphibians and reptiles. *Comp Biochem Physiol A* **55**, 407–411.

Spotila, J. R., O'Connor, M. P. and Bakken, G. S. (1992). 4. Biophysics of heat and mass transfer. In: *Environmental Physiology of the Amphibians*, eds W. Burggren and M. Feder, pp. 59–80. University of Chicago Press.

Spruyt, P., Y. Ghafir, T. Art and P. Lekeux. (1995). Infrared thermography in the study of thermoregulation: Bibliographical study. Annales de Medecine Veterinaire 139: 413–418.

Srivastav, A. K. and Rani, L. (1989). Influence of calcitonin administration of serum calcium and inorganic phosphate level of the frog *Rana tigrina*. *Gen Comp Endocrinol* **74**, 14–17.

St-Pierre, J. and Boutilier, R. G. (2001). Aerobic capacity of frog skeletal muscle during hibernation. *Physiol Biochem Zool* **74**, 390–397.

St-Pierre, J., Brand, M. D. and Boutilier, R. G. (2000a). Mitochondria as ATP consumers: cellular treason in anoxia. *PNAS* **97**, 8670–8674.

St-Pierre, J., Brand, M. D. and Boutilier, R. G. (2000b). The effect of metabolic depression on proton leak rate in mitochondria from hibernating frogs. *J Exp Biol* **203**, 1469–1476.

St-Pierre, J., Tattersall, G. J. and Boutilier, R. G. (2000c). Metabolic depression and enhanced O_2 affinity of mitochondria in hypoxic hypometabolism. *Am J Physiol Regul Integr Comp Physiol* **279**, R1205–1214.

St. Aubain, M. L. and Wingstrand, K. G. (1979). A spincter in the pulmonary artery of the frog *Rana temporaria* and its influence on blood flow in skin and lungs. *Acta Zoologica* **60**, 163–172.

Stanier, M. W., L. E. Mount, J. Bligh. (1984). Energy Balance and Temperature Regulation. Cambridge: Cambridge University Press.

Stark-Vancs, V., Bell, P. B. and Hutchison, V. H. (1984). Morphological and pharmacological basis for pulmonary ventilation in *Amphiuma tridactylum*. *Cell Tissue Res* **238**, 1–12.

Starrett, P. H. (1973). Evolutionary patterns in larval morphology. In: *Evolutionary Biology of the Anurans*, ed. J. L. Vial, pp. 257–271. University of Missouri Press: Columbia, MO.

Stary, C. M. and Hogan, M. C. (2000). Impairment of Ca^{2+}release in single *Xenopus* muscle fibers fatigued at varied extracellular P_{O2}. *J Appl Physiol* **88**, 1743–1748.

Stary, C. M. and Hogan, M. C. (2005). Intracellular pH during sequential, fatiguing contractile periods in isolated *Xenopus* skeletal muscle fibers. *J Appl Physiol* **99**, 308–312.

Stary, C. M., Mathieu-Costello, O. and Hogan, M. C. (2004). Resistance to fatigue of individual *Xenopus* single skeletal muscle fibres is correlated with mitochondrial volume density. *Exp Physiol* **89**, 617–621.

Stebbins, R. C. (1954). Natural history of the salamanders of the plethodontid genus *Ensatina*. *University of California Publications in Zoology* **54**, 377–512.

Stebbins, R. C. and Cohen, N. C. (1995). *A Natural History of Amphibians*. Princeton University Press: New Jersey.

Steele, W. K. and Louw, G. N. (1988). Caecilians exhibit cutaneous respiration and high evaporative water loss. *SA J Zool* **23**, 134–135.

Steen, W. B. (1929). On the permeability of the frog's bladder to water. *Anat Rec* **43**, 215–220.

Stefano, F. J. E. and Donoso, A. O. (1964). Hypophyso-adrenal regulation of moulting in the toad. *Gen Comp Endocrinol* **4**, 473–480.

Stefanski, M., Gatten, R. E. and Pough, F. H. (1989). Activity metabolism of salamanders: tolerance to dehydration. *J Herpetol* **23**, 45–50.

Steiner, A.A., Petenusci, S.O., Brentegani, L.G. and Branco, L.G.S. (2000). The importance of glucose for the freezing tolerance/intolerance of the anuran amphibians Rana catesbeiana and Bufo paracnemis. Rev.Bras. Biol. SciELO Brasil

Sten-Knudsen, O. and Ussing, H. H. (1981). The flux ratio equation under non-stationary conditions. *J Membrane Biol* **63**, 233–242.

Stevens, E. D. (1988). Acclimation temperature markedly alters the motivation to feed in the toad *Bufo Amus*. *J Thermal Biol* **13**, 73–78.

Stevenson, R. D. (1985). Body size and limits to the daily range of body temperature in terrestrial ectotherms. *Am Naturalist* **125**, 102–117.

Stewart, E. R., Reeses, S. A. and Ultsch, G. R. (2004). The physiology of hibernation in Canadian leopard frogs (*Rana pipiens*) and bullfrogs (*Rana catesbeiana*). *Physiol Biochem Zool* 77, 65–73.

Stewart, M. M. (1995). Climate driven population fluctuations in rain forest frogs. *J Herpetol* **29**, 437–446.

Stewart, M. M. and Woolbright, L. L. (1996). Amphibians. In: *The Food Web of a Tropical Rain Forest*, eds D. P. Reagan and R. B. Waide, pp. 363–398. University of Chicago Press: Chicago.

Steyermark, A. C., Miamen, A. G., Feghahati, H. S. and Lewno, A. W. (2005). Physiological and morphological correlates of among-individual variation in standard metabolic rate in the leopard frog *Rana pipiens*. *J Exp Biol* **208**, 1201–1208.

Stiffler, D. F. (1981). The effects of mesotocin on renal function in hypophysectomized *Ambystoma tigrinum* larvae. *Gen Comp Endocrinol* **45**, 49–55.

Stiffler, D. F. (1991). Partitioning of acid-base regulation between renal and extrarenal sites in the adult, terrestrial stage of the salamander *Ambystoma tigrinum* during respiratory acidosis. *J Exp Biol* **157**, 47–62.
Stiffler, D. F. (1993). Amphibian calcium metabolism. *J Exp Biol* **184**, 47–61.
Stiffler, D. F. (1995). Active calcium transport in the skin of the frog *Rana pipiens*: kinetics and seasonal rhythms. *J Exp Biol* **198**, 967–974.
Stiffler, D. F. and Alvarado, R. H. (1974). Renal function in response to reduced osmotic load in larval salamanders. *Am J Physiol* **226**, 1243–1249.
Stiffler, D. F., DeRuyter, M. L. and Talbot, C. R. (1990). Osmotic and ionic regulation in the aquatic caecilian *Typhlonectes compressicauda* and the terrestrial caecilian *Ichthyophis kohtaoensis*. *Physiol Zool* **63**, 649–668.
Stiffler, D. F., Eskandari, S. and Dejbakhsh, S. (1997). Cutaneous transport of Ca^{2+} in the frog *Rana pipiens*: significance and specificity. *J Exp Zool* **277**, 371–381.
Stiffler, D. F., Hawk, C. T. and Fowler, B. C. (1980). Renal excretion of urea in the salamander, *Ambystoma tigrinum*. *J Exp Zool* **213**, 205–212.
Stiffler, D. F. and Manokham, T. (1994). Partitioning of nitrogen excretion between urea and ammonia and between skin and kidneys in the aquatic caecilian *Typhlonectes natans*. *Physiol Biochem Zool* **67**, 1077–1086.
Stiffler, D. F. and Talbot, C. R. (2000). Exchanges of oxygen, carbon dioxide, nitrogen and water in the caecilian *Dermophis mexicanus*. *J Comp Physiol B* **170**, 505–509.
Stiffler, D. F., Yee, J. C. and Tefft, J. D. (1998). Responses of frog skin, *Rana pipiens*, calcium ion transport to parathyroid hormone, calcitonin, and vitamin D3. *Gen Comp Endocrinol* **112**, 191–9.
Stille, W. T. (1951). The nocturnal amphibian fauna of the southern Lake Michigan beach. *Ecology* **33**, 149–162.
Stille, W. T. (1958). The water absorption response of an anuran. *Copeia* 1958, 217–218.
Stinner, J. and Hartzler, L. K. (2000). Effect of temperature on pH and electrolyte concentration in air-breathing ectotherms. *J Exp Biol* **203**, 2065–2074.
Stinner, J., Newlon, D. L. and Heisler, N. (1994). Extracellular and intracellular carbon dioxide concentration as a function of temperature in the toad *Bufo marinus*. *J Exp Biol* **195**, 345–360.
Stinner, J., Zarlinga, N. and Orcutt, S. (1994). Overwintering behavior of adult bullfrogs, *Rana catesbeiana*, in northeastern Ohio. *Ohio J Sci* **94**, 8–13.
Stinner, J. N. and Shoemaker, V. H. (1987). Cutaneous gas exchange and low evaporative water loss in the frogs *Phyllomedusa sauvagei* and *Chiromantis xerampelina*. *J Comp Physiol B* **157**, 423–427.
Stoner, L. C. (1977). Isolated, perfused amphibian renal tubules: the diluting segment. *Am J Physiol* **233**, F438–F444.
Storey, J. M. and Storey, K. B. (1985a). Triggering of cryoprotectant synthesis by the iniation of ice nucleation in the freeze tolerant frog, *Rana sylvatica*. *J Comp Physiol* **156B**, 191–195.
Storey, J. M. and Storey, K. B. (1985b). Adaptations of metabolism for freeze tolerance in the gray tree frog, *Hyla versicolor*. *Can J Zool* **63**, 49–54.
Storey, J. M. and Storey, K. B. (1996). β-adrenergic, hormonal, and nervous influences on cryoprotectant synthesis by liver of the freeze-tolerant wood frog *Rana sylvatica*. *Cryobiology* **33**, 186–195.
Storey, K. B. (1985). Freeze tolerance in terrestrial frogs. *Cryo-Letters* **6**, 115–134.

Storey, K. B. (1987a). Organ-specific metabolism during freezing and thawing in a freeze-tolerant frog. *Am J Physiol* **253**, R292–297.

Storey, K. B. (1987b). Glycolysis and the regulation of cryoprotectant synthesis in liver of the freeze tolerant wood frog. *J Comp Physiol* **157B**, 373–380.

Storey, K. B. (1988). Suspended animation: The molecular basis of metabolic depression. *Can J Zool* **66**, 124–132.

Storey, K. B. (1997). Organic solutes in freezing tolerance. *Comp Biochem Physiol* **117A**, 319–326.

Storey, K. B. (1999). Living in the cold: freeze-induced gene responses in freeze-tolerant vertebrates. *Clin Exp Pharmacol Physiol* **26**, 57–63.

Storey, K. B. (2002). Life in the slow lane: Molecular mechanisms of estivation. *Comp Biochem Physiol A* **133**, 733–754.

Storey, K. B. (2004). Strategies for exploration of freeze responsive gene expression: advances in vertebrate freeze tolerance. *Cryobiology* **48**, 134–145.

Storey, K. B., Dent, M. E. and Storey, J. M. (1999). Gene expression during estivation in spadefoot toads, *Scaphiopus couchii*: upregulation of riboflavin binding protein in liver. *J Exp Zool* **284**, 325–333.

Storey, K. B., Mosser, D. D., Douglas, D. N., Grundy, J. E. and Storey, J. M. (1996). Biochemistry below 0 degrees C: nature's frozen vertebrates. *Braz J Med Biol Res* **29**, 283–307.

Storey, K. B. and Storey, J. M. (1984). Biochemical adation for freezing tolerance in the wood frog, *Rana sylvatica*. *J Comp Physiol* **155B**, 29–36.

Storey, K. B. and Storey, J. M. (1986a). Freeze tolerance and intolerance as strategies of winter survival in terrestrially-hibernating amphibians. *Comp Biochem Physiol A* **83**, 613–617.

Storey, K. B. and Storey, J. M. (1986b). Freeze tolerant frogs: cryoprotectants and tissue metabolism during freeze-thaw cycles. *Can J Zool* **64**, 49–56.

Storey, K. B. and Storey, J. M. (1987). Persistence of freeze tolerance in terrestrially hibernating frogs after spring emergence. *Copeia* 1987, 720–726.

Storey, K. B. and Storey, J. M. (1988). Freeze tolerance in animals. *Physiol Rev* **68**, 27–84.

Storey, K. B. and Storey, J. M. (1990). Facultative metabolic rate depression: Molecular regulation and biochemical adaptation in anaerobiosis, hibernation, and estivation. *Quart Rev Biol* **65**, 145–174.

Storey, K. B., Storey, J. M. and Chruchill, T. A. (1997). De novo protein biosynthesis responses to water stresses in wood frogs: freeze-thaw and dehydration-rehydration. *Cryobiology* **34**, 200–213.

Straus, C., Wilson, R. J. A. and Remmers, J. E. (2001). Oxygen sensitive chemoreceptors in the first gill arch of the tadpole, *Rana catesbeiana*. *Can J Pharmacol* **79**, 959–962.

Strawinski, S. (1956). Vascularization of respiratory surfaces in ontogeny of the edible frog *Rana esculenta*. *Zool Polon* 7, 327–365.

Strübing, H. (1954). Über vorzugstemperaturen von Amphibien. *Zeitschrift fuer Morphologie und Ökologie der Tiere* **43**, 357–386.

Strum, J. M. and Danon, D. (1973). Fine structure of the urinary bladder of the Bullfrog (*Rana catesbeiana*). *Anat Rec* **178**, 15–40.

Stuart, L. C. (1951). The distributional implications of temperature tolerance and hemoglobin values in the toads *Bufo marinus* (Linnaeus) and *Bufo bocourti* Brocchi. *Copeia* 1951, 220–221.

Stuart, S. N., Chanson, J. S., Cox, N. A., Young, B. E., Rodrigues, A. S. L., Fichman, D. L. and Waller, R. W. (2004). Status and trends of amphibian declines and extinctions worldwide. *Science* **306**, 1783–1786.

Sullivan, B. K. (1982). Sexual selection in Woodhouse's toads (*Bufo woodhousei*): 1. Chorus organization. *Anim Behav* **30**, 680–686.

Sullivan, B. K. and Hinshaw, S. H. (1992). Female choice and selection of male calling behavior in the grey treefrog (*Hyla versicolor*). *Anim Behav* **44**, 733–744.

Sullivan, B. K. and Walsberg, G. E. (1985). Call rate and aerobic capacity in Woodhouse's toad (*Bufo woodhousei*). *Herpetologica* **41**, 404–407.

Sullivan, L. C., Daniels, C. B., Philips, I. A., Orgeig, S. and Whitsett, J. A. (1998). Conservation of surfactant protein A: Evidence for a single origin for vertebrate pulmonary surfactant. *J Mol Evol* **46**, 131–138.

Suzuki, M., Hasegawa, Y., Ogushi, Y. and Tanaka, S. (2007). Amphibian aquaporins and adaptation to terrestrial environments: A review. *Comp Biochem Physiol* **148A**, 72–81.

Swanson, D. L. and Graves, B. M. (1995). Supercooling and freeze intolerance in overwintering juvenile spadefoot toads (*Scaphiopus bombifrons*). *J Herpetology* **29**, 280–285.

Swanson, D. L., Graves, B. M. and Koster, K. L. (1996). Freezing tolerance/ intolerance and cryoprotectant synthesis in terrestrially overwintering anurans in the Great Plains, USA. *J Comp Physiol* **166B**, 110–119.

Szarski, H. (1964). The respiratory organs in relation to body size in amphibia. *Evolution* **18**, 118–126.

Tachiki, K. H. and Baxter, C. F. (1980). Role of carbon dioxide fixation, blood aspartate and glutamate in the adaptation of amphibian brain tissue to a hyperosmotic internal environment. *Neurochem Res* **5**, 993–1010.

Taigen, T. L. and Beuchat, C. A. (1984). Anaerobic threshold of anuran amphibians. *Physiol Zool* **57**, 641–647.

Taigen, T. L., Emerson, S. B. and Pough, F. H. (1982). Ecological correlates of anuran exercise physiology. *Oecologia* **52**, 49–56.

Taigen, T. L. and Pough, F. H. (1983). Prey preference, foraging behavior and metabolic characteristics of frogs. *Am Nat* **122**, 509–520.

Taigen, T. L. and Pough, F. H. (1985). Metabolic correlates of anuran behavior. *Amer Zool* **25**, 987–997.

Taigen, T. L. and Wells, K. D. (1985). Energetics of vocalization by an anuran amphibian (*Hyla versicolor*). *J Comp Physiol* **155**, 163–170.

Taigen, T. L., Wells, K. D. and Marsh, R. L. (1985). The enzymatic basis of high metabolic rates in calling frogs. *Physiol Zool* **58**, 719–726.

Takada, M. and Hokari, S. (2007). Prolactin increases Na(+) transport across adult bullfrog skin via stimulation of both ENaC and Na(+)/K(+)-pump. *Gen Comp Endocrinol* **151**, 325–331.

Takada, M. and Kasai, M. (2003). Prolactin increases open-channel density of epithelial Na^+ channels in adult frog skin. *J Exp Biol* **206**, 1319–1323.

Takada, M., Shimomura, T., Hokari, S., Jensik, P. J. and Cox, T. C. (2006). Larval bullfrog skin expresses ENaC despite having no amiloride-blockable transepithelial Na^+ transport. *J Comp Physiol B* **176**, 287–293.

Takada, M., Yai, H. and Takayama-Arita, K. (1995a). Corticoid-induced differentiation of amiloride-blockable active Na^+ transport across larval bullfrog skin in vitro. *Am J Physiol* **268**, C218–C226.

Takada, M., Yai, H. and Takayama-Arita, K. (1995b). Prolactin inhibits corticoid-induced differentiation of active Na+ transport across cultured frog tadpole skin. *Am J Physiol* **269**, C1326–C1331.

Takada, M., Yai, H., Takayama-Arita, K. and Komazaki, S. (1996). Prolactin enables normal development of Ach-stimulated current in cultured larval bullfrog skin. *Am J Physiol* **271**, C1059–C 1063.

Takahashi-Iwanaga, H. (2002). Comparative anatomy of the podocyte: a scanning electron microscopic study. *Microsc Res Tech* **57**, 196–202.

Talbot, C. R. (1992). Regional variation and control of cutaneous gas exchange in bullfrogs. *Respir Physiol* **89**, 261–72.

Tamori, Y. and Wahahara, M. (2000). Conversion of red blood cells (RBCs) from the larval to the adult type during metamorphosis in *Xenopus*: specific removal of mature larval-type RBCs by apoptosis. *Int J Dev Biol* **44**, 373–380.

Tanaka, Y. (1979). Whole body transvascular filtration coefficients and interstitial space capacitance. *Jpn J Physiol* **29**, 181–193.

Tanii, H., Hasegawa, T., Hirakawa, N., Suzuki, M. and Tanaka, S. (2002). Molecular and cellular characterization of a water-channel protein, AQP-h3, specifically expressed in the frog ventral skin. *J Membr Biol* **188**, 43–53.

Tasman, K. (1956). 'Amphibian underground' (or ground amphibians). *Proceedings Transactions the Rhodesia Scientific Association* **44**, 22–35.

Tattersall, G. J. and Boutilier, R. G. (1997). Balancing hypoxia and hypothermia in cold-submerged frogs. *J Exp Biol* **200**, 1031–1038.

Tattersall, G. J. and Boutilier, R. G. (1999). Behavioural oxy-regulation by cold-submerged frogs in heterogeneous oxgen environments. *Can J Zool* 77, 843–850.

Tattersall, G. J., Eterovick, P. C. and de Andrade, D. V. (2006). Skin colour and body temperature changes in basking *Bokermannohyla alvarengai* (Bokermann 1956). *J Exp Biol* **209**, 1185–1196.

Taugner, R., Schiller, A. and Ntokalo-Knittel, S. (1982). Cells and intercellular contacts in glomeruli and tubules of the frog kidney. *Cell Tissue Res* **226**, 589–608.

Taylor, A. E. and Townsley, M. I. (1987). Evaluation of the Starling fluid flux equation. *News Physiol Sci* **2**, 48–52.

Taylor, E. H. (1968). *The Caecilians of the World*. University of Kansas Press: Lawrence, KS.

Taylor, E. N., Denardo, D. F. and Malawy, M. A. (2004). A comparison between point and semi-continuous sampling for assessing body temperature in a free-ranging ectotherm. *J Thermal Biol* **29**, 91–96.

Taylor, S. and Ewer, D. W. (1956). Moulting in the Anura: The normal moluting cycle of *Bufo regularis* Reuss. *Proc Zool Soc London* **127**, 461–478.

Tazawa, H. and Johansen, K. (1987). Comparative model analysis of central shunts in vertebrate cardiovascular systems. *Comp Biochem Physiol A* **86**, 595–607.

Tazawa, H. and Mochizuki, M. (1977). Analysis of oxygen transport in bullfrogs. *Adv Exp Med Biol* **94**, 199–204.

Tazawa, H., Mochizuki, M. and Piiper, J. (1979a). Respiratory gas transport by the incompletely separated double circulation in the bullfrog, *Rana catesbeiana*. *Respir Physiol* **36**, 77–95.

Tazawa, H., Mochizuki, M. and Piiper, J. (1979b). Blood oxygen dissociation cutrve of the frogs *Rana catesbeiana* and *Rana brevipoda*. *J Comp Physiol* **129**, 111–114.

Tenney, S. M. and Tenney, J. B. (1970). Quantitative morphology of cold-blooded lungs: Amphibia and Reptilia. *Respir Physiol* **9**, 197–215.

Tester, J. R. and Breckenridge, W. J. (1964a). Winter behavior patterns of the Manitoba toad, *Bufo hemiophrys*, in northwestern Minnesota. *Ann Acad Sci Fenn (Biol)* **71**, 421–431.

Tester, J. R. and Breckenridge, W. J. (1964b). Population dynamics of the Manitoba toad, *Bufo hemiophrys*, in nortwestern Minnesota. *Ecology* **45**, 592–601.

Thompson, G. G., Withers, P. C., McMaster, K. A. and Cartledge, V. A. (2005). Burrows of desert-adapted frogs, *Neobatrachus aquilonius* and *Notaden nichollsi*. *J Royal Soc Western Aust* **88**, 17–23.

Thorson, T. B. (1955). The relationship of water economy to terrestrialism in amphibians. *Ecology* **36**, 100–116.

Thorson, T. B. (1961). The partitioning of body water in Osteichthyes: phylogenetic and ecological implications in aquatic vertebrates. *Biol Bull* **120**, 238–254.

Thorson, T. B. (1964). The partitioning of body water in amphibia. *Physiol Zool* **37**, 395–399.

Thorson, T. B. (1968). Body fluid partitioning in Reptilia. *Copeia* 1968, 592–601.

Thorson, T. B. and Svihla, A. (1943). Correlation of the habitats of amphibians with their ability to survive the loss of body water. *Ecology* **24**, 374–381.

Toews, D. P., Boutilier, R. G., Todd, L. and Fuller, N. (1978). Carbonic anhydrase in the amphibia. *Comp Biochem Physiol* **59A**, 211–213.

Toews, D. P. and Macintyre, D. (1977). Blood respiratory properties of a viviparous amphibian. *Nature* **266**, 464–465.

Toews, D. P., Mallet, M. and Maclatchy, L. (1989). Renal responses to continuous and short-pulse infusion of arginine vasotocin in the toad *Bufo marinus*. *J Exp Biol* **142**, 177–189.

Toews, D. P., Shelton, G. and Randall, D. J. (1971). Gas tensions in the lungs and major blood vessels of the urodele amphibian, *Amphiuma tridactylum*. *J Exp Biol* **55**, 47–61.

Toews, D. P. and Wentzell, L. A. (1995). The role of the lymphatic system for water balance and acid-base regulation in the Amphibia. *Adv Comp Environ Physiol* **21**, 201–213.

Toft, C. A. (1980). Seasonal variation in populations of panamanian litter frogs and their prey: a comparison of wetter and drier sites. *Oecologia* **47**, 34–38.

Toledo, R. C. and Jared, C. (1993). Cutaneous adatations to water balance in amphibians. *Comp Biochem Physiol* **105A**, 593–608.

Tonosaki, Y., P. Cruijsen, K. Nishiyama, H. Yaginuma, E. W. Roubos. (2004). Low temperature stimulates alpha-melanophore-stimulating hormone secretion and inhibits background adaptation in *Xenopus laevis*. *J Neuroendocrinology* **16**, 894–905.

Toolson, E. C. (1978). Diffusion of water through the arthropod cuticle: thermodynamic consideration of the transition phenomenon. *J Thermal Biol* **3**, 69–73.

Toolson, E. C. (1980). Thermodynamic and kinetic aspects of water flux through arthropod cuticle. *J Thermal Biol* **5**, 1–6.

Toop, T. and Donald, J. A. (2004). Comparative aspects of natriuretic peptide physiology in non-mammalian vertebrates: a review. *J Comp Physiol* **174B**, 189–204.

Torday, J. S. and Rehan, V. K. (2004). Deconvuluting lung evolution using functional/comparative genomics. *Am J Respir Cell Mol Biol* **31**, 8–12.

Townson, R. (1795). *Physiological Observations on the Amphibia. Dissertation the Second. Respiration continued; with a fragment upon the subject of absorption*. Gottingen.

Tracy, C. R. (1975). Water and energy relations of terrestrial amphibians: Insights from mechanistic modelling. In: *Perspectives of Biophysical Ecology*, eds D. M. Gates and R. B. Schmerl, pp. 325–346. Springer-Verlag.

Tracy, C. R. (1976). A model of the dynamic exchanges of water and energy between a terrestrial amphibian and its environment. *Ecolog Mono* **46**, 293–326.

Tracy, C. R. (1979). Further thoughts on anuran thermoregulation. Discussion. In: *The Behavioral Significance of Color*, ed. E. H. Burtt, pp. 63–67. Garland Press: New York.

Tracy, C. R., Betts, G., Tracy, C. R. and Christian, K. A. (2007). Plaster models to measure opertaive temperature and evaporative water loss of amphibians. *J Herpetology* **41**, 597–603.

Tracy, C. R. and Christian, K. A. (2005). Preferred temperature correlates with evaporative water loss in hylid frogs from northern Australia. *Physiol Biochem Zool* **78**, 839–846.

Tracy, C. R., Christian, K. A., Betts, G. and Tracy, C. R. (2008). Body temperature and resistance to evaporative water loss in tropical Australian frogs. *Comp Biochem Physiol* A 150: 102–108.

Tracy, C. R., Christian, K. A., O'Connor, M. P. and Tracy, C. R. (1993). Behavioral thermoregulation by *Bufo americanus*: The importance of the hydric environment. *Herpetologica* **49**, 375–382.

Tracy, C. R. and Porter, W. P. (1973). *Technical Report 1. Properties of Air.* University of Wisconsin, Laboratory of Biophysical Ecology.

Tracy, C. R., Tracy, C. R. and Turner, J. S. (1992). Contrasting physiological abilities for heating and cooling in an amphibian (*Rana pipiens*) and a reptile (*Sauromalus obesus*). *Herpetologica* **48**, 57–60.

Tracy, C. R., Welch, W. R. and Porter, W. P. (1980). *Properties of Air: A Manual for use in Biophysical Ecology. Technical Report 1.* University of Wisconsin: Madison.

Tufts, B. L., Mense, D. C. and Randall, D. J. (1987). The effects of forced activity on circulating catecholamines and pH and the water content of erythrocytes in the toad. *J Exp Biol* **128**, 411–418.

Tufts, B. L. and Toews, D. P. (1985). Partitioning of regulatory sites in *Bufo marinus* during hypercapnoea. *J Exp Biol* **119**, 199–209.

Turner, R. J. (1988). Amphibians. In: *Vertebrate Blood Cells*, eds A. F. Rowley and N. A. Ratcliffe, pp. 129–209. Cambridge University Press.

Turney, L. D. and Hutchison, V. H. (1974). Metabolic scope, oxygen debt, and the diurnal oxygen consumption cycle of the leopard frog, *Rana pipiens*. *Comp Biochem Physiol* **49A**, 583–601.

Twitty, V. C. (1966). *Of Scintists and Salamanders.* WH Freeman and Co: San Francisco, CA.

Tyler, M. J. (1982). *Frogs*, 2nd edn. Australian Naturalist Library. Collins: Sydney.

Tyler, M. J. (1999). Distribution patterns of amphibians in the Australo-Papuan region. In: *Patterns of Distribution of Amphibians A Global Perspective*, ed. W. E. Duellman, pp. 541–563. Johns Hopkins University Press: Baltimore and London.

Tyler, M. J., Roberts, J. D. and Davies, M. (1980). Field observations on *Arenophryne rotunda* Tyler, a leptodactylid frog of coastal sandhills. *Australian Wildlife Research* 7, 295–304.

Tyml, K. (1987). Red cell perfusion in skeletal muscle at rest and after mild and severe contractions. *Am J Physiol* **252**, H485–493.

Tyml, K. and Groom, A. C. (1980). Regulation of blood flow in individual capillaries of resting skeletal muscle in frogs. *Microvasc Res* **20**, 346–357.

Uchiyama, M. (1980). Hypocalcemic factor in the ultimobranchial gland of the newt, *Cynops pyrrhogaster*. *Comp Biochem Physiol* **66A**, 331–334.

Uchiyama, M. and Konno, N. (2006). Hormonal regulation of ion and water transport in anuran amphibians. *Gen Comp Endocrinol* **147**, 54–61.

Uchiyama, M., Takeuchi, T. and Matsuda, K. (1998). Effects of homologous natriuretic peptides in isolated skin of the bullfrog, *Rana catesbeiana*. *Comp Biochem Physiol C Pharmacol Toxicol Endocrinol* **120**, 37–42.

Ultsch, G. R. (1973). A theoretical and experimental investigation of the relationships between metabolic rate, body size, and oxygen exchange capacity. *Respir Physiol* **18**, 143–160.

Ultsch, G. R. (1976). Respiratory surface area as a factor controlling the standard rate of O_2 consumption in aquatic salamanders. *Respir Physiol* **26**, 357–369.

Ultsch, G. R., Brainerd, E. L. and Jackson, D. C. (2004). Lung collapse among aquatic reptiles and amphibians during long-term diving. *Comp Biochem Physiol* **139A**, 111–115.

Ultsch, G. R., Reeses, S. A. and Stewart, E. R. (2004). Physiology of hibernation in *Rana pipiens*: metabolic rate, critical oxygen tension, and the effects of hypoxia on several plasma variables. *J Exp Zool* **301A**, 169–176.

Uranga, J. (1969). The hepatic production of a glomerular pressure substance in the toad (*Bufo arenarum*). *Gen Comp Endocrin* **13**, 179–184.

Uranga, J. (1971). The characteristics of hepatic glomerular pressure substance. *Am J Physiol* **220**, 1617–1620.

Urbani, E. (1962). Comparative biochemical studies on amphibian and invertebrate development. In: *Advances in Morphogenesis,* Vol. 2, eds M. Abercrombie and J. Brachet, pp. 61–108. Academic Press: New York.

Usher-Smith, J. A., Xu, W., Fraser, J. A. and Huang, C. L. H. (2006). Alterations in calcium homeostasis reduce membrane excitability in amphibian skeletal muscle. *Pflugers Arch—Eur J Physiol* **453**, 211–221.

Ussing, H. H. (1949). The active ion trasport through the isolated frog skin in the light of tracer studies. *Acta Physiol Scand* 17, 1–37.

Ussing, H. H. (1978). Interpretation of tracer fluxes. In: *Membrane Transport in Biology,* Vol. 1 *Concepts and Models,* eds. G. Giebisch, D. C. Tosteson and H. H. Ussing, pp. 115–140. New York: Springer Verlag.

Ussing, H. H. and Fuhrman, F. A. (1951). Characteristic response of the isolated frog skin potential to neurohypophysial principles and its relation to the transport of sodium and water. *J Cell Physiol* **38**, 109–130.

Ussing, H. H. and Windhager, E. E. (1964). The nature of the shunt path and active transport through frog skin epithelium. *Acta Physiol Scand* **61**, 484–504.

Ussing, H. H. and Zerhan, K. (1951). Active transport of sodium as the source of electric current in the short-circuited isolated frog skin. *Acta Physiol Scand* **23**, 110–127.

Vanatta, J. C. and Frazier, L. W. (1981). The epithelium of *Rana pipiens* secretes H^+ and NH_4^+ in acidosis and HCO_3^- in alkalosis. *Comp Biochem Physiol A* **68**, 511–513.

van Beurden, E. K. (1980). Energy metabolism of dormant Australian water-holding frogs (*Cyclorana platycephalus*). *Copeia* 1980, 787–799.

Van Beurden, E. K. (1984). Survival strategies of the Australian water-holding frog, *Cyclorana platycephalus*. In: *Arid Australia: Proceedings of a Symposium on the Origins, Biota and Ecology of Australia's Arid Regions*, eds H. G. Cogger and E. E. Cameron, pp. 223–234. Australian Museum: Sydney.

van der Laarse, W. J., des Tombe, A. L., van Beck-Harmsen, B. J., Lee-de Groot, M. B. E. and Jaspers, R. T. (2005). Krogh's diffusion coefficient for oxygen in isolated *Xenopus* skeletal muscle fibers and rat myocardial trabeculae at maximum rates of oxygen consumption. *J Appl Physiol* **99**, 2173–2180.

van der Laarse, W. J., Diegenbach, P. C. and Elzinga, G. (1989). Maximum rate of oxygen consumption and quantitative histochemistry of succinate dehydrogenase in single muscle fibres of *Xenopus laevis*. *J Musc Res Cell Motil* **10**, 221–228.

van der Laarse, W. J., Lannergren, J. and Diegenbach, P. C. (1991). Resistance to fatigue of single muscle fibers from *Xenopus* related to myofibrillar succinate dehydrogenase and ATPase activities. *Exp Physiol* **76**, 589–596.

Van Driessche, W. (1984). Physiological role of apical potassium ion channels in frog skin. *J Physiol* **356**, 79–95.

Van Driessche, W. (1994). Noise and impedance analysis. In: *Methods in Membrane and Transporter Research*, eds Shafer, J. A., Giebisch, G., Kristensen, P., and Ussing, H. H. , pp. 22–80. Landes: Georgetown, TX, USA.

Van Driessche, W. and Lindemann, B. (1978). Low-noise amplification of voltage and current fluctuations arising in epithelia. *Rev Sci Instruments* **49**, 52–57.

Van Driessche, W. and Zeiske, W. (1985). Ca^{2+}-sensitive, spontaneously fluctuating cation channels in the apical membrane of the adult frog skin. *Pflugers Archiv* **405**, 250–259.

van Gelder, J. J., Olders, J. H. J., Bosch, J. W. G. and Starmans, P. W. (1986). Behaviour and body temperature of hibernating common toads *Bufo bufo*. *Holarctic Ecol* **9**, 225–228.

Vanhooydonck, B. and Van Damme, R. (1999). Evolutionary relationships between body shape and habitat use in lacertid lizards. *Evol Ecol Res* **1**, 785–805.

van't Hoff, J. H. (1887). The role of osmotic pressure in the analogy between solutions and gasses. *Zeitschrift fur Physikalische Chemie* **1**, 481–508.

Vazquez Illanes, M. D. and Storey, K. B. (1993). 6-Phosphofructo-2-kinase and control of cryoprotectant synthesis in freeze tolerant frogs. *Biochim Biophys Acta* **1158**, 29–32.

Vences, M., Galán, P., Vieites, D. R., Puente, M., Oetter, K. and Wanke, S. (2002). Field body temperatures and heating rates in a montane frog population: the importance of black dorsal pattern for thermoregulation. *Ann Zoologici Fennici* **39**, 209–220.

Vences, M., Vietes, D. R., Glaw, F., Brinkmann, H., Kousch, J., Vieth, J. and Meyer, A. (2003). Multiple overseas dispersal in amphibians. *Proc R Soc Lond* B **270**, 2435–2442.

Vernberg, F. J. (1953). Hibernation studies of two species of salamanders, *Plethodon cinereus cinereus* and *Eurycea bislineata bislineata*. *Ecology* **34**, 55–62.

Vezzoli, A., Gussoni, M., Greco, F. and Zetta, L. (2003). Effects of temperature and extracellular pH on metabolites: kinetics of anaerobic metabolism in resting muscle by ^{31}P- and ^{1}H-NMR spectroscopy. *J Exp Biol* **206**, 3043–3052.

Vezzoli, A., Gussoni, M., Greco, F., Zetta, L. and Cerreteli, P. (1997). Quantitative analysis of anaerobic metabolism in resting anoxic muscle by ^{31}P and ^{1}H MRS. *Biochim Biophys Acta* **1322**, 195–207.

Viborg, A. L. and Hillyard, S. D. (2005). Cutaneousl blood flow and water absorption by dehydrated toads. *Physiol Biochem Zool* **78**, 394–404.

Viborg, A. L. and Rosenkilde, P. (2001). Angiotensin II elicits water seeking behavior and the water absorption response in the toad *Bufo bufo*. *Hormones Behavior* **39**, 225–231.

Viborg, A. L. and Rosenkilde, P. (2004). Water potential receptors in the skin regulate blood perfusion in the ventral pelvic patch of toads. *Physiol Biochem Zool* 77, 39–49.

Viborg, A. L., Wang, T. and Hillyard, S. D. (2006). Cardiovascular and behavioral changes during water absorption in toads, *Bufo alvarius* and *Bufo marinus*. *J Exp Biol* **209**, 834–844.

Viertel, B. (1999). Salt tolerance of *Rana temporaria*: spawning site selection and survival during embryonic development (Amphibia, Anura). *Amphibia-Reptilia* **20**, 161–171.

Voituron, Y., Eugene, M. and Barre, H. (2003). Survival and metabolic responses to freezing by the water frog (*Rana ridibunda*). *J Exp Zoolog A Comp Exp Biol* **299**, 118–126.

Voituron, Y., Joly, P., Eugene, M. and Barre, H. (2005). Freezing tolerance of the European water frogs: the good, the bad, and the ugly. *Am J Physiol Regul Integr Comp Physiol* **288**, R1563–1570.

Vondersaar, M. E. and Stiffler, D. F. (1989). Renal function in amphibians: a comparison of strictly aquatic and amphibious species, with observations on the effects of anesthesis. *Comp Biochem Physiol A* **94A**, 243–248.

Wake, D. B. (1987) Adaptive radiation of salamanders in Middle American cloud forests. *Ann Missouri Bot Garden* **74**, 242–264.

Wake, M. H. (1970). Evolutionary morphology of the caecilian urogenital system: II. The kidneys and urogenital ducts. *Acta Anatomica* **75**, 321–358.

Wake, M. H. (2003). Reproductive modes, ontogenies and the evolution of body form. *Anim Biol* **53**, 209–223.

Wake, M. H. and Dickie, R. (1998). Oviduct structure and function an reproductive modes in amphibians. *J Exp Zool* **282**, 477–506.

Walker, A., M, and Hudson, C. L. (1937). The reabsorption of glucose from the renal tubule in Amphibia and the action of phlorizin upon it. *Am J Physiol* **118**, 140–143.

Walker, J. C. G. (1977). *Evolution of the Atmosphere.* MacMillan: New York.

Walls, G. L. (1942). The vertebrate eye and its adaptive radiation. *Bull Cranbrook Inst Sci* **19**.

Walsberg, G. E. (1986a). Comparison of two techniques for estimating the maximum aerobic capacity of amphibians. *Herpetologica* **42**, 389–394.

Walsberg, G. E. (1986b). A simple technique for estimating operative environmental temperature. *J Thermal Biol* **11**, 67–72.

Walsberg, G. E. (1996). An appraisal of operative temperature mounts as tools for studies of ecological energetics. *Physiol Zool* **69**, 658–681.

Walsberg, G. E. and Hoffman, T. C. M. (2006). Using direct calorimetry to test the accuracy of indirect calorimetry in an ectotherm. *Physiol Biochem Zool* **79**, 830–835.

Walsberg, G. E., Lea, M. S. and Hillman, S. S. (1986). Individual variation in maximum aerobic capacity: cardiovascular and enzymatic correlates in *Rana catesbeiana*. *J Exp Zool* **239**, 1–5.

Walters, P. J. and Greenwald, L. (1977). Physiological adaptations of aquatic newts (*Notopthalmus viridescens*) to a terrestrial environment. *Physiol Zool* **50**, 88–98.

Walton, M. (1988). Relationships among metabolic, locomotory and field measures of organismal performance in the Fowler's toad (*Bufo woodhousei fowleri*). *Physiol Zool* **61**, 107–118.

Walton, M. (1993). Physiology and phylogeny: the evolution of locomotor energetics in hylid frogs. *Am Nat* **141**, 26–50.

Walton, M. and Andersen, B. D. (1988). The aerobic cost of saltatory locomotion in the Fowler's toad (*Bufo woodhousei americanus*). *J Exp Biol* **136**, 273–288.

Walton, M., Peterson, C. C. and Bennett, A. F. (1994). Is walking costly for anurans? The energetic cost of walking in the northern toad *Bufo boreas halophilus*. *J Exp Biol* **197**, 165–178.

Wang, T., Abe, A. S. and Glass, M. L. (1998). Temperature effects on lung and blood gases in *Bufo paracnemis*: consequences of bimodal gas exchange. *Respir Physiol* **113**, 231–238.

Wang, T., Burggren, W. W. and Nobrega, E. (1995). Metabolic, ventilatory, and acid-base responses associated with specific dynamic action in the toad *Bufo marinus*. *Physiol Zool* **68**, 192–205.

Wang, T., Busk, M. and Overgaard, J. (2001). The respiratory consequences of feeding in amphibians and reptiles. *Comp Biochem Physiol A Mol Integr Physiol* **128**, 535–549.

Wang, T., Hedrick, M. S., Ihmied, Y. M. and Taylor, E. W. (1999). Control and interaction of the cardiovascular and respiratory systems in anuran amphibians. *Comp Biochem Physiol* **124A**, 393–406.

Wang, T. and Hicks, J. W. (2002). An integrative model to predict maximum O_2 uptake in animals with central vascular shunts. *Zoology* **105**, 45–53.

Warburg, M. R. (1988). Adaptation of amphibians to life in xeric habitats. In: *Ecophysiology of Desert Vertebrates*, ed. P. K. Ghosh and I. Prakash, pp. 59–98. Scientific Publishers: Jodhpur.

Warburg, M. R. and Degani, G. (1979). Evaporative water loss and uptake in juvenile and adult *Salamandra salamandra* (L.) (Amphibia: Urodela). *Comp Biochem Physiol A* **62**, 1071–1075.

Warburg, M. R. and Goldenberg, S. (1978). The changes in osmoregulatory effects of prolactin during the life cycle of two urodeles. *Comp Biochem Physiol* **61A**, 321–324.

Warburg, O. (1924). Verbesserte methode zur messung der atmung und glykolyse. Biochemische Zeitscrift 152: 51–63.

Waringer-Loschenkohl, A. and Schagerl, M. (2001). Algal exploitation by tadpoles: an experimental approach. *Int Rev Hydrobiol* **86**, 105–125.

Warkentin, K. M. (2000). Environmental and developmental effects on external gill loss in the red-eyed tree frog, *Agalychnis callidryas. Physiol Biochem Zool* **73**, 557–565.

Warkentin, K. M. (2002). Hatching timing, oxygen availability, and external gill regression in the tree frog, *Agalychnis callidryas. Physiol Biochem Zool* **75**, 155–164.

Warren, D. E. and Jackson, D. C. (2005). The role of mineralized tissue in the buffering of lactic acid during anoxia and exercise in the leopard frog *Rana pipiens. J Exp Biol* **208**, 1117–1124.

Watson, P. A. (1991). Function follows form: generation of intracellular signals by cell deformation. *FASEB J* **5**, 2013–2019.

Weathers, W. W. (1975). Circulatory responses of *Rana catesbeiana* to temperature, season and previous thermal history. *Comp Biochem Physiol* **51A**, 43–52.

Weathers, W. W. (1976a). Influence of temperature on the optimal hematocrit of the bullfrog (*Rana catesbeiana*). *J Comp Physiol* **105**, 173–184.

Weathers, W. W. (1976b). Influence of temperature acclimation on oxygen consumption, haemodynamics and oxygen transport in bullfrogs. *Aust J Zool* **24**, 321–330.

Weber, R. E., Ostojic, H., Fago, A., Dewilde, S., Van Hauwert, M., Moens, L. and Monge, C. (2002). Novel mechanism for high-altitude adaptation in the hemoglobin of the Andean frog *Telmatobius peruvianus. Am J Physiol* **283**, R1052–R1060.

Weber, R. E., Voelter, W., Fago, A., Echner, H., Campanella, E. and Low, P. S. (2004). Modulation of red cell glycolysis: interactions between vertebrate hemoglobins and cytoplasmic domains of band 3 red cell membrane proteins. *Am J Physiol* **287**, R454–R464.

Weed, I. G. (1933). Cytological studies of the epidermis of *Rana pipiens* and *Rana climitans* tadpoles with special references to the figures of Eberth. *J Morphol* **56**, 213–229.

Weibel, E. R., Tayloe, C. R. and Hoppler, H. (1991). The concept of symmorphosis: a testable hypothesis of structure-function relationship. *PNAS* **88**, 10357–10361.

Weldon, P. J., Demeter, B. J. and Rosscoe, R. (1993). A survey of shed skin-eating (dermatophagy) in amphibians and reptiles. *J Herpetology* **27**, 219–228.

Wells, K. D. and Taigen, T. L. (1989). Calling energetics of a neotropical treefrog, *Hyla microcephala. Behav Ecol Sociobiol* **25**, 13–22.

Welsch, U. (1983). Phagocytosis in the Amphibian lung. *Anat Anz, Jena* **154**, 319–327.

Weng, L., Wong, W. P., Chew, S. F. and Ip, Y. K. (2004). Excretory nitrogen metabolism in the Chinese fire-belly newt *Cynops orientalis* in water, on land, or in high concentrations of environmental ammonia. *J Comp Physiol* **174B**, 113–120.

Wentzell, L. A., McNeil, S. A. and Toews, D. P. (1993). The role of the lymphatic system in water balance processes in the toad *Bufo marinus*. *Physiol Zool* **66**, 307–321.

Wernecke, L., Withers, P. C., Schleucher, E. and Maloney, S. K. (2007). Body temperature variation of free-ranging and captive southern brown bandicoots *Isoodon obesulus* (Marsupialia: Peramelidae). *J Thermal Biol* **32**, 72–77.

West, G. B., Woodruff, W. H. and Brown, J. H. (2002). Allometric scaling of metabolic rate from molecules and mitochondria to cells and mammals. *PNAS* **99**, 2473–2478.

West, N. H. and Burggren, W. W. (1982). Gill and lung ventilation responses to steady-state aquatic hypoxia and hyperoxia in the bullfrog tadpole. *Respir Physiol* **47**, 165–176.

West, N. H. and Burggren, W. W. (1983). Reflex interactions between aerial and aquatic gas exchange organs in larval bullfrogs. *Am J Physiol* **244**, R770–777.

West, N. H. and Burggren, W. W. (1984). Factors influencing pulmonary and cutaneous arterial blood flow in the toad, *Bufo marinus*. *Am J Physiol* **247**, R884–894.

West, N. H., Kimmel, P. B., Topor, Z. L. and Evered, M. D. (1998). The role of angiotensin in arterial blood pressure regulation in the toad *Bufo marinus*. *J Exp Biol* **201**, 2219–2224.

West, N. H. and Smits, A. W. (1994). Cardiac output in conscious toads (*Bufo marinus*). *J Exp Biol* **186**, 315–323.

West, N. H. and Van Vliet, B. N. (1992). Sensory mechanisms regulating the cardiovascular and respiratory systems. In: *Environmental Physiology of the Amphibians*, eds W. Burggren and M. Feder, pp. 151–182, University of Chicago Press.

West, N. H. and Van Vliet, B. N. (1994). The role of arterial baroreceptors in the undivided circulation of anuran amphibians. *Physiol Biochem Zool* **67**, 1305.

Westerblad, H., Allen, D. G. and Lannergren, J. (2002). Muscle fatigue: lactic acid or inorganic phosphate the major cause? *News Physiol Sci* **17**, 17–21.

Westerblad, H., Bruton, J. D., Allen, D. G. and Lannergren, J. (2000). Functional significance of Ca^{2+} in long-lasting fatigue of skeletal muscle. *Eur J Appl Physiol* **83**, 166–174.

Westerblad, H., Lannergren, J. and Allen, D. G. (1997). Slowed relaxation in fatigued skeletal muscle fibers of *Xenopus* and mouse. *J Gen Physiol* **109**, 385–399.

White, B. A. and Nicol, C. S. (1979). Prolactin receptors in *Rana catesbeiana* during development and metamorphosis. *Science* **204**, 851–853.

White, C. R., Philips, N. F. and Seymour, R. S. (2006). The scaling and temperature dependence of vertebrate metabolism. *Biol Lett* **2**, 125–127.

White, H. L. (1929a). Some measurements of ciliary activity. *Am J Physiol* **88**, 282–285.

White, H. L. (1929b). The question of water reabsorption by the renal tubule and its bearing on the problem of tubular secretion. *Am J Physiol* **88**, 267–281.

White, H. L. (1929c). Observations on the nature of glomerular activity. *Am J Physiol* **90**, 689–704.

Whitford, W. G. (1968). Physiological responses to temperature and desiccation in the endemic New Mexico plethodontids, *Plethodon neomexicanus* and *Aneides hardii*. *Copeia* 1968, 247–251.

Whitford, W. G. and Hutchison, V. H. (1965). Gas exchange in salamanders. *Physiol Zool* **38**, 228–242.

Whitford, W. G. and Hutchison, V. H. (1967). Body size and metabolic rate in amphibians. *Physiol Zool* **40**, 127–133.

Whitney, C. L. and Krebs, J. R. (1975). Mate selection in Pacific tree frogs. *Nature* **255**, 325–326.

Whittouck, P. J. (1972a). Modification du retention du sodium chez *Ambystoma mexicanum* (axolotl) intact et hypophysectomise sous l'effect du prolactine. *Archives International du Physiologie et Biochemie* **80,** 373–381.

Whittouck, P. J. (1972b). Intensification par la prolactine de l'absorption d'ions sodium au niveau des branchies isolees de larves d' *Ambystoma mexicanum. Archives International du Physiologie et Biochemie* **80,** 825–827.

Wiens, J. J. (2007). Global patterns of diversification and species richness in amphibians. *Am Naturalist* **170,** S86–S106.

Wilkinson, M. and Nussbaum, R. A. (1997). Comparative morphology and evolution of the lungless caecilian *Aretochoana eiselti* (Amphibia:Gynophiona:Typhlonectidae). *Biol J Linn Soc* **62,** 39–109.

Wilkinson, M., Sheps, J.A., Oommem, O.V. and Cohen, B.L. (2002). Phylogenetic relationships of Indian caecilians (Amphibia: Gymniophiona) inferred from mitochondrial rRNA gene sequences. *Mol Phylogen Evol* **23,** 401–407.

Williams, A. A. and Wygoda, M. L. (1993). Dehydration stimulates behavioral hypothermia in the Gulf Coast Toad, *Bufo valliceps. J Thermal Biol* **18,** 223–227.

Williams, P., Humphries, C., Araujo, M. B., Lampinen, R., Hagemeijer, W., Gasc, J. P. and Mitchell-Jones, T. (2000). Endemism and important areas for conserving European biodiversity: a preliminary examination of atlas data for plants and terrestrial vertebrates. *Belgian J Entomol* **2,** 21–46.

Willis, Y. L., Moyle, D. L. and Baskett, T. S. (1956). Emergence, breeding, hibernation, movements, and transformation of the bullfrog, *Rana catesbeiana. Copeia* 1956, 30–41.

Willumsen, N. J., Amstrup, J., Mobjerg, N., Jespersen, A., Kristensen, P. and Larsen, E. H. (2002). Mitochondria rich cells as experimental model in studies of epithelial chloride channels. *Biochim Biophys Acta* **1566,** 28–43.

Willumsen, N. J., Amstrup, J., Nejsum, L. M., Larsen, E. H., Nielsen, S. and Hillyard, S. D. (2003). Differential localization of aquaporins 1–3 in amphibian skin (abstract). *FASEB J* **17,** A919.

Willumsen, N. J. and Larsen, E. H. (1986). Membrane potentials and intracellular Cl^- activity of toad skin epithelium in relation to activation and deactivation of the transepithelial conductance. *J Membrane Biol* **94,** 173–190.

Willumsen, N. J., Viborg, A. L. and Hillyard, S. D. (2007). Vascular aspects of water uptake mechanisms in the toad skin; Perfusion, diffusion, confusion. *Comp Biochem Physiol* **148A,** 55–63.

Wilson, R. S. (2001). Geographic variation in thermal sensitivity of jumping performance in the frog *Limnodynastes peronii. J Exp Biol* **204,** 4227–4236.

Wind-Larsen, H. and Jorgensen, C. B. (1987). Hormonal control of seasonal growth in a temperate zone toad *Bufo bufo. Acta Zoologica* **68,** 49–56.

Windmiller, B. (1996). Tracking techniques useful for field studies of anuran orientation and movement. *Herpetological Review* **27,** 13–14.

Withers, P. C. (1977). Measurement of V_{O_2}, V_{CO_2}, and evaporative water loss with a flow-through mask. *J Applied Physiol* **42,** 120–123.

Withers, P. C. (1980). Oxygen consumption of plethodontid salamanders during rest, activity and recovery. *Copeia* 1980, 781–787.

Withers, P. C. (1992). *Comparative Animal Physiology.* Saunders College Publishing: Philadelphia.

Withers, P. C. (1993). Metabolic depression during aestivation in the Australian frogs, *Neobatrachus* and *Cyclorana. Aust J Zool* **41,** 467–473.

Withers, P. C. (1995a). Cocoon formation and structure in the aestivating Australian desert frogs, *Neobatrachus* and *Cyclorana*. *Aust J Zool* **43**, 429–441.
Withers, P. C. (1995b). Evaporative water loss and colour change in the Australian desert tree frog *Litoria rubella* (Amphibia: Hylidae). *Records Western Australian Museum* **17**, 277–281.
Withers, P. C. (1998). Evaporative water loss and the role of cocoon formation in Australian frogs. *Aust J Zool* **46**, 405–418.
Withers, P. C. (2001). Design, calibration and calculation for flow-through respirometry systems. *Aust J Zool* **49**, 445–461.
Withers, P. C. and Cooper, C. E. (2008). Dormancy. In: *Encyclopedia of Ecology*, ed. S. E. Jorgensen. Elsevier, B. V: Amsterdam, in press.
Withers, P. C. and Guppy, M. (1996). Do Australian desert frogs co-accumulate counteracting solutes with urea during aestivation? *J Exp Biol* **199**, 1808–1816.
Withers, P. C. and Hillman, S. S. (1981). Oxygen consumption of *Amphiuma means* during forced activity and recovery. *Comp Biochem Physiol A* **69**, 141–144.
Withers, P. C. and Hillman, S. S. (1983). The effects of hypoxia on pulmonary function and maximal rates of oxygen consumption in two anuran amphibians. *J Comp Physiol* **152**, 125–129.
Withers, P. C. and Hillman, S. S. (1988). A steady-state model of maximal oxygen and carbon dioxide transport in anuran amphibians. *J Appl Physiol* **64**, 860–868.
Withers, P. C. and Hillman, S. S. (2001). Allometric and ecological relationships of ventricle and liver mass in anuran amphibians. *Funct Ecol* **15**, 60–69.
Withers, P. C., Hillman, S. S. and Drewes, R. C. (1984). Evaporative water loss and skin lipids of anuran amphibians. *J Exp Zool* **232**, 11–17.
Withers, P. C., Hillman, S. S., Drewes, R. C. and Sokol, O. M. (1982a). Water loss and nitrogen excretion in sharp-nosed reed frogs (*Hyperolius nasuta*: Anura: Hyperoliidae). *J Exp Biol* **97**, 335–343.
Withers, P. C., Hillman, S. S., Hedrick, M. S. and Kimmel, P. B. (1991). Optimal hematocrit theory during activity in the bullfrog (*Rana catesbeiana*). *Comp Biochem Physiol A* **99**, 55–60.
Withers, P. C., Hillman, S. S. and Kimmel, P. B. (1988). Effects of activity, hemorrhage, and dehydration on plasma catecholamine levels in the marine toad (*Bufo marinus*). *Gen Comp Endocrinol* **72**, 63–71.
Withers, P. C., Hillman, S. S., Simmons, L. A. and Zygmunt, A. (1988). Cardiovascular adjustments to enforced activity in the anuran amphibian, *Bufo marinus*. *Comp Biochem Physiol* **89A**, 45–49.
Withers, P. C., Lea, M., Solberg, T. C., Baustian, M. and Hedrick, M. (1988). Metabolic fate of lactate during recovery from activity in an anuran amphibian, *Bufo americanus J Exp Zool* **246**, 236–243.
Withers, P. C., Louw, G. and Nicholson, S. (1982b). Water loss, oxygen consumption, and colour change in "waterproof" reed frogs (*Hyperolius*). *S A J Sci* **78**, 30–32.
Withers, P. C. and Roberts, J. D. (1993). Water balance response in the terrestrial Western Australian frog, *Myobatrachus gouldii* (Anura: Myobatrachidae). *J Royal Soc Western Australia* **76**, 49–52.
Withers, P. C. and Richards, S. J. (1996). Cocoon formation by the treefrog *Litoria alboguttata* (Amphibia : Hylidae): a waterproof. taxonomic tool? *J Royal Soc Western Aust* **78**, 103–106.

Withers, P. C. and Thompson, G. G. (2000). Cocoon formation and metabolic depression by the aestivating hylid frogs *Cyclorana australis* and *Cyclorana cultripes* (Amphibia: hylidae). *J Royal Soc Western Aust* **83**, 39–40.

Wittenberg, J. B. (1970). Myoglobin-facilitated oxygen diffusion: role of myoglobin in oxygen entry into muscle. *Physiol Rev* **50**, 559–636.

Wittle, L. W. (1983). Effect of parathroidectomy on plasma calcium levels in the seal salamander, *Desmognathus monticola*. *Gen Comp Endocrin* **51**, 159–162.

Wittle, L. W. and Dent, J. N. (1979). Effects of parathroid extract on levels of calcium and phosphate in the blood and urine of the red-spotted newt. *Gen Comp Endocrin* **37**, 428–439.

Woinarski, J. C. Z., Fisher, A. and Milne, D. (1999). Distribution patterns of vertebrates in relation to an extensive rainfall fradient and soil texture in the tropical savannas of the Northern Territory, Australia. *J Trop Ecol* **15**, 381–398.

Wolanczyk, J. P., Storey, K. B. and Baust, J. G. (1990). Ice nucleating activity in the blood of the freeze-tolerant frog, *Rana sylvatica*. *Cryobiology* **27**, 328–335.

Wood, C. M., Munger, R. S. and Toews, D. P. (1989). Ammonia, urea and H^+ distribution and the evolution of ureotelism in amphibians. *J Exp Biol* **144**, 215–233.

Wood, S. C. (1971). Effects of metamorphosis on blood respiratory properties and erythrocyte adenosine triphosphate level of the salamander *Dicamptodon ensatus*. *Respir Physiol* **12**, 53–65.

Wood, S. C. (1995). Oxygen as a modulator of body temperature. *Brazilian J Med Biol Res* **28**, 1249–1256.

Wood, S. C. and Hicks, J. W. (1985). Oxygen homeostasis in vertebrates with cardiovascular shunts. In *Cardiovascular Shunts: Phylogenetic, Ontogenetic, and Clinical Aspects*, eds K. Johannsen and W. W. Burrgren, pp. 354–366. Munksgaard: Copenhagen.

Wood, S. C. and Malvin, G. M. (1991). Physiological significance of behavioral hypothermia in hypoxic toads (*Bufo marinus*). *J Exp Biol* **159**, 203–215.

Wood, S. C., Malvin, G. M. and Riedel, C. (1989). Effect of hematocrit on behavioural temperature regulation of the toad, *Bufo marinus*. *FASEB J* **3**, A234.

Wood, S. C., Weber, R. E., Maloiy, G. M. O. and Johansen, K. (1975). Oxygen uptake and blood respiratory properties of the caecilian *Boulengerula taitanus*. *Respir Physiol* **24**, 355–363.

Woody, A. and Justus, J. T. (1974). Comparative studies of hemoglobins from the clouded tiger salamander before and after metamorphosis. *J Exp Zool* **188**, 215–224.

Woolbright, L. L. (1985). Patterns of nocturnal movement and calling by the tropical frog, *Eleutherodactylus coqui*. *Herpetologica* **41**, 1–9.

Word, J. (2008). Physiological adjustments to aestivation and activity in the cocoon-forming frogs *Cyclorana platycephala* and *Cyclorana maini*. PhD, School of Animal Biology, University of Western Australia, Perth.

Word, J. M. and Hillman, S. S. (2005). Osmotically absorbed water preferentially enters the cutaneous capillaries of the pelvic patch in the toad *Bufo marinus*. *Physiol Biochem Zool* **78**, 40–47.

Wright, P. A. (2004). Nitrogen excretion: three end products, many physiological roles. *J Exp Biol* **198**, 273–281.

Wygoda, M. (1984). Low cutaneous evaporative water loss in arboreal frogs. *Physiol Zool* **57**, 329–337.

Wygoda, M. (1988a). Adaptive control of water loss resistance in an arboreal frog. *Herpetologica* **44**, 251–257.

Wygoda, M. (1988b). A comparative study of time constants of cooling in green tree frogs (*Hyla cinerea*) and southern leopard frogs (*Rana sphenocephala*). *Herpetologica* **44**, 261–265.

Wygoda, M. (1989). A comparative study of heating rates in arboreal and nonarboreal frogs. *J Herpetol* **23**, 141–145.

Wygoda, M. L. (1984). Low cutaneous evaporative water loss in arboreal frogs. *Physiol Zool* **57**, 329–337.

Wygoda, M. L., Garman, R. H. and Howard, C. E. (1987). Cutaneous and subcutaneous adipose tissue in anuran amphibians. *Copeia* 1987, 1031–1035.

Yamane, M. (1990). The arterial baroreceptor-lymph heart inhibitory reflex and its pathway in the bullfrog, *Rana catesbeiana*. *Kawasaki Med J* **16**, 93–100.

Yancey, P. H. and Somero, G. N. (1979). Counteraction of urea destabilization of protein structure by methylamine osmoregulatory compounds of elasmobranch fishes. *Biochem J* **183**, 317–323.

Yancey, P. H. and Somero, G. N. (1980). Methylamine osmoregulatory solutes of elasmobranch fishes counteract urea inhibition of enzymes. *J Exp Zool* **212**, 205–213.

Yano, K., Hicks, J. W., Vaudry, H. and Conlon, J. M. (1995). Cardiovascular actions of frog urotensin II in the frog, *Rana catesbeiana*. *Gen Comp Endocrin* **97**, 103–110.

Yaoi, Y., Suzuki, M., Tomura, H., Sasayama, Y., Kikuyama, S. and Tanaka, S. (2003). Molecular cloning of otoconin-22 complementary deoxyribonucleic acid in the bullfrog endolymphatic sac: effect of calcitonin on otoconin-22 messenger ribonucleic acid levels. *Endocrinology* **144**, 3287–3296.

Yetka, N. and Blackburn, D. (1992). Sexual dimorphism in mass and protein content of the forelimb muscles of the northern leopard frog *Rana pipiens*. *Can J Zool* **70**, 670–674.

Yokota, S. D., Benyajati, S. and Dantzler, W. H. (1985). Comparative aspects of glomerular filtration in vertebrates. *Renal Physiol* **8**, 193–221.

Yokota, S. D. and Hillman, S.S. (1984). Adrenergic control of the anuran cutaneous hydroosmotic response. *Gen Comp Endocrinol* **53**, 309–314.

Young, J. E., Christian, K. A., Donnellan, S. and Parry, D. (2005). Comparative analysis of cutaneous evaporative water loss in frogs demonstrates correlation with ecological habits. *Physiol Biochem Zool* **78**, 847–856.

Zancanaro, C., Merigo, F., Digito, M. and Pelosi, G. (1996). Fat body of the frog *Rana esculenta*: an ultrastuctural study. *J Morphol* **227**, 321–334.

Zardoya, R. A. and Meyer, A (2001). On the origin of and phylogenetic relationships among living amphibians. *PNAS* **98**, 7380–7383.

Zeiske, W. and Van Driessche, W. (1980). Spontaneous fluctuations of potassium channels in the apical membrane of frog skin. *J Physiol* (London) **299**, 101–116.

Zerella, M. J. and Stiffler, D. F. (1999). Active transport of calcium ions in the skin of the salamander *Ambystoma tigrinum*. *J Comp Physiol B* **169**, 481–486.

Zhao, S., Fang, J., Peng, C. and Tang, Z. (2006). Relationships between species richness of vascular plants and terrestrial vertebrates in China: analysis based on data of nature reserves. *Diversity Distributions* **12**, 189–194.

Zimmerman, S. L., Frisbie, J., Goldstein, D. L., West, J., Rivera, K. and Krane, C. M. (2007). Excretion and conservation of glycerol, and expression of aquaporins and glyceroporins, during cold acclimation in Cope's gray tree frog *Hyla chrysoscelis*. *Am J Physiol* **292**, R544–R555.

Zimmitti, S. J. (1999). Individual variation in morphological, physiological, and biochemical features associated with calling in spring peepers (*Pseudacris crucifer*). *Physiol Biochem Zool* **72**, 666–676.

Zipser, R. D., Licht, P. and Bern, H. A. (1969). Comparative effects of mammalian prolactin and growth hormone on growth in the toad *Bufo boreas* and *Bufo marinus*. *Gen Comp Endocrinol* **13**, 382–391.

Zug, G. R. and Zug, P. B. (1979). The marine toad, *Bufo marinus*: a natural history resume of native populations. *Smithsonian Contributions to Zoology* **284**.

Zwemer, R. L. and Foglia, U. G. (1943). Fatal loss of plasma after lymph heart destruction in toads. *Proc Soc Exp Biol Med* **53**, 14–17.

Index